Translated Texts

This series is designed to meet the need ̅ ̅ ̅ ̅ ̅ ̅ ̅ ̅
eval history and others who wish to ̅ ̅ ̅ ̅ ̅ ̅ ̅ ̅ ̅ ̅
source material, but whose knowledge ̅ ̅ ̅ ̅ ̅ ̅ ̅ ̅ ̅ ̅ ̅ ̅ ̅ ̅ ̅ ̅ ̅ ....ient
to allow them to do so in the original ̅ ̅ ̅ ̅ ̅ ̅ ̅ ̅ ̅ ̅ ̅ ̅ ̅. Many important Late
Imperial and Dark Age texts are currently unavailable in translation
and it is hoped that TTH will help to fill this gap and to complement the
secondary literature in English which already exists. The series relates
principally to the period 300–800 AD and includes Late Imperial,
Greek, Byzantine and Syriac texts as well as source books illustrating a
particular period or theme. Each volume is a self-contained scholarly
translation with an introductory essay on the text and its author and
notes on the text indicating major problems of interpretation, including
textual difficulties.

Editorial Committee
Sebastian Brock, Oriental Institute, University of Oxford
Averil Cameron, Keble College, Oxford
Henry Chadwick, Oxford
John Davies, University of Liverpool
Carlotta Dionisotti, King's College, London
Peter Heather, University College, London
William E. Klingshirn, The Catholic University of America
Michael Lapidge, Clare College, Cambridge
Robert Markus, University of Nottingham
John Matthews, Yale University
Claudia Rapp, University of California, Los Angeles
Raymond Van Dam, University of Michigan
Michael Whitby, University of Warwick
Ian Wood, University of Leeds

General Editors
Gillian Clark, University of Liverpool
Mary Whitby, Oxford

Front cover drawing: Bede writing *The Reckoning of Time*, after an initial in a 12th-century version of *De temporum ratione*, Glasgow U. L. Hunter T.4.2, fol. 35r (Durham Cathedral Priory S.XII[1]) (Drawn by Gail Heather)

A full list of published titles in the Translated Texts for Historians series is available on request. The most recently published are shown below.

Pseudo-Dionysius of Tel-Mahre: *Chronicle*, **Part III**
Translated with notes and introduction by WITOLD WITAKOWSKI
Volume 22: 192 pp., 1995, ISBN 0–85323–760–3

Venantius Fortunatus: Personal and Political Poems
Translated with notes and introduction by JUDITH GEORGE
Volume 23: 192 pp., 1995, ISBN 0–85323–179–6

Donatist Martyr Stories: The Church in Conflict in Roman North Africa
Translated with notes and introduction by MAUREEN A. TILLEY
Volume 24: 144 pp., 1996, ISBN 0–85323–931–2

Hilary of Poitiers: Conflicts of Conscience and Law in the Fourth-Century Church
Translated with introduction and notes by LIONEL R. WICKHAM
Volume 25: 176 pp., 1997, ISBN 0–85323–572–4

Lives of the Visigothic Fathers
Translated and edited by A. T. FEAR
Volume 26: 208 pp., 1997, ISBN 0–85323–582–1

Optatus: Against the Donatists
Translated and edited by MARK EDWARDS
Volume 27: 220 pp., 1997, ISBN 0–85323–752–2

Bede: A Biblical Miscellany
Translated with notes and introduction by W. TRENT FOLEY and
ARTHUR G. HOLDER
Volume 28: 240 pp., 1998, ISBN 0–85323–683–6

Bede: The Reckoning of Time
Translated with introduction, notes and commentary by FAITH WALLIS
Volume 29: 582 pp., 1999, ISBN 0–85323–693–3

Ruricius of Limoges and Friends: A Collection of Letters from Visigothic Gaul
Translated with notes and introduction by RALPH W. MATHISEN
Volume 30: 272 pp., 1998, ISBN 0–85323–703–4

For full details of Translated Texts for Historians, including prices and ordering information, please write to the following: **All countries, except the USA and Canada:** Liverpool University Press, Senate House, Abercromby Square, Liverpool, L69 3BX, UK (*Tel* +44–[0]151–794 2233, *Fax* +44–[0]151–794 2235, *Email* J.M.Smith@liv.ac.uk, http://www.liverpool-unipress.co.uk). **USA and Canada:** University of Pennsylvania Press, 4200 Pine Street, Philadelphia, PA 19104–6097, USA (*Tel* +1–215–898–6264, *Fax* +1–215–898–0404).

Translated Texts for Historians
Volume 29

Bede:
The Reckoning of Time

translated, with introduction, notes
and commentary by
FAITH WALLIS

Liverpool
University
Press

First published 1999
Liverpool University Press
Senate House, Abercromby Square
Liverpool, L69 3BX

British Library Cataloguing-in-Publication Data
A British Library CIP Record is available
ISBN 0–85323–693–3

Set in Monotype Times by
Wilmaset Ltd, Birkenhead, Wirral
Printed in the European Union by
The Cromwell Press, Trowbridge, Wiltshire

To Kendall,
who keeps the mice at bay

TABLE OF CONTENTS

ABBREVIATIONS

Augustine, *DCD* = Augustine, *De civitate Dei.* Ed. B. Dombart and
 A. Kalb. CCSL 47–48 (1955).
Bede, *De tem.* = *De templo.* Ed. D. Hurst. CCSL 119A (1969):143–234.
 Trans. S. Connolly. Translated Texts for Historians 21. Liverpool:
 Liverpool University Press, 1995.
 HE = *Historia ecclesiastica gentis Anglorum.* Ed. and trans. Bertram
 Colgrave and R.A.B. Mynors. Oxford: Clarendon Press, 1969.
 Hom. = *Homeliae.* Ed. D. Hurst. CCSL 122 (1955):1–384. Trans.
 Lawrence T. Martin and David Hurst, *Bede the Venerable: Homilies
 on the Gospels*, 2 vols. Cistercian Studies Series 110–111. Kalama-
 zoo: Cistercian Publications, 1991.
 In Gen. = *Libri quatuor in principium Genesis usque ad nativitatem
 Isaac et eiectionem Ismahelis adnotationum.* Ed. C.W. Jones. CCSL
 118A (1967).
 Letter to Plegwin = *Epistola ad Pleguinam*, in *BOD* 615–626. Trans. in
 Appendix 3.1.
 Letter to Wicthed = *Epistola ad Wichthedum*, in *BOD* 634–642. Trans.
 in Appendix 3.3.
 The Nature of Things = *De natura rerum*, in *BOD* 174–234.
 On Times = *De temporibus*, in *BOD* 580–611.
 The Reckoning of Time = *De temporum ratione*, in *BOD* 241–544.
BOD = *Bedae opera didascalica.* Ed. C.W. Jones. 3 vols. CCSL 123A–C
 (1975–1980).
BOT = *Bedae opera de temporibus.* Ed. with intro. by C.W. Jones. Cam-
 bridge, Mass.: Mediaeval Academy of America, 1943.
CCCM = *Corpus christianorum continuatio medievalis.* Turnhout:
 Brepols, 1966–
CCSL = *Corpus christianorum series latina.* Turnhout: Brepols, 1953–
CSEL = *Corpus scriptorum ecclesiasticorum latinorum.* Vienna, Prague,
 Leipzig: Temsky, etc., 1866–
Isidore of Seville, *DNR* = *De natura rerum.* Ed. Jacques Fontaine, *Traité
 de la nature.* Bordeaux: Feret, 1960.
 Etym. = *Isidori Hispaliensis episcopi Etymologiarum sive originum
 libri XX.* Ed. W.M. Lindsay. 2 vols. Oxford: Clarendon Press, 1911.

Krusch, *Studien I* = Bruno Krusch, *Studien zur christlich-mittelalter-liche Chronologie. Der 84jahrige Ostercyclus.* Leipzig: Von Veit, 1880.

Studien II = *idem.*, *Studien zur christlich-mittelalterliche Chronologie. Die Entstehung unserer heutigen Zeitrechnung.* Abhandlungen der preussischen Akademie der Wissenschaften. Jahrgang 1937. Philo-sophisch-historische Klasse 8. Berlin: Akademie der Wissenschaft-en, 1938.

MGH = *Monumenta Germaniae historica.* Leipzig, Hannover, Berlin: [various publishers], 1826–

AA = Auctores antiquissimi.

Ep. = Epistolae.

Leg. = Leges.

Poetae = Poetae latini medii aevi.

SS = Scriptores.

Script. rer. mer. = Scriptores rerum merovingicarum.

PG = *Patrologia cursus completus.* Series graeco-latina. Ed. J.-P. Migne. 167 vols. Paris: Garnier Frères, 1857–1864.

PL = *Patrologia cursus completus.* Series latina. Ed. J.-P. Migne. 221 vols. Paris: J.-P. Migne and Garnier Frères, 1841–1880, with volumes reissued by Garnier to 1905.

Pliny, *HN* = Pliny the Elder, *Historia naturalis.* Ed. H. Rackham, W.E.S. Jones, and D.E. Eichholz. 10 vols. Cambridge, Mass.: Harvard Uni-versity Press and London: Heinemann, 1938 (rpt. 1961).

ACKNOWLEDGEMENTS

There are many people to whom I owe debts of gratitude for the encouragement, advice, and salutary criticism they have given me in the course of preparing this book. My thanks go to Dr Rosamond McKitterick, who first suggested that *Translated Texts for Historians* might be interested in *The Reckoning of Time*, and to Dr Gillian Clark, who gave me valuable suggestions for recasting the presentation of the entire work. The anonymous reader who evaluated this book for Liverpool University Press proposed many significant and useful improvements, which I have been happy to incorporate. I salute that small and intrepid band of *computus*-scholars whose support, direct and (through their published works) indirect, have made my own work possible, particularly Dr Daniel McCarthy for the generous gift of offprints, Prof. Bruce Eastwood for plain talk about style, and Dr Charles Burnett for (among other things) giving me the opportunity to present my reconstruction of the technique of finger-reckoning at the Warburg Institute. Dr Mary Whitby did much more than copy-edit the manuscript; she rescued me from public shame a hundred times over by catching errors, inconsistencies and infelicities, and she did so with gentle courtesy and good humour. Both her erudition and her attention to detail are exceptional. Finally, I thank my beloved husband, Kendall Wallis, a reference librarian without peer, who tracks down obscure publications, prunes and proof-reads my prose, and tolerates the obtrusive presence of that other man in my life, "Mr Bede".

INTRODUCTION

1. "OUR LITTLE BOOK ABOUT THE FLEETING AND WAVE-TOSSED COURSE OF TIME ..."

At the close of his *Ecclesiastical History of the English People*, the great Northumbrian scholar and monk Bede (*ca.*673–735) appended a sort of autobio-bibliography. He told of how he was born on the lands of the double monastery of Wearmouth-Jarrow, and how he was offered as an oblate there at the age of seven. Educated under Benedict Biscop, the founder of the house, and under Biscop's successor, Ceolfrith, he was in due course ordained deacon and then priest by Bishop John of Hexham. That ordination was the last "event" in his life. From that time forward, Bede's life was one of sacred sameness, ruled not by change and chance, but by the stable rhythms of monastic time.

> From then on I have spent all my life in this monastery, applying myself entirely to the study of the Scriptures; and amid the observance of the discipline of the Rule and the daily task of singing in the choir, it has always been my delight to learn or to teach or to write.[1]

Bede defines his life's main achievement as the study of Scripture, and so it is fitting that his bibliography of his own writings begins with his biblical exegesis. In its wake come the lives of the saints – Felix, Anastasius and Cuthbert – and then the biographies of the "confessors" so to speak, the abbots of Wearmouth-Jarrow. His *Ecclesiastical History* follows. Then there comes his liturgical writing: a martyrology and hymns. After that, poetry on sacred subjects. The list closes with two books on the secular subjects of grammar and rhetoric: one on orthography and the other on schemes and tropes. Just below the poetry, and just above the grammar texts, Bede lists "[t]wo books, one on the nature

---

1 *cunctumque ex eo tempus uitae in eiusdem monasterii habitatione peragens, omnem meditandis scripturis operam dedi, atque inter obseruantiam disciplinae regularis, et cotidianam cantandi in ecclesia curam, semper aut discere aut docere aut scribere dulce habui.* Bede, *HE* 5.24 (566); trans. Colgrave and Mynors 567.

of things and the other on chronology (*de temporibus*): also a longer book on chronology (*item de temporibus librum unum maior*)".[2]

This book is a translation of that "longer book on chronology", which Bede also called his "little book about the fleeting and wave-tossed course of time".[3] Medieval readers knew it as *De temporibus* or *De temporibus liber secundus*, to distinguish it from Bede's earlier,[4] briefer treatment of time-reckoning, also entitled *De temporibus*. Its conventional medieval and modern title, *De temporum ratione* (*The Reckoning of Time*), is apparently derived from the incipit of the book's prologue: *De natura rerum et ratione temporum duos quondam stricto sermone libellos discentibus ut rebar necessarios composui.* ("Some time ago I wrote two short books in a summary style which were, I judged, necessary for my students; these concerned the nature of things and the reckoning of time.") The "two short books" mentioned by Bede were *De natura rerum* (*On the Nature of Things*), an introduction to cosmology, and *De temporibus* (*On Times*).

The Reckoning of Time is about measuring time and constructing a Christian calendar, or what later medieval writers called *computus*.[5] It

---

2 *Ibid.* 571. Besides the specialized studies cited below, this introduction draws on the major essay collections and monographs about Bede, notably: *Bede: His Life, Times and Writings*, ed. A.H. Thompson (Oxford: Clarendon Press, 1935); *Famulus Christi. Essays in Commemoration of the Thirteenth Centenary of the Birth of the Venerable Bede*, ed. Gerald Bonner (London: S.P.C.K., 1976); *Saints, Scholars and Heroes*, ed. Margot King and Wesley Stevens (Collegeville, Minn.: Hill Monastic Manuscript Library, 1979), esp. section entitled "The World of Bede"; *Beda Venerabilis: Historian, Monk and Northumbrian*, ed. L.A.J.R. Houwen and A.A. MacDonald (Groningen: Egbert Forsten, 1996); and the collected edition of the annual lectures delivered at St Paul's Church, Jarrow, from 1958 onwards, ed. Michael Lapidge, *Bede and His World*, 2 vols. (Aldershot: Variorum, 1994), hereafter referred to as "Jarrow Lectures"; Peter Hunter Blair, *The World of Bede* (London: Secker and Warburg, 1970); G.H. Brown, *Bede the Venerable*, Twayne's English Authors Series 443 (Boston: Twayne, 1987); and Benedicta Ward, *The Venerable Bede* (London: Geoffrey Chapman, 1990).

3 *noster libellus de uolubili ac fluctiuago temporum lapsu*: *The Reckoning of Time*, ch. 71 (544.91–92).

4 *On Times* is commonly dated to 703, and *The Reckoning of Time* to 725, based on the final events recorded in the chronicles of these works. Illustrative calculations in *On Times* 14 and 22 also indicate an *annus praesens* of 703; those in *The Reckoning of Time* 49, 52, 54, and 58 point to 725. However, Bede seems to have begun work on *The Reckoning of Time* at least as early as 722: see *The Reckoning of Time*, ch. 11, n. 121.

5 For a discussion of the history of this term, see Appendix 4, "A note on the term *computus*".

is the earliest comprehensive treatment of this subject, for though there was an abundant calendar literature before Bede's day, it was both fragmented and partisan in character. Most of it was not even in the form of texts, but rather, of tables – notably Paschal tables, which projected the dates of future Easters, calculated according to various theological criteria, and according to different notions of how such dates might be arranged to form a cyclic pattern. These tables attracted into their orbit a technical apparatus in the form of operating instructions (called *canones*) or formulae (called *argumenta*). But each of these early Paschal tables also faced a formidable array of rival tables, based on different criteria for what constituted a valid Easter date, or different projections of cyclic recurrence, or both. Hence most explanations of *computus* prior to Bede were in the form of polemical letters or prologues attached to the tables. Some were authentic, many others anonymous or pseudonymous. Few aimed beyond defending or promoting one set of principles against those of other systems.

Bede's book is very different in form and content. Although he is also the partisan of one form of Paschal table – the Alexandrian 19-year cycle, as elaborated by Dionysius Exiguus – he sought to establish its credibility by making it the basis of a comprehensive manual of time-reckoning. It was a gamble that paid off. So lucid, thorough and well-organized was Bede's exposition, so easy was it to teach from and learn from, that it can be said to have not only guaranteed the ultimate success of Dionysius' system, but to have made *computus* into a science, with a coherent body of precept and a technical literature of its own.

In making such a statement, we are also making two assumptions about *The Reckoning of Time*. The first assumption is that it is a "textbook"; the second is that its content is "science". Both assumptions need to be explained and defended. If it is a textbook, it must have been designed for programmed teaching and learning, as distinct from speculation, meditation, exhortation, etc. Behind the book, then, lies a person who is teaching, another person who is learning, and something being taught. None of these three elements can be defined apart from the others, and all must be seen in the context of the special kind of knowledge that *computus* was.

2. *COMPUTUS* AS PROBLEM-BASED SCIENCE AND *DOCTRINA CHRISTIANA*

Computus is not an easy subject to categorize.[6] It is hard to claim that it is a science, because it is essentially an application of other sciences, particularly astronomy and mathematics. It does not seek to establish universal principles, and it boasts no theory. It has no ancestor in the ancient canon of the sciences, and no posterity in the modern one.[7] On one level, it can be described as nothing more than a complicated mathematical problem: how to find the date of Easter. Solving this problem, however, involves elaborate co-ordination of lunar and solar data; it also demands some delicate interpretation of conflicting biblical texts.[8]

Easter commemorates Christ's death and Resurrection. According to the Gospels, this took place during the Jewish feast of Passover. Passover, following the divine injunctions recorded in Exodus, was to be celebrated at the first full Moon of "the first month" of the Hebrew lunar calendar. Patristic writers interpreted this to mean the first full Moon to follow the spring equinox. Because the Julian calendar is solar, the spring equinox has a fixed date each year. When this calendar was first established in 46 BC, that date was considered to be 25 March, and so it remained in popular lore for many centuries thereafter. However, the true length of the tropical solar year is slightly less than $365\frac{1}{4}$ days; 365.2422 days, to be precise. This means that inserting an extra day every four years overcompensates for the shortfall of the three preceding years. Unless some adjustment is made, the date of the true astronomical equinox will slowly creep backwards in the calendar. Indeed, at the time

6 For an overview of secondary literature on *computus*, see the Bibliography at the end of this volume, section 1 "Bibliographic Note: Literature on *computus*".

7 For this reason, *computus* has been largely neglected by historians of science. The older school is represented by George Sarton, who admitted that *computus* treatises might be "interesting from the point of view of the history of civilization and popular education, but they hardly concern the historian of science", *Introduction to the History of Science* (Baltimore: Williams and Wilkins, 1927–1947):2.19, and by Dirk Struik, *A Concise History of Mathematics*, 3rd ed. (New York: Dover, 1967):84. Though more sympathetic to the subject, David Lindberg mentions *computus* only once in his *The Beginnings of Western Science* (Chicago: University of Chicago Press, 1992):159. The context, significantly, is the work of Bede.

8 For the following summary of the *computus* problem, and the more detailed discussion below, I am indebted to a large body of technical literature on the history of the calendar: see Bibliography, section 1 "Bibliographic Note: Literature on *computus*".

of the Council of Nicaea in 325, the astronomical equinox was considered by experts to fall on 21 March. The Alexandrian Church, whose
Paschal *computus* became generalized in the Greek East after this
Council, fixed the equinox on this day, though other Churches, especially
in the West, were confused as to whether the equinox occurred on 21
March or 25 March. As we shall see, Bede pays particularly close attention to dating the equinox, for on it hangs the validity of the Alexandrian
type of Easter formula he is promoting.[9]
    Their assumption that the Jewish Passover was supposed to be cele-
brated on the first full Moon after the spring equinox also committed
Christian computists to correlating the lunar and solar cycles. This
means tracking the phases of the Moon against the background of the
Julian solar calendar commonly used in late-antique and medieval socie-
ties.[10] What the computist wants to do is to predict the *age* of the Moon
on a particular *date*, namely the next vernal equinox, so that he will
know whether that lunation qualifies as the lunation of Easter or not. If
the Moon is full on or after the equinox, it does; if not, one will have to
wait until the following lunation. But pinning down the lunar month to
the dates of a solar calendar is not easy to do, because the synodic lunar
month or lunation (the period between one new Moon and the next) is
29.5306 days, while the solar calendar is based on the tropical year of
365.2422 days. This means that the solar year does not contain a whole
number of lunar months. Twelve lunar months is roughly 11 days
shorter than the solar year, which means that the Moon on any given
calendar date will be about 11 days older next year on the same date,
than it is this year. Every two or three years, this annual advance of 11
days results in the accumulation of an entire lunar month. "Embolismic"
years, then, will have 13 lunations, but they still do not fit exactly into
solar years, for 13 lunations is 384 days. The solution is to discover a
luni-solar cycle, that is, a whole number of solar years into which a
whole number of lunar months can be inserted, so that the lunar phases
will fall on the same calendar dates after the end of the cyclic period.

    9 See section 3 below, and the Commentary on *The Reckoning of Time*, chs. 6, 16,
22, 30, 35 and 42; also Appendix 3.3, the *Letter to Wicthed*.
    10 This calendar was used throughout the Empire, though the names and some-
times the lengths of the months varied with local tradition. Cf. Alan E. Samuel, *Greek and
Roman Chronology. Calendars and Years in Classical Antiquity*, Handbuch der Altertumswissenschaft I,17 (Munich: Beck, 1972): ch. 6, esp. discussion of *hemerologia* (concordances
of the Julian calendar with local calendars), 171–178.

To complicate matters further, the Christian Church from early times celebrated Easter only on Sunday, to commemorate the historic weekday of the Resurrection. This adds a third criterion into the calculation. The Julian solar year of 365 days is 52 weeks and 1 day long. This means that if 1 January falls on Monday this year, it will fall on Tuesday next year, and on Wednesday the year following. However, the steady progress of calendar dates through the weekdays is interrupted by leap years, which introduce an extra day into the solar year in every fourth year. Hence if the present year is a leap year, and 1 January falls on Monday, next year 1 January will fall, not on Tuesday, but on Wednesday. The calendar date/weekday cycle is played out in 28 years, which is the product of the seven days of the week and the four years of the leap-year cycle. But the critical full Moon of Easter could fall on any day of the week, and so the computist had to devise a formula for correlating this calendar date to its next following Sunday.

Christians do not have the option of calculating the date of Easter year by year, using direct observation of the astronomical equinox and full Moon, in the manner in which Muslims determine Ramadan. The date of Easter must be known well in advance, for on it depend the dates of some two months of pre-Paschal observances. Hence, the Church from at least the early third century was in quest of a cycle which would permit Easter to be determined for many years in the future. But a cycle had at least two additional advantages: it diminished the possibility of recurrent conflicts over the correct date, and it meant that clergy everywhere could be assured that they were celebrating Easter in unity with their co-religionists without being dependent upon annual announcements from some distant authority. Such a system of annual announcements from the Pope in Rome had functioned in late antiquity, but to an ever increasing extent, deteriorating communications made local responsibility for the calendar a necessity.

Making a cycle of Easters inevitably involves doing some violence to astronomical realities. The periods of the Moon and of the Sun are, strictly speaking, incommensurable, so a luni-solar cycle involves artificially adjusting the lunar period to the solar period over a certain number of years.[11] *Computus*, then, is not an observational science, or a physics of time, but a technique of patterning time into repeating cycles according to certain conventions. It is closer to engineering than

---

11 The theory of luni-solar cycles is described below, section 3.

science, at least in the ancient sense of the term "science" – that branch of philosophy which investigates the natural world. *Computus* starts with a problem, not with curiosity or speculation; it ends with a product, not a hypothesis. The end-products of the Paschal disputes of the patristic period were the first Easter tables, which projected the dates of Easter for a discrete number of years into the future.[12] Nonetheless, there were historic forces at play in the evolution of the early medieval *computus* which made it more than a problem-solving skill, something which was, in fact, a vehicle for and stimulus to "science". These forces differentiate the history of *computus* from that of other technical subjects like surveying, and bring it closer to domains like medicine. Both medicine and *computus* were essentially applied sciences, but distinguished by the nobility of their respective subject-matters, and by their pretentions to encyclopaedic scope and philosophical resonance.

The factors in question centre around the breakup of the ancient liberal arts encyclopaedia, and the emergence of a new definition of erudition, *doctrina christiana*. *De doctrina christiana* is the title of St Augustine's manifesto for a new approach to the role of learning in the Christian economy, an approach already elaborated by a number of patristic writers, including Origen, the Cappadocian fathers, and Jerome. *Doctrina christiana* sees Christian erudition as a means to an end: the training of exegetes and preachers – men who could understand the Word of God and convey its message accurately and persuasively.[13] Augustine began with a radical division of knowledge into two categories: useful and useless. Christians did not need to know anything that was not useful to salvation. On the other hand, the Bible was such a rich, complex and mysterious text that its study demanded formidable erudition of a philological, historical and scientific nature. Augustine invited the Christian intellectual to pillage the "useful" knowledge accumulated by the ancients, and rearrange it in forms more pertinent to the Christian project. For instance, he thought it would be very useful if someone could compile a kind of dictionary of biblical science and mathematics, in which every number, creature, and geographical loca-

---

12 Discussed in greater detail below, section 3.

13 *De doctrina christiana* 1.1, ed. Joseph Martin, CCSL 32 (1961):6.1–3; see also Thomas L. Amos, "Augustine and the Education of the Early Medieval Preacher", in *Reading and Wisdom: The De doctrina christiana of Augustine in the Middle Ages*, ed. E. English, Notre Dame Conferences in Medieval Studies 6 (Notre Dame: University of Notre Dame Press, 1995):23–40.

tion in the Bible would be defined, explained, and arranged for handy reference.[14] In short, *doctrina christiana* involved dismantling and re-arranging ancient erudition so that it would solve specific problems of biblical interpretation or exposition.

Bede inherited a body of Christian scientific and computistical litera-ture in which this dismantling and rearrangement was already well advanced. By his day, the manuscript *computus* anthology had assumed a characteristic shape which would endure beyond the Middle Ages itself in the form of the popular almanac.[15] The anthology was the major form of *computus* literature. At the core of these anthologies were the two essential tables of the computists' art: the Julian solar calendar (a perpetual calendar, often accompanied by tables which translated calendar dates into weekdays or lunar phases) and the Paschal table, possibly equipped with *canones* and *argumenta*. But *computus* manu-scripts also became convenient "filing cabinets" for fragments of ancient scientific erudition which impinged on the issue of time. Around the nucleus of *computistica* gathered a variable halo of other subjects. Some could be considered background materials: mathematics, cosmology, astronomy. Others were associated with the calendar by analogy, like medicine (diagnostics and therapeutics being closely regu-lated by astronomical time)[16] or even prosody, which is the science of the measurement of speech in time. In short, a number of scattered leaves of the ancient encyclopaedia had drifted to rest around the calendar, attracted by association rather than by disciplinary affiliation. Bede reflects and encourages this trend by including much of this ency-clopaedic material in *The Reckoning of Time*: the excursus on medicine in ch. 35 is a typical example.

Bede also drank from a broad stream of Christian encylopaedism,

---

14 *De doctrina christiana* 2.39.58 (72.1–23).

15 On the forms and evolution of medieval *computus* anthologies, see Faith Wallis, "The Church, the World and the Time", in *Normes et pouvoirs à la fin du moyen âge*, ed. Marie-Claude Deprez-Masson, Inedita et rara 7 (Montreal: Ceres, 1990):15–29, and my 1985 University of Toronto doctoral dissertation, "MS Oxford St John's College 17: A Medieval Manuscript in its Contexts". The varieties of computistical anthology will be the theme of my forthcoming *Computus: Manuscripts, Texts and Tables*, for the series "Typo-logie des sources du Moyen Age occidental". On the *computus* anthology used by Bede, see section 5 below.

16 Cf. Faith Wallis, "Medicine in Medieval Computus Manuscripts", in *Manu-script Sources of Medieval Medicine*, ed. Margaret Schleissner (New York: Garland, 1995):105–143.

beginning with the hexaemeral literature of the patristic period, and continuing in his own day with the work of Isidore of Seville and the Irish cosmographers.[17] The influence of the Irish is especially important, because they were the first to suggest that cosmological and computistical lore might join forces in a new type of systematic *computus* treatise. One Irish text well known to Bede, entitled *De computo dialogus*,[18] takes a rather elevated view of *computus*. It opens by sketching a sort of Christian *quadrivium* of *canon divinus*, *historia*, *numerus*, and *grammatica*. By *numerus*, the *Dialogus* plainly means *computus*: *numerus* is the skill "by which future events and divine celebrations are reckoned up" (*facta futurorum et solemnitates diuinae dinumerantur*), and its meaning is illustrated by a quotation from Isidore of Seville:

> In praise of *computus*, Isidore says: "The calculation [*ratio*] of numbers is not to be scorned, for it reveals the mystery contained in many passages of Holy Scripture. Not in vain is it said of God [Wisdom 11.21]: 'Thou hast made all things in measure, number and weight'. The number six, perfect in its factors, proclaims the perfection of the cosmos by a certain numerical significance.

17 On Bede's debt to the Irish authors of *De mirabilibus sacrae scripturae* and *Liber de ordine creaturarum*, see D. Ó Cróinín, "The Irish Provenance of Bede's Computus", *Peritia* 2 (1983):238–242, and Marina Smyth, *Understanding the Universe in Seventh-Century Ireland*, Studies in Celtic History 15 (Woodbridge: Boydell, 1996): *passim*. The *De mirabilibus* (AD 655) (Smyth 168, n. 218) is printed in PL 35.2149–2200 (amongst works of Augustine of Hippo). On the *De mirabilibus*, see J.F. Kenney, *Sources for the Early History of Ireland: Ecclesiastical. An Introduction and Guide*, 2nd ed. (1968) (rpt. Dublin: Four Courts Press, 1993):275–277; and G. MacGinty, "The Irish Augustine: *De mirabilibus sacrae scripturae*", in *Irland und die Christenheit*, ed. P. Ní Chatháin and M. Richter (Stuttgart, 1985):70–83. MacGinty edited this text in his unpublished 1971 doctoral dissertation at the National University of Ireland: I have not been able to consult this work. The critical edition of the *Liber de ordine creaturarum* is that by Manuel C. Diaz y Diaz, *Liber de ordine creaturarum: Un anónimo irlandés del siglo VII* (Santiago de Compostela: Universidad de Santiago, 1972).

18 PL 90.647–652. This treatise and its companion in the PL, *De divisionibus temporum*, form part of a larger work, composed in Ireland in the 7th century. The prologue and *capitula* were edited by Jones, *BOT* 393–395, and by Alfred Cordoliani, "Une encyclopédie carolingienne de comput: les *Sententiae in laude computi*", *Bibliothèque de l'École des Chartes* 104 (1943):237–242. Dr Dáibhí Ó Cróinín of University College, Galway, is preparing an edition of this work which in its reconstructed form will be entitled *De ratione temporum uel de computo annali*. The treatise appears in the "Sirmond" manuscript of Bede's *computus* anthology, Oxford, Bodleian Library, Bodley 309, fols. 62v–64v (see section 5 below).

Likewise the forty days wherein Moses, Elijah and our Lord himself fasted cannot be understood without a knowledge of number. There are other numbers in Holy Scripture whose symbolism [*figuras*] can only be unravelled by those knowledgeable in this science. Using the science of numbers, we have an ability to stand fast [*consistere*] to some degree, when through this science we discuss the course of the months or learn the span of the revolving year. Indeed through number we are taught so that we do not fall into confusion. Take number away, and everything lapses into ruin. Remove *computus* from the world, and blind ignorance will envelop everything, nor can men who are ignorant of how to calculate be distinguished from other animals."[19]

But *numerus* or *ratio numerorum* are terms charged with metaphysical import. *Ratio numerorum* was established at the beginning of the world when God made evening and morning into the first day, and created the Sun and Moon to mark the seasons, days and years.

PUPIL: Tell me when this reckoning [*ratio (numerorum)*] was first invented. TEACHER: At the time when the creatures were made, that is, in the beginning of the world. For that was when number first began, as we read in Genesis, 'And the evening and the morning were the first day' ... It also speaks of number when it says 'the first day, the second day, the third day ...' Again, God

19 *Isidorus in computi laude dixit: Ratio numerorum contemnenda non est. In multis locis sacrarum Scripturarum, quantum mysterium habet, elucet. Non enim frustra in laudibus Dei dictum est: Omnia in mensura et numero et pondere fecisti. Senarius namque numerus, qui partibus suis perfectus est, perfectionem mundi quadam numeri sui significatione declarat. Similiter et quadraginta dies, quibus Moses et Elias et ipse Dominus jejunaverunt, sine numerorum cognitione non intelliguntur. Sic et alii in Scripturis sacris numeri existunt, quorum figuras non nisi noti hujus artis scientiae solvere possunt. Datum etiam nobis est, ex aliqua parte sub numerorum consistere disciplina, quando mensium curricula disputamus, quando anni spatium redeuntis per numerum agnoscimus. Per numerum siquidem ne confundamur, instruimur. Tolle numerum a rebus omnibus, et omnia pereunt. Adime saeculo computum, et omnia caeca ignorantia complectitur. Nec differi possunt a caeteris animalibus, qui calculi nesciunt rationem: De computo dialogus* 647; my translation. The quotation is from *Etym.* 3.4. Isidore's immediate source for both the phrasing and sense of this passage is Cassiodorus, *Institutiones* 2.4.7; ed. R.A.B. Mynors (Oxford: Clarendon Press, 1937):141.1–7. The final phrase may ultimately derive from Augustine, *De libero arbitrio* 2.16.42, ed. W.M. Green, CCSL 29 (1970):265.25–26: *Formas habent creaturae, quia numeros habent, adime illis haec, nihil erunt.* ("Created things have forms because they have numbers; take this away from them, and they are nothing.")

spoke of number when He said of the Sun and Moon, 'And let them be as signs for seasons and days and years'. Who can understand days and seasons and years, save by number?[20]

Number is God's privileged instrument of order in his cosmos. As the Teacher says, "All things are fashioned from the first nature of things, and are seen to be shaped by the *ratio* of numbers. For this was the principal exemplar in the Creator's mind."[21] *Numerus* is an "art", and in its widest sense is coterminous with nothing less than *philosophia*,[22] but it specifically pertains to that branch of philosophy called *physica*, which in turn is divided into the *quadrivium* of arithmetic, geometry, music and astronomy.[23]

I think it exceedingly likely that this Irish text was a direct inspiration for Bede's project in *The Reckoning of Time*. Besides sketching out an order and system for *computus*, it endowed the lore of the calendar with two additional qualities: an encyclopaedic dimension, and a theological resonance. These qualities were certainly implicit in the pre-Bedan *computus* tradition, but that a text contained in a manuscript undoubt-

---

20 *D[ISCIPULUS]. Dic ergo quando primum inventa est ista ratio? M[AGISTER]. Ex illo tempore ex quo factae sunt creaturae, hoc est ab origine saeculi. Tunc enim primum numerus initiavit: sicut in Genesi legitur: 'Et factum est vespere et mane dies unus'. . . De numero autem dixit, quando dixit, dies unus, sic secundus, sic tertius . . . Iterum de numero dixit Deus, quando dixit de sole et luna: 'Et sint in signa et tempora et dies et annos'. Quis enim potest intelligere dies et tempora et annos, nisi per numerum?*: De computo dialogus 649A. For echoes in Bede's works, see *In Gen.* 1.1.17 (18.510–527). In *Hom.* 1.6, Bede refers to God as "maker of time" (*temporum conditor*) (38.36), and in *Hom.* 2.4 as "author and controller of time" (*auctor et ordinator temporum*) (226.51–52).

21 *Omnia quaecumque a prima rerum natura constructa sunt, numerorum videntur ratione formata. Hoc enim fuit principale exemplar in animo Conditoris . . .*: ibid. 649 A. Cf. *Hom.* 1.6., where Bede states that Christ deliberately chose to be born at a time of universal peace. This peace is demonstrated by the census itself, a symbol of the order that number and computation could impose on the world: "For what could be a greater indication of peace in this life than for the entire world to be enrolled by one man and to be included in a single coinage?", trans. Martin and Hurst 52 (*Quod enim maius in hac uita potuit esse pacis indicium quam ab uno homine orbem describi uniuersum atque unius census numismate concludi?*: 37.12–15).

22 *D[ISCIPULUS]. Haec igitur ars, hoc est numerus, quod nomen generale habet? M[AGISTER]. Philosophia scilicet, quia omnis sapientia, philosophia nominatur.* ("PUPIL: This art, that is, number – what is its general name? TEACHER: Philosophy, to be sure, because all wisdom is called philosophy.") *De computo dialogus* 649B.

23 *Ibid.* 649D.

edly consulted by Bede sets these elements forth in an explicit way, argues for a more direct dependence.

Yet even in a systematic treatise such as *The Reckoning of Time*, Bede does not attempt, as later writers would, to define *computus* as a "discipline" or "art". Taking our cue once again from *De computo dialogus*, we would perhaps understand his approach better if we defined *computus* as one of the "four divisions of Scripture", or "four things necessary in the Church of God".[24] Bede seems to treat *computus* more as a "function" than as a "subject"; what it is, and what it is permitted to do or explain, are defined by its variable associations with its subject-matter, time. Hence *The Reckoning of Time* can extend its elastic boundaries to include astronomy and cosmology, but also moral theology and biblical exegesis, as applied to time. By the same token, Augustine's concept of *doctrina christiana* permitted Scriptural commentary to incorporate *computus* as a strategy for interpreting a passage about time, without any sense of embarrassment at having transgressed some genre boundary, or having illegitimately used one kind of authority or knowledge to explain something else completely different.[25] A common object, namely time, permits different modes of discussion to operate simultaneously and without conflict.

Computus, then, hovers somewhere between technique and a kind of science-in-progress, tentatively combing the flotsam and jetsam of ancient erudition in search of some as yet undefined identity. What can we make of such indeterminacy? The tendency to abandon scientific theory in favour of technique, and the fragmentation and rearrangement of ancient science into new encyclopaedic formats, have in the past been seen as signs of the cultural poverty of early medieval Europe.[26] However, newer scholarship, especially the work of Brigitte Englisch, is

---

24 *Augustinus dixit de quatuor divisionibus scripturae: Quatuor necessaria sunt in Ecclesia Dei . . .* ("Augustine says concerning the four divisions of Scripture: four things are necessary in the Church of God . . .") *ibid.* 647.

25 Augustine, *De doctrina christiana* 2.29.46 (64.28–65.45). For examples of Bede's use of *computus* themes in his exegetical works, see C.W. Jones, "Some Introductory Remarks on Bede's *Commentary on Genesis*", *Sacris Erudiri* 19 (1969–1970):115–198. Parallels between *The Reckoning of Time* and *In Gen.*, *Hom.*, and other exegetical works are signalled in the notes to the translation.

26 See in particular William H. Stahl, *Roman Science: Origins, Development, and Influence to the Later Middle Ages* (Madison: University of Wisconsin Press, 1962), and his introduction and commentaries to his translations of Macrobius' *Commentary on the Dream of Scipio* (New York: Columbia University Press, 1962) and Martianus Capella, in

questioning this consensus. What was once scorned as "decline" may, in fact, be "liberation". Englisch points out that ancient science was strictly circumscribed in its range of inquiry and in its applications, because the classical conception of the mathematical sciences – the *quadrivium* of the *artes liberales* – was essentially speculative. Studying mathematics was a mental and spiritual propaedeutic to the abstract reasoning required of philosophy, a way of familiarizing oneself with the non-material world of ideas. Consequently, the ancients had a low opinion of "experimentation" in any sense of the word; scientific questions were resolved, ideally, by metaphysical reflection.[27] By dismantling the liberal arts and valorizing such instrumental applied sciences as *computus*, Englisch argues, the early Middle Ages fundamentally reoriented Western science towards the resolution of problems. This in turn changed the definition of science itself. She sees Bede's way of combining elements from the traditional *artes liberales* with the technical, problem-oriented literature of the Christian calendar as essentially scientific in that it is structured, empirical and rational. Bede's task necessitated such a synthesis. He was obliged not only to harmonize scientific data with theological implications, but to transform the calendar into a reproducible system, logically consistent and hence capable of resolving doubts and refuting criticism.[28] But in so doing, Bede also invited the *artes liberales* back in on fresh terms, as partners in the making of answers to problems.

This hypothesis is very suggestive, and sheds fresh light on the genesis of scholastic science. Far from being merely a recovery of ancient learning, scholastic science innovated from its very beginnings by making room within its institutional and conceptual framework for

---

Martianus Capella and the Seven Liberal Arts, 2 vols. (New York: Columbia University Press, 1971–77).

27 Brigitte Englisch, *Die Artes liberales im frühen Mittelalter (5.–9. Jh.): Das Quadrivium und der Komputus als Indikatoren für Kontinuität und Erneuerung der exacten Wissenschaften zwischen Antike und Mittelalter*, Sudhoffs Archiv, Beiheft 33 (Stuttgart: Franz Steiner, 1994): section 2.1, esp. p. 23.

28 Brigitte Englisch, "Realitätsorientierte Wissenschaft oder praxisferne Traditionswissen? Inhalte und Probleme mittelalterlicher Wissenschaftsvorstellungen am Beispiel von *De temporum ratione* des Beda Venerabilis", in *Dilettanen und Wissenschaft. Zur Geschichte und Aktualität eines wechselvollen Verhältnisses*, ed. Elisabeth Strauss, Philosophie und Repräsentation/Philosophy and Representation 4 (Amsterdam: Rodopi, 1996):19–21.

applied sciences, of which medicine was the most significant. Medicine and the sciences proved highly susceptible to the influence of another problem-based discipline, law. Law, natural science and medicine redis-covered the ancient rhetorical exercise of *quaestio*, and transformed it into an instrument of dialectical analysis and even speculative research. The "experimental" strain within high medieval science indicates how deeply rooted the problem-solving dimension of scientific thought had become. Though detailed exploration of this theme exceeds the mandate of this brief introduction, it suggests that the particular problem-centred approach of *computus* had a long-term importance in the history of Western culture, as part of a fundamental reorientation of science between antiquity and the rise of the universities.

But at least at this phase of its history, it is more appropriate to define *computus* as *doctrina christiana*. Bede would have appreciated this desig-nation, for not only was he well acquainted with Augustine's *De doctrina christiana*,[29] but he was brought up in an educational regime based on its principles. Not only did this new paedagogy dispense with the ancient mathematical sciences of the *quadrivium* and ancient natural history as such, but it also taught the young in a manner which was very different from that of the ancient school. In Bede's world there were no professional teachers who systematically imparted a formalized syllabus of subjects to groups of people assembled only for the purpose of learning. There were, in fact, no schools, that is, no institutions created for and exclusively dedicated to teaching. Instead, the monastery trained monks, and the episcopal *familia* clergy, by initiating them into the functions, duties and mores of their calling. These were not absorbed through "lessons" on "subjects", with set texts expounded by a master, but rather through the socializing force of the *vita communis*, through the self-instruction of *lectio*, and through unstructured and informal *confabulatio mutua*.[30] Its goal was perfection in the practice of a religious

29 On Bede's use of *De doctrina christiana*, see Roger Ray, "Bede, the Exegete, as Historian", in *Famulus Christi*, 125–140.

30 See especially C.W. Jones, "Bede as Early Medieval Historian", *Mediaevalia et Humanistica* 4 (1946):26–36 – a sketch for his later monograph, *Saints' Lives and Chroni-cles* (Ithaca: Cornell University Press, 1947) –, his introduction to *BOD* 1.vi, and his final word on the subject, "Bede's Place in Medieval Schools", in *Famulus Christi*, 261–285, esp. p. 263. In the latter essay, he draws heavily on recent scholarship on early medieval educa-tion, esp. Pierre Riché's *Education et culture dans l'Occident barbare, 6e–8e siècles* (Paris: Ed. du Seuil, 1962), and Henri-Irénée Marrou, *Histoire de l'éducation dans l'Antiquité*, 3rd

vocation, its method was imitation, and its medium was the relationship between *seniores* and *iuniores*. Monastic education had content, but no curriculum. Moral ascesis, scripture study, musical drill, grammar and *computus* were not disciplines learned separately and according to a staged syllabus, but rather organically connected reference points within an integrated *conversatio.*[31]

This was certainly the type of teaching to which Bede seems to have been accustomed. He himself names only three men as his teachers: Abbots Benedict Biscop and Ceolfrith, and the monk Trumberht. Bede does not say what Benedict Biscop and Ceolfrith taught him: apparently, they taught him virtually everything, including *computus*.[32] Trumberht, who trained under Chad at Lastingham, instructed Bede in the Scriptures,[33] but given the dominance of Scripture in monastic erudition, this is

---

ed. (Paris: Ed. du Seuil, 1964). These methods of monastic education are analysed by Detlef Illmer, *Formen der Erziehung und Wissenvermittlung im frühen Mittelalter,* Münchner Beiträge zur Mediävistik und Renaissance-Forschung 7 (Munich: Arbeo-Gesellschaft, 1971): ch. 2 fleshes out the broader assertions of Marrou and Riché with detailed documentary evidence. See also Laetitia Boehm, "Die wissenschaftstheoretische Ort der Historia im früheren Mittelalter: die Geschichte auf dem Wege zur 'Geschichtswissenschaft'", in *Speculum historiale: Geschichte im Spiegel von Geschichtsschreibung und Geschichtsdeutung. Festschrift für Johannes Spörl* (Freiburg and Munich, 1965):666, 681; Jan Davidse, "On Bede as Christian Historian", in *Beda Venerabilis: Historian, Monk and Northumbrian*, 8–9.

31 C.W. Jones, "Bede's Place", 263, 267, and his introduction to *BOD*, 1:vi. See also *BOT* 135–136, where he argues that Bede's students may only have been expected to master a few chapters of *The Reckoning of Time* each year. Jones further suggests that Bede wrote for a very restricted and local audience, and that he saw *computus* as auxiliary or propaedeutic to biblical exegesis. However, in "Bede's Place" he claims that Bede and his colleagues had to abandon this monastic approach for "classrooms" and "structured texts". He ascribes this change to the rising numbers of students, but does not furnish evidence, or pursue the implications for *computus*. On the non-programmed and imitative nature of monastic training, see Jean Leclercq, "Les études dans les monastères du Xe au XIIe siècles", in *Los monjes y los estudios*, IV Semana de estudios monasticos, Poblet, 1961 (Poblet: Abadia de Poblet, 1961):105–117, and "Pédagogie et formation spirituelle", in *La scuola nell'occidente latino dell'alto medioevo*, Settimane di studio 4 (Spoleto: Centro italiano di studi sull'alto medioevo, 1956):255–290. The argument advanced by Jones in *Saints' Lives and Chronicles* that Bede might have been choirmaster at Jarrow, with attendant responsibilities for teaching and the library, has not met with acceptance: see Brown, *Bede the Venerable*, 19.

32 In the *Letter to Plegwin* 14, Bede reveals that he was studying *computus* as a boy. Dorothy Whitelock argues that his teacher was Ceolfrith: "Bede's Teachers and Friends", *Famulus Christi*, 24; cf. Ward, *The Venerable Bede* 7, and Stevens "Bede's Scientific Achievement", *Jarrow Lecture* 1985, 659.

33 *HE* 4.3 (342).

equivalent to a general education. Bede clearly states that he himself taught, but he does not tell us that he taught a specific subject. Monastic teaching seems to have no predetermined schedule or sequence; it has no courses, no one ever "graduates", and its rhythm is set by the monk's own reading, more particularly, the text of Scripture he might be studying or commenting upon at the time. The description of Bede's last days provided by his fellow monk, Cuthbert, fills in the details of this picture:[34] between his Psalms and prayers, the ailing Bede gave lessons, taught and dictated, but Cuthbert does not specify what he taught, or what the lessons were. The heterogeneous character of his two final projects – a translation of St John's Gospel into the vernacular, and the preparation of an anthology of excerpts from Isidore of Seville's survey of cosmology, *De natura rerum* – ought perhaps to suggest that these final lessons were not necessarily programmed elements of a formalized curriculum, but generic monastic instruction.

This distinctively monastic praxis of vocational education calls into question any easy definition of *The Reckoning of Time* as a "textbook", for there seem to be no courses or classrooms in which it could be *used*. C.W. Jones goes so far as to reject this label entirely. *The Reckoning of Time*, he argues, was really an elaborate defence of Bede's own orthodoxy. The background is this. Shortly after the appearance of *On Times* in 703, Bede heard that the chronology of the World Ages which he had proposed in that work had been the subject of discussion at the dinner-table of Bishop Wilfred and his entourage. Someone accused Bede of heresy, for by dating the Incarnation to *annus mundi* 3952, instead of the conventional date found in Eusebius' chronicle of AM 5199, Bede had (it was claimed) displaced the Incarnation from the Sixth Age of the World. Did not the Apostle Peter declare that for God, a thousand years were as a day?[35] If the six World-Ages correspond to the six days of Creation, they should each endure at least a thousand years. Bede defended himself in his *Letter to Plegwin* (AD 708) both by demonstrating that the World-Ages were not, as the vulgar thought, each one thousand years in length, and by justifying the revised Old Testament chronology of Jerome's Vulgate translation, the *hebraica veritas*.[36]

---

34 *Epistola de obitu Bedae*, in Colgrave and Mynors' ed. of *HE*, 582.
35 II Peter 3.8.
36 For a translation of the *Letter to Plegwin*, see Appendix 3.1 of the present volume.

Jones speculates that malicious rumours continued to circulate nonetheless, and that Bede felt obliged to expand *On Times* to clarify and defend his views. Proof is that he summarized the *Letter to Plegwin* in the preface of *The Reckoning of Time*. *The Reckoning of Time* is therefore not a textbook, but an extended *pièce justificative* for the world-chronicle in chapter 66, where Bede reiterates and fleshes out the controversial chronology.[37]

Bede certainly smarted under this insult, but to suggest that he brooded over it for twenty years, only to defend himself by restating his original position at greater length, seems somewhat contrived. After all, there is no contemporary evidence that the charge of heresy was taken seriously by anyone save Bede himself, or that it persisted after the publication of the *Letter to Plegwin*. Chiliasm is a theme in *The Reckoning of Time*, but only one amongst many, and confined to the final chapters of the work, which are essentially a reprise of the *Letter to Plegwin*. Ironically, given that he thought them so central to the mission of the book, Jones did not even include these chapters in his first edition of *De temporum ratione*. Finally, Jones' own observation that Bede's texts furnished the materials for a true formal classroom education in the Carolingian period[38] seems to undermine the assertion that they could not possibly have been so used by Bede himself. In fact, the form and content of *The Reckoning of Time* reveal that programmed instruction was exactly what Bede had in mind.

In the preface to *The Reckoning of Time*, Bede states that he not only "gave" his earlier works on *computus* to his fellow monks, but also that he "began to expound" these books to them.[39] This suggests that he expected these works to be used not only for private study, but also as the basis for some kind of instruction. At several points in *The Reckoning of Time*, Bede actually lets us glimpse him at work, teaching *computus*. Surprisingly, the scene does not look much like *confabulatio mutua*; on the contrary, it resembles, of all things, the schoolrooms of classical Rome or scholastic Paris. To begin with, *The Reckoning of Time* is a *lectio* on a text, only in this case the "text" is two tables: the solar calendar and the Paschal table. Like the ancient *grammaticus* with his Vergil or the scholastic *magister* with his Aristotle, Bede teaches by

---

37 "Some Introductory Remarks", 194–195; "Bede's Place", 268.
38 "Bede's Place", 263–264.
39 *. . . cum fratribus quibusdam dare atque exponere coepissem . . .* (263.3–4).

commenting, line by line. Within that framework, he proceeds in logical order, starting with a technical prologue: finger calculation and the names of the units of time. He tackles the solar calendar first, as it is the frame of reference within which all computistical calculations take place. He conceives the Julian calendar as a hierarchy of units of time: the day (chs. 5–7), the week (chs. 8–10), the month (chs. 11–29), and the solar year itself (chs. 30–43). Then he turns to the Paschal table of Dionysius Exiguus, which is a cycle of 19 years (chs. 44–46). Bede explains each of its eight columns in turn: the *annus Domini* (ch. 47), indictions (chs. 48–49), lunar epacts (chs. 50–52), solar epacts or concurrents (chs. 53–55), the lunar cycle (chs. 56–58), the Easter terminus (chs. 59–60), Easter Sunday (ch. 61), and the age of the Moon on Easter Sunday (ch. 62). But the 19-year cycles of Dionysius can be joined together to form an even larger unit of time, the Great Paschal Cycle of 532 years (ch. 65). Straddling the Great Cycles are the World-Ages chronicled in ch. 66; beyond them lie the unmeasured tracts of future time, and the very end of time itself (chs. 67–71). In short, the book is based on two overlapping patterns: the ascending hierarchy of the units of time, and the graphic forms of the calendar and Paschal table. The former derives its purpose and meaning from its integration with the latter, for the business of *computus* is to make such tables, and make them right. But the tables also transcend the merely technical and utilitarian by being visualized as a grand scheme of time, stretching from the briefest atom of duration to the eternity of the age to come. This in itself would not be sufficient to prove that Bede intended *The Reckoning of Time* to be a textbook, or actually taught *computus* according to such a plan. But it does suggest that he was beginning to conceive of *computus* as more than just a problem; it was an "art".

But there is further evidence that Bede actually taught *computus* in a programmed manner. To begin with, he frequently complains that it is much easier to teach *computus* face to face than to commit it to writing,[40] which suggests that his book was an effort to transcribe what he did in the classroom. Moreover, when he explains a formula in *The*

40 E.g. *The Reckoning of Time*, ch. 4: "These things can be both learned and taught more easily through speech than by the pen of a writer"; ch. 16: "Much can be said about this, but to better effect by someone speaking than by a writer"; ch. 55: "But many aspects of this discipline, just as of the other arts, are better conveyed by the utterance of a living voice than by the labour of an inscribing pen."

Reckoning of Time, Bede works in a characteristically paedagogical manner. He starts with a simple worked example, proceeds to a more complicated one, and finally inverts the process, i.e. begins with the answer and works backward to the formula, in order to corroborate its validity.[41] He wants not only to drill the student in the formula, but also to teach him why it works. Moreover, Bede explicitly "streams" his readership according to the degree of their background preparation. For example, he provides a number of methods for finding the zodiac sign in which the Moon is located, depending on whether the pupil has or has not learned the names and sequence of the signs, and does or does not know how to perform sophisticated arithmetical operations (chs. 16–19).[42] At one point, he begs the more advanced students to explain the zodiac signs to the less well-prepared (ch. 16).[43] Bede constructs an ingenious experiment with hanging lamps in a darkened church to prove that a heavenly body which appears to an earthly observer to be higher up in the sky than another heavenly body may in fact be closer to Earth (ch. 26). Apostrophes to the reader are frequent, and at times we can catch echoes of classroom dialogues, and Bede's own somewhat salty schoolmaster's language.[44] Finally, Bede likes to illustrate his points with familiar examples which summon up the experience of life in Anglo-Saxon England, a world lit and warmed by fire: torches advancing in the night are like stars in the distant heavens (ch. 7); an open-air fire-pit demonstrates the theory of climatic zones (ch. 34).

A case can be made, then, that Bede did indeed teach *computus* in a formal, programmed manner, and possibly to purposefully constituted groups of clergy. If he did, others probably did as well; this might explain in part why *The Reckoning of Time* was considered so useful, for all the disadvantages of text over direct pupil–teacher interaction. To be sure, teaching in this way does not mean that Bede regarded intellectual training, let alone *computus*, as other than *doctrina christiana* in the Augustinian sense: vocational formation for the exegete, preacher,

---

41 For a good example of this method, see ch. 17.

42 Bede did not, however, entirely approve of computistical tables as a crutch for the lazy or ignorant: see Commentary on chs. 19 and 23.

43 For other illustrations of Bede's classroom manner, see the Commentary on *Reckoning of Time*, chs. 16 and 46.

44 See Commentary on ch. 46.

and man of prayer.[45] He did not teach in the manner of the Carolingian masters, who turned *computus* into a platform for reconstructing the classical *quadrivium*:[46] significantly, when he adapted Book 5 of Isidore of Seville's *Etymologies* for his own *On Times*, he stripped out the references to classical writers and substituted Scriptural quotations.[47] Like the Irish scholars of his age, Bede saw *computus* as a "division of Scripture", that is, as a conjunct to exegesis. It was part of an undifferentiated body of Christian erudition that could be shaped as needed into exegesis, homily, or hagiography. But that, apparently, did not impede him from composing a work which handled *computus* as if it were a "subject", to be taught integrally and systematically, according to *ratio*.

3. A BRIEF HISTORY OF THE CHRISTIAN CALENDAR BEFORE BEDE

The Reckoning of Time confronts an array of technical and theological issues which can only be understood in the context of the "Easter question" which vexed the Church for the first seven centuries of its existence. This Easter question is in fact two questions: first, what *criteria* determine a suitable date for Easter? and secondly, how can such a date be *calculated* in advance? The first question is essentially theological. The second is mathematical and astronomical.

Three interconnected issues surrounded the criteria for a valid Easter date:

45 On the clerical and vocational ethos of Bede's scholarship, see H. Mayr-Harting, "The Venerable Bede, the Rule of St Benedict, and Social Class" (Jarrow Lecture, 1976):421; and H.-J. Diesner, "Das christliche Bildungsprogramm des Beda Venerabilis (672/3–735)", *Theologische Literatur-Zeitung* 106 (1981):865–872. On the pastoral role of monks in early Anglo-Saxon England, see T.R. Eckenrode, "The Venerable Bede and the Pastoral Affirmation of the Christian Message in Anglo-Saxon England", *Downside Review* 99 (1981):258–278; and T.L. Amos, "Monks and Pastoral Care in the Early Middle Ages", in *Religion, Culture and Society in the Early Middle Ages: Studies in Honor of Richard E. Sullivan*, ed. T.F.X. Noble and J.J. Contreni, Studies in Medieval Culture 23 (Kalamazoo: Medieval Institute Publications, 1987):165–180.

46 See section 6 below. The most striking example of this process is the so-called "Encyclopaedia of 809", a *computus* anthology with considerable cosmological expansions drawn from ancient sources: see Arno Borst, "Alkuin und die Enzyklopädie von 809", in *Science in Western and Eastern Civilization in Carolingian Times*, ed. P.L. Butzer and D. Lorhmann (Basel: Birkhäuser Verlag, 1993):53–78.

47 "Bede's Place", 267–268 and 281 n. 44. This "purging" procedure is paralleled in Bede's grammatical works, notably *De schematibus et tropis.*

1. The annual celebration of Easter is supposed to coincide with the Jewish Passover, the time of the historic Passion and Resurrection of Christ. But if Christians, at least from the second century on, chose to celebrate Easter only on a Sunday, even though Passover (14 Nisan in the Jewish lunar calendar) can fall on any of the seven days of the week, what is the range of lunar dates – i.e. dates in Nisan – on which it is permissible to celebrate Easter?

2. Nisan is "the first month",[48] and falls in "spring". But spring, unlike Nisan, is a solar season. When does "spring" begin, and when must Nisan begin if it is to be "the first month"?

3. Given the above, within what range of dates in the Julian calendar can Easter fall?

Easter involves, then, three interlocking, but distinct criteria. Sunday is a day of the week, which is an arbitrary count of seven days, unconnected to any seasonal or astronomical phenomenon. Nisan is a lunar month; it begins with the first thin crescent of the first lunation of spring. But spring is a phenomenon produced by the Sun's annual journey of $365\frac{1}{4}$ days around the zodiac. To keep Nisan in the spring, the Jews had to adopt a luni-solar calendar, which adjusted the lunations to the solar year. Potentially, such a calendar could match up dates in a solar calendar, like the Julian calendar, with phases of the Moon.

Primitive Jewish Christians, and many early Christian communities especially in Syria and Asia Minor, chose to coincide their celebration of Easter with the Jewish Passover. These were the "Quartodecimans", or those who celebrated on the *fourteenth* day of Nisan, with the Jews, regardless of weekday. Other Churches, especially in Egypt and the West, insisted that Easter be celebrated on the historic weekday of the Resurrection, Sunday. Both groups had strong feelings about the theological symbolism of their choice. For the Quartodecimans, Christ was the true Paschal Lamb, slain to liberate God's people from the bondage of sin. For the Sunday group, Easter was the anniversary of the Resurrection, not of the Passion, the "first day" of the new age.[49] Eventually

48 Exodus 12.2.
49 There are two convenient anthologies of patristic texts on the significances accorded to Easter, and the controversies these entailed. These are Rainero Cantalamessa, *Easter in the Early Church: An Anthology of Jewish and Early Christian Texts*, ed. and trans. James M. Quigley and Joseph T. Leinhard (Collegeville, Minn.: Liturgical Press,

Quartodecimanism was officially rejected as a "Judaizing" heresy. But that did not solve the Easter question: if anything, it complicated it.

Rejecting Quartodecimanism raised two fresh difficulties. Was it *never* permissible to celebrate on 14 Nisan, even if it fell on a Sunday? At least from the beginning of the third century, the Roman Church was clear that it was not. If Easter commemorated the Resurrection, it could not be celebrated until 16 Nisan; this is the third day, counting inclusively, after the Passover, for the Resurrection took place on the third day after the Passion. But other Christian communities were not so certain. Many were unwilling to abandon the link with Passover; the very name of the Christian Easter, *Pascha*, attests to this. For them, 14 Nisan was still acceptable. Others held that since the historic Passion took place on 15 Nisan, at least according to the Synoptic Gospels, this should be the earliest date.

The second difficulty concerned the determination of Nisan itself. Should Christians follow Jewish rules? The issue was, to be sure, highly political, as Christians increasingly resisted any hint of dependence on the Jews. But that resistance was expressed in "scientific" terms. The Jews, it was claimed, were not observing the equinox in determining Nisan. Since the equinox determined the beginning of the year in which Nisan was supposed to be the first month, the Jewish *computus* was in error. Christians ought therefore to develop their own *computus*, based on the equinox. The earliest known Easter table based on this principle was published by Dionysius, bishop of Alexandria, between 257 and 265.[50]

Western Christians, however, took a different approach to "spring". Hippolytus of Rome devised a Paschal table in 222, which operated on the assumption that the earliest date of 14 Nisan was 18 March, the day

---

1993), and August Strobel, *Texte zur Geschichte des frühchristlichen Osterkalendars* (Münster in Westfalen: Achendorffsche Verlagsbuchhandlung, 1983). For discussion, see Strobel, *Ursprung und Geschichte des frühchristlichen Osterkalendars*, Texte und Untersuchungen 121 (Berlin: Akademie Verlag, 1977), and Odon Casel, *La Fête de Pâques dans l'église des Pères*, Lex orandi 37 (Paris: Cerf, 1963). The "Easter controversy" is also summarized concisely by Jones, *BOT* 6–77, though a number of his conclusions have been revised by newer evidence.

50 V. Grumel, "Le problème de la date paschale aux IIIe et IVe siècle", *Revue des études byzantines* 18 (1960):161–178; Marcel Richard, "Le comput pascal par octaétéris", *Muséon* 87 (1974):311. Dionysius' cycle and criteria are mentioned in Eusebius, *HE* 7.20.

the Sun entered the constellation of Aries in the old Julian calendar, and hence the astronomical beginning of spring. The earliest possible date for the beginning of Nisan was therefore 5 March. Evidently this definition of "spring" roused controversy, for shortly afterwards, an elaborate effort was launched to justify it on theological, rather than merely astronomical, grounds. This was the aim of a revised version of the Hippolytan table which appeared in 243, falsely ascribed to Cyprian of Carthage.[51] The criteria it invoked were intriguing, and as we shall see, found an echo in *The Reckoning of Time*, ch. 6. The world, said "Cyprian", must have been created at the spring equinox, which in the old Roman calendar was fixed at 25 March (ch. 4). The Moon was created on the fourth day, i.e. 28 March, at dusk, and it must have been created "perfect", i.e. full (ch. 5). Thus 29 March would have been 14 Nisan, the first full Moon of spring.[52] At the time of Creation, however, there was no 1 Nisan, because the Moon was created full; hence, in Jones' phrase "the first month of the lunar year was *virtually* inaugurated on [the evening of] March 16",[53] and the first 1 Nisan would have been 17 March. The first *real* 1 Nisan occurred in the following year, *anno mundi* 2. Since the Moons fall eleven days earlier with respect to the solar calendar with each passing year, the first historic 1 Nisan began on 5 March, or more precisely, on the evening of 4 March (ch. 6). But this Roman solution to the question of "spring" and the calendar limits for Easter was not allowed to go unchallenged. It may have been in response to the Romans, as much as to the "Judaizers", that Dionysius of Alexandria sent out his new 8-year Easter table, accompanied by a clear statement, in the form of a Paschal letter, that it was not licit to celebrate Easter before the equinox, the astronomical marker for spring.[54]

Contrary to later legend, the Council of Nicaea in 325 laid down no

51 Ed. Wilhelm Hartel, CSEL 3.1 (1871):248–271; trans. George Ogg, *The Pseudo-Cyprianic De pascha computus* (London: S.P.C.K., 1955).

52 Bede points out in *The Reckoning of Time*, ch. 5, that the Christian computistical day begins at sunset, and thus the full Moon of the evening of 28 March "belongs" to 29 March. If the Paschal 14th Moon appeared on Sunday evening, it "belonged" to Monday, and one would have to wait until the following Sunday to celebrate Easter. Hence the computist of Bede's Alexandrian persuasion looked for the 14th Moon, but celebrated on the 15th day.

53 Jones, *BOT* 12.

54 On Paschal letters, see "Paschal Letters", in *A Dictionary of Christian Antiquities*, ed. Sir William Smith and Samuel Cheetham (London: John Mullary, 1908):1462–1464.

criteria for Easter save one: that it could not be celebrated on 14 Nisan, "with the Jews", even if that day fell on a Sunday.[55] The Alexandrian Church interpreted this to mean that if 14 Nisan fell on a Sunday, Easter should be postponed to the following Sunday – in other words, the lunar limits were effectively 15–21. This meant that Easter could still be the antitype of Passover without ever being celebrated on Passover.[56] The Romans held firm to their lunar limits of 16–22, though they found them increasingly difficult to defend, and in practice allowed celebration on the 15th.[57]

According to Eusebius, however, the Emperor Constantine interpreted the decrees of Nicaea as forbidding the celebration of Easter "twice in the same year".[58] The *Apostolic Constitutions* declares that the only way to avoid this is by observing the equinox – the computistical year being defined as the period from one spring equinox until the following one.[59] But this brought no resolution to the Easter question, for there was considerable confusion as to what the equinox was supposed to be the limit *of*. For the Romans, who thought in terms of their solar calendar, the equinox was the early limit of Easter itself. The Alexandrians, on the other hand, thought in terms of the Jewish lunar calendar.

---

55 Those who hesitated to adopt the "traditional" equinox-based Nisan were held to keep Easter "with the Jews" because they followed contemporary Jewish practice in determining the date of Nisan. These "Judaizers" are not to be confused with true Quartodecimans, for they celebrated Easter on Sunday only: Grumel, "Le problème", 168, 172. On the decrees of Nicaea concerning the criteria for Easter, see Eusebius, *De vita Constantini* 3.18–19, ed. I.A. Heikel, *Eusebius Werke* 1 (Leipzig: Hinrichs'sche Buchhandlung, 1902):85–87; cf. Cassiodorus, *Historia tripartita* 9.38, ed. W. Jacob and R. Hanslik, CSEL 71 (1952):557–564. For discussion, see Paul Grosjean, "La date de Pâques et le Concile de Nicée", *Académie royale de Belgique, Bulletin de la classe des sciences*, 5th ser., 48 (1962):55–66, and Kenneth Harrison, *The Framework of Anglo-Saxon History to A.D. 900* (Cambridge: Cambridge University Press, 1975):30.

56 This is the line of argumentation followed in the first major Alexandrian *computus* tract, the *Prologue* of Theophilus (ed. Krusch, *Studien I* 223–224), of which more below; cf. Jones, *BOT* 19.

57 Paul Grosjean argues that the objective of Nicaea was to replace the Jewish Nisan with a Christian one, and that Rome's peculiarities were of little interest to the Fathers. He disagrees with Grumel ("Le problème", 169) that Rome fell into line after Nicaea, precisely on the grounds that Nicaea enunciated no specific criteria: "La date de Pâques et le Concile de Nicée".

58 Eusebius, *De vita Constantini* 3.18 (85.26–27); cf. Jones, *BOT* 20; Grumel, "Le problème", 169.

59 *Apostolic Constitutions* 17.

For them, the equinox was the early limit of 14 Nisan, not of Easter. They were already familiar with an Easter table devised by Bishop Anatolius of Laodicea (AD. 258) which incorporated this principle.

Another source of contention was the fact that there was no agreement about the date of the equinox. By the time of Nicaea, the Alexandrian Church had fixed the equinox at the date established by Ptolemy's astronomical tables, namely 21 March. Most Romans, however, thought that the equinox fell on 25 March, the day Julius Caesar had marked in their calendar, but which, as we have seen, was now obsolete.

The result was that for the next hundred years or so, Rome and Alexandria played a curious shadow-boxing game over the date of Easter. In principle, their systems diverged greatly. Alexandria dated the equinox to 21 March, which was the *terminus a quo* for the Paschal full Moon, or 14 Nisan. The lunar limits for Easter Sunday were 15–21, and therefore the calendar limits were 22 March–26 April. The latter date was one full lunar month after the former, plus seven days to accommodate the following Sunday, for if 15 Nisan fell on 21 March, that whole lunation had to be disregarded, and only the following lunation would count as the first one of spring. The Romans, on the other hand, held that 25 March was the equinox, and that it was the *terminus a quo* for Easter itself. Their lunar limits were 16–22, but their upper calendar limit was 21 April. This was insufficient to accommodate the necessary second lunation, but Rome clung stubbornly to this date, for the popes wished to avoid any coincidence of Holy Week with the boisterous celebrations of the feast of Rome's foundation on 21 April.[60] Sometime in the middle of the fourth century, Rome quietly abandoned the 25 March equinox and moved its lower calendar limit accordingly,[61] but it refused to compromise on the upper calendar limit.

These fundamental differences ought to have produced very divergent dates for Easter, but in fact, the records of actual Easter dates in fourth- and fifth-century Rome and Alexandria show that the Romans almost always celebrated on the Alexandrian dates.[62] Alexandria,

---

60 Jones, *BOT* 28; cf. Prosper, *Epitoma chronicorum* 479, ed. Th. Mommsen, MGH AA 11 (1895).

61 Cf. André van de Vyver, "L'évolution du comput alexandrin et romain du IIIe au Ve siècle", *Revue d'histoire ecclésiastique* 52 (1952):5–25.

62 These records are (a) an Alexandrian list of Easters from 328 to 373 (surviving only in a Syriac version, ed. William Cureton, *The Festal Letters of Athanasius* (London: Society for the Publication of Oriental Texts, 1848); for Angelo Mai's Latin translation,

however, was not content with *de facto* victory. Its aim was to promote universal acceptance of its Easter system. This seems to have been the motivation behind the Easter table created by Bishop Theophilus of Alexandria shortly before his death in 395, and dedicated with much effusive praise to the Emperor Theodosius. Theophilus was essentially lobbying the emperor to endorse Alexandrian reckoning, and bring Rome into conformity.[63] The explanatory material or *quaestiones* attached to the table[64] lay out the Alexandrian criteria very completely. The Bible dictates that Passover must be in the "month of first fruits", which means spring, which means after the equinox. Therefore the 14th day of Nisan must never fall before 21 March. Should 14 Nisan coincide with a Sunday, Easter must be postponed until the following week. Finally, Easter must sometimes be celebrated in the "second month", by which Theophilus means the second *solar calendar month* after the equinox (i.e. after 22 April) rather than the second lunar month – in short, he is targetting the Roman calendar limits here.

Though Theodosius never made the Alexandrian criteria law, the letter and table of Theophilus signal a new phase in the Paschal controversy, because its strategy marks a change in the rules of the contest. Henceforward, the winner would be the party best capable of designing and promoting an authoritative and reliable Easter *table*. The validity of the criteria would be as much proved by the table, as the table by the criteria. We must, therefore, now turn to the question of tables and cycles for the calculation of Easter.

Christians, like Jews, must begin their search for a prospective

---

see PG 36.1351 *sqq.*) and (b) the Roman *Chronograph of 354*. The latter contains two lists of Easter dates: one covering AD 312–342 records the actual dates of Easter, the second from 343 to 354 is a list of calculated dates, according to the Roman system. The *Chronograph* is edited by Theodore Mommsen, MGH AA 9:13–148. For discussion, see Henri Stern, *Le Calendrier de 345. Étude sur son texte et ses illustrations*, Institut français d'archéologie de Beyrouth, Bibliothèque archéologique et historique 60 (Paris: Imprimerie nationale, 1953), and Michele Renee Salzman, *On Roman Time. The Codex-Calendar of 345 and the Rhythms of Urban Life in Late Antiquity*, The Transformation of the Classical Heritage 17 (Berkeley: University of California Press, 1990):39–41. Cf. Jones, *BOT* 25–26.

63　Jones, *BOT* 29–30.

64　Ed. Krusch, *Studien I*, 220–226; there are lacunae in Krusch's text, which may be supplied from older editions, e.g. those of Denis Petau, *De doctrina temporum* (Paris: Sebastien Cramoisy, 1727), for the Greek, and Gilles Bouchier, *De doctrina temporum* (Antwerp: ex officina Plantiniana Balthasaris Moreti, 1633), for the Latin.

Paschal table by establishing a luni-solar cycle, that is, a system for adjusting the lunation to the solar year. Unlike Jews, however, they use a solar calendar. That means that they need a system which will assign to each lunation a date in the solar calendar.[65] But matching up individual *days* within each successive lunation to *dates* in the solar calendar presents numerous problems. The Julian calendar (or its Eastern cognates) is a list of named solar days, but neither the lunation nor the solar year cover a whole number of such days. Both therefore must be artificially normalized, because calendars cannot handle units of time smaller than the day. The lunation is slightly more than $29\frac{1}{2}$ days, and so the fiction was adopted that lunations were alternately 29 and 30 days long: medieval computists called these "hollow" and "full" lunar months. The solar year is a quarter-day longer than the solar calendar year; this excess is cumulated into an additional day, added once every four years. In antiquity and the Middle Ages, the leap-year day was inserted on 24 February (the 6th kalends of March). In leap years, this day "happened" twice: hence the term for leap year was *annus bissextus* – a year with two 6th kalends of March.

A second fundamental problem in constructing a luni-solar cycle is that the solar year does not contain a whole number of lunations. A true lunar month averages 29.5306 days – and it is important to stress that this is an average, which can be affected from one month to another by the gravitational pull of Earth and Sun. Moreover, when any lunar months "begins" – i.e. when the Moon first appears as a thin crescent at dusk after conjunction – depends on time of year and latitude. The solar year now measures 365.2422 days. If one divides

---

65 For useful summaries and discussions of the astronomical basis of cycles, see Kristian Peter Moesgaard, "Basic Units in Chronology and Chronometry", in *The Gregorian Reform of the Calendar. Proceedings of the Vatican Conference to Commemorate its 400th Anniversary, 1582–1982*, ed. G.V. Coyne, M.A. Hoskin and O. Pedersen (Vatican City: Pontificia Academia Scientiarun; Specola Vaticana, 1983):3–13; Vénance Grumel, *La Chronologie*, Traité des études byzantines 1 (Paris: Presses universitaires de France, 1968), esp. ch. 3 for the history of the 19-year cycle, and ch. 9 for the solar 28-year cycle and the 532-year Paschal cycle; Kenneth Harrison, "Luni-solar Cycles: Their Accuracy and Some Types of Usage", in *Saints, Scholars and Heroes*, 2:65–78; O. Schissel and Maria Ellend, "Berechnung des Sonnen-, Mond- und Schaltjahrszirkels in der griechisch-christlichen Chronologie", *Byzantinische Zeitschrift* 22 (1943):150–157; van Wijk, Preface, section 1; Jones, *BOT* 11; *idem*, "A Legend of St Pachomius", *Speculum* 18 (1943):204–205; Krusch, *Studien II* (1938):14–15; A.N. Zélinsky, "Le calendrier chrétien avant la réforme grégorienne", *Studi medievali* ser. 3, 23 (1982):549.

365.2422 by 29.5306 the result is 12.3683, which is the number of lunations in a solar year. Over about three years, that decimal excess of .3683 will amount to slightly more than one whole lunation. This means that in every third year (more or less), an extra lunar month will be inserted, and the solar year will contain 13 whole lunations. This inserted lunar month can be explained in a different way. Because the Julian year is 365 days long, and 12 lunar months of $29\frac{1}{2}$ days total 354 days, the Moon is about 11 days older at the end of the Julian year than it is at the beginning. Thus the age of the Moon on any given calendar date will rise by 11 each year. But this running total cannot go over 30, the maximum length of a calculated lunar month. When it does, the computist subtracts 30 from the running total, and thereby intercalates an additional lunar month.

The aim of a luni-solar cycle is to find a whole number of solar years which can accommodate a whole number of lunations, so that dates within any lunation (e.g. "the fifteenth day of the Moon") can be plotted against a solar calendar, with its 365 successive dates, and so that the numbering of lunations can begin again at the close of the cycle.

Strictly speaking, such a cycle is impossible, since the decimal fraction excess of lunar months (.3683) is an irrational number. That means that it cannot be translated into a common fraction, and therefore no number of lunations can ever be fitted evenly into any number of whole solar years. Approximations, however, are possible. The decimal excess can be expressed fairly closely by the following common fractions:

$\frac{3}{8}$ (.3750)

$\frac{4}{11}$ (.3636)

$\frac{7}{19}$ (.3684)

$\frac{31}{84}$ (.3690)

The options, then, are to intercalate 3 lunar months over 8 years, or 4 lunar months over 11 years, or 7 over 19, or 31 over 84. These are the bases of the 8-year cycle, or *octaëteris*, the various 84-year cycles, and the 19-year cycle.[66] It should be stressed that in every case, the lunar

66 Jones, *BOT* 11. The 19-year cycle is sometimes regarded as a fusion of the cycles of 8 and 11 years, the error of the first being cancelled out by the opposing error of the second.

month is the calculated lunar month, and the solar year is the regularized solar year described above. What the ancient computists were aiming for, then, was a mathematical system to correlate two artificially regularized astronomical cycles, so as to do the minimum amount of violence to the natural phenomena, while still generating a cycle. All the cycles mentioned above were used in antiquity, though the 19-year cycle evidently is the most accurate.

The interpolated lunar months were known as *embolisms*, and a year containing 13 lunar months was *embolismic*. But in any given embolismic year, where (in relation to the solar calendar) ought one to insert the extra lunation? This was an important issue, because embolismic lunations were always counted as 30 days long, so they disrupted the regular alternation of "full" (30-day) and "hollow" (29-day) lunar months, requiring the addition of an extra day to the lunar count. The alternation of full and hollow months was supposed to march in step with the months of the solar calendar. Every lunation "belonged" to a solar month, the month in which it ended: January's lunation was full, February's hollow, etc. Therefore the ideal solution would be to make the embolism "invisible" by inserting it in a calendar month whose proper lunation ended on its first or second day. A whole embolismic month could then be inserted, which would end on the first or second day of the next month. This would allow this second month still to absorb its "own" lunation, and for the pattern to continue undisturbed. But as we shall see, even if it is "invisibly" inserted, an embolism can nonetheless throw off the calculation of 14 Nisan by a day.

Finally, the 19-year cycle ends with one more *calculated* lunar day than necessary. At some point within the cycle, then, the lunar count has to "jump" a day to get in phase with reality. This is what medieval computists called the *saltus lunae*, or "leap of the Moon". The 84-year cycles also incorporated *saltus lunae*, though they required more than one. Like the embolism, the positioning of the *saltus* was critical, for it would change the calculated age of the Moon by a day.

Any of these cycles might produce defensible dates for the Easter full Moon, but only some of them could actually produce a true cycle of Easters, namely, those which can also accommodate a repeating cycle of Sundays. To do so, a cycle would have to be divisible by 28 – the number of years in the weekday cycle described above. Only two cycles offered this possibility: the 112-year cycle of Hippolytus (which was a double cycle of 8 years, repeated four times) and the 84-year cycle. This was a major argument in their favour, but in the end, it could not compensate for these cycles' lunar inaccuracies.

The 8-year cycle or *octaëteris* and its 112-year variant form need not detain us here, as they were obsolete well before Bede's day. However, the literature created by Hippolytus and ps.-Cyprian to defend this cycle did have a long-term effect on Bede's conception of the relationship of history to *computus*. This is discussed below, in the Commentary to ch. 6. More significant was the 84-year cycle which emerged in the fourth century. This was a cycle designed with Rome in mind. Its lunar limits for Easter were 16–22, and its lunar calculations were based on the age of the Moon (in computist's jargon, the *epact*) on 1 January, the Roman New Year. This was such a distinctive feature of 84-year cycles that early chronicles written in Ireland, where the 84-year cycle long held sway, identified years by the weekday and lunar phase of 1 January, even after the Irish had switched to the Paschal reckoning of Dionysius Exiguus.[67]

The early history of the 84-year cycle is very obscure,[68] but it behoves us to try to summarize it, because the Celtic *computus* which Bede confronted in *The Reckoning of Time* was a special variation on the 84-year system. For our purposes here, it suffices to note that by the beginning of the fourth century, an 84-year table called the "Roman *Supputatio*" was circulating in the West. This table prescribed the Roman lunar limits of 16–22, and the 21 March equinox. Moreover, it inserted a *saltus lunae* every 12 years. The difference between 84 solar and 84 lunar years is 924 days (11 × 84), or 30 intercalated lunar months, plus 24 days. If this remainder is treated as an extra embolismic month, the cycle will end 6 days behind the astronomical reality. To deal with this, the calculated Moon will have to "jump back" six days in order to get in phase with reality. There were two ways of doing this: one could insert a one-day "jump" (*saltus*) every 14 years, or one could

67 Kenneth Harrison, "Epacts in Irish Chronicles", *Studia Celtica* 12–13 (1977–8):17–32; "Episodes in the History of Easter Cycles in Ireland", in *Ireland in Early Medieval Europe. Studies in Memory of Kathleen Hughes*, ed. Dorothy Whitelock, Rosamond McKitterick and David Dumville (Cambridge: Cambridge University Press, 1982):314–315.

68 The foundational account is Krusch, *Studien I*, which Jones follows on most points. Important revisions are offered by Eduard Schwartz, *Christliche und jüdische Oster-tafeln*, Abhandlungen der königlichen Gesellschaft der Wissenschaften zu Göttingen, phil.-hist. Kl. n.f. 8,6 (Berlin: Weidmannsche Buchhandlung, 1905):66 *sq.*: see n. 69.

insert a *saltus* every 12 years, and omit it in the last year of the cycle. The Romans used the 12th-year *saltus* system; the Celts, as we shall see, used the 14th-year system.[69] Finally, the *Supputatio* acknowledged that the curtailed Roman calendar limits made it impossible at some times to find a fully "orthodox" Easter date. When this happened, the two nearest dates were listed, and the reader advised to consult the pope for a decision.[70]

Meanwhile, between the time of Bishop Dionysius and the Council

69 The major revision to be made to Krusch's account of the history of the 84-year cycle concerns this issue of the *saltus*. The Easter dates recorded in the *Chronograph of 354* differ from those of the *Supputatio* on a number of occasions. The *Supputatio* uses a 21 March equinox, and lunar limits 16–22. Krusch concluded that the *Chronograph* must have based its dates on a lost Roman table, which he called the "older Roman *Supputatio*". He argued that this "older *Supputatio*" differed from the "newer" one by fixing the equinox on 25 March, and using lunar limits of 14–20. The older *Supputatio* was a continuation of an 84-year table devised by one Augustalis, and described in the "Carthaginian *Computus* of 455". Augustalis' table supposedly began in 213, and ran for 100 years, so the older *Supputatio* would have taken over in 312, in time to be used by the *Chronograph*. According to Krusch, the Carthaginian text describes Augustalis' table as an 84-year cycle, where the *saltus lunae* comes every fourteen years, while the "newer Roman *Supputatio*" (as preserved in MS Milan Ambrosiana H 150 inf.) has the *saltus* in the twelfth year. A table for an 84-year cycle with 14th-year *saltus* survives in a 9th-century codex, Munich CLM 14456. In this Munich manuscript, the lunar limits are 14–20, and the earliest possible date for Easter is 25 March. Therefore, Krusch concluded that the Munich manuscript contained a copy of the *laterculus* or table of Augustalis, which in turn was the basis of the "older Roman *supputatio*", which was the table used by the *Chronograph of 354* (*Studien I*, 5–23). This theory has been attacked by Eduard Schwartz, on a number of grounds. First, the *Chronograph of 354* is not a prescriptive table, but a record of actual Easters, and whenever it departs from the Milan table, it does so in agreement with the Alexandrian Easters. In short, the *Chronograph* is evidence of Rome's compromises with Alexandria, not a relic of an earlier type of 84-year cycle. Secondly, the lunar limits of 14–20 in Krusch's "older Roman *Supputatio*" fly in the face of Roman tradition, which ever since Hippolytus had staunchly defended 16–22. Finally, Schwartz demonstrated that the Carthaginian *computus* was describing a system with a 12-year, not a 14-year *saltus*. Schwartz concluded that there never was an "older Roman *Supputatio*", and that the *laterculus* of Augustalis was never in use in Rome. What Krusch thought was a primitive Roman 84-year cycle was, in fact, the eccentric version of this cycle used by the Celtic churches of the British Isles: see Schwartz 66 *sq.* For a summary of these arguments, see D.J. O'Connell, "Easter Cycles in the Early Irish Church", *Journal of the Royal Society of Antiquaries of Ireland* 66 (1936):67–106. Unfortunately, Jones at the time of *BOT* was unacquainted with the work of Schwartz and O'Connell; this is also the case with other scholars of the immediate post-war era, such as Richard and Cordoliani. The major point to retain here, is that no Roman 84-year cycle ever used a 14th-year *saltus*; only the Celtic 84-year cycle did.

70 Jones, *BOT* 27.

of Nicaea, the Alexandrian Church had switched from the *octaëteris* to the 19-year cycle, devised in the fifth century BC by Meton of Athens, and long used by professional astronomers.[71] As this cycle eventually won out over its rivals, it is important to understand how it is constructed.

Within 19 solar years, one can accommodate 228 lunar months of $29\frac{1}{2}$ days (or 6,726 days), plus 7 intercalated lunar months of 30 days (or 210 days), plus $4\frac{3}{4}$ additional lunar days for the leap years. The total number of lunar days is $6,940\frac{3}{4}$, or 1 day more than 19 Julian years ($6,939\frac{3}{4}$ days). For this reason one lunar day – the *saltus lunae* – is omitted at the end of the 19th year.

The 19 years are arranged in patterns of common (C) and embolismic (E), as follows:

> *ogdoas* or first 8 years: CCECCECE
> *hendecas* or remaining 11 years: CCECCECCECE

As explained above, embolismic months are ideally positioned in a such a way that they are absorbed within a calendar month, while not interrupting the rhythm of full and hollow lunar months. In the 19-year cycle, the embolismic months are inserted as follows:

> year 3: 2 December
> year 6: 2 September
> year 8: 6 March
> year 11: 4 December
> year 14: 2 November
> year 17: 2 August
> year 19: 5 March

Note that in years 8 and 19, the embolism occurs in March, at a time when it *will* affect the Paschal reckoning.[72]

The 19-year cycle could be used to calculate the first full Moon after the spring equinox (in Alexandria, 21 March) in the Julian calendar. All

---

71 For a summary history of the Alexandrian *computus*, see W.E. van Wijk, *Le Nombre d'or. Étude de chronologie technique suivi du texte de la Massa compoti d'Alexandre de Villedieu* (La Haye: Martinus Nijhoff, 1936): Preface, sections 2–3. On the date of this transition, see Richard, "Le comput pascal", 314; on the ascription of the cycle to Meton, and the disputed dating of his life, see van de Vyver, "L'évolution", 5–6.

72 Cf. *The Reckoning of Time*, ch. 43. Year 11 also presents problems, albeit for a different reason: see Commentary on *The Reckoning of Time*, ch. 20.

one would have to do is memorize these 19 dates, and one would have the Easter full Moons forever.[73]

The table of Anatolius, bishop of Laodicea, composed in 258, is the first recorded Paschal table based on the Metonic cycle to be used in the Greek East. We know of this table only through Eusebius' *Historia ecclesiastica*[74] and the criteria governing its construction are not clear,[75] but Anatolius insisted that the equinox[76] was the *terminus a quo* for the Easter full Moon, and he provided rules and reasons for this choice, which were apparently used at Nicaea.[77] The success of Anatolius' table was reinforced by the fact that the expertise was available at Alexandria to construct new 19-year tables when the old ones expired; indeed, Athanasius of Alexandria did so very shortly after the Council of Nicaea,[78] as did others, up to the time of Theophilus.

The Easter table of Theophilus contained an interesting feature: a column containing the numbers 1–7, standing for the weekdays from

73 To assist the memorizing of these 19 dates, a famous 19–verse poem was composed which gave the date of the full Moon and the *ferial regular* (a number which, when added to the *concurrent* – a number representing the weekday of 24 March – will give the day of the week on which 14 Nisan falls): *Nonae aprilis norunt quinos . . .*, etc. This poem was probably written in the late 5th century: see Jones, "Legend of St Pachomius". For ed., see P. de Winterfeld, "Rhythmi computistici", MGH Poetae 4.1 (1899):670–671.

74 7.32.14, or in Rufinus' Latin translation, 8.28. Michael Whitby and Mary Whitby, trans. of *Chronicon Paschale 284–628 AD*, Translated Texts for Historians 7 (Liverpool: Liverpool University Press, 1989) *s.a.* 285 (n. 2) say that Anatolius' table was probably 95 years long, and its notional beginning was AD 258, the year under which it is recorded in the *Chronicon*.

75 Reconstructions of the table have not been notably convincing, largely because of confusions between Anatolius and the *Liber Anatolii*, an Irish computistical "forgery": see below, n. 117; cf. van de Vyver, "L'évolution", 11.

76 Debate continues on which day Anatolius thought the equinox was. Van de Vyver ("L'évolution", 9) argues that it was 20 March, because Anatolius identifies 22 March as the day when the Sun is in the fourth degree of Aries. Schwartz, however, thinks that Anatolius did not identify the equinox with the entry of the Sun into Aries (15–16). It is also unclear when Anatolius inserted his *saltus*. One feature of Anatolius' cycle which is clear is that it began with the year in which a new Moon fell on 26 Phamenoth, or 22 March; this ever after became the locus of the epacts in Alexandrian tables: cf. *Reckoning of Time*, ch. 50.

77 Anatolius' arguments are preserved in a fragment of a treatise by his contemporary Peter of Alexandria included in the computistical preface to the *Chronicon Paschale*. For echoes of Anatolius in Eusebius' account of Nicaea in *Historia ecclesiastica* and *De vita Constantini*, see Jones, *BOT* 22, n. 3.

78 For a reconstruction, see Schwartz 24–25.

Sunday to Saturday, and representing the weekday of 24 March in the year in question. Using this column, one could count forwards or backwards to locate the weekday of the Easter full Moon recorded in the table; the following Sunday would be Easter.[79] This clarified the principles on which the table was constructed, and suggested the manner in which it might be transformed into a fully cyclical Easter table, where all the lunar and solar criteria would recur – namely, by repeating the 19-year cycle 28 times. This did not actually take place until the time of Bede, but Theophilus' table did establish the 19-year Alexandrian table on a permanent footing, from which it never substantially deviated thereafter. It was now poised to begin its conquest of the Christian world.

That conquest, however, was very slow, especially in the West, where the Roman 84-year table was well entrenched. Alexandrian 19-year tables were hard to adapt to Western usage, because they were based on the Egyptian calendar, which began in September, and which inserted its leap-year day in August. Moreover, the Egyptian months did not coincide with the Roman ones.[80] To have a bridgehead in the West, the 19-year table needed to be modified for users of the Julian calendar. This was accomplished by the table ascribed (falsely) to Bishop Cyril of Alexandria, nephew and successor of Theophilus. It covered 95 years (AD 437–531) or five 19-year cycles,[81] and was certainly a feat of computistical skill. For the first time, the Alexandrian Easter system was successfully laid over the framework of the Julian calendar, with its year beginning on 1 January, its leap-year day inserted on 24 February, and its specific grid of months.

The ps.-Cyrillan table comprised eight columns: the number of the year as reckoned from the beginning of the reign of Diocletian,[82] the

---

79 For further details, see my Commentary on *The Reckoning of Time*, ch. 53.

80 Cf. *The Reckoning of Time*, ch. 11.

81 An early version of this 95-year table survives in MS Paris, Bibliothèque nationale lat. 10319 (s. VIII), a Spanish codex whose contents point towards a North African exemplar: van de Vyver, "L'évolution", 21–25.

82 The choice of the "era of Diocletian" was totally fortuitous. To establish the equinox on 21 March, a computistical reform took place in Alexandria in the early 4th century. The revised cycle began in 303, and picked up on the previous cycle, which began in 284. The latter date was the first year of Diocletian's reign, and regnal years were commonly used for dating in Egypt; the new cycle simply incorporated this as its era: see *Chronicon Paschale s.a.* 285, n. 2.

indiction,[83] the lunar epact for 22 March,[84] the concurrent or weekday of 24 March,[85] the year of the "lunar cycle", i.e. a 19-year cycle beginning in the year when the Moon is one day old on 1 January,[86] the Julian date of 14 Nisan, the Julian date of Easter, and finally, the age of the Moon on Easter Sunday. When Dionysius Exiguus adapted and continued the ps.-Cyrillan table, his only change was to alter the first column to conform to his own *annus Domini* chronology.

The arrival of the Alexandrian 19-year cycle in the West made the deficiencies of the 84-year table with respect to lunar reckoning very evident. An 84-year cycle is essentially four successive 19-year cycles, with an *octaëteris* tacked on the end, but since 84 is divisible by seven, the dates of the new Moons will recur on the same days of the week, and hence Easters will fall on the same dates when the cycle repeats. However, the cycle is not accurate in its prediction of the lunations: after 84 years there is an error of two days, which is compounded if the cycle is repeated.[87] The 19-year cycle too was not perfect, and unfortunately was not cyclic for Easter itself, but only for the Easter full Moon; however, it was by far the closest match to the astronomical realities. In the face of the challenge posed by the Alexandrian table, proponents of the 84-year cycle started tinkering with their tables to fix the lunar anomaly, either by adjusting the epact or age of the Moon on 1 January, or by shifting the beginning of the cycle to a different year. The year selected to inaugurate the cycle was typically one of theological and computistical significance, such as the one corresponding to the year of the Exodus, or the year of Christ's Passion.[88] The result was a multipli-

---

83 See *The Reckoning of Time*, ch. 48.

84 *Ibid.* ch. 50.

85 *Ibid.* ch. 53.

86 *Ibid.* ch. 56.

87 Kenneth Harrison, "Episodes", 308.

88 E.g. the fragmentary marble "Zeitz table" of 447 (ed. Mommsen, *Chronica minora* 1.503–510) starts in AD 29, the traditional year of the Passion, and corrects the age of the Moon in line with Alexandrian reckoning (O'Connell 74–75; Krusch, "Die Einführung der griechischen Paschalritus im Abendland", *Neues Archiv* 9 (1884):106; *Studien I*, 116 *sq.*). The Carthaginian *computus* of 455 (Krusch, *Studien I*, 279–297) describes two "reformed" 84-year cycles, one beginning in AD 29 and another which began in AD 439, a year corresponding to 2100 BC, the traditional date of the Exodus and first Passover. In both cases, the effect was to cancel out the accumulated error in lunar reckoning of the 84-year cycle.

city of different versions of the 84-year cycle, and even greater confusion amongst Western computists.

Rome showed no interest in formally endorsing any of these cycles, or even in consistently supporting any principle save the 21 April limit. In practice, it followed Alexandria, except when the Alexandrian date of Easter fell after 21 April. In 444 and again in 455, however, this is exactly what happened, and the occurrence of such a crisis twice in little more than a decade forced Rome to acknowledge the need for a definitive solution. In both cases, Pope Leo yielded to Alexandria, not on principle, but as a concession for the sake of unity. But the Papacy seemed at last convinced of the necessity for a decision on the Paschal question, a decision which seemed inevitably to lean towards accepting the Alexandrian system. In 444, Bishop Paschasinus of Lilybaeum had responded to Pope Leo's query for advice on the differences between the Alexandrian and Roman reckoning by endorsing the former, apparently in the form of the ps.-Cyrillan table.[89] To prove that the Alexandrian Easter was the right one, Paschasinus resorted to an anecdote about a baptistry with a miraculous font which filled automatically at the Easter Vigil – but only the Vigil reckoned according to the Alexandrian system. Such legends speak volumes about the difficulty of persuading people, either through authority or through reason, that any particular reckoning was infallible: Bede sometimes felt constrained to invoke them.[90]

As the 455 crisis approached, Pope Leo cast about for help once again, first to Paschasinus, whose reply has not survived, and then to the Emperor Marcian.[91] Marcian referred the matter to Proterius, bishop of Alexandria, who wrote a firm, almost condescending response to Leo, defending the Alexandrian reckoning, including its calendar limits.[92] This time Alexandria would not bow to Rome's claims to superiority. Leo gave in again, though only, as he said, to preserve peace.

To avoid such embarassments in the future, Leo's archdeacon Hilarius commissioned Victorius of Aquitaine to investigate the reasons for the discrepancies between the Roman and Alexandrian systems, and suggest a way to resolve them. Victorius is a somewhat shadowy figure,

---

89 Paschasinus' letter is edited by Krusch, *Studien I*, 245–250; cf. Jones, *BOT* 55–56.

90 Cf. *The Reckoning of Time*, ch. 43.

91 Leo's letters are edited by Krusch, *Studien I*, 257–264.

92 The Latin translation of Proterius' letter is edited by Krusch, *Studien I*, 269–278.

a "calculator of infinitesimals" (*scripulorum calculator*) according to Gennadius,[93] and author of a *calculus* or set of multiplication tables with an appendix on fractions.[94] The extent of his training in computistics is unclear, but Hilarius evidently felt confident in assigning him the job. In 457, Victorius issued his new tables.[95] These abandoned the 84-year cycle in favour of the 19-year cycle, which Victorius knew in the form of Theophilus', and perhaps ps.-Cyril's, table. Victorius also set aside without comment the sacrosanct Roman calendar limits. His table dated years according to the *annus Passionis*, assumed to coincide with *annus mundi* 5229, according to the reckoning of Eusebius, as translated by Jerome, or AD 28. This chronology was reinforced by a list of consuls down to 457. In one sense, and one sense only, his effort was actually an improvement over ps.-Cyril's cycle: Victorius projected his cycles for 532 years, and he knew that his table would repeat after this period – although he did not know *why*. Victorius saw his 532-year cycle as 4×133 years, not 19×28, as Bede correctly did.[96]

On every other ground, however, Victorius' table was riddled with problems. First, he placed his *saltus* in the 6th year of his (and the Alexandrian) cycle, which meant that his cycle was out of phase with the Alexandrian one for years 7–19.[97] Secondly, he placed the earliest beginning of Nisan at 5 March, and 14 Nisan as early as 18 March, whereas the

93 *De viris illustribus* 88, ed. W. Herdin (Leipzig: Teubner, 1879):108.9–25; cf. *The Reckoning of Time*, ch. 66, *s.a.* 4427. See Krusch, *Studien II*, 4–15; Jones, *BOT* 60–68; M. Walsh & D. Ó Cróinín, *Cummian's Letter 'De controversia paschali' and the 'De ratione conputandi'*, Studies and Texts 86 (Toronto: Pontifical Institute of Mediaeval Studies, 1988):42.

94 Ed. G. Friedlein, "Der Calculus des Victorius", *Zeitschrift für Mathematik und Physik* 7 (1871):42–79.

95 These tables, and the correspondence with Hilarius, are edited by Krusch, *Studien II*, 16–52.

96 Bartholomew MacCarthy, *Annals of Ulster* (Dublin: His Majesty's Stationery Office, 1901):4.lxxxv; cf. Jones, *BOT* 64.

97 Why Victorius did this may seem baffling, since his letter to Hilarius acknowledges that the insertion of the *saltus* at the end of nineteen years is preferable (*quod est verius*): Krusch, *Studien II*, 19.8. The reason was the epoch Victorius chose for his table. Victorius held that his Paschal cycle should follow nature by beginning with Creation. His epoch, though, was the *annus Passionis*, which he set at *annus mundi* 5229, following Eusebius' chronology. 5229 divided by 19 leaves 4 as a remainder, so *annus Passionis* 1 is year 4 of Victorius' cycle. In order to insert the *saltus* in year 19 of his cycle, Victorius had to place it at *annus Passionis* 16, and every 19th year following. But *annus Passionis* 16 was year 6 of the Alexandrian cycle: cf. Jones, *BOT* 65. This error is discussed by Bede in *The Reckoning of Time*, ch. 42.

Alexandrians, to avoid a full Moon before the equinox on the 21st, began Nisan at the latest on 8 March.[98] Thirdly, he clung to the old Roman lunar limits of 16–22, rather than switching to the Alexandrian limits of 15–21.[99] Finally, to make matters even worse, Victorius did not actually fulfil his mandate to produce an unambiguous, trouble-free Paschal table. He set out his own table of Easters, listed beside them the "Greek" Easters, and invited the Pope to choose in cases of conflict. But Victorius' "Greek" Easters were not, in fact, the Alexandrian dates; they were dates arrived at through Victorius' own rules, and not observed anywhere in the Church. Ironically, his "Latin" Easter dates were often identical with the Alexandrian dates, because his insertion of the *saltus* in year 6 cancelled out the discrepancy between his lunar limits for Easter, and those of the Alexandrian Church. Therefore if the Pope chose to celebrate the "Latin" Easter as recorded in Victorius' table, he would, in years 7–19 of the cycle, in fact be choosing the Alexandrian Easter date.[100]

But for all its flaws, the fortunes of the Victorian table were assured by the fact that Archdeacon Hilarius, shortly after commissioning the table, became pope himself. In some manuscripts, his title in the covering letter from Victorius is even altered from "Archidiaconus" to "Papa",[101] thereby lending unmerited authority to Victorius' tables. This was reinforced by the coincidence that whenever the Pope announced an Alexandrian date for Easter, it looked as if he was endorsing Victorius' "Latin" dates. Nonetheless, the success of Victorius' system was mitigated, save in Gaul, where in 541, a synod at Orleans formally adopted it as authoritative.

Victorius' strange dating of Nisan and his eccentric positioning of the *saltus*, to say nothing of the problem of double dates, drew down considerable criticism. At the Pope's request, Victor of Capua wrote a merciless critique of Victorius' tables, and called for new tables.[102] Even at Rome,

98 Cf. *ibid.* ch. 51.

99 Jones, "The Victorian and Dionysiac Tables in the West", *Speculum* 9 (1934):409; significantly, Bede did not address this error, for reasons which will shortly become evident.

100 *Ibid.* 413.

101 Jones, "Legend of St Pachomius", 205.

102 Victor's treatise has not survived, but passages from it are quoted by Bede in *The Reckoning of Time*, ch. 51 and *Letter to Wicthed*. These and other recovered fragments were printed by J.B. Pitra in *Spicilegium Solesmense* (Paris: Firmin Didot, 1852):1.296–301.

support for the uncanonical Easter Moons generated by Victorius' *saltus* was tepid.

Dionysius Exiguus, a monk of Scythian origin resident in Rome, and apparently friendly with (among other people) Cassiodorus, felt that the only way to resolve the problem of the Victorian tables was to scrap them in favour of thoroughly Alexandrian tables. He translated ps.-Cyril of Alexandria's table into Latin, constructed a continuation for another 95 years (AD 532–627), attached to it a prologue and translation of the computistical rules or *argumenta* current in Alexandria,[103] and sent it with a covering letter to an otherwise unidentified bishop named Petronius, possibly an African.[104] The Prologue states that the Alexandrian system described by Dionysius was endorsed by the Council of Nicaea. Earlier commentators regarded this as an outrageous falsehood, but as Jones observes, Dionysius was far from the first to conclude that the bishops at Nicaea had drawn up a table, and certainly correct in his inference that they approved the Alexandrian system.[105] Despite these claims, however, Dionysius' work attracted little attention at first; he was a foreigner in Rome, and apparently not altogether popular with the popes,[106] and his work was addressed to an obscure bishop.

In 525, however, it looked as if Dionysius might get a hearing. In the following year, there would be a serious discrepancy between the Victorian tables and the Alexandrian Easter, due to Victorius' misplacement of the *saltus*. To explain this discrepancy, Dionysius addressed a letter to Bonifatius and Bonus, *primicerii* of Pope John I, outlining the proper Alexandrian sequence of common and embolismic years. Bonifatius conveyed this explanation to Pope John himself.[107] The Pope might

---

103 Ed. Krusch, *Studien II*, 75–81. As Jones pointed out, this edition of Dionysius' *argumenta titulorum paschalium*, as well as the older one by J.W. Jan, reprinted in PL 67.453 *sq.*, stand in need of revision, as only *argumenta* 1–9 are authentic: *BOT* 70–71. For a draft of such a new edition, see Joan Gómes Pallarés, "Hacia una nueva edición de los 'Argumenta titulorum paschalium' de Dionisio el Exiguo", *Hispania sacra* 46 (1994):13–31.

104 Jones, *BOT* 68 *sqq.*

105 Jones, *BOT* 71. Dionysius may have derived the notion of Nicene endorsement of the 19-year cycle from the prefatory matter of Cyril's tables: Harrison, "Easter Cycles and the Equinox in the British Isles", *Anglo-Saxon England* 7 (1978):4.

106 Jones, "Victorian and Dionysiac Tables", 414.

107 Bruno Krusch, "Ein Bericht der päpstlichen Kanzlei an Papst Johannes I. von 526 und die Oxforder HS Digby 63 von 814", in *Papstum und Kaisertum. Forschungen zur politischen Geschichte und Geisteskultur des Mittelalters Paul Kehr zum 65. Geburtstag dargebracht*, ed. Albert Brackmann (Munich: Verlag der Münchner Drucke, 1926):48–58.

have endorsed Dionysius' tables had he not died shortly thereafter; his successors were in no position, thanks to the Gothic Wars, to deal with the Paschal question. By comparison with Victorius' "orthodox" and "papally-sanctioned" table, then, the Dionysian solution had only very marginal prestige.

But though it had no official status, Dionysius' solution was not without its admirers. A *computus* ascribed to his friend Cassiodorus is essentially a copy of Dionysius' *argumenta*,[108] and Isidore of Seville plainly knew of Dionysius' table, though his attempt to reproduce it in *Etym.* 6.17 was something of a botch. When the table expired in 626, a continuation until 721 was constructed by a certain "Felix of Ghylli-tanus", apparently a north African.[109] However, we have no idea when exactly the Roman Church began to adapt its Easter calculations to the Dionysian table.[110] It is interesting to observe that while the Irish, by Columbanus' day, had heard of Victorius, they seem not to have known of Dionysius.

The Dionysian Paschal table forms the backbone of the second half of *The Reckoning of Time*, and will be discussed in detail in the Commentary on this section of Bede's book. A few notable features of the table should, however, be signalled here. The old Roman 84-year cycles were based on the epact of 1 January; in other words, the cycle began in a year when the Moon was one day old on 1 January. The Alexandrian epacts, on the other hand, began on 1 September, the Egyptian New

---

108 O. Neugebauer, "On the Computus Paschalis of 'Cassiodorus'", *Centaurus* 25 (1982):292–302. Edited in PL 69.1249–1250, and by Paul Lehmann, "Cassiodorusstudien II: Die Datierung der Institutiones und der Computus paschalis", *Philologus* 71 (1912):278–299 (text is on 297–299).

109 Jones, *BOT* 73–74; Krusch, *Studien I*, 301, n. 2, argues that the continuator was a Spanish monk named Leo, but Jones rejects this argument, since the continuation was apparently unknown to Isidore of Seville. A table has survived in a 9th-c. Fleury MS, Harley 3017 fols. 50r–51r, which also projects Dionysius' cycle forward for another 95 years, from 627 to 721, but it is not the one by "Felix", since (contradicting "Felix's" statement) it does not follow the format of Dionysius' table. In particular, it does not contain the column for the *annus Domini*. But it is evidently based on Dionysius' formulae, and it agrees with the Alexandrian reckoning. Its existence demonstrates "that there were computists capable of adapting the rules of Dionysius with more success than did Isidore; it shows the spread of the Alexandrian reckoning in the West; it shows that in at least one center the methods of Dionysius were used without adoption of his *annus Domini* era; and it is, to my knowledge, the only extant cycle for the missing period between the cycle of Dionysius and the cycle of Bede": C.W. Jones, "Two Easter Tables", *Speculum* 13 (1938):204.

110 Ó Cróinín and Walsh 38.

Year. Dionysius attached the epacts to 22 March, because this date falls on the same weekday as the Roman leap-year day of 24 February, and because it marks the earliest possible date for Easter. In this way, he adapted epacts both to Roman calendar usage, and the exigencies of Paschal reckoning. Dionysius' cycles begin in a year when the Moon is new on 22 March. Secondly, Dionysius chose a different era for his table, the year of the Incarnation.[111]

The final crisis of the Paschal controversy took place, not in the Papal court, but in the British Isles. Though commonly referred to as the "Irish" Paschal controversy, it would be more exact to characterize it as "Insular" or "Celtic" rather than "Irish". The 84-year cycle was used by the British and Picts as well as the Irish. In fact, by the time of Bede, most of the Irish had already conformed to Rome, at least by adopting Victorian tables, and when *The Reckoning of Time* was composed, they all had. Ongoing resistance came from the British.[112]

The Celtic Paschal cycle was an 84-year cycle, but several features distinguish it from the Roman *Supputatio*. Its lunar limits are very archaic – 14–20 – and so is its 25 March equinox. Finally, unlike any other documented version of the 84-year cycle, it inserts its *saltus* in the 14th year.[113] How or when this 84-year cycle reached the British and Irish Churches is unknown, nor has it been established whether the peculiar criteria were introduced with the cycle, or already used before the cycle arrived. The Celtic system was already a firmly entrenched tradi-

111 For a useful summary, see Gustav Teres, "Time Computations and Dionysius Exiguus", *Journal of the History of Astronomy* 15 (1984):177–188; van Wijk, Preface, section 4.

112 This point is well expressed by Daniel McCarthy, "Easter Principles and a Fifth-Century Lunar Cycle used in the British Isles", *Journal of the History of Astronomy* 24 (1993):204–224.

113 Much of the earlier literature on the Celtic Easter has been preoccupied with finding, or attempting to reconstruct, an actual Celtic Paschal table which would prove that these were indeed its criteria. Milestones in this research are: Krusch, "Einführung", 101–169, esp. 140–169; Bartholomew MacCarthy, *Annals of Ulster* 4.lxv *sqq*; and O'Connell, "Easter Cycles". Thus the discovery of an authentic Irish Easter table in MS Padua, Bibl. Antoniana I.27, fols. 76r–77v is an important breakthrough: see D. Ó Cróinín and D. McCarthy, "The 'Lost' Irish 84-year Easter Table Rediscovered", *Peritia* 6–7 (1987–88):227–242. For corrections to the technical basis of their reconstruction, see McCarthy, "Easter Principles", 204–224, and "The Origin of the *Latercus* Paschal Cycle of the Insular Celtic Churches", *Cambrian Medieval Celtic Studies* 28 (1994):25–49. This table establishes beyond doubt the criteria traditionally ascribed to the Celtic system.

tion by the turn of the seventh century, when Columbanus complained to Pope Gregory I that the Frankish bishops were harassing his monks for celebrating Easter according to their ancestral custom. The bishops insisted that he follow the Victorian *computus*, which the learned men of his island had inspected, and found ludicrous.[114] In his response to the bishops, Columbanus added that the Victorian table was a mere modern novelty, without authority, especially when compared to the Celtic 84-year cycle, which dated from "the times of the great Martin and great Jerome and Pope Damasus".[115] What this claim means remains unclear,[116] but it is certain that only a few decades after Columbanus' time, the Irish Church itself was profoundly split by the Easter issue.

As usual, the debate hinged on two interconnected issues: the criteria for a canonical Easter, and the computing of a reliable Easter table. On the issue of criteria, the Irish waters were muddied not only by the variety of apparently authoritative systems available – Cummian lists ten in his *Epistola de controversia paschali* (AD 632) – but by the rich literature of Alexandrian pseudepigrapha which collected in Ireland. The label "Irish forgeries", though conventional, is however unfortunate and misleading in that it suggests wilful deception. *Computus* was a matter of profound importance, and medieval Christians deeply believed that such issues had to be decided on the authority of tradition. But if tradition failed to guide, then reason and reckoning would have to disguise themselves as tradition. The motivation was less deception than desperation.

The most important of these "Irish forgeries" was the *Liber Anatolii* or canon of ps.-Anatolius, the primary document debated at the Council of Whitby in 664.[117] It has additional significance for us

114 Columbanus' letters on the Easter question are edited by W. Gundlach, MGH Epp. 3 (1892):156 (to Gregory I); 160 (to the synod of Gallic bishops); 164 (to Pope Sabinian?); 177 (to Pope Boniface IV); see also ed. and trans. by G.S.M. Walker, *Sancti Columbani Opera* (Dublin: Dublin Institute for Advanced Studies, 1970). On the letter to Boniface, see Bruno Krusch, "Chronologisches aus Handschriften", *Neues Archiv* 10 (1885):83–88. On the dating of Columbanus' letters, see P. Grosjean, "Recherches sur les débuts de la controverse paschale chez les Celtes", *Analecta Bollandiana* 64 (1946):206–215.

115 Trans. G.S.M. Walker 19.

116 The arguments presented by McCarthy, "Origin", that the 84-year cycle with Celtic criteria was created by Sulpicius Severus are not convincing.

117 Bede, *HE* 3.25; cf. Jones, *BOT* 82–85. The most recent edition of the work is that by Krusch, *Studien I*, 311–327; for earlier editions, and criticisms of Krusch's, see

because Bede cites it frequently in *The Reckoning of Time*,[118] and in the *Epistle to Wicthed*. It was included in the *computus* anthology of Irish provenance which served as one of Bede's major sources of computistical knowledge, and which is represented today by the manuscripts known collectively as the "Sirmond Group".[119]

Everyone involved in the Celtic Paschal controversy accepted the *Liber Anatolii* as an authentic work by Anatolius, though many, including Bede, claimed that the text had been corrupted, perhaps deliberately. Their conviction rested on the fact that the *Liber Anatolii* includes a passage from a genuine work of Anatolius found in Rufinus' translation of Eusebius' *Ecclesiastical History*.[120] However, the Paschal criteria in the genuine text are 22 March equinox, and lunar limits 15–21; in the *Liber Anatolii*, they are the Celtic principles of 25 March equinox, and lunar limits of 14–20.

Presumably the "forger" lifted the excerpt from Rufinus, changed the dates of the Paschal criteria, and added all the rest of the material to confect what we now have as the *Liber Anatolii*. And yet no adequate explanation has been offered for the most curious aspect of the *Liber Anatolii*, namely that it advocates Celtic criteria *with a 19-year Alexandrian cycle*. In fact, the table included with the treatise purports to be Anatolius' 19-year Alexandrian table, but it is constructed according to the criteria of a Celtic 84-year table, in other words with lunar limits 14–20. That the "forger" tried to marry a 19-year cycle to non-Alexandrian criteria had the ironic effect of convincing many readers of the *Liber Anatolii* that "Anatolius" *supported*

Jones, *BOT* 82, n. 4. For analysis of the text, see MacCarthy 4.cxxiii *sqq.* For recent bibliography, see Ó Cróinín and Walsh 33, nn. 123–124. The text is corrupt in the MSS, but the table even more so, and attempts to reconstruct it, such as MacCarthy's (4.cxix–cxxvii) have not met with success, since they were based on inadequate older editions, and an incomplete knowledge of the MSS; see in particular the criticisms levelled by C.H. Turner, "The Paschal Canon of 'Anatolius of Laodicea' ", *English Historical Review* 10 (1895):699–710, against the efforts of A. Anscombe, "The Paschal Canon attributed to Anatolius of Laodicea", *ibid.* 515–535.

    118 Chs. 6, 14, 22, 30, 35, 42.

    119 See below, section 5.

    120 Rufinus' trans. of Eusebius, *HE* 7.32.14, ed. Th. Mommsen, *Eusebius Werke* 2.1–3, Griechischen christlichen Schriftsteller 9.1 (Leipzig: Hinrichs'sche Buchhandlung, 1903–1904):723.

the 19-year cycle against the 84-year cycle. Cummian was one,[121] and so were Aldhelm[122] and Bede.

What exactly the motivation was behind this curious production is quite obscure. Ó Cróinín and Walsh argue that its main target was the Alexandrian 19-year system, and that the forgery included a 19-year table in order to demonstrate that 19 years was an insufficiently long period to incorporate the solar data necessary to produce a true Easter cycle.[123] If that was the case, then the plan backfired badly. More persuasive are the suggestions of Alfred Cordoliani, who argues that the real point of the forgery was to defend the Celtic *lunar limits* by cloaking them in the authority of the great Alexandrian computist; the author really did not care that the *cycle* differed from the Celtic one.[124] This might seem improbable at first blush, but if the Celtic criteria historically antedated the arrival of the 84-year cycle, then the defence of the former at the expense of the latter seems less bizarre.[125] The idea that the Irish forgers saw the issues of Paschal criteria and Paschal cycle as separate – the criteria to be defended at all costs, the cycle as less relevant – fits with other evidence about the early Irish computists. They seem, for instance, to have been fairly indifferent to astronomy, and only vaguely aware of their cycle's technical defects.[126] Hence they were happy to adopt the Alexandrians as their advocates, regardless of the differences between the two cycles, as long as the criteria could be modified to conform to Celtic tradition. It is noteworthy that at the Council of

121 Ó Cróinín and Walsh 35.

122 In his letter to King Geraint, Aldhelm refers to the Celts as Quartodecimans who use a 19-year table. This description fits ps.-Anatolius, which presents a 19-year table, with lunar limits of 14–20: ed. R. Ehwald, *Aldhelmi Opera*, MGH AA 13 (1919):483; trans. Michael Lapidge and Michael Herren, *Aldhelm: The Prose Works* (Cambridge: D.S. Brewer and Totowa, N.J.: Rowman and Littlefield, 1979):157.

123 Ó Cróinín and Walsh 34–35.

124 "Les computistes insulaires et les écrits pseudo-alexandrins", *Bibliothèque de l'École des Chartes* 106 (1945–1946):7.

125 This possibility is proposed by O'Connell 92–93, and Schwartz 103.

126 Smyth 147–151. The author of *De mirabilibus sacrae scripturae* seems unaware of the concept of the regularized lunar and solar cycles, and assumes that the Easter table is a literal description of astronomical reality (164); when he discusses "the solar cycle" for instance, he is not referring to an astronomical event, but to Victorius' 532-year table (151–2). The *De ordine creaturarum* understands the importance of the equinox for Paschal reckoning, but does not seem to know what the word "solstice" means (152–3), etc.

Whitby, Colman defended the Celtic *criteria*, while Wilfred attacked the Celtic *cycle*.[127]

It was against this backdrop of confusion and contention that a meeting of southern Irish clergy was held at Mag Léne in anticipation of a major conflict over the date of Easter in AD 631,[128] and in response to a letter from Pope Honorius (probably written in 628/9) summoning them to conform.[129] The Synod was divided over what to do, and decided to send envoys to Rome to ascertain the practice of the universal Church. When the envoys returned to report that the Alexandrian reckoning was observed everywhere but in the Western islands, the southern Irish converted to the new system. However, the northern Irish communities, and especially the *paruchia* of St Columba, centred at Iona, strongly disapproved of the decision. It was in defence of the new way that Cummian wrote his *Epistola de controversia paschali*.

As Cummian's letter was not, as far as we are aware, known to Bede, we shall refer the reader to Ó Cróinín and Walsh's excellent edition for further details, and confine ourselves here to what Cummian reveals about a source directly relevant to Bede. Though Cummian employs many arguments, biblical, patristic, and commonsensical in favour of the change, his major ally in advocating the Alexandrian reckoning is Cyril – or rather pseudo-Cyril. The *Epistola*

---

127 Bede, *HE* 3.25 (304): "Colman replied, 'Did Anatolius, a man who was holy and highly spoken of in the history of the Church to which you appeal, judge contrary to the law and the Gospel when he wrote that Easter should be celebrated between the fourteenth and twentieth day of the Moon?'... Wilfred replied, 'It is true that Anatolius was a most holy and learned man, worthy of all praise; but what have you to do with him since you do not observe his precepts? He followed a correct rule in celebrating Easter, basing it on a cycle of nineteen years, of which you are either unaware or, if you do know of it, you despise it, even though it is observed by the whole Church of Christ.'" (trans. Colgrave and Mynors 305).

128 On the problems of dating this meeting, and consequently, the Letter of Cummian, see Ó Cróinín and Walsh, Introduction, section A. The choice of 631 is based on the statement in the letter that the Celtic Easter of that year diverged from the Roman by a month (88.238). Since the Celtic 84-year cycle retained the old Roman limits of 25 March–21 April, this could only happen when the Roman (i.e. Alexandrian) Easter fell on 22–24 March or 22–25 April. This happened in 631, when the Celtic Easter fell on 21 April, and the Roman on 24 March. When they published their edition, Ó Cróinín and Walsh were using MacCarthy's reconstruction of the 84-year table, but the rediscovered 84-year table since published by Ó Cróinín and McCarthy (see above, n. 113) confirms that these data are correct for 631.

129 Summarized by Bede, *HE* 2.19.

Cyrilli,[130] like the *Liber Anatolii,* presents a doctored version of a genuine text (a letter from Cyril of Alexandria to the Council of Carthage of 419)[131] to which a forgery had been attached. It was certainly composed after Dionysius Exiguus, since it incorporates Dionysius' "signature", namely the assertion that the Council of Nicaea approved the 19-year cycle,[132] and seems to have been provoked by a problem in determining the Easter of 607.[133] There is where certainty ends, for no known table will produce the data "Cyril" describes. Jones argues that dissatisfaction with the Victorian table resulted in ill-informed tinkering, which "Cyril" corrected, and for good measure, backed up by copies of Dionysius' table; O'Connell, on the other hand, sees it as an attack on Victorius by a champion of Dionysius.[134]

In obeying the papal behest, the southern Irish in 633 adopted "Alexandrian reckoning" generically, and seem to have used Victorius' or Dionysius' tables interchangeably, or perhaps even a fusion of the two.[135] In fact, the Letter of Cummian seems to indicate that the Victorian system, or a hybrid Victorian-Dionysian system, was the preferred one, because it was thought – and perhaps correctly – that this was the one used in Rome and specifically recommended by Pope Honorius.[136] Certainly, the author of *De mirabilibus sacrae scripturae* was using Victorian tables in 654, and was probably unaware that they were incompatible with Dionysian ones.[137] That is undoubtedly why Bede's *Reckoning of Time* assumes that there are *two* Paschal systems

130 Ed. Krusch, *Studien I,* 344–9; the treatise was first edited by Gilles Bouchier, *De doctrina temporum* 72–74.

131 Cf. Denis Petau, *De doctrina temporum* 1.220–221, who noted that the authentic letter was published by Dionysius Exiguus in his *Collectio canonum* and who corrected Bouchier's dating of the original letter accordingly: cf. Cordoliani, "Les computistes", 25; Jones, *BOT* 72 n. 3; Paul Grosjean, "Recherches", 227.

132 Jones, *BOT* 94; see above, p. liii.

133 Cordoliani, "Les computistes", 25–27; Grosjean, "Recherches", 226 *sqq.* AD 607 presents the same problem as AD 410, about which the real Cyril wrote to the Council of Carthage – namely the problem of an embolismic year where the embolism occurs in the spring, and therefore affects the Easter calculation.

134 O'Connell, "Easter cycles", 79. The suggestion by Jones (*BOT* 95) that "Cyril" is really Pope Boniface IV, sporting a disguise contrived by Bishop Laurentius of Canterbury, is unconvincing.

135 Ó Cróinín and Walsh 46.

136 Ó Cróinín and Walsh 28 and n. 99; 38; Harrison, *Framework,* 59.

137 Kenney, *Sources,* 276–277; Jones, *BOT* 66; Harrison, "Episodes", 310, n. 14; Smyth 147. There is evidence that Victorius' cycle continued in use in Ireland until the 11th

which require refutation: the old 84-year cycle, still used by most of the British, and the Victorian cycle, now favoured by some of the Irish. The origin of the *Epistola Cyrilli*, then, remains obscure, but its purpose seems clear: to promote the Alexandrian system, without invoking either the contentious name of Victorius (whose system, as Columbanus observed, commanded little respect in Ireland), or the obscure name of Dionysius.

After the southern Irish had converted to the Alexandrian reckoning, they evidently wrote to Pope Severinus explaining that their northern confrères had refused to follow suit. A response came from Severinus' successor, the Pope-elect John IV, who roundly condemned the recalcitrant Irish as Quartodecimans.[138] This is something of an exaggeration, of course, and Bede tends to avoid this allegation in *The Reckoning of Time*.[139] He even excises the reference to Quartodecimanism in his transcription of Pope John's letter in *HE* 2.19.[140] But the accusation stuck: it was raised by Aldhelm, as we have seen, and also used by Wilfred at Whitby.

In interpreting the Council of Whitby, the historian is faced with the problem that both sources – Bede and Eddius Stephanus – have a strong vested interest in enlarging the Council's importance. This has perhaps distracted attention from an equally significant issue. Why was the meeting summoned in 664? The Celtic and Alexandrian tables had been producing discordant Easters in Northumbria for twenty years and nobody had seen fit to remedy the situation.[141] On the other hand, one of the rare disagreements between the the Dionysian and the Victorian tables was looming in 665. It has therefore been suggested that the bone of contention at Whitby was not only the 84-year cycle, but the super-

---

century: see Daniel McCarthy, "The Chronological Apparatus of the Annals of Ulster AD 431–1131", *Peritia* 8 (1994):46–97.

138 Kenneth Harrison, "A Letter from Rome to the Irish Clergy", *Peritia* 3 (1984):222–229.

139 However, see *The Reckoning of Time*, ch. 69, and ch. 66, *s.a.* 4591, and note. Bede, like most computists, often confused Quartodecimanism with accepting the 14th Moon as a legitimate date for Easter: see *The Reckoning of Time*, ch. 66, *s.a.* 4146.

140 Alan Thacker, "Bede and the Irish", in *Beda Venerabilis* 39; for the missing passage, see K. Harrison, "A Letter from Rome", 228. Bede also deletes a reference to the link between celebrating on 14 Nisan and Pelagianism in his account of the papal letter of 640 (*HE* 2.19), though he leaves it in Ceolfrith's letter to Nechtan (5.21, p. 544): Thacker 40.

141 Jones, *BOT* 103; Harrison, *Framework*, 59–60.

iority of Dionysius' over Victorius' table as its major rival. In short, Whitby marks the arrival of Dionysius' table in England.

The fact that Wilfred learned the *computus* of Dionysius under Archdeacon Boniface in Rome in 645 indicates that the popes had adopted Dionysius' system by this time, but when exactly they switched is uncertain. Harrison argues that if Cummian was correct that Rome was using Victorian tables, the switch must have taken place between 631 (the date of Cummian's letter) and 640 (John IV's letter).[142] However, this is not certain, since the Victorian and Dionysian tables produced for the most part identical dates. Since Cummian's letter mentions the Dionysian tables, they would seem to have been known in the British Isles before Whitby, and the southern Irish *computus* manuscript used by Bede contained both a Victorian and a Dionysian cycle.[143] So argue Krusch and Poole; but Jones and Harrison demur.[144] Harrison points out that since Bede claims that Wilfred was unable to learn the correct *computus* in England, it would follow that the Dionysian system was unknown there before 645.[145] It was therefore Wilfred who introduced the Dionysian system into England, and the Council of Whitby was really a promotion of that system, at the expense of the Victorian one, as the only valid alternative to the Celtic 84-year system.

This argument fails to convince me, not because I have any fresh chronological evidence to offer, but because it does not take into account the considerable difference between being able to *consult* a table, and knowing how to *construct* one. The evidence of Cummian's letter seems incontrovertible: the Irish had access to the tables of Dionysius before 631 – but they probably did not know how they were constructed, any more than they knew how to test their own 84-year cycle against the astronomical phenomena. Even the *Epistola de ratione conputandi*, probably written in Cummian's own circle, and the most sophisticated of the Irish *computus* treatises yet to come to light, does not offer a coherent explanation of the Dionysian system's principles, and why they are different from and superior to those of Victorius. Wilfred did not learn of the *existence* of the correct *computus* in Rome,

---

142 *Framework*, 161.

143 See below, section 5.

144 Krusch, "Einführung", 150 *sqq.*; Poole, "The Earliest Use of the Easter Cycle of Dionysius", in *Studies in Chronology and History* (Oxford: Clarendon Press, 1934):28–37; Jones, "Victorian and Dionysiac Paschal Tables", 415; Harrison, *Framework*, 62.

145 *Framework*, 64.

but rather, how it *operated*. Indeed, Bede's statement that Wilfred was discontented with the customs of Lindesfarne and wanted to investigate the ecclesiastical practices of Rome[146] would suggest that prior acquaintance with the Dionysian table was what *motivated* Wilfred's study-tours. What made *The Reckoning of Time* so popular, and such a significant milestone in the Insular Paschal controversy, was that it made a trip to Rome unnecessary. In short, Whitby may have secured the victory of the Alexandrian system in Northumbria, but it may not have struck a decisive blow for the Dionysian version of that system. Bede, certainly, continued to see Victorius as a threat.

The chronology of the subsequent conversion of the Irish is clear. The southern Irish followed the Roman *computus* by 631. The northern Irish, except for Iona, were converted for the most part by Adomnan, *ca.*686; their agreement was sealed at the Synod of Birr in 697.[147] The Picts converted under their king Nechtan about 710. Iona and its dependencies were won over by Egbert in 715 or 716. No serious attempt to convert the British to the Roman Easter was made until the days of Theodore and Hadrian; in 705, Aldhelm of Malmesbury, at the request of a Wessex synod, wrote to King Geraint in Wales, recommending the Dionysian version of the Alexandrian system.[148] In sum, at the time when Bede wrote *The Reckoning of Time* there were three competing *computus* systems: the old Celtic 84-year cycle, now abandoned by all save the British; the Victorian system, still used by the southern Irish, in Gaul, and perhaps also in England; and the *computus* of Dionysius Exiguus, known in the British Isles since at least 631, and actively promoted, if not exactly introduced, by Wilfred of Hexham. It was Bede's refutation of Victorius' errors concerning Nisan and the *saltus*, and his transformation of Dionysius' system into a perpetual Paschal table, which effectively won the day for Dionysius.

4. STRUCTURE AND CONTENT OF *THE RECKONING OF TIME*

St Augustine's appeal for a Christian encyclopaedia went unanswered for nearly two centuries, and when it finally appeared, it was something

---

146 *HE* 5.19 (518).

147 Harrison, "Episodes", 309.

148 For ed., see above, n. 122. We know that Aldhelm is referring to the Dionysian, and not the Victorian system because he alludes to the "318 Fathers" at the Council of Nicaea, a tag from Dionysius' preface.

of a disappointment from the perspective of *doctrina christiana*. Isidore of Seville's *Etymologies* was not structured by the Bible, but according to the Liberal Arts and the traditional divisions of philosophy. Likewise, Isidore's cosmological handbook, *De natura rerum*, was based on the hierarchy of the four elements, not on the six days of Creation. Isidore's treatment of time reveals something of the difficulty experienced by early medieval thinkers in finding a format to match this subject. In the *Etymologies*, Isidore deals with the movements of Sun and Moon in Book 3, the units of time-reckoning (including a world-chronicle) in Book 5, and the Paschal *computus* in Book 6 (under the rubric of "Scripture and liturgy"). This dispersal reflects the division of genres: cosmology, historiography, *computus*. In *De natura rerum*, Isidore attempts an explicitly Christian cosmology which included time, and yet he does not discuss *computus* or the relation of time to history.

What Bede aimed for in *The Reckoning of Time*, on the other hand, was nothing less than a new genre of writing which would integrate *computus*, its astronomical and cosmological context, and its relation to historical time. Moreover, this new science of time would be an explicitly Christian one, based on Christian sources and useful for Christian purposes. This achievement can best be analysed in three stages: (1) the integration of *computus* and cosmology, (2) the integration of *computus* and history, and (3) the theological reframing of the science of time, and particularly, the purifying of eschatology. Concretely, Bede accomplished all this by fusing three of his earlier scientific writings: *The Nature of Things*, a Christian cosmology; *On Times*, a computistical manual with a chronicle; and the *Letter to Plegwin*, a defence of *computus* against the more dangerous speculations which the calculation of time might inspire.

Computus and cosmology

In the preface of *The Reckoning of Time*, Bede identifies his new book as a sequel to two others: *On Times*, his first work on the calendar, and *The Nature of Things*, a treatise on cosmology. *On Times* was composed in 703, and so is one of the earliest works to come from Bede's pen. *The Nature of Things* followed shortly thereafter.[149] The two works drew on quite separate models, and were conceived by Bede as having distinctive

149 Jones, *BOD* 174.

missions. *On Times* was strictly about the calendar, but not modelled on any pre-existing computistical genre. Bede was certainly indebted to Irish models, notably to *De computo dialogus*; he also drew heavily on Isidore, both *De natura rerum* 1–7 and *Etymologies*, Book 5. Both his Irish and Isidorean sources suggested to him the idea of basing his discussion on the units of time, ranked from smallest to largest. From Book 5 of the *Etymologies*, he adapted the notion of inserting a chronicle under the rubric of "world-ages". But unlike Isidore's works or the Irish treatises, Bede's *On Times* meshes this hierarchy of time-units with the primary documents of the computist: calendar and Paschal table. Bede describes the units of time-measurement up to the year, the span of time covered by the calendar. Then he turns to the Paschal table, explaining the 19-year cycle, including its *saltus lunae*, the various columns of the Dionysian Paschal table, and the significance of Easter. Finally, he closes with a skeletal world-chronicle. Isidore, by contrast, deals with the Paschal table completely separately from the units of time, in *Etymologies* 6.7.

Like Isidore, however, Bede avoided mixing cosmology and *computus*. *On Times* is a concise and focused course in basic computistics, with no discussion of the Sun and Moon, or the zodiac, solstices and equinoxes, seasons and variations of daylight, or tides. Isidore had also dealt with these matters apart from the context of *computus*, in *Etymologies* 3 (astronomy) and 13 (cosmology), and in *De natura rerum*. So when Bede wrote his own *Nature of Things*, he simply took Isidore's framework in *De natura rerum*, and subtracted the section on the units of time, which he had already shifted to *On Times*. To all intents and purposes, *The Nature of Things* is Isidore's *De natura rerum*, minus the materials on the units of time, and improved by revisions based on Pliny, an author Isidore did not know. Bede replicates Isidore's top-down arrangement, beginning with the universe as a whole, and then descending from the highest heaven down to earth, with its atmosphere, oceans and rivers, and land masses. The pattern is that of the four elements, with the heavens standing for fire, the atmosphere for air, the seas for water, and the land for earth.[150]

However, Bede's tentative experiment in redeploying parts of *De natura rerum* as part of a *computus* textbook suggested the possibility of using even more cosmological material to elucidate the calendar. An

---

150 Cf. Alessandra di Pilla, "Cosmologia e uso delle fonte nel *De natura rerum* di Beda", *Romanobarbarica* 11 (1991):128–147.

epigram at the head of *The Nature of Things* declares that "I Bede, the servant of God, have in these brief chapters touched on the various natures of things *and of this fragile age and on extended aeons* (*labentis et aeui ... tempora lata*) ...".[151] This epigram does not describe *The Nature of Things* as it stands, for the treatise does not touch upon time at all, let alone "extended aeons". It does not even allude to the Isidorean model, which discussed the units of time from the day up to the year, but not ages or aeons. The epigram does, however, match the profile of a hypothetical fusion of *On Times* and *The Nature of Things*. It presages Bede's most striking innovation in composing his second work on *computus*, namely the fusion of natural history and calendar science.

Bede reflected on this move for nearly twenty years, and doubtless drew courage from Irish models like *De computo dialogus* and *De divisionibus temporum*. These treatises liked to stretch the framework of *computus* to absorb other kinds of erudition, largely philological and historical: the names of the various parts of the day or night, the origins of month-names, and the like. This philological bent is visible in much of the first part of *The Reckoning of Time*, especially chs. 5–15. But Bede was alert to the dangers of *curiositas*, and not interested in philology or etymology for its own sake. Where the Irish *De ratione conputandi* analyses the word for "day" in every known ancient language,[152] Bede is content with Latin. The Irish tracts also show an interest in pure mathematics which is foreign to Bede. The author of *De ratione conputandi* makes his students work problems involving the minutest divisions of time[153] – divisions which Bede introduces briefly in chs. 3 and 4, but then passes over as irrelevant for computistical purposes. Indeed, as we shall see, he may have considered such rarified calculations as dangerously close to astrology.[154]

This policy of controlled integration also governed Bede's more innovative project of systematically incorporating cosmology into *computus*, something neither Isidore nor the Irish had done. Especially in chs. 16–36 of *The Reckoning of Time*, we find Bede redistributing the information recorded in *The Nature of Things* under computistical rubrics, namely the lunar month and the solar year. The discussion on tides which in *The*

---

151 Ed. Jones, *BOD* 189.
152 Ed. Ó Cróinín and Walsh, ch. 22 (130).
153 E.g. chs. 108–109 (209–210).
154 See *The Reckoning of Time*, ch. 3 and Commentary.

Nature of Things is filed under "oceanography" (ch. 39) is now to be found amongst the chapters on the month, the unit of time defined by the Moon, whose power creates the tides. The chapters on climates in *The Nature of Things*, classed as "geography" (chs. 47–48), are integrated into the discussion of the solar year in *The Reckoning of Time* (chs. 31–34) because latitude determines the duration of daylight at different seasons of the year. Significantly, not all of *The Nature of Things* is ransacked in this way: Bede omits the discussion of the planets (chs. 12–13), atmospheric phenomena and meteorology (chs. 24–37), earthquakes and volcanoes (chs. 49–50). He is determined not to turn *computus* into an encyclopaedia at the expense of its primary calendrical mission. But this does not detract from the originality of his project. No *computus* text before Bede attempted to integrate cosmographical explanations into its argumentation. That Bede did so bespeaks a certain breadth of scientific vision, grounded in a Christian conception of time. God created time and the universe simultaneously. The Sun and Moon do not make time; there were days and nights before they were created. Rather, they are the *signs* of time, and the time which they mark is God's time, which is more than just the calendar. It involves rhythms which the calendar does not record: seasons, tides, the cycles of growth and decay (ch. 28), the pattern of eclipses (ch. 27), and "natural years" (ch. 36). In fusing *computus* and chronology, Bede re-envisions time in a theological light, encompassing both liturgy and providential history. In fusing *computus* and cosmology, Bede likewise re-envisions *computus* in a perspective which can best be described as "scientific", anchored to the calendar and its demands, but open to the other "times" of which the Sun and Moon were signs. In our chapter-by-chapter commentary on *The Reckoning of Time*, we will show how he enlarges the scope of calendar science, without losing sight of its fundamental structure and problems.

Computus and history

It has often been remarked that Bede inaugurates his *Ecclesiastical History of the English People* by situating the island of Britain in geographical space. It has not often been noticed that he also endows Britain with a special kind of time:

> Because Britain lies almost under the North Pole, it has short nights in summer, so that often at midnight it is hard for those

who are watching to say whether it is evening twilight or whether morning dawn has come ... On the other hand the winter nights are also of great length, namely eighteen hours, doubtless because the Sun has then departed to the region of Africa. In summer too the nights are extremely short; so are the days in winter, each consisting of six standard equinoctial hours, while in Armenia, Macedonia, Italy and other countries in the same latitude the longest day or night consists of fifteen hours and the shortest of nine.[155]

The parallel in *The Reckoning of Time* is chs. 32–33, on geographical latitude and its influence on the length of daylight. It seems not unlikely, then, that Bede had his textbook on *computus* in mind when he set out to describe the situation of the island of Britain.[156]

For Bede, a strong link between *computus* and historiography seems to have been self-evident. Indeed, his second major innovation as a *computus* writer was to incorporate a full world-chronicle into his computistical writings. But it is important to understand what *kind* of history Bede incorporated into *On Times* and *The Reckoning of Time*. These chronicles are not a history like the *Ecclesiastical History*; they are universal, not national, and based on *annus mundi*, rather than *annus Domini* reckoning. But neither do they resemble the annals commonly recorded on the margins of Easter tables: these also tend to be local, not universal, and Easter tables in the West rarely use *annus mundi* as their era.[157] Though Bede uses the Dionysian Paschal table to formally intro-

155 *HE* 1.1 (15).

156 This temporal element is overlooked by J.M. Wallace-Hadrill, *Bede's Ecclesiastical History of the English People: A Historical Commentary* (Oxford: Clarendon Press, 1988):6, and by the major commentators on this passage, e.g. Calvin B. Kendall, "Imitation and the Venerable Bede's *Historia ecclesiastica*", *Saints, Scholars and Heroes* 1.161–190.

157 In *Saints' Lives and Chronicles* and in his essay on "Bede as Early Medieval Historian", C.W. Jones argues that medieval historiography was shaped by the form of the Paschal table, whose sequence of years provided orderly storage for annalistic notes which could later be elaborated into narrative. It was supplemented by the calendar, whose chain of saints' days spawned biographical notices. Both documents were maintained by one man, the monastic schoolmaster, computist, martyrologist and liturgical expert – the protean historian. Bede's own *History*, in which edifying biographical vignettes are threaded on a narrative chain, is a typical product of this encounter of manuscript form and monastic function. The chronicle of ch. 66 is a totally different kind of production from the *Ecclesiastical History*, and based on a totally different source, namely Eusebius' world-chronicle.

duce the great chronicle in ch. 66 of *The Reckoning of Time*, the chronicle really has little to do with the Paschal table as such. Bede is not demonstrating the usefulness of Paschal tables for historians; rather, he is demonstrating how the World-Ages, like every other kind of time, are ordered by a *ratio* of Divine providence. Moreover, Bede's models – Isidore of Seville and the Irish treatises – saw the chronicle as the capstone of an ascending series of units of time, not as an appendix to the Paschal table. Nonetheless, Bede explicitly connects the Paschal table to his great chronicle,[158] so it is worth asking what he thought that connection was.

The ancestor of Bede's chronicle is in fact Eusebius' great world-chronicle, the *Kanones*. This world-chronicle starts with Abraham, and weaves the histories of various peoples together into a single chronological framework.[159] In the words of Brian Croke, Christian world-chronicle "takes for granted . . . a fixed starting point in the human story and that the history of all known kingdoms and countries can be fitted into a single calculated sweep".[160] The aim is to synchronize the histories of many peoples, and the intention is to facilitate reference: the location of events and their relation to other events.[161] The world-chronicle was designed for finding dates and events, just as the *computus*, with its formulas and tables, was designed for locating dates and feasts. Like *computus*, chronicles conceal theological meaning beneath a utilitarian surface. Chronicles seem to sacrifice narrative coherence for the sake of

---

However, Jones does not explain why Bede did not choose a historical form based on the Paschal table to illustrate a treatise on *computus*. Jones' overall argument contains other weaknesses, and has increasingly come under attack, most notably on historiographical grounds: see Peter Hunter Blair, "Bede's *Ecclesiastical History of the English Nation* and its Importance Today" (Jarrow Lecture, 1959):22; Jan Davidse, "The Sense of History in the Works of the Venerable Bede", *Studi medievali*, ser. 3, 23 (1982):647–695, and "On Bede as a Christian Historian", 10.

158 See Commentary on ch. 65.

159 On the distinctive character of the world-chronicle as a genre of Christian historiography, see Anna-Dorothee von den Brincken, *Studien zur lateinischen Weltchronistik bis in die Zeitalter Ottos von Freising* (Düsseldorf: Michael Triltsch, 1957): esp. 38–49; Hildegard L.C. Tristram, *Sex aetates mundi. Die Weltzeitalter bei den Angelsachsen und den Iren. Untersuchungen und Texte* (Heidelberg: Carl Winter, 1985): esp. 12–15; Brian Croke, "The Origins of the Christian World Chronicle", in *History and Historians in Late Antiquity*, ed. Brian Croke and Alanna M. Emmett (Sydney and Oxford: Pergamon Press, 1983):116–131. For further discussion, see Commentary on ch. 66.

160 Croke 120.

161 Croke 125.

chronological schematism, but this is only on the level of appearances. Though episodic, disjointed and paratactic, they are well suited to conveying certain kinds of messages. In the patristic period, world-chronicle was a sub-genre of *adversus paganos* apologetic, which aimed at proving that Christianity was not a "novelty", because it was derived from Judaism, a religion older than the cultures of Greece or Rome.[162] It also served to refute pagan accusations that the Bible contained absurdities or contradictions: Augustine in *De doctrina christiana* claims that chronology is a useful study for Christians because errors concerning the consulship of the Lord's birth and Passion have caused some to suppose that he was forty-six years of age at his death.[163] In the case of the seventh-century Byzantine *Chronicon Paschale*, the aim was to celebrate the new age inaugurated by Heraclius' defeat of the Persians in 630, and to reinforce the orthodox calendar in forthcoming talks with Monophysites and Persian Nestorians.[164] Above all, chronicles were composed to prove or disprove contentions about the duration and approaching end of the world: John Malalas' chronicle was written to convince contemporaries that the dreaded year AM 6000 had passed without the expected apocalypse,[165] and the chronicles of Eusebius and even Bede himself were in part intended to revise the age of the world to demonstrate that the end was still far off.[166]

The historiographical intention in the *Reckoning of Time* is, then, quite different from that of the *Ecclesiastical History*. The theme of the latter is the particular providence of God with regard to the English; that of the former is the continuity and pattern of general providence throughout time. Bede uses *annus mundi* throughout the chronicle because it underscores this continuity; he uses *annus Domini* reckoning in the *Ecclesiastical History* because all its action takes place within the Sixth Age of the world, and because the Dionysian *computus* associated with *annus Domini* era plays such an important role in his overall

---

162 Croke 122.
163 2.28.43 (63.36–39).
164 *Chronicon Paschale* xii–xiii.
165 *Ibid.* xxiv.
166 Richard Landes, "Lest the Millenium Be Fulfilled: Apocalyptic Expectations and the Pattern of Western Chronography 100–800 CE", in *The Use and Abuse of Eschatology in the Middle Ages*, ed. Werner Verbeke, Caliel Verhelst and Andries Welkenhuysen, Mediaevalia Lovaniensia Series I/Studia XV (Leuven: Leuven University Press, 1988):137–209. This theme is discussed further in the Commentary on ch. 66.

story.[167] Like his choice of genre, Bede's choice of era was deliberate and strategic: what is irrelevant to the *Ecclesiastical History*, is all-important to *The Reckoning of Time*, and *vice versa*.

A purified eschatology

On Times ends with the six World-Ages, but *The Reckoning of Time* ventures beyond to explore the end of time, and the transition from time to eternity. This innovation grew out of the Plegwin episode, where Bede's chronology had been criticized by "rustics" who believed that each Age of the world would last 1,000 years. The implication of such a belief is that the world will end precisely in the year 6000. In the commentary on ch. 66, we shall explain how *computus* and chronology were enlisted from early in the Church's history to combat this heresy. Here we would simply note that in electing to face this issue squarely, Bede is acknowledging the occupational dangers computists run. To project the dates of Easter is to project the future, and to give names to years which have not yet been. This was a very unusual project for early medieval people. Neither Romans nor Germans had any prospective era; they could only name the years in the present and past, by reference to consuls or kings. Computists not only thought about the years to come, but also counted and named them in the columns of their Paschal tables. But in so doing, they opened up prospects for millenarian speculation which the Fathers had desperately sought to bar.

But the purified eschatology of chs. 68–71 of *The Reckoning of Time* also gives Bede an opportunity to do what no previous computist had ever attempted: to turn the reckoning of time into a *figura* of eternity. The calculation of Easter merges into a meditation upon the last things, a spiritual exercise whose purpose was to rise through the contemplation of time to the perception of eternity. Chapter 71 brings *The Reckoning of Time* to a close by underscoring the book's essential character as a vision of eternity through time.

---

167 One might add that there are other, practical reasons for adopting AD reckoning in the *HE*. The sources used for the *HE*, especially the papal letters dated by indiction, could be readily dated using a Dionysian Paschal table, which correlates AD era and indiction. AD was also being used in charters by Bede's day: Harrison, *Framework*, 97–98. This is not to claim that there are no "world history" elements in *HE*: see section on "the English as Chosen People" in Jones, "Some Introductory Remarks", 125–127.

5. BEDE'S SOURCES

The technical literature of *computus*

Analysis of the sources cited in *The Reckoning of Time* shows that Bede had access to a manuscript *computus* anthology, such as we have described above. It was of Irish provenance, and a copy made in the eleventh century in Vendôme survives as Oxford, Bodleian Library Bodley 309. This codex is called the "Sirmond manuscript", after the French scholar Jacques Sirmond (1559–1651), who owned it and who loaned it to the Jesuit chronographer Denis Petau (1583–1652). Gilles Bouchier also borrowed the volume for his edition of and commentary on Victorius.[168]

According to Petau, Sirmond's volume contained the prologue and complete tables of Victorius of Aquitaine, the prologues of Theophilus and Cyril of Alexandria, the letter of ps.-Cyril, the *Liber Anatolii*, the two letters by Dionysius Exiguus, as well as much miscellaneous computistical information.[169] The Bodleian manuscript is Sirmond's, for it contains all the items listed by Petau, including (uniquely) the Victorian cycles.[170] But the Sirmond volume is part of a larger manuscript cousinage. Three other manuscripts share much of the same material, namely Vatican City Rossiana lat. 264 s. XI [R], Paris lat. 16361 s. XI [P], and Geneva, Bibl. de l'Université 50 s. IX [G]. Jones also identified a number of other codices containing similar configurations of material. All these codices are linked to Britain.[171]

The Sirmond manuscript is divided into two "books". The exemplar of Book 1 was compiled in Ireland before 718; it contained the following items, almost all of which were known to Bede:

168 C.W. Jones, "The 'Lost' Sirmond Manuscript of Bede's Computus", *English Historical Review* 51 (1937):204–205.

169 Petau 217–219, 225–228.

170 *Ibid.* 204–213; see also *BOT*, ch. 6.

171 The other MSS include [L] Cotton Caligula A.XV, s. VIII; [O] Bodl. Digby 63, s. IX;[C] Cologne Dombibliothek 83(ii), *ca.*805; [M] Milan Ambrosiana H 150 inf., *ca.*810 (the "Bobbio *computus*"); [D] Leiden Scaliger 28, s. IX; [B] Bern 610, s. X; [Be] Besançon 186, s. IX; [Ba] Basel F III 15 k, s. IX [V]; Vat. lat. 642, s. XI[1]. Evidence for Insular provenance includes the following: M contains a unique Irish Easter cycle; D, C, G show Insular palaeographical symptoms; O was written at Winchester or Canterbury, AD 867, and contains no works of Bede, but many works Bede used; O and C are the only MSS to contain the "Cologne Prologue", edited by Krusch (*Studien I*, 227–244), which Bede describes in his letter to Plegwin.

A. *The "Irish computus".* Fol. 62r–v of the Sirmond manuscript contains the prologue and capitula of an anonymous *computus* text (*De numero igitur . . . De Victorio et dionisio. De Boetio. De calculo.*) which probably belongs to the following item, namely *De computo dialogus* (Sirmond, fols. 62v–64v). Together with the next item, *De divisionibus temporum liber*[172] (Sirmond, fols. 64v–73v), it probably formed a single work.[173] As we have already seen, this work (or works) strongly influenced Bede's conception of the scope and content of *computus.* Like other Irish *computus* treatises, notably the *De ratione conputandi,*[174] this work is organized according to divisions of time in ascending order of magnitude. *The Reckoning of Time* adopts and expands upon this strategy. *De divisionibus temporum* is cited directly in chs. 2, 3, 5 and 11 of *The Reckoning of Time,* and *De computo dialogus* in ch. 6.

B. *Short tracts and formulae.* These include a tract on the *bissextus* (Sirmond fols. 74r–76r),[175] and one on the *saltus lunae* (Sirmond fols. 76r–78r),[176] as well as formulae for the *saltus lunae* (Sirmond fol. 78r–v),[177] the *bissextus* (Sirmond fol. 78v),[178] and one to find the number of hours of moonlight on any night of the lunation (Sirmond fols. 78v–79r). None of these are quoted directly by Bede, but none teach anything different from what is found in *The Reckoning of Time.* Such formulae and tracts, being anonymous, were freely adapted and excerpted.

C. *Computistical poem:* (Sirmond fols. 79r–80r) *Annus solis continetur . . . de soli secula.*[179] This item is not cited by Bede; however

172 PL 653–664, but the text in the Sirmond MS diverges from the PL text at 657B and contains all 14 divisions of time.

173 C.W. Jones, *Bedae pseudepigrapha: Scientific Works Falsely Attributed to Bede* (Ithaca and London: Cornell University Press, 1939):48–51. See above, n. 18.

174 Bede did not, apparently, know this work.

175 Published amongst the works of Alcuin in PL 101.993 (based on the edition of Froben, Regensburg, 1760).

176 PL 101.984–9.

177 PL 101.989–90.

178 PL 101.998–9.

179 Published (a) as authentic work of Bede by Giles, *Venerabilis Bedae opera quae supersunt omnia* (London: Whittaker and Co., 1843–1844):1.54–5, whence it was reprinted in PL 94.605–6 and (b) anonymously in PL 129.1369–72 from the "Bobbio *computus*" in Milan, Ambrosiana H 150 inf. Critical ed. by Karl Strecker in MGH Poetae 4.2 (1923): no. 114, pp. 682–686.

Jones, following the lead of some medieval manuscripts, thought it might actually be by Bede.[180]

The items in "Book 2" of the Sirmond manuscript are also found, and in the same order, in G and (for the most part) L, which was written before the end of the eighth century, and probably copied from a pre-Bedan exemplar. The strong Insular affiliations of these manuscripts argue for an Irish origin. So do the contents of Book 2: the *Epistola Cyrilli*, ps.-Anatolius, the *Disputatio Morini*, and the Acts of the spurious Synod of Caesarea are all Irish forgeries; Cummian's *Epistola* furnishes proof that the letters of Paschasinus and Dionysius, the *Prologus Cyrilli* (a critique of Victorius composed in Spain in the seventh century), and the Victorian and Dionysian cycles were also available in Ireland before the time of Bede. Book 2 contains the following items:

D. *Computistical formulae.* After three anonymous *argumenta* (fols. 80v–81r) *De annis domini, De indictione, De Pascha,*[181] the Sirmond manuscript reproduces the authentic *Argumentum titulorum paschalium* of Dionysius Exiguus (Sirmond fols. 81r–82r).[182] These Dionysian rules of thumb for discovering various computistical data form the backbone of the whole second half of *The Reckoning of Time*, since they cover each of the eight columns of the Paschal table. Bede cites the *Argumenta* directly in chs. 47–49, 52, 54, and 58. The formula for finding the Paschal Moon which follows on fol. 82r–v (*Incipit calculatio quomodo repperiri possit quota feria in singulis annis xiiii luna paschalis id est circuli decennouenalis. A primo anno . . . luna paschalis xiiii. Haec argumenta hic finitur.*) is not by Dionysius, but circulated in the Middle Ages as part of the expanded *Argumenta titulorum paschalium.*[183]

E. *Dossier of letters in defence of the 19-year Alexandrian cycle.* The dossier opens with ps.-Jerome, *Disputatio de sollemnitatibus*

---

180 *BOT* 270 and 283, n. 69. It is attributed to Bede e.g. in Leiden, Scaliger 38 (s. XI). Jones may have abandoned this idea later, however, for he does not include the poem in the *opuscula fortassis genuina* in *BOD.*

181 Jones says all three are frequently found in MSS. The first is printed in PL 67 from Reims 298.

182 In the Sirmond MS, the *argumenta* stop with no. 10 in Jan's and Krusch's editions: see n. 103 above.

183 Cf. PL 67.505.

paschae (Sirmond fols. 82v–84r), a tract in favour of the Alexandrian 19-year cycle.[184] Bede does not directly cite this work, but he definitely makes use of the remainder of the dossier, namely the letter of Paschasinus to Pope Leo (Sirmond fols. 84r–86r),[185] cited in chs. 9, 11, 43, 66 (though not at all in *On Times*); the letters of Dionysius Exiguus to Bonifatius and Bonus (Sirmond fols. 85r–86r),[186] and to Bishop Petronius (Sirmond fols. 86r–87v)[187] (the letter of Dionysius to Petronius appears in chs. 11, 16, 30, 38 and 47, while the one to Bonifatius and Bonus is cited in chs. 42, 45 and 56); Proterius of Alexandria's letter on the Easter controversy of 455 addressed to Pope Leo I (Sirmond fols. 88r–89v)[188] (cf. chs. 6, 16 and 25, as well as the *Letter to Wicthed*); and the *Epistola Cyrilli* (Sirmond fols. 89v–90v),[189] quoted in ch. 44, as well as in the *Letter to Wicthed*.

F. *The "Irish forgeries"*. These include the *Liber Anatolii* (Sirmond fols. 90v–93v)[190] which Bede, like so many others, accepted as genuine, and which is cited extensively in *The Reckoning of Time*,[191] as well as a puzzling oddity entitled *Disputatio Morini Alexandrini episcopi de ratione paschali* (Sirmond fol. 94r–v). This Insular forgery seems to be pro-Alexandrian, but the text is very obscure and its motivations not entirely intelligible. More a theological tract than a computistical essay, it seems to argue that while the equinox is on 25 March, one must start to celebrate Easter on 22 March, since (a) Christ was crucified on 21 March and (b) 21 March was the anniversary of the Creation of the

---

184 The rubric (*Exemplum suggestionis boni sci. primice. De sollemnitatibus et sabbatis... ora pro me venerabilis papa*) belongs to the *Exemplum Boni* (ed. Krusch, "Einführung", 109) but the text is ps.-Jerome, *Disputatio de sollemnitatibus paschae*, ed. Krusch *Neues Archiv* 10 (1885):84–89 from P, pp. 212–217. Also printed in PL 22.1220. Cf. *BOT* 108–109.

185 See above, p. l.

186 See above, p. liii.

187 See above, p. liii.

188 See above, p. l.

189 See above, pp. lix–lx.

190 See above, pp lvi–lviii.

191 Chs. 6, 14, 16, 17, 22, 25, 30, 42, 50, 66 as well as the *Letter to Wicthed*. In the Sirmond manuscript, this is followed by extracts from Eusebius, Jerome et al. concerning Anatolius (Sirmond fols. 93v–94r).

world, which occurred four days before the equinox. Since the conclusion follows that it is legitimate to celebrate before the equinox, it is little wonder that this tract fell into obscurity.[192] Bede does not cite Morinus, but he had better authorities at his disposal. Finally, there are the spurious Acts of the Synod of Caesarea (Sirmond fols. 94v–95v), here entitled *Epistola Philippi de pascha*,[193] another Insular tract composed in favour of the Alexandrian *computus*. It is cited in chs. 47 and 56 of *The Reckoning of Time*.

G. *Formula for finding the age of Moon on the first of the month.* (Sirmond fol. 96r–v: *Incipit calculatio quomodo reperire ... errore sublato reperies.*) Under this rubric are found three lengthy formulae using the Victorian system of calculating from the kalends of January, later adapted by Bede in *The Reckoning of Time*, ch. 20 for use with the Dionysian system.[194]

H. *Pope Leo I's first letter to Emperor Marcian concerning the Easter controversy of 455* (Sirmond fols. 96v–97r)[195] appears in ch. 44 of *The Reckoning of Time*.

I. *A tract on the mystical significance of Easter.* (Sirmond fol. 97r–v: *De pascha autem tanquam maximo sacramento ... illuminante comedamus.*) Not cited by Bede, and not related to his own chapter 47 on the "allegorical interpretation of Easter".

J. *Romana computatio.* (Sirmond fols. 97v–98r: *Romana computatio ita digitorum ... aures retro respicientes.*) A tract on finger reckoning, probably the source of *The Reckoning of Time*, ch. 1.[196]

K. *Theophilus' prologue, dedicated to Theodosius* (Sirmond fols. 98r–99r)[197] is frequently quoted by Bede in *The Reckoning of Time*, notably in chs. 6, 11, 43, 59 and 61.

192 Jones, "Lost Sirmond Manuscript", 216; Grosjean, "Recherches", 225 *sqq.* Cordoliani ("Les computistes", 28–34) provides an edition of the text from MS Tours 334, in parallel with Muratori's edition from Milan Ambrosiana H 150 Inf. (PL 129.1357–1358).

193 Ed. Krusch, *Studien II*, 303–310, and from a different recension by A. Wilmart, "Un nouveau texte du faux concil de Césarée sur le comput pascal", in his *Analecta reginensia*, Studi e testi 59 (Vatican City: Bibliotheca Apostolica Vaticana, 1933):19–27.

194 Jones, "Lost Sirmond Manuscript", 217.

195 See above, p. l.

196 Ed. Jones, BOD 3.671–2; see Commentary on *The Reckoning of Time*, ch. 1.

197 See above, p. xl.

L. Fol. 99r–v. Computistical *argumenta*.

M. *Prologus Cyrilli* (Sirmond fols. 90v–101r)[198] is quoted in chs. 11, 30 and 43.

N. *Extracts from Macrobius' Saturnalia* (Sirmond fols. 101r–105v), known to Bede (though not so identified in the Sirmond manuscript) as *Disputatio Hori [or Chori] et Praetextati*.[199] The *Disputatio Chori et Praetextati* is ubiquitous, but especially significant for chs. 11–13 of *The Reckoning of Time*, and also cited in chs. 16, 36, 46. Bede used it much more sparsely in *On Times*, and not at all in *The Nature of Things*.

O. *Miscellany of computistica*: including computistical extracts, some from Isidore (fols. 105v–107v), the rubric of Dionysius Exiguus' cycles (fol. 107v), and a rota showing lunar and solar months and number of days in seasons (fol. 108r).

P. *Victorius of Aquitaine*. Following a biographical note on Victorius, based on Gennadius (Sirmond fol. 108r), the Sirmond manuscript reproduces Archdeacon Hilarius' letter to Victorius, as well as Victorius' prologue (Sirmond fols. 108r–110v). Hilarius' letter to Victorius appears in ch. 43 of *The Reckoning of Time*, and Victorius' prologue in chs. 50, 51 and 61. A short chronicle intervenes (Sirmond fols. 111r–113r), comprising selections from Eusebius/Jerome, from Olympiad 157 to AD 32, and then follows Victorius' 532-year cycle (Sirmond fols. 113r–120r). The Sirmond manuscript's version has no double dates. Bede refers directly to this cycle in chs. 50, 51 and 61.

Q. *Dionysius' 19-year cycles.* (Sirmond fols. 120r–131v). The Sirmond manuscript's version runs from AD 532–1421, and contains annals.

What is most intriguing about the discovery of the Sirmond manuscript, is that it reveals that Bede's knowledge of the Dionysian *computus* was derived, not from Rome or even from the Continent, but from Ireland. The Sirmond *computus* anthology is plainly an Irish production, behind which lies a Spanish or African chrestomathy. Ó Cróinín argues that the

---

198 Ed. Krusch, *Studien I*, 337–343.
199 See notes and Commentary on *The Reckoning of Time*, ch. 12.

exemplar of the Sirmond group was compiled in southern Ireland in AD 658, and passed from there shortly after to Northumbria and Jarrow.[200] The itinerary of this collection was probably from southern Ireland to south or south-western England, and thence north. A southern Irish provenance is likely precisely because the collection contains *both* Victorian and Dionysian material, but *not* an 84-year Irish Paschal cycle. The south of Ireland had switched to the Alexandrian reckoning after Mag Léne, but was uncertain whether "Alexandrian" meant "Victorian" or "Dionysian"; the north, however, was loyal to the 84-year cycle until 715, so the collection could not have derived from Iona.[201]

Did Bede know that he was using an Irish *computus* anthology? Would it have made any difference to him? Jones thinks it would: he points out that Bede never mentions an Irish teacher or school or even the word *Scotti* because he "feared these sources unless they were carefully checked".[202] Ó Cróinín even accuses Bede of deliberately suppressing information about the Irish origin of his material, presumably out of dislike for the Irish.[203] Indeed, it has become almost fashionable now to accuse Bede of anti-Hibernian prejudice. Alan Thacker, however, refutes these allegations. Bede rarely refers to "the Irish" in general terms, but when he does, it is in a complimentary light: they are the *gens innoxa*, friendly to the English, whom Ecgfrith of Northumbria unjustly attacks (*HE* 4.26). If Bede suffers from ethnic prejudice, it is against the British, not the Irish.[204] Bede was aware that a majority of the Irish followed Roman Paschal reckoning; in his account of Whitby, an Irishman named Ronan argues on the Roman side. In fact, Thacker suggests that the Insular Paschal "controversy" may have been less a real conflict than a struggle in the mind of Bede himself. For Bede, Lindisfarne and Iona are the cradles of Northumbrian Christianity, and the resistance of Iona to Roman *computus* was painful and embarrassing; that is why the conversion of Iona to the orthodox reckoning is the climax of the *Historia Ecclesiastica*.[205] Thacker makes the sensible

---

200 Dáibhí Ó Cróinín, "Irish Provenance", 233.

201 *Ibid.* 242. Ó Cróinín suggests that Aldhelm also had access to a Sirmond-type MS. A further argument in favour of a southern Irish origin is the fact that Cummian apparently had a similar anthology: cf. Ó Cróinín and Walsh 31.

202 *BOT* 131.

203 "Irish Provenance", 246–247.

204 Thacker 32–36.

205 *Ibid.* 41–42.

suggestion that if Bede did not always credit his Irish computistical sources, it was probably because most of them were either anonymous, like the *De computo dialogus*, or pseudonymous.[206] In the *Historia ecclesiastica* he certainly expresses admiration for Irish learning in general.[207]

In the case of *The Reckoning of Time*, it is very unlikely that Bede was targeting "the Irish" *en bloc*; in fact, by the time he wrote *The Reckoning of Time* Bede would not associate the 84-year cycle so much with the Irish as with the British, who still refused to convert. If anything, Bede may have felt that the problem with the present-day Irish was that many of them still favoured the cycle of Victorius, but they certainly were not peculiar in that, as the Gaulish churches did too.[208] There was no need for Bede to single out "the Irish" by name in *The Reckoning of Time*.

The Sirmond manuscript is uniquely important for Bede's project, but it is certainly not the only *computus* collection available to him. Much computistical literature known to Bede is not accounted for by the Sirmond volume. In the *Letter to Plegwin*, for example, he mentions that when he was a boy, he saw a *computus* treatise "written by some heretic". His description of the contents would suggest that this suspect tract is the "Cologne Prologue", an Irish description of an 84-year cycle with *saltus* in the 12th year.[209] He knew a number of anonymous tracts, such as the *De saltu lunae* falsely ascribed to Columbanus (ch. 42), the anonymous *De causis quibus nomina acceperunt duodecim signa* (chs. 16, 17), and the *De ratione embolismorum* (ch. 45); earlier computistic literature like Victor of Capua's *De pascha* (chs. 27, 50, 51 and 64); as well as other calendar material such as the *Laterculus* of Polemius Silvius (chs. 11, 14).

Secular scientific, technical and general literature

As we have seen, Bede was indebted to the Irish computists and cosmographers for much of his "scientific programme",[210] but in one respect

---

206 *Ibid.* 50.

207 E.g. *HE* 3.3.27; Thacker 50. For further evidence of Bede's interest in the Irish, and indifference to the issue of ethnic identification, see T.M. Charles-Edwards, "Bede, the Irish, and the Britons", *Celtica* 15 (1983):42–52, and C.A. Ireland, "Boisil: An Irishman Hidden in the Works of Bede", *Peritia* 5 (1986):400–403.

208 Thacker 53–4.

209 Ed. Krusch, *Studien I*, 227–235; cf. Jones, *BOT* 91. However, this text is found in other MSS of the "Sirmond group": see n. 171 above.

210 See Introduction, sections 2 and 5.

he differed from them markedly. The Irish cosmographers of the seventh century, especially the "Irish Augustine" who wrote *De mirabilibus sacrae scripturae* and the anonymous author of the *Liber de ordine creaturarum*, drew their lore about the natural world entirely from the Bible, patristic exegesis, and works of grammar. They did not have access to any secular scientific literature, neither Isidore's *The Nature of Things*, nor the scientific books of the *Etymologies*, nor Pliny.[211] Bede knew all three of these works, and others as well. Indeed, he is one of the rare early medieval writers to make systematic use of such writings. In this respect he not only goes beyond the Irish cosmographers, but departs from Isidore himself, whose science largely came from grammars and literary compilations.

Isidore of Seville is a major source for both *The Nature of Things* (which even borrows its title from Isidore) and *On Times*, which depends heavily on the fifth book of the *Etymologies*. As we have seen, Bede is accepting and diplomatic about his Irish sources, even when he disagrees with them. This accords with his usual non-confrontational attitude towards the learned tradition: when he disagrees with received wisdom, he passes over the matter in silence. Not so in the case of Isidore. Bede not only cuts down considerably on his use of Isidore's works in *The Reckoning of Time*, by comparison with *On Times* and *The Nature of Things*,[212] but displays his impatience with Isidore more openly. He is discretely critical of Isidore's dating of the equinox (ch. 16) – discretion being called for because St Ambrose made the same error. But in ch. 24, his refutation of Isidore's views on weather prognostication by the Moon is less restrained. In ch. 35 he is even more forthright:

> Bishop Isidore the Spaniard said that winter begins on the 9th kalends of December [23 November], spring on the 8th kalends of March [22 February], summer on the 9th kalends of June [23 May], and autumn on the 10th kalends of September [23

---

211 Smyth 30–33.

212 Isidore's *De natura rerum* is cited directly or indirectly 28 times in *The Reckoning of Time*, compared with 50 times in *The Nature of Things* and 12 times in *On Times*. The *Etymologiae* are also cited more often in *The Nature of Things* (46 times) than in the much longer *Reckoning of Time* (34 times). Bede may have thought that the Irish *De ordine creaturarum* was by Isidore: it is cited only once in *The Reckoning of Time* and 13 times in *The Nature of Things*, as well as in *In Gen*. Isidore's chronicle forms the backbone of the world-chronicle in *On Times*, but is used only for the first two World-Ages in *The Reckoning of Time*, ch. 66.

August].[213] But the Greeks and Romans, whose authority on
these matters, rather than that of the Spaniards, it is generally
preferable to follow, deem that winter begins on the 7th ides of
November [6 November], spring on the 7th ides of February [6
February], summer on the 7th ides of May [8 May], and autumn
on the 7th ides of August [7 August].

The issue of Bede's attitude towards Isidore has been the matter of much
recent debate. The consensus opinion is that while Bede may have been
dependent upon Isidore in the early phases of his career as a scholar, he
gradually became disillusioned with the bishop of Seville, and at the end
of his life, spent his final hours compiling a list of the errors in *De natura
rerum* on the grounds that he did not want his "children learning what is
not true, and losing their labour on this after I am gone".[214] Roger Ray
argues that Bede's quarrels with Isidore on a wide variety of issues
(including the nature of history) intensified over the course of Bede's lifetime.[215] Jones acknowledges that Bede actively disliked certain Isidorean
postures, particularly his "pretentiously humanistic approach",[216] but
points out that on the issues of natural science and *computus*, Bede also
had positive reasons for wishing to revise Isidore. He had access to
superior materials: the Irish *computistica*, and above all, Pliny, for
whom Bede felt evident admiration. It is by calling on Pliny and ps.-
Anatolius that Bede refutes the Isidorean teaching on the seasons
quoted above. What seems perhaps less likely is that Bede would have
occupied himself upon his deathbed with refuting errors which he had
already implicitly, and occasionally explicitly, corrected in *The Reckoning of Time*. Nor is it entirely plausible that Cuthbert, who wrote the
account of Bede's last moments, would have chosen to present his
master in such a pugnacious pose. Perhaps what Bede was compiling in
his final days was not *exceptiones* (errors) but *excerptiones* (extracts),[217]

---

213 Isidore, *DNR* 7.5 (203.46–51).

214 Cuthbert, *Epistola de obitu Bedae*, in Colgrave and Mynors, pp. 582–583.

215 "Bede's *vera lex historiae*".

216 "Bede's Place", 266; see also Thomas Eckenrode, "The Growth of a Scientific
Mind: Bede's Early and Late Scientific Writings", *Downside Review* 94 (1976):208–209, and
the literature cited by Max Lejbowicz, "Postérité médiévale de la distinction isidorienne *as-
trologia/astronomia*: Bède et le vocabulaire de la chronometrie", in *Documents pour l'histoire du vocabulaire scientifique*, no. 7 (Paris: Editions du CNRS, 1985):26 and 41, n. 47.

217 William McCready, "Bede, Isidore, and the *Epistola Cuthberti*", *Traditio* 50
(1995):88. McCready, however, concludes that what Bede was doing was correcting a

i.e a florilegium or edition of Isidore which weeded out pernicious errors such as those Bede refutes in *The Reckoning of Time*. Be that as it may, the fact that the manuscripts of Cuthbert's letter prefer *exceptiones* suggests that Bede's coolness towards Isidore was no secret, and not necessarily regarded as a stain on his character.

Bede generally prefers not to cite pagan writers if he can avoid it, and certainly avoids commending them.[218] But in ch. 27 of *The Reckoning of Time* he is not ashamed to express open admiration for Pliny's "delightful book, the *Natural History*"; in ch. 31, he boldly compares the authority of Pliny with that of Basil and Ambrose. Yet Bede's use of Pliny is oddly spotty. Jones' index of sources indicates that Bede drew on Books 2 (heavily), 3–7, 28, 30, 35 and possibly 37 of the *Historia naturalis*, and elsewhere he suggests that Bede knew Pliny only through a florilegium of extracts.[219] Such an anthology did circulate in eighth-century England, from whence it was transported to the continent.[220] However, it contains excerpts only from Books 2 and 18; moreover, distinctive readings from this anthology are found in four citations in *The Nature of Things*, but not in any of the Pliny passages in *The Reckoning of Time*, nor in the other Pliny citations in *The Nature of Things* or *On Times*. Furthermore, Bede never cites Book 18 of the *Historia naturalis*, which would have been very useful to him in his researches on time, and which does appear in the

---

corrupt manuscript of *De natura rerum*, whose textual aberrations might deceive his students, or even translating the work into English, or correcting a faulty translation. McCready feels that Bede's hostility to Isidore has been exaggerated, but does not mention these passages from *The Reckoning of Time*.

218 For example, Bede makes considerable use of Vegetius' *Epitoma rei militaris*, but never cites the author or title of the book. He may not have known who Vegetius was (no writer prior to the Carolingian period mentions Vegetius by name). But Jones also observes that Bede "often concealed the source from which he drew if the orthodoxy of the source was questionable" ("Bede and Vegetius", *The Classical Review* 46 (1932):248). Vegetius was a pagan writer, and war is certainly an unedifying topic.

219 *BOT* 366.

220 Ed. Karl Rück, *Auszüge aus der Naturgeschichte des C. Plinius Secundus in einem astronomisch-komputistischen Sammelwerke des achten Jahrhunderts*, Programm des Königlichen Ludwigs-Gymnasiums für das Studienjahr 1887–88 (Munich: F. Straub, 1888). I have been unable to consult V. King, "An Investigation of Some Astronomical Excerpts from Pliny's Natural History found in Manuscripts of the Earlier Middle Ages", D.Litt. thesis, Oxford, 1969.

anthology.[221] In short, we can conclude that Bede did not use the usual English Plinian anthology, but that he did not have access to a full edition of the *Historia naturalis* either.[222]

Besides Pliny, Bede cites Solinus in ch. 31 (the same passage is found in *The Nature of Things* ch. 9), and Vegetius in chs. 25, 27, and 29.[223] Vegetius seems to have been a major classical discovery of Bede's adult life, for he also used him (albeit without naming him) in his commentary on the Acts of the Apostles, and in the *Historia ecclesiastica*.[224] His use of Macrobius (whom he discovered in his "Irish *computus*") is discussed above. Bede's little library of secular literature also included some medical works: ps.-Hippocrates' *Letter to Antigonus*, cited in ch. 30, and the *Epistola ad Pentadinum* of St Augustine's friend Vindicianus Afer, used in ch. 35.[225] A few lines from Vergil and Ausonius, and some excerpts from grammarians like Priscian (ch. 4) and Julian of Toledo (ch. 3), round out the secular literature.

Patristic writings

Bede may have differed from the Irish cosmographers in his use of secular scientific literature, but he thoroughly resembled them in the manner in which he employed the Fathers as sources of information about the natural world.[226] Hexaemeral literature and commentaries on Genesis furnish the lion's share: Ambrose's *Hexaemeron* is used more frequently in *The Reckoning of Time* than in any of Bede's other scientific writings, including *The Nature of Things*; *The Reckoning of Time* is the only scien-

---

221 M.L.W. Laistner, "The Library of the Venerable Bede", in *Bede: His Life, Times and Writings*; rpt. in *Intellectual Heritage of the Early Middle Ages*, ed. C.G. Starr (Ithaca: Cornell University Press, 1957):124.

222 Bede's situation seems to have been the normal one in the pre-Carolingian and Carolingian period: see Bruce Eastwood, "The Astronomies of Pliny, Martianus Capella and Isidore of Seville in the Carolingian World", in *Science in Western and Eastern Civilization*, 162–168.

223 Vegetius is also used in *The Nature of Things* chs. 19, 27, 36; these are not the same passages as those in *The Reckoning of Time*.

224 Jones, "Bede and Vegetius", 249.

225 For editions and discussion, see *Bibliographie des textes médicaux latins: antiquité et haut moyen âge*, ed. Guy Sabbah, Pierre-Paul Corsetti and Klaus-Dieter Fischer (Saint-Étienne: Publications de l'Université, 1987):96–99, 154.

226 Smyth 18–20; compare her list of patristic sources exploited by Irish cosmographers (23 *sq.*) with those used by Bede and discussed below.

tific work where Bede cites Basil's *Hexaemeron*. The works of Augustine which form the backbone of Bede's explanation of Creation in *In Genesim* reappear here as well: *De Genesi ad litteram* and *De Genesi contra Manichaeos*. It is interesting to observe, however, that Bede uses Augustine rather less in *The Reckoning of Time* than in his earlier works. There is only one citation from *De Genesi ad litteram* in *The Reckoning of Time*, compared to ten in *The Nature of Things*; there are no quotations from *De Genesi imperfectus liber*, a work which Bede uses frequently in his earlier cosmological book. Bede's relationship to Augustine was a rather complex one. In ch. 5, he discreetly distances himself from Augustine's philosophical allegorizing of the Creation narrative in Genesis, in favour of the more literal approach of Ambrose and Basil. This may represent a fundamental ideological shift for Bede, or it may simply reflect his concern not to confuse students with irrelevant subtleties. The Irish loved to juxtapose differing patristic opinions,[227] but Bede on the whole preferred to soft-pedal differences in the interests of clarity and orthodoxy.

Besides these hexaemeral sources, Bede drew on his wide reading in patristic literature for a vast array of scientific lore and theological commentary. Works like Augustine's *De civitate Dei* are heavily mined (chs. 8, 27–28, 34, 43, 66, 68, 70), as are his *Letters* 60 (chs. 25, 27, 64) and 199 (chs. 5, 9, 27, 68), both on subjects connected with Easter. Bede's doctrine of the six World-Ages (chs. 10, 66) is thoroughly Augustinian, and might have come from *Contra Adimantum Manichaei discipulum*, *De Genesi contra Manichaeos*, *De catechizandis rudibus*, the *Tractatus in evangelium Ioannis* or *De Trinitate*. Incidental use is made of Augustine's *De consensu evangelistarum* (ch. 4) and *De sermone Domini in monte* (ch. 4), Ambrose's *Explanatio super Psalmos* (ch. 71), the letters of Jerome, his *Adversus Jovinianum* (ch. 1), *In Danielem* (ch. 69) and commentaries on Isaiah (ch. 5), Amos (ch. 11) and Matthew (ch. 27), the pseudo-Clementine *Recognitiones* (chs. 5, 7), Cassian's *De incarnatione Domini contra Nestorium* (ch. 5), and Philippus Presbyter's commentary on Job (chs. 4, 29).[228] Echoes of other patristic works are abundant.

---

227 Smyth 35 *sq.*

228 Philippus was a disciple of Jerome who died AD 455/6. His commentary on Job was published as a work of Bede in the folio edition of Hervagius (Basle, 1563):IV, 602. It appears amongst the works of Jerome in PL 26.619–802. There are a number of excellent

In contrast to secular works, patristic literature is cited openly and specifically. Bede is proud to engage the Fathers in his exploration of the cosmos and time, for in his view, the main source of truth about the natural world is the Bible itself, and the main reason for studying nature is to understand what the Bible is saying. He knew the Bible intimately and in detail, and its verses served as a flexible storehouse for the information he would need.[229] Bede was also heir to an Irish intellectual culture that regarded the study of nature as a key to validating the miracles of the Bible (cf. *De mirabilibus sacrae scripturae*), or of penetrating God's own creative processes (cf. *De ordine creaturarum*). His heavy debt to the Fathers is not merely *faute de mieux* – he had Isidore and Pliny, after all – but a declaration of allegiance to the ideal of *doctrina christiana*. How he deployed this erudition in the service of this ideal is the theme of our commentaries on the individual chapters of *The Reckoning of Time*.

6. MANUSCRIPTS, GLOSSES, EDITIONS, AND PRINCIPLES OF TRANSLATION

Manuscripts

> ... for God, the orderer of natures (*naturarum dispositor*) who raised the Sun from the east on the fourth day of Creation, in the Sixth Age of the world, has made Bede rise from the West as a new Sun to illuminate the whole earth.[230]

Though Notker the Stammerer commended Bede primarily as a Scriptural exegete, the computistical flavour of this eulogy is not without significance. Bede was a very popular writer not only in his own time and country, but for generations thereafter, and across

---

8th-c. MSS, and a critical edition is sorely needed: cf. E.A. Lowe, *English Uncial* (Oxford: Clarendon Press, 1960): Pl. XXXVII = MS Leningrad F.v.I.3, fol. 38. Jones notes that this work was nearly as popular as Gregory's *Moralia* in Bede's day: *BOT* 334–335.

229 Jones, *BOT* 128.

230 *...quem naturarum dispositor deus, qui quarta die mundanae creationis solem ab oriente produxit, in sexta aetate saeculi nouum solem ab occidente ad inluminationem totius orbis direxit.* Notker Balbulus, *Notatio de illustribus viris*, ed. Erwin Rauner, "Notkers des Stammlers 'Notatio de illustribus viris': Teil I: kritische Edition", *Mittellateinisches Jahrbuch* 21 (1986):60. The translation of this passage is by Michael Idomir Allen, "Bede and Frechulf at Medieval St Gallen", *Beda Venerabilis* 65.

Europe,[231] and *The Reckoning of Time* was the most popular of Bede's textbooks. 240 manuscripts of all or part of the book survive, compared with 93 of *On Times*, 134 of *The Nature of Things*, and 96 of *De schematibus et tropis*. The fortunes of *The Reckoning of Time* are traced through its manuscripts, their glosses, and later, the printed editions.

Whitelock argues that Bede's computistical works were the first of his writings to find an audience on the Continent, where they served to stimulate annalistic writing as well as *computus* studies.[232] *The Reckoning of Time* entered the European mainstream through two channels: the Irish, and the Anglo-Saxon. The two streams seem to have been contemporary, and their earliest witnesses are fragmentary manuscripts whose state attests to heavy use and copying.[233] Two fragments – Bückeburg, Niedersachsiche Staatsarchiv Dept.3/1 fols. i–viii and Münster in Westphalen, Staatsarchiv Misc. I. 243 fols. 1r–2v, 11r–12v – are linked through their north German provenance with the missions of Boniface. Boniface appealed to English supporters to send him copies of the works of that "candle of the Church", Bede, and one of those works was very probably *The Reckoning of Time*. The Bückeburg and Münster fragments are from a single codex written in Northumbrian uncial script, probably about 746–750 and possibly in Wearmouth-Jarrow. The third fragment, Darmstadt, Hessische Landes- und Hochschulbibliothek 4262, also in Northumbrian uncial, may have been written during Bede's lifetime.[234] These are the only English manuscripts of *The Reckoning of Time* before the age of Dunstan.

231 Dorothy Whitelock, "After Bede" (Jarrow Lecture, 1960); Jones, "Bede's Place"; J.E. Cross, "Bede's Influence at Home and Abroad: An Introduction", in *Beda Venerabilis* 17–29; Allen, "Bede and Frechulf"; J.M. Clark, *The Abbey of St Gall as a Centre of Literature and Art* (Cambridge: Cambridge University Press, 1926):55–70; W. Levison, *England and the Continent in the Eighth Century* (Oxford: Clarendon Press, 1946): *passim*; Wolfgang Viereck, "Beda in Bamberg", in *Einheit in der Vielfalt: Festschrift für Peter Lang zum 60. Geburtstag*, ed. Gisela Quast (Bern: Peter Lang, 1988):556–569.

232 Whitelock, "After Bede", 41–42, 47; on the impact of *The Reckoning of Time* on Carolingian annalistic writing, see also Landes 178–181.

233 Jones, "Bede's Place", 265.

234 Wesley Stevens, "Bede's Scientific Achievement", 39. For discussion of the Bückeburg and Münster fragments, see Jürgen Petersohn, "Neue Bedafragmente in Northumbrischer Unziale Saec. VIII", *Scriptorium* 20 (1966):215–247 and Pl. 17–18, and *idem*, "Die Bückeburger Fragments von Bedas *De temporum ratione*", *Deutsches Archiv für Erforschung des Mittelalters* 22 (1966):587–597. The fragments are reproduced in E.A. Lowe, *Codices latini antiquiores* (Oxford: Clarendon Press, 1934–71):9, no. 1233 and Supplement

The Irish stream[235] is represented by the fragments of a late eighth- or early ninth-century codex, Vienna Nationalbibliothek Supp. 2698. These four leaves contain chs. 7–9, 11–16, and 21–22, written and glossed in Ireland.[236] The glosses themselves are in Old Irish,[237] and raise the intriguing question of whether *computus* was taught in the vernacular. It is not impossible that even Bede himself taught *computus* in Old English. Irish scholars on the Continent were also responsible for one of the best manuscripts of *The Reckoning of Time*, Karlsruhe Aug. 167 (s. IX med.), containing *The Nature of Things*, *On Times*, and *The Reckoning of Times*, with Latin and Old Irish glosses closely related to those of the Vienna fragments.

Jones examined 104 manuscripts for his edition, 45 of which were written within 70–100 years of the completion of the work. This shows how important *The Reckoning of Time* was for the Carolingian schools.[238] The text is astoundingly stable, despite occasional scribal perplexity over the computistical jargon.[239] The wide dissemination of *The Reckoning of Time* in the Carolingian period has made any meaningful stemma out of the question but there seem to be two avenues of dissemination: (a) the upper Rhine group (St Gall, Reichenau, Bobbio) perhaps *via* Boniface's mission, and (b) the northern France/lower

---

4 (1233), and in his *English Uncial*, Tab. XVIIIa–c. The Darmstadt fragments are described, transcribed and illustrated in Kurt Hans Staub, "Ein Beda-Fragment des 8. Jahrhunderts in der Hessischen Landes- und Hochschulbibliothek Darmstadt", *Bibliothek und Wissenschaft* 17 (1983):1–7.

235 The Irish use of *The Reckoning of Time*, as well as other works of Bede, has been investigated by Steven B. Killion, "Bede's Irish Legacy: Knowledge and Use of Bede's Works in Ireland from the Eighth through the Sixteenth Century", PhD diss., University of North Carolina at Chapel Hill, 1992.

236 Lowe, *Codices latini antiquiores* 10, p. 21, no. 1511.

237 Killion 141–154; cf. John Strachan, "The Vienna Fragments of Bede", *Revue celtique* 23 (1902):40–49 and Miles Dillon, "The Vienna Glosses on Bede", *Celtica* 3 (1956):340 *sqq.*

238 For a full list of manuscripts of *The Reckoning of Time*, see Jones, *BOD* 242–256. For a handlist of 8th- and 9th-century codices, see Stevens, "Bede's Scientific Achievement", 39–42. On early MSS of *The Reckoning of Time* and early mentions of the work in library catalogues, see Dorothy Whitelock, "After Bede", 42–43.

239 Jones, *BOT* 140–141. But see Juan Gómez Pallarés, "Los *excerpta* de Beda (*De temporum ratione*, 23–25) en el MS. ACA, Ripoll 225", *Emerita* 59 (1991):101–122; the readings in this 11th-c. MS suggest that the text tradition may be more complicated.

Rhine group.[240] These two regional groups correspond to the two main families of glosses discussed below.

Manuscripts of *The Reckoning of Time* crest in the Carolingian period, and then taper off somewhat, though the book continued to be copied, and later printed, up to the calendar reform of 1582 and even beyond. Most manuscripts written after 1050 tend to come from England, Spain, or Bavaria-Austria.[241]

The Carolingian enthusiasm for *The Reckoning of Time* entailed certain changes to the way the text was treated. When Charlemagne legislated educational reforms for his realm, he envisaged the curriculum of the new monastic and cathedral schools as a programmed version of the old monastic vocational curriculum: *psalmos, notas, cantus, compotum, grammaticam . . . libros catholicos . . .*[242] The term he used for reading was *psalmos*, for memorizing the Psalms was at once the first stage in learning to read and the first initiation into the Bible.[243] *Compotum*, like *notas* (writing), *cantus* (chant) and *libros catholicos* (religious literature), is redolent of the programme's vocational roots.

The Carolingian adoption of *computus* into its official educational policy made it a requisite element of every educated man's mental equipment. The emperor himself was interested in *computus*[244] and exchanged letters with Alcuin on the subject.[245] Groups of *compotistae* met for debate and questioning: a report of one such convocation in 809,

240 Jones, *BOT* 142.

241 Jones, "Bede's Place", 270–271.

242 *Admonitio Generalis*, ed. A. Boretius, *Capitularia regum francorum*, MGH Leges (Quarto), 2 (1883):1.59–60.

243 Illmer 150–156; Pierre Riché, "Le psautier, livre de lecture élémentaire d'après les vie des saints merovingiens", in *Études merovingiennes. Actes des journées de Poitiers* (Paris: Picard, 1953):253–256.

244 *Discebat artem conputandi et intentione sagaci siderum cursum curiosissime rimabatur.* ("He learned the art of computing and by intelligent application explored with great inquisitiveness the course of the stars", Einhard, *Vita Caroli*, ed. Louis Halphen, 3rd ed. (Paris: Les belles lettres, 1947). The reference to astronomy suggests that the *ars conputandi* was *computus*. Einhard also records that Charlemagne gave German names to the months, perhaps consciously inspired by Bede's discussion of the Anglo-Saxon months in *The Reckoning of Time*, ch. 15.

245 *Alcuini epistolae*, ed. Ernst Dümmler, *Epistolae Carolini aevi*, MGH Ep. 4 (1895):2.231–235 (Ep. 145, to Charles on the *saltus lunae*) and 237–241 (Ep. 148, to Charles on the calculation of the zodiacal signs). Cf. Dietrich Lohrmann, "Alkuins Korrespondenz mit Karl dem Grossen über Kalender und Astronomie", in *Science in Western and Eastern Civilization in Carolingian Times*, 79–114. On the second letter, see Wolfgang Edelstein,

perhaps drawn up by Abbot Adelard of Corbie, survives.[246] Diocesan synods were admonished to ensure that the clergy knew chant and *computus*,[247] and *missi* were to report on the level of *computus* expertise in the religious establishments they visited.[248] The memorandum *Quae a presbyteris discenda sint* ("What priests should learn") stipulates that they should know *computus*[249] and they were expected to own at least a calendar.[250] The study of *computus* was enjoined at the Council of Aachen in 789.[251] Whether every Carolingian priest was expected to master *The Reckoning of Time* is somewhat doubtful,[252] but official pressure helps to explain the high levels of production of manuscripts of *The Reckoning of Time* in the eighth and ninth centuries. Carolingian manuscripts of the full text of *The Reckoning of Time* certainly bear witness to its active use in teaching. In Bede's formulae and worked examples, the *annus praesens* was modified to bring it up to date,[253] and sometimes additional formulae were interpolated.[254]

While Bede's great treatise on the calendar was becoming a cornerstone of the Carolingian curriculum, other forces were at work which changed the way that curriculum was conceived. Carolingian school-

---

Eruditio und sapientia. Weltbild und Erziehung in der Karolingerzeit. Untersuchungen zu Alkuins Briefe (Freiburg im Breisgau, 1965):102–103.

246 C.W. Jones, "An Early Medieval Licensing Examination", *History of Education Quarterly* 3 (1963):19–29. Jones' text is from Oxford, Bodleian Library 309, fol. 141. He thought this MS was unique, and was not aware of a previous edition, based on Paris, Bibliothèque nationale nouv. acq. lat. 1613 fols. 20–22 (s. IX[1]) and Brussels, Bibl. Royale 9590 fols. 55v–56r (s. XI) by E. Dümmler, *Epistolae Carolini aevi* 2, MGH Ep. 4 (1895):565–567. I have noticed another copy in Paris, Bibl. nationale lat. 1615 fols. 143v–144v, and Wesley Stevens informs me that there is yet another in Paris, Bibl. nationale lat. 2796 fols. 98r–99r (s. IX).

247 *Capitularia regum francorum* 2.237.

248 *Ibid.* 2.121.

249 *Ibid.* 2.235.

250 Regino of Prüm, *De synodalibus causis* 92 (PL 132.191C).

251 MGH Leg. 2.1, p. 60.

252 For a rather negative assessment of Carolingian expertise in *computus*, see Alexander Murray, *Reason and Society in the Middle Ages* (Oxford: Clarendon Press, 1978):153.

253 E.g. in Würzburg Theol. fol. 46 (Salzburg 792–807) the *annus praesens* in ch. 49 was changed by the original scribe to 800.

254 E.g. Munich CLM 14725 (Regensberg, s. IX[1]) adds additional formulae to chs. 49, 52, 54 (cf. Jones, *BOT* 147).

masters, especially those attached to cathedrals, discovered long-
neglected works like Martianus Capella's *De nuptiis Philologiae et
Mercurii*, which introduced them to the idealized ancient curriculum
known as the Seven Liberal Arts. Many of these schoolmasters glossed
both Martianus and the works of Bede: men like Martin the Irishman
(817–875), head of the cathedral school of Laon,[255] and his followers
Manno of Laon and Heiric of Auxerre.[256] None of these men were
computists, and when they read Bede, they were looking for material to
fill the category of *astronomia* in their new taxonomy of learning. As we
shall see shortly, this had a considerable impact on the way in which *The
Reckoning of Time* was glossed in the Carolingian age. It also entailed
an approach to *computus* which differed significantly from Bede's. Bede
never mentions the Liberal Arts, and as we have seen, his monastic
conception of *doctrina christiana* encouraged the dismantling of ancient
genres of scientific and didactic literature and the incorporation of their
contents into new Christian formats. This tide began to reverse in the
Carolingian period. The works of astronomy and natural science which
Bede had pillaged on behalf of *computus* were now copied and studied
for their own sake.[257] In consequence, Bede himself was mined for astronomical information that could be rearranged in more "classical"
formats. For example, the *Aratus latinus* is a Latin translation of a
group of texts, including the astronomical poem of Aratus of Soli, a
commentary on the same, and other associated texts. It survives in two
versions. The original version was put together in the first half of the
eighth century, probably at Corbie. The revised version, prepared in the
latter half of the same century, likewise came from northern France.
This version abbreviated the poem drastically, and rewrote the commentary in better Latin, with some parts replaced by new texts borrowed
from Hyginus, Isidore of Seville, Fulgentius, Pliny, etc. The Pliny

---

255 On Martin's glosses to *The Nature of Things* and *The Reckoning of Time*, see
John J. Contreni, *The Cathedral School of Laon from 850 to 930: Its Manuscripts and
Masters*, Münchner Beiträge zur Mediävistik und Renaissance-Forschung 29 (Munich:
Arbeo-Gesellschaft, 1978):124–126, and Jones, *BOD* 257–261.

256 Contreni, ch. 10, esp. 139, 150–151.

257 Cf. Stephen C. McCluskey, *Astronomies and Cultures in Early Medieval
Europe* (Cambridge: Cambridge University Press, 1998):ch. 8, and "Astronomies in the
Latin West from the Fifth to the Ninth Centuries", in *Science in Western and Eastern Civi-
lization*, 139–160; Bruce Eastwood, "The Astronomies of Pliny, Martianus Capella and
Isidore of Seville in the Carolingian World" (see n. 222).

excerpts are in fact derived from Bede's *The Nature of Things*, and the *Aratus latinus* also borrows other passages from Bede.[258]

An index of this change of direction with regard to *computus* is Rabanus Maurus' manual of *doctrina christiana*, *De clericorum institutione*. Rabanus' model was Cassiodorus, who had no place for *computus* in either his "divine" or his "human" categories of learning. So Rabanus elected to equate *computus* with one of the Liberal Arts, specifically what Isidore would call *astrologia naturalis*, or astronomy:[259]

> It behoves God's cleric by skilful art to learn about this part of astronomy which follows natural inquiry and which prudently searches out the courses of the Sun, Moon and stars and the sure distinctions of time, so that through sure interpretation of the rules, and established and true appraisal of the formulae, he may not only inspect in a trustworthy manner the past courses of the years but also know how to reason about the time to come, and so that he may find out for himself the beginning of the Paschal feast, and the true places [i.e. in the calendar] which ought to be observed for all solemnities and celebrations and be able to proclaim their lawful celebration to the people of God.[260]

In terms of the manuscripts of *The Reckoning of Time*, the most striking result of this change of perspective was the wholesale dismemberment of the treatise into excerpts which could be included in a new type of anthology, containing *computus*, astronomy and natural science. The most prominent example of this new genre is the "Aachen encyclopaedia of 809", whose most outstanding examples are the twin manuscripts

---

258 Vernon King, "An Unreported Early Use of Bede's *De natura rerum*", *Anglo-Saxon England* 22 (1993):85–91.

259 Isidore, *Etym.* 3.27.1–2, quoted in Rabanus, *De clericorum institutione* 25, PL 107.403C–D, cf. Lejbowicz, "Postérité", 24, and "Théorie et pratique astronomiques chez Isidore de Seville", in *L'Homme et son univers au moyen âge*, ed. Christian Wenin, Philosophies médiévaux 27 (Louvain-la-Neuve: Éditions de l'Institut supérieur de philosophie, 1986):622–630.

260 *Hanc quidem partem astrologiae quae naturali inquisitione exsequitur, solis lunaeque cursus atque stellarum, et certas temporum distinctiones caute rimatur, oportet a clero Domini solerti meditamine disci, ut per certas regularum conjecturas, et ratas ac veras argumentorum aestimationes, non solum praeterita annorum curricula veraciter investiget, sed et de futuris noverit ratiocinari temporum atque Paschalis festi exordia, et certa loca omnium solemnitatum atque celebrationum, sibi sciat intimare observanda, et populo Dei rite valeat indicare celebranda*: *De clericorum institutione* 25 (403D–404A).

Munich CLM 210 and Vienna Nationalbibl. 387, and the *de luxe* edition in Madrid Bibl. nac. 3307. Similar encylopaedic anthologies can be found in Vatican Reg. lat. 123, Cologne Dom- und Diözesanbibliothek 83(II) and elsewhere.[261] *The Reckoning of Time* appears in these chrestomathies as part of an array of cosmological and astronomical excerpts, including passages from Pliny, Chalcidius, Martianus, Macrobius, Isidore, and the star-catalogues of the *Aratea*. Bede forms the backbone of these works, but to do so, he had to be fundamentally rearranged. Typical of such reconfigurations is Reg. lat. 123, a single *computus* divided into four books (I. *De sole*, II. *De luna*, III. *De natura rerum*, IV. *De astronomia*), and composed of rearranged passages from Pliny, Isidore, Bede, Boethius, Hyginus and others. What is striking is the substitution of a cosmological for a computistical framework: while *The Reckoning of Time* is ordered by the the units of time-reckoning, as they appear in the fundamental documents of *computus*, the great Carolingian encyclopaedias reconfigure the material in terms of the categories of natural science.

Of course, *The Reckoning of Time* continued to be copied *in extenso* in the Carolingian period. Moreover, not all excerpted versions were in natural science encyclopaedias like those described above: many were in conventional *computus* anthologies of tables, rules, short notices and excerpts, and mnemonics.[262] Excerpts appear even in the same codex as the full text of *The Reckoning of Time*, e.g. Geneva, Bibliothèque de l'Université 50 (Massai, 805?). They are often incorporated into practical *computus* handbooks,[263] a pattern which would persist throughout the Middle Ages.

---

261 Cf. Arno Borst, "Alkuin und die Enzyklopädie von 809", in *Science in Western and Eastern Civilization*, 53–75, and Anton von Euw, "Die künstlerische Gestaltung der astronomischen und komputistischen Handschriften des Westens", in *ibid.* 251–269.

262 For an analysis of a typical *computus* anthology of this type, see Faith Wallis, "The Church, the World and the Time".

263 Early examples include Berlin, Geheimes Staatsarchiv R 94 IX A 3 (prov. Werden, s. X^2); Bern, Bürgerbibliothek 417 (s. IX); St Gall 397 (St Gall, s. IX); Vatican City BAV Pal. lat. 1448 (Trier/Mainz *ca.*810); Vat. lat. 645 (St Quentin or Noyon? 825?); London British Library Harley 3017 (Fleury, s. IX). The earliest instance of this seems to be Bern Bürgerbibliothek 207 (Fleury?, s. VIII) which contains only ch. 1 (Jones, *BOT* 149); but see also Paris 7530 (Monte Cassino, s. VIII^2); and Karlsruhe Reichenau 229 (Reichenau 821?) where it appears as part of *computus* anthology based on an archetype of 780, and possibly copied by a schoolboy (Jones, *BOT* 151–152). Vatican City, Reg. lat. 309 (St

Glosses

Like the manuscript tradition of *The Reckoning of Time*, the glossing tradition is twofold: an Irish tradition which antedates the arrival of the texts on the continent, and a later Carolingian tradition. The Irish tradition is essentially computistic, while the Carolingian tradition incorporates elements of the new "liberal arts" approach to *computus* discussed above.

The Irish glossing tradition itself comprises two discernible streams: the Karlsruhe-Vienna glosses, and those associated with Martin the Irishman, *scholasticus* of Laon. The first family stems from the upper Rhine region of St Gall and Reichenau, and extends to Bobbio; it is represented by Karlsruhe Reichenau 167.[264] The glosses in this manuscript are in both Latin and Old Irish, and share a common source with those of the Vienna fragments. Karlsruhe's Latin glosses are essentially translations of Vienna's Irish glosses.[265] These glosses stick very closely to computistical issues, apart from some lexical and etymological notes, mainly on Greek terms. What is remarkable about them is the sophistication of the Old Irish computistical terminology, which suggests that the Irish may have taught *computus* in their vernacular. The Irish glosses are also interesting in that they draw upon a developed body of native Irish computistical literature, much of which predates Bede, and because they aim not only to explain *The Reckoning of Time*, but also to present contested or alternative points of view. They bespeak a lively and rather advanced computistical culture, but one which is grounded in the monastic and biblical curriculum with which Bede was familiar.

The second family is from north-western Gaul, the region of Auxerre and Metz. These codices contain anthologies of didactic matter, for example the selection of excerpts from Pliny's *Historia naturalis* traceable to Northumbria, and discussed above (p. lxxxii). Its outstanding representatives are Melk 370 G. 32 (St Germain d'Auxerre, 840–*ca.*876)

Denis, s. X) is a *computus* in seven books composed of extracts from Bede's works; Rome, Biblioteca Vallicelliana E 26 (Lyon, s. IX[1]) contains "possibly a half of [*The Reckoning of Time*] rearranged in an accurate and careful text" (Jones, *BOT* 156).

264 Killion 154–174; the Reichenau origin of this MS has been contested by Marc Schneiders, "Zur Datierung und Herkunft des Karlsruher Beda (*Aug.* CLXVII)", *Scriptorium* 43 (1989):247–252, who argues that the presence of Irish glosses precludes any origin other than Ireland.

265 For a detailed comparison, see Killion 279 *sqq.*

written in Tironian notes, the autograph of Heiric of Auxerre; and Berlin Staatsbibliothek 130 (Phillipps 1832) from Metz. The glosses in the Berlin manuscript were edited by Jones and included in his *Bedae opera didascalica* edition of *The Reckoning of Time*. Berlin 130, once united with Berlin 129, is from St Vincent in Metz, but not copied there: Jones argues that the glosses were composed in Laon.[266] The *computistica* refer to the *annus praesens* as 873, and Jones deduces that it was written about then. On the basis of early annals in Paschal tables, and other evidence that the exemplar of Berlin 130 was written in Laon, I shall refer to them hereafter as the Laon-Metz glosses. The Berlin manuscript also contains glosses to *The Nature of Things*, possibly by the same author who glossed *The Reckoning of Time*.

A third category of Carolingian glosses is the so-called "Byrhtferth Glosses". Volume 2 of the *Opera* of Bede, published by J. Herwagen Jr. in Basel in 1563,[267] contains *The Reckoning of Time*, *The Nature of Things*, and *On Times*; in addition, there are the *scholia* of Bede's earlier editor Noviomagus, glosses by an *incertus auctor*, and a *Vetus commentarius* which is essentially the computistical works of Abbo of Fleury (*ca.*940–1004).[268] This gloss material all comes from Noviomagus, although Herwagen's text of Bede – which is quite corrupt – does not. But the manuscript from which Herwagen obtained his text was probably the source of the additional glosses which he appended to each chapter, and which he ascribed to Byrhtferth of Ramsey. The ascription gained currency because Herwagen's edition was reprinted by Migne in the *Patrologia latina*.

In 1932, C.W. Jones demonstrated that the so-called "Byrhtferth glosses" appear in a number of *computus* manuscripts antedating Byrhtferth, and of Continental origin. The most notable of these are Berlin 130 and Melk 370. It is not impossible that some of them originated with the glosses of Martin the Irishman, a schoolmaster active at Laon in the mid-ninth century, who was known to have glossed Bede's

---

266 Jones, *BOD* 260.

267 This and other printed editions of Bede are discussed below, pp. xcvii–xcix.

268 Noviomagus's *Vetus commentarius* is essentially Abbo's computistical writings, as found in Berlin Staatsbibliothek 138 and Bern 250, but divided up and arranged as a commentary. Noviomagus also incorporated Abbonian matter into his own *scholia*. According to Jones ("The Byrhtferth Glosses", *Medium aevum* 7 (1938):82), Noviomagus' main sources for all save the *Vetus commentarius* were Cologne Dombibliothek 102 and 103.

computistica, and who was the teacher of Heiric of Auxerre.[269] In short, some of the "Byrhtferth" glosses are related to the Laon-Metz group. However, the same "Byrhtferth" glosses are not found in both manuscripts, which led Jones to conclude that Herwagen conflated his "Byrhtferth" glosses from two sources. The attribution to Byrhtferth was probably inspired by Bale's *Scriptorum illustrium*, published in Basel in 1559 by Herwagen's stepfather, Oporinus.[270] In short, the "Byrhtferth" glosses represent a second stream of Irish commentary on *The Reckoning of Time*, this time filtered through the cathedral schools of northern France, and considerably more influenced by the humanistic strains of the Carolingian renaissance.

The transmutation of *computus* from vocational *doctrina christiana* into something very close to "science" is reflected in these early glosses to *The Reckoning of Time*. Jones observes that "[t]he diocesan masters who had founded and popularized the new cathedral schools that were the wave of the future (at Auxerre, Laon, Liège, Reims, Paris) had been educated through Bede's texts and curriculum. But now, partly by glossing him away, they were preparing first for Martianus, then for Boethius, and finally for Aristotle".[271] One indicator of the fact that these glosses were for secular students is the introduction of an *accessus ad auctorem* for Bede, about whom the glossators seem to have known very little.[272] Another indicator is the fact that *The Reckoning of Time* tended to be glossed together with Bede's cosmographical treatise, *The Nature of Things*. Frances Randall Lipp suggests that the two were used as "companion texts" in the Carolingian schools,[273] another sign that

269 Contreni 124–126. The Melk glosses also appear in Reg. lat. 755 (s. IX–X), Paris BN lat. 5329 (s. IX–X), Vat. lat. 643 (s. XII); cf. Jones *Bedae pseudepigrapha* 30, and appendix, where these glosses are edited from the MSS, with the passages omitted by Herwagen supplied.

270 "The Byrhtferth Glosses", 88–97. In his article "The Glosses on Bede's *De temporum ratione* Attributed to Byrhtferth of Ramsey", *Anglo-Saxon England* 25 (1996):209–232, Michael Gorman argues that there is no reason why the "Byrhtferth glosses" could not be by Byrhtferth, but admits that there is also no positive evidence that they are by him. The lack of any Insular MS tradition for the "Byrhtferth glosses" would seem to undermine this hypothesis. Moreover, Gorman does not demonstrate any connection between the contents of the glosses and Byrhtferth's authentic computistical writings that cannot be explained by mutual use of commonly available computistical material.

271 "Bede's Place", 275–276.

272 This *accessus* is edited by Jones, *BOD* 701–702.

273 "The Carolingian Commentaries on Bede's *De natura rerum*", PhD diss., Yale University, 1961, p. ii.

Bede's policy of incorporating science into *computus* was now being reversed. By the eleventh century, it was not uncommon to find glosses like those in Munich CLM 18158 (prov. Tegernsee, s. XI), which Jones describes as "largely philosophical with little relation to the text".[274]

Though the two families of glosses discussed here are outstanding for their content and long-range influence, they do not by any means exhaust the glossing history of *The Reckoning of Time*. Of the 240 manuscripts of *The Reckoning of Time*, 66 contain a commentary of two or more glosses, and most of these glosses are surprisingly independent.[275] They vary from simple lexical notes, to extensive digressions on philological and scientific matters. Many contain additional worked examples, and one English family of eleventh-century manuscripts contains a unique body of "graphic glosses" framed as tables or diagrams.[276]

Use by later computists

As the computistical system espoused by Bede became more widely accepted, and as his own perpetual Paschal tables made the study of computistical theory less urgent, one would expect *The Reckoning of Time* to have faded in popularity and importance. Although the number of manuscripts peaked in the Carolingian period, there was, in fact, no diminution of interest in either *computus* or Bede's treatise. In the mid-eleventh century, as Bede's 532-year cycle was coming to an end, there was considerable debate over the *annus Domini* chronology, and many alternatives, some openly critical of Bede, were proposed.[277] In the twelfth century, the new urban schools substituted the study of Arabic

274 *BOT* 153.

275 Jones, "Bede's Place", 271 *sqq.*

276 I will discuss this group of manuscripts in a forthcoming article, "An English Family of Graphic Glosses on Bede's *De temporum ratione*".

277 See Alfred Cordoliani, "Abbon de Fleury, Heriger de Lobbes et Gerland de Besançon sur l'ère de l'Incarnation de Denys le Petit", *Revue d'histoire ecclésiastique* 44 (1949):463–487; *id.*, "L'activité computistique de Robert, évêque de Hereford", in *Mélanges offerts à René Crozet*, ed. Pierre Gallais & Yves-Jean Rion (Poitiers: Société d'études médiévales, 1966):1.333–340; Anna-Dorothee von den Brinken, "Die Welt- und Inkarnationsära bei Heimo von St Jakob. Kritik an der christlichen Zeitrechnung durch Bamberger Komputisten in der ersten Hälfte des 12. Jahrhunderts", *Deutsches Archiv zur Erforschung des Mittelalters* 16 (1960):155–194; *id.*, "Beobachtungen zum Aufkommen der retrospektiven Inkarnationsära", *Archiv für Diplomatik, Schriftgeschichte, Siegel- und Wappenkunde* 25 (1982 for 1979):1–20.

mathematics and astronomy for the old *computus*, which in turn was reduced to a subordinate study of *astronomia*. The growing dissatisfaction with the failings of the traditional calendar in the thirteenth to the fifteenth centuries made Bede rather obsolete, and the rise of "natural *computus*", i.e. time-reckoning in the service of astronomy and astrology, made his *computus ecclesiasticus* seem crude and artifical. Yet none of these movements spelled the end of *The Reckoning of Time*. Manuscripts continued to be copied up to the sixteenth century, and the book was quickly put into print.

One of the more interesting manifestations of continuing interest in *The Reckoning of Time* is the number of adaptations and translations of the work made in the Middle Ages. Portions of *The Reckoning of Time* were turned into poetry,[278] and even set to music.[279] Aelfric turned Bede into Old English in *De temporibus anni*, but there was an anonymous versified version as well.[280] The vigorous tradition of vernacular Old English *computus* owes much to Bede's book,[281] but its influence on Old English literature in general has only begun to be studied.[282]

Printed editions

The first printed appearance of *The Reckoning of Time* [283] was typically medieval in that it was an excerpt: it contained only the chronicle in ch. 66. This edition appeared in Venice in 1505 (reprinted in Paris in 1506, and again in Venice in 1509), and was edited by Petrus Marenus Aleander

278 C.J. Fordyce, "A Rhythmical Version of Bede's *De ratione temporum*", *Archivum latinitatis medii aevi* 3 (1927):59–73, 129–141.

279 Alma Colk Santosuosso, "Music in Bede's *De temporum ratione*: An 11th-century Addition to MS London, British Library, Cotton Vespasian B. VI", *Scriptorium* 43 (1989):255–259.

280 Ed. by Thomas Wright, *Popular Treatises on Science Written During the Middle Ages* (London: Taylor, 1841):1–19.

281 See Heinrich Henel, *Studien zum altenglischen Computus*, Beiträge zur englischen Philologie 26 (Leipzig: Tauchnitz, 1934).

282 On the use of *The Reckoning of Time* in the *Old English Martyrology*, for example, see Cross, "Bede's Influence", 25–26, and "On the Library of the Old English Martyrologist", in *Learning and Literature in Anglo-Saxon England*, ed. Michael Lapidge and Helmut Gneuss (Cambridge: Cambridge University Press, 1985):239–242.

283 On printed editions of *The Reckoning of Time*, see Jones, *BOD* 256–257, and Stevens, "Bede's Scientific Achievement", Appendix 1.

of Padua.[284] The world-chronicle was also printed with John Smith's critical edition of the *Historia ecclesiastica* in Cambridge in 1722, along with the preface of *The Reckoning of Time*. The standard modern edition is that of Theodore Mommsen in *Chronica minora* 3, MGH AA 13 (1898):247–327, which is reproduced without change in Jones' *BOD* edition.

Another favourite *excerptum* was ch. 1, on finger-reckoning. This was published in Regensburg in 1532 as *Abacus vetustissima, veterum latinorum per digitos manusque numerandi (quinetiam loquendi) consuetudo, ex beda ...*, edited by Iohannes Aventinus; it was reprinted in Leipzig by Johann Friedrich Braun in 1710. Chapters 1 and 2 also appeared as part of an omnibus volume dealing with, amongst other things, ancient weights and measures (*Hoc in volumine haec continentur. M. Val. Probus de notis Romanorum ...*) printed by Ioannes Tacuinus in Venice, 1525.

The first complete text version of *The Reckoning of Time* was that of Johannes Sichardus, published in Basel in 1529,[285] which also contained *The Nature of Things, On Times*, and Bede's 532-year cycle. Johannes Bronchorst of Neumagen, alias Noviomagus, prefaced his edition of Bede's computistical works (*Opuscula quamplurima de temporum ratione, diligenter castigata atque illustrata ueteribus quibusdam annotationibus, una cum scholiis ... authore* (Cologne: Johannes Prael and Petrus Quentel, 1537),[286] which includes the *Epistle to Wicthed*, with some anonymous tracts on Paschal computation, and included the medieval glosses discussed above, as well as original annotations.

The collection of Bedan and pseudo-Bedan *computistica* printed in Giles' 1843 edition of the *Opera omnia*, and in Migne's *Patrologia latina* 90, basically reproduces the version in the folio edition of Bede's *Opera omnia* printed by Johann Herwagen (Hervagius) in Basel in 1563.[287] In his edition of *The Reckoning of Time*, Herwagen copied Noviomagus' *Vetus commentarius* and *scholia*, but as was mentioned above, his texts of Bede's works, as well as the mysterious "Byrhtferth" glosses, are

284 *Venerabilis Bedae ... De temporibus sive De sex aetatibus huius saeculi liber...* (Venice: per Ioan. de Tridino alias Tacuino, 1505).

285 *Bedae ... De natura rerum et temporum ratione libri duo* (Basel: Heinricus Petrus, 1529). *The Reckoning of Time* is on fols. 48–64v.

286 Described by Jones, *Bedae pseudepigrapha*, 6–13. *The Reckoning of Time* is on fols. XXXI–XCVIv.

287 *The Reckoning of Time* appears in vol. 2, pp. 49–173.

from another, unidentifiable manuscript. To these he appended a vast body of miscellaneous treatises on the *computus* and related topics, such as are frequently found in *computus* manuscripts. While Herwagen had no scruples about assigning them to the pen of Bede, the editors of *Patrologia latina* correctly designated them as *spuria et dubia*.[288]

The two modern critical editions of *The Reckoning of Time* are both the work of C.W. Jones. The first appeared in 1943 as *Bedae opera de temporibus* (Cambridge, Mass.: Mediaeval Academy of America), and contained only chs. 1–65; the second was published in 1977 as vol. 2 of *Bedae opera didascalica* in the series *Corpus christianorum series latina*, 123B, and contains the full text, with ch. 66 supplied from Mommsen's edition. It is this edition which is the basis for our translation of *The Reckoning of Time*.

Principles of translation

This is the first complete translation of this work into any language. C.W. Jones translated chs. 1–29 as part of his 1932 Cornell dissertation, "Materials for an edition of Bede's *De temporum ratione*". Parts of ch. 1 have been translated in a number of studies of finger-reckoning[289] and the section of ch. 66 devoted to the Sixth Age has been translated by Judith McClure and Roger Collins as part of the "World's Classics" edition of *The Ecclesiastical History of the English People* (Oxford: Oxford University Press, 1994):307–340. I did not see the Jones translation until well after my own was completely terminated, and have relied on it only for slight modifications. The McClure-Collins translation of ch. 66 unfortunately reproduces some dates incorrectly, omits passages, and in numerous instances renders the original incorrectly or incoherently.

My translation indicates the corresponding pages of Jones' critical edition in *BOD*, to enable the reader to refer readily to the Latin text. I have not always been able to indicate the page transition with perfect precision, due to the transposition of phrases from the original in the translation.

---

288 See Jones, *Bedae pseudepigrapha* 14–18, and Bernhard Bischoff, "Zur Kritik der Heerwagenschen Ausgabe von Bedas Werken (Basel 1563)", *Mittelalterliche Studien* (Stuttgart: Hiersemann, 1966–1981):1.112–117.

289 See Commentary on ch. 1.

Not every literal translation is a faithful translation, and with no author is this more true than Bede. His grammar is impeccable and his style lucid, but long, complex sentences frequently need to be broken into more digestible morsels for the English reader. I have also attempted to reproduce Bede's technical language faithfully, while at the same time avoiding jargon. Hence I have elected to retain Bede's Latin terms for units of time like *momenta*, because it would be misleading to render this exact unit of measurement as "moment". On the other hand, I have translated *cyclus decemnovennalis* as "19-year cycle" rather than "decennovenal cycle". Bede uses the word *Pascha* to mean both Easter and Passover. I have chosen one or other term to fit the context, and in ambiguous cases used phrases like "the Pasch" or "the Paschal season".

I have given Bede's dates in the Roman calendar form he used, with modern calendar dates in brackets following. My purpose in so doing is to give the reader some impression of what it was like to think about calendar problems using such a counter-intuitive calendar schema.

For the sources and analogues cited in the notes, I have drawn freely on Jones' own *apparatus fontium*, and on his notes in the *BOT* edition of *The Reckoning of Time*. However, I have checked all his references, eliminating some as inconsequential or misleading, adding many others, and correcting and expanding throughout. To indicate Bede's use of his sources with greater precision, particularly when he is not quoting directly, I have italicized all passages taken verbatim from the source text.

Quotations from Scripture for the most part follow the wording of the Authorized Version, which itself was profoundly influenced by the Vulgate Bede knew. In some cases, minor changes have been introduced to conform to the Latin. Where Bede is quoting from the *Vetus latina* used in the liturgy, the modifications are sometimes major. Names of biblical personages and books of the Bible also follow the Authorized Version. In the case of references to the Psalter, the Vulgate numbering follows that of the Authorized Version, in parentheses.

To help the reader to understand the broad architecture of Bede's text, I have divided it into six thematic sections, following the schema outlined above (p. xxxii). As these are my divisions, not Bede's, I enclose them within square brackets.

Finally, this translation is followed by an extensive, chapter-by-chapter Commentary. The reader, especially if he or she is approaching *computus* with little prior knowledge, is advised to read the text and

Commentary in parallel. Sources and issues regarding details of the text are handled in the text footnotes; the content and broader significance of each chapter is reserved for the Commentary. A brief glossary of frequently-used computistical terms is found in Appendix 5.

BEDE:
THE RECKONING OF TIME

PREFACE

Some time ago I wrote two short books in a summary style which were, I judged, necessary for my students; these concerned the nature of things, and the reckoning of time. When I undertook to present and explain them to some of my brethren, they said that they were much more concise than they would have wished, especially the book on time, which was, it seems, rather more in demand because of the calculation of Easter. So they persuaded me to discuss certain matters concerning the nature, course, and end of time at greater length. I yielded to their enthusiasm, and after surveying the writings of the venerable Fathers, I wrote a longer book on time. I was enabled to do so by the largesse of Him who, abiding eternal, established the seasons when it pleased Him, and who knows the limits of the ages; indeed, when He sees fit, He himself shall decree an end to the unstable cycles of time.

Lest anyone be shocked that in this work I have preferred to follow the Hebrew Truth[1] rather than the version of the Seventy Translators as to the sequence of the unfolding ages, I have introduced it in every instance where there seemed to be a discrepancy, so that the reader, whoever he might be, could see both [versions] at the same time and select whichever he thinks preferable to follow. /**264**/ But it is my firm judgment (which I dare say is not countered by any of the wise) that, just as the most reverend translator of this same Hebrew Truth said to those who cavilled at his work, *I neither condemn nor reprove the Seventy, but I prefer the Apostles to all of them,*[2] so also shall I proceed with confidence. For I do not reprove the old chronographers who sometimes followed the translation of the Seventy and sometimes disregarded it, as their fancy took them[3] (this will be demonstrated in this little work of ours), but I prefer to all of these the integral purity of the Hebrew Truth, which the foremost men of learning – Jerome in his book *On Hebrew Questions*, Augustine in his book *On the City of God*, the chronographer Eusebius himself in the third book of his *Ecclesiastical*

---

1 The Vulgate translation of the Bible.
2 Jerome, *Apologia aduersus libros Rufini* 2.25 (PL 23.449–450).
3 Bede is referring to Eusebius; cf. *Letter to Plegwin* 6 (Appendix 3.1).

History, where he cites the historian Josephus, writing against Apion the grammarian[4] – agree contains a shorter span of time than is commonly conveyed in the edition of the Seventy. Even those who laud the Seventy to the skies with great and divine praises do not doubt that this [shorter span] should be followed. Should anyone examine what these men have written, I am sure that he will immediately cease to criticize our efforts, provided he does not regard them with envious eyes, which God forbid.

But however /265/ they who read my writings take them, I offer this little book, now completed to the best of my ability, to you first of all, my most beloved Abbot Hwætbert, that you may read it through and examine it, earnestly beseeching you that should you find anything reprehensible in it, you make it known to me immediately so that I can correct it. But if you perceive that something has been done reasonably and properly, then devoutly thank God with me, Who gave this gift, and without Whom we can do nothing.

Should anyone be annoyed that I have presumed to try my hand at this subject, since I have laboured to confect a new work out of what can be found scattered here and there in the writings of the ancients, then let him listen to what St Augustine says: *It is necessary that many men make many books, in a different style, but not in a different faith, and even concerning the same questions, so that the subject-matter itself might be available to the greatest number: one way to some, another way to others.*[5] Let him also listen to me as I reply with simplicity on my own behalf to him who is displeased or to whom it seems superfluous that I have compiled this work from a number of sources at the request of my brethren and enclosed it in the bounds of a single book: let him discard this book, if that is what he wants. Let him drink in my company from the common wellspring of the Fathers whatever he deems adequate for himself and his, but nonetheless guard uncorrupted the duty he owes to fraternal feeling.

4 For citations of all these works, see *Letter to Plegwin.*
5 Augustine, *De Trinitate* 1.3.25–28, ed. W.J. Mountain and F. Glorie, CCSL 50 (1978): 33.25–28.

[TABLE OF CONTENTS]

[I. TECHNICAL PREPARATION (CHS. 1–4)]

1. CALCULATING OR SPEAKING WITH THE FINGERS

/268/ Before discussing the basics of the calculation of time, we have decided to demonstrate a few things, with God's help, about that very useful and easy skill of flexing the fingers, so that when we have conveyed maximum facility in calculation, we may then, with our readers' understanding better prepared, attain equal facility in investigating and explaining the sequence of time through calculation. For one ought not to despise or treat lightly that rule with which almost all the exegetes of Holy Scripture have shown themselves well acquainted, no less than they are with verbal expressions. Many have said other things [on this topic], and even Jerome, that translator of the sacred narrative, says in his treatise on the evangelical precept[1] (and [Jerome] did not hesitate to take up the aid of its discipline): *The thirty-fold, sixty-fold and hundred-fold fruit, though born /269/ of one earth and one seed, nevertheless differ vastly as to number. Thirty refers to marriage, for this conjunction of fingers depicts husband and wife, wrapped and linked (as it were) in a tender kiss. Sixty refers to widows, because their position is one of confinement and tribulation; hence they are pressed down against the upper finger, for the more the will of a [sexually] experienced person suffers in abstaining from sin, the greater the reward. Finally the hundred-fold number (pay careful attention, reader, I pray!) is transferred from the left hand to the right, and symbolizes the crown of virginity by making a*

---

1 Jones argues that this phrase (*evangelicae tractatus sententiae*) shows that Bede was unaware that the passage from Jerome he is about to cite comes from *Adversus Jovinianum*, and confused it with one of Jerome's commentaries on the Gospels. He speculates that Bede may, in fact, have used a florilegium of excerpts from Jerome (*BOT* 330–331). Such a conclusion is unnecessary. The manner in which Bede proceeds to qualify this phrase ("and [Jerome] did not hesitate to take up the aid of its discipline") indicates that *evangelica sententia* is not a synonym for "Gospels", but an allusion to the "counsels of perfection" in Matthew 19. Here, Christ describes a more perfect way, for those with a special calling: a way particularly characterized by the renunciation of worldly goods, and of marriage and sexual relations, in order to follow Christ more single-heartedly. *Adversus Jovinianum*, being a radical argument against marriage, can be seen as an exposition of one such counsel. As Bede observes, Jerome followed his own advice, and never married.

*circle with the same fingers, but not on the same hand, by which marriage
and widowhood are signified on the left hand.*[2]

So when you say "one", bend the little finger of the left hand and fix it
on the middle of the palm. When you say "two", bend the second from
the smallest finger and fix it on the same place. When you say "three",
bend the third one in the same way. *When you say "four", lift up the little
finger again. When you say "five", lift up the second from the smallest in
the same way.* When you say "six", you lift up the third finger, while
only the finger in between, which is called *medicus*,[3] is fixed in the
middle of the palm. When you say "seven", *place the little finger only
(the others being meanwhile raised), on the base of the palm. When you
say "eight", put the medicus beside it.* When you say /**270**/ "nine", *add
the middle finger.* When you say "ten", touch the nail of the index finger
to the middle joint of the thumb. When you say "twenty", you insert the
tip of the thumb between the middle joints of the index and middle
fingers. When you say "thirty", you join the tips of the index and middle
fingers in a gentle embrace. When you say "forty", you pass the under
side of the thumb over the side or top of the index finger while holding
both erect. When you say "fifty", you rest the thumb, bent at the last
joint into the shape of the Greek letter *gamma*, against the palm. When
you say "sixty", *you carefully encircle the thumb, bent as before, by
curving the index finger forward.* When you say "seventy", you fill the
index finger, bent as before, by inserting the thumb, with its nail upright,
through the middle joint of the index finger. When you say "eighty",
you fill the index finger, curved as before, with the thumb extended full
length and its tip placed against the middle joint of the index finger.
When you say "ninety", you place the tip of your bent index finger
against the base of your upright thumb. So much for the left hand. You
make one hundred on the right hand the way you make ten on the left,
*two hundred on the right the way you make twenty on the left, three
hundred on the right the way you make thirty on the left*, and the rest in
the same manner up to nine hundred. You make *one thousand on the
right hand the way you make one on the left, two thousand on the right
hand the way you make two on the left, three thousand on the right hand*

---

2 Jerome, *Adversus Jovinianum* 1 (PL 23.213B–214A); cf. *Ep.* 48, ed. I. Hilberg,
2nd ed., CSEL 54.1 (1996):353.15–354.9; *In Evangelium Matthaei* 2.13, ed. D. Hurst and
M. Adriaen, CCSL 77 (1959):106.805–814.

3 The ring finger.

the way you make three on the left, and so forth up to nine thousand. Then when you say "ten thousand", you place your left hand flat on the middle of your chest, /271/ but with the fingers pointing upwards to the neck. When you say "twenty thousand", place the same hand, spread out sideways, on your chest. When you say "thirty thousand", place it flat but upright with the thumb on the breastbone. When you say "forty thousand", turn it on its back upright against the belly. When you say "fifty thousand", lay it flat but upright, with your thumb against your belly. When you say "sixty thousand", grasp your left thigh with your flattened hand. When you say "seventy thousand", turn [your hand] on its back on your thigh. When you say "eighty thousand", lay it flat on your thigh. When you say "ninety thousand", grasp your hip with your thumb turned towards the groin. One hundred thousand, two hundred thousand and so forth up to nine hundred thousand, you perform in the same manner as we said, but on the right side of the body. When you say "one million", cross your two hands, linking your thumbs together.[4]

There is also a second type of computation worked on the joints of the fingers which, since it pertains to the reckoning of Easter, will be more conveniently explained when we have arrived at that point.[5] From the kind of computation I have just described, one can represent a sort of manual language, whether for the sake of exercising one's wits, or as a game. /272/ By this means one can, by forming one letter at a time, transmit the words contained by those letters to another person who knows this procedure, so that he can read and understand them even at a distance. Thus one may either signify necessary information by secret intimation, or else fool the uninitiated as if by magic. The method of this game or language is as follows. When you wish to show the first letter of the alphabet, hold up "one" with your hand; for the second, "two"; for the third, "three" and so on in that order. For example, if you wish to warn a friend who is among traitors to act cautiously, show with your fingers 3, 1, 20, 19, 5 and 1, 7, 5; in this order, the letters signify *caute age* ["act cautiously"]. It can be written down in this manner, if greater secrecy is demanded. But this can be more easily learned and manipulated using the letters and numbers of the Greeks, who do not, like the Latins, express numbers by a few

---

4 *Romana computatio*, ed. Jones, *BOD* 671–672. On the history of this text, see Commentary.

5 See below, ch. 55.

letters and their duplicated forms;[6] rather, they depict the figures of
numbers with individual signs, by means of all the letters of the
alphabet, plus three additional numbers,[7] /273/ as follows:

A	1
B	2
Γ	3
Δ	4
E	5
Ϛ	6
Z	7
H	8
Θ	9
I	10
K	20
Λ	30
M	40
N	50
Ξ	60
O	70
Π	80
φ	90
P	100
Σ	200
T	300
Y	400
Φ	500
X	600
Ψ	700
Ω	800
↑	900

Thus whoever has learned to signify numbers with his fingers knows
without hesitation how to shape letters with them as well. So much for

6 That is, the Latins use multiples and combinations of the letters I, V, X, L, C, D,
and M to form their numbers.

7 These "three additional numbers" are 6, represented by the *stigma*; 90, represented by the archaic Greek letter *koppa*; and 900, represented by a symbol called *ennacosi*
(i.e. ἐναχόσιοι, 900) by the Laon-Metz glossator (297, *ad lineam* 101).

this. Now let us proceed to the explanation of time, in so far as He who created and imposed order upon time will deign to assist us.

2. THREE WAYS OF RECKONING TIME

Times (*tempora*) take their name from "measure" (*temperamentum*),[8] either because every unit of time is separately measured (*temperatum*), or because all the courses of mortal life are measured (*temperentur*) in moments, hours, days, months, years, ages and epochs.[9] We shall explain each in turn, as the Lord grants us the power to do so.

First, we point out to the reader that there are three kinds of time-reckoning: it operates either according to nature, or according to custom, or according to authority. This authority is itself twofold. It is either human authority – for example, holding Olympics in a cycle of four years, markets in a cycle of eight days, and indictions in a fifteen-year cycle; likewise the Romans, Greeks and Egyptians ordered the [leap-year] day formed out of quarter-days to be intercalated in the month of February or August as it pleased them[10] – or divine authority, as the Lord in the Law commands that the sabbath be kept on the seventh day,[11] that one refrain from agriculture in the seventh year,[12] or that the fiftieth year be called a Jubilee.[13] For although it is true that barbarian nations are believed to have weeks, it is nonetheless obvious that they borrowed this from the people of God.

Now it is by human custom /**275**/ that the month is considered as having 30 days, even though this does not match the course of either the Sun or the Moon. Those who probe with subtlety into these matters

---

8 The source of this etymology is unknown. Isidore in *Etym.* 5.35.1 states that the seasons are called *tempora* because of "the tempering of commonality, by which each tempers the other with respect to moisture, dryness, heat and cold" (*Dicta sunt autem tempora a communionis temperamento, quod invicem se humore, siccitate, calore et frigore temperent*). But Bede's explanation concerns measurement, not tempering, and he speaks of times in general, not seasons in particular. Jones' apparatus cites the "Merovingian Computus of 727", ch. 7 (ed. Krusch, *Studien II*, 54), but this work also refers to seasons, not times in general, and was evidently not Bede's source.

9 Cf. *De divisionibus temporum* 1 (PL 90.633A); Isidore, *Etym.* 5.29.1; Bede, *On Times* 1.

10 Cf. *Letter to Wicthed.*

11 Exodus 23.11–12.

12 Leviticus 25.1–7.

13 Leviticus 25.8 *sqq.*, esp. 10.

confirm that, in fact, the Moon has – setting aside the calculation of the "leap of the Moon"[14] – 12 hours less [than 30 days],[15] and the Sun has $10\frac{1}{2}$ hours more. Thus with nature as our guide we discover that the solar year is made up of $365\frac{1}{4}$ days, but the lunar year is finished in 354 days if it is common and 384 days if it is embolismic.[16] The complete course of the Moon is encompassed in a nineteen-year cycle. But each of the planets as well is borne around the zodiac at its own rate.[17] This Nature was created by the one true God when He commanded that the stars which He had set in the heavens should be the signs of seasons, days and years;[18] it is not, as the folly of the pagans asserts, a creating goddess, one amongst many.[19]

3. THE SMALLEST INTERVALS OF TIME

An hour is one-twelfth of a day, and a day is made up of twelve hours, as the Lord testifies, saying, *Are there not twelve hours in a day? If any man walk in the day, he stumbleth not*.[20] Here, although He spoke of Himself allegorically as the "day", and of His disciples (who received the light from Him) as "hours", He nonetheless defined the number of hours according to the usual manner of human computation. Now it should be noted that if all the days of the year were thought to have twelve hours, summer days would of necessity be composed of longer hours and winter days of shorter hours. But if we wish to make all hours

14 See below, ch. 42.

15 Bede is referring here to the synodic lunar month, or the period between one new Moon and the next. He points out that this period is slightly more than $29\frac{1}{2}$ days, but that this slight excess is disregarded until the end of the 19-year cycle, when its cumulative effect will be to put the calculated Moon one day ahead of the astronomical Moon. This is corrected by "leaping over" one day in the lunar calculation.

16 See below, ch. 45. The $10\frac{1}{2}$ hours by which the Sun exceeds the 30 days of the notional month, when multiplied by 12 months, makes 180 hours, or $5\frac{1}{4}$ days. Added to 30×12, or 360 days, this produces the solar year of $365\frac{1}{4}$ days. The lunar "year" is 12 months of $29\frac{1}{2}$ days, or 354 days; in embolismic years, an additional month of 30 days is added, bringing the total to 384 days.

17 Cf. ch. 8 below, and *The Nature of Things* 13.

18 Genesis 1.14.

19 Jones' apparatus cites Lucretius, *De rerum natura* 1.629, ed. Joseph Martin (Leipzig: Teubner, 1969):24. It is, however, extremely unlikely that Bede knew Lucretius at first hand, save perhaps through an excerpt in a grammar text. Where he heard about the "goddess Natura" remains uncertain.

20 John 11.9.

equal, that is, to have equinoctial [hours], we must then give fewer hours to the winter day, and more to the summer day.

An hour has four *puncti*, 10 *minuta*, 15 *partes*, 40 *momenta*, and in some lunar calculations, five *puncti*.[21] These divisions of time are not natural, but apparently are agreed upon by convention. For since it was necessary for calculators to divide the day into 12, or the hour into 4 or 10 or 15 or 40 or other segments, whether larger or smaller, they sought out terminology for themselves by which they might designate what they wished, and might denote one thing or another. What [constitutes] the margin [*ora*] of a certain [span of] time, they call an "hour" [*hora*], even as we are accustomed to call the boundaries *of garments, rivers, or of the sea "margins"* [*oras*].[22] *Puncti* they name after the swift passage of the point [*punctus*] /277/ on a sundial, *minuta* after an even smaller [*minore*] interval, and *partes* from the partition of the zodiacal circle, which they divided into thirty days for each month. Then they name *momenta after the swift motion [motu] of the stars*, when it was observed *that something moved and succeeded itself* in a very brief space of time.[23] The smallest time of all, and one which cannot be divided by any reckoning, they call by the Greek word "atom", that is, "indivisible" or "that which cannot be cut".[24] Because of its tiny size, it is more readily apparent to grammarians than to computists, for when they divide a verse into words, words into feet, feet into syllables, and syllables into quantities [*tempora*], and give double quantity to the long [foot] and single to the short, they are pleased to call this an *atomus*, as they had nothing more beyond this which they could divide.[25] In exploring the nativities of men, astrologers likewise claim to arrive at the atom when they divide the zodiacal circle into 12 signs, /278/ each sign into 30 *partes*, each *pars* into 12 *puncti*, each *punctus* into 40 *momenta*, and each *momentum* into 60 *ostenta*, so that by carefully observing the position of the stars they might learn, virtually without error, the fate of the

21 *De divisionibus temporum* 3–6 (654D–655D). On the "lunar calculations", see *The Reckoning of Time*, ch. 17.

22 *De divisionibus temporum* 6 (655D); Isidore, *Etym.* 5.29.2.

23 Isidore, *Etym.* 5.29.1.

24 *Ibid.* 13.2.3.

25 Julian of Toledo, *Excerpta*, ed. Heinrich Keil, *Grammatici latini* (Leipzig: Teubner, 1855–1880):5.321–323; *De divisionibus temporum* 2 (643A–D); Bede, *De arte metrica* 2 (*BOD* 86.4); MS Bern 207, fol. 112 (ed. Keil, 8.xxiv).

newborn.[26] Let us see to it that these things are avoided, because such observance is futile and alien to our faith.

The Apostle uses the term for this kind of time in a better sense, to suggest the swiftness of the Resurrection, stating, *We shall all rise, but we shall not all be changed, in a moment, in the twinkling of an eye, at the last trumpet.*[27] This deserves our attention, because although computists make a strict distinction [between these terms], many writers indiscriminately call that tiniest interval of time in which the lids of our eyes move when a blow is launched [against them], and *which* cannot *be divided or distributed*,[28] either a *momentum*, a *punctus* or an atom.

4. THE RECKONING OF DUODECIMAL FRACTIONS

To know how to divide duodecimal fractions [*unciae*] is no unworthy ambition, for one can apply it to the calculation not only of coinage, but of time and of other things. /**279**/ Because from time to time various historical narratives, and even Holy Scripture itself, will use the names and symbols [of fractions] we have taken the trouble to record them in summary form here.

✕	pound or *as* or *assis*	12 *unciae*
⟨⟨⟩	*deunx* or *iabus*	11 *unciae*
⟨⟨⟨	*decunx* or *dextans*	10 *unciae* [$\frac{5}{6}$]
⟨⟨	*dodrans* or *doras*	9 *unciae* [$\frac{3}{4}$]
⟨⟩	*bes* or *bisse*	8 *unciae* [$\frac{2}{3}$]
⟨	*septunx* or *septus*	7 *unciae*
⟨	*semis*	6 *unciae* [$\frac{1}{2}$]
⟨⟨	*quincunx* or *cingus*	5 *unciae*
⟨⟩	*triens* or *treas*	4 *unciae* [$\frac{1}{3}$]

26 Cf. Basil, *Hexaemeron* 6.5, trans. Eustathius (PL 53.925C–927B); Ambrose, *Hexaemeron* 4.4, ed. C. Shenkl, CSEL 32.1 (1897):120.9–122.10. Since the passage which follows is from Jerome's *Letter* 119.5 (see below), this is probably Bede's immediate source, though Jones (*BOT* 333) opts for Ambrose or Basil.

27 I Corinthians 15.51–52; cf. Augustine, *Ep.* 205, ed. W. Goldbacher, CSEL 57 (1911):355.1–4. This translation may sound strange to English readers familiar with the Authorized Version or the libretto of Handel's *Messiah*, but it renders Bede's Vulgate: *Omnes quidem resurgemus, sed non omnes immutabimur . . .* This was a difficult crux, and Augustine devotes considerable time to its resolution in *Ep.* 205.

28 Jerome, *Ep.* 119.5, ed. I. Hilberg, CSEL 55 (1996):450.8–20.

⋎	*quadrans* or *quadras*	3 *unciae* [$\frac{1}{4}$]
⌐	*sextans* or *sextas*	2 *unciae* [$\frac{1}{6}$]
ⱬ	*sescunx* or *sescuncia*	$1\frac{1}{2}$ *unciae* [$\frac{1}{8}$]
⌐	*uncia*	24 *scripuli*
⌐	*semuncia*	12 *scripuli*
ⱴ	two *sextulae* or *sesclae*, that is, $\frac{1}{3}$ of an *uncia*	8 *scripuli*
7	*sicilicus*	6 *scripuli* [$\frac{1}{4}$ *unciae*]
⌐	*sextula* or *sescla*	4 *scripuli*
⌣	$\frac{1}{2}$ *sextula* or *sescla*	2 *scripuli*
℔	*scripulus*	6 *siliquae*[29]

These names and symbols for weights apply, I say, not only to measuring money, but to quantifying any body or length of time. Hence it is both logical and customary that in reciting their computations, boys often change "one" and "two" to "*assis*" and "*dipondium*"; likewise they say "*tressis*", "*quartus*", "*quinquis*", "*sexis*", "*septis*" and so forth in this fashion, instead of "three *asses*", "four *asses*", and a sequence of however many numbers in the same manner.[30]

Therefore, should you wish to divide one hour, or a whole day, /**281**/ or a month, or a year, or any other specific length of time, greater or less, into twelve, that twelfth part is an *uncia*; the remaining eleven are called *deunx* [$\frac{11}{12}$]. If you divide into six, the sixth part is called *sestans* [$\frac{1}{6}$]; the remaining five are called *dextans*, or according to others, *distas* [$\frac{5}{6}$]. If into four, the fourth part is called *quadrans* [$\frac{1}{4}$]; the residual three take the name of *dodrans* [$\frac{3}{4}$]. Something which perturbs many beginners is solved by the rule of this discipline; for Philip, in his commentary on the blessed Job, describing how the ocean tide comes in twice each day, adds that it arrives, *whether by day or by night, later [than the previous day's tide] by an interval of a dodrans [i.e. $\frac{3}{4}$] of an equinoctial hour, with no break in*

29 Priscian, *De figuris numerorum* 10–11, ed. Heinrich Keil, *Grammatici latini* (Leipzig: Teubner, 1855–1880):3.408–409; cf. 15–16 (410–411); 31 (416). Cf. Victorius of Aquitaine, *Calculus*, ed. G. Friedlein, *Zeitschrift für Mathematik und Physik* 7 (1871):42–97.

30 If this passage reflects Bede's own experience, and is not copied from some untraced classical source, it suggests two things: first, that boys in monasteries learned computational tables by rote, and hence in some kind of programmed way; secondly, that numbers were commonly understood, not as abstractions, but as concrete numbers of things.

continuity.[31] Now if you want to divide by three, you will call the third part *triens*, and the remaining two parts *bisse* [⅔]. If by two, you will call the half *semis*; and so on for the rest. These things can be both learned and taught more easily through speech than by the pen of a writer. /**282**/

If your task is to divide miles or *stadia* or acres or perches, or even cubits, feet or palms, proceed according to the method given above, and likewise for bodies. Thus in Exodus a half-cubit is called *semis*, for Moses tells that the Ark of the Covenant was two and one-half [*duos semis*] cubits long and a cubit and a half high,[32] and in the Gospel the fourth part of the universe, that is, earth, is symbolically designated by the term *quadrans*. It is said to the sinner as he is sent to his punishment, *You shall not go hence until you have paid the last farthing [quad-rantem],*[33] that is *until you atone for your earthly sins,* as St Augustine explains: *For the sinner heard this: You are earth and to earth you shall go.*[34] *Now the fourth and last of the distinct divisions of the universe is found to be earth, so that you begin with heaven, count air as coming second, water third, and earth fourth.*[35] He also alludes to this branch of study in his commentary on the Gospel, where it is written, *And it was about the sixth hour of the preparation of the Passover,*[36] saying, *For it was not fully the sixth hour, but about the sixth hour, that is, the fifth hour was finished* /**283**/ *and the sixth somewhat begun. For these men would never have said five and a quarter [quadrans] or five and a third [triens] or five and a half [semis] or anything of that kind.*[37]

---

31 Philippus Presbyter, *Commentarii in librum Iob* 38 (PL 26.752D). Jones argues that this passage is a criticism of Philippus: "Bede may have believed that if Philippus had known the proper names for fractions he would not have said *dodrans*, which is forty-five minutes. Bede's calculation for the retardation of the tides is forty-eight minutes" (*BOT* 334–335). But Bede is merely stating that the term *dodrans* is unfamiliar to many readers, not that Philippus is misusing it. In the translation included in his doctoral dissertation, Jones misconstrues the first part of this sentence as "The rules of this discipline are in disorder, because many untrained persons use them inaccurately" ("Materials" 160). This is undoubtedly the source of the misunderstanding.

32 Exodus 25.10.

33 Matthew 5.26.

34 Genesis 3.19.

35 Augustine, *De sermone Domini in monte* 1.11.30, ed. A. Munzenbecker, CCSL 35 (1967):31.664–665, 660–663.

36 John 19.14.

37 Augustine, *De consensu Evangelistarum* 3.13.41, ed. F Weihrich, CSEL 43 (1904):324.21–325.3.

[II. THE JULIAN CALENDAR (CHS. 5–41)]

5. THE DAY

Day is air which is lit up by the Sun, and it derives its name from the fact *that it separates and divides the darkness.*[1] Because, *at the very beginning* of Creation, *darkness was upon the face of the deep, and God said: Let there be light. And there was light, and God called the light "day".*[2]

This *word is defined in two ways, that is, according to common parlance, and according to its proper [meaning].*[3] On the whole, ordinary folk call the Sun's presence above the Earth "day". But properly speaking, a day comprises 24 hours, that is, a circuit of the Sun lighting up the entire globe.[4] [The Sun] always and everywhere carries the daylight around with itself, and it is believed to be borne aloft at night under the Earth by no less a space of air than it is by day above the Earth. This claim rests on the authority of much Christian as well as secular literature, but we need present the testimony of only one Father, Augustine. In the second book of questions on the Gospels, explaining the sum of the seventy-two disciples[5] /**284**/ according to its figural meaning, he says, *Just as the whole globe is traversed and lit up in 24 hours, so the mystery of the illumination of the globe through the Gospel of the Trinity is intimated by the 72 disciples.*[6] For 24 times 3 is 72. The same [Augustine] says in the first book of *The Literal Interpretation of Genesis*: *Can it be said that although this work of God's was swiftly concluded, light remained without a night to follow it so long as the diurnal interval ran its course, and that the night which succeeded upon the light remained until the interval of nocturnal time had passed and the dawn of the second day came, when the first day was over? But were I to*

---

1 Basil, *Hexaemeron* 6.8 (431A–B); Isidore, *Etym.* 5.30; *De divisionibus temporum* 8 (656A).

2 Genesis 1.1–4.

3 Cf. *De divisionibus temporum* 8 (656B).

4 Augustine, *Ep.* 199.18, CSEL 57 258.13–16. Cf. ch. 3 above, where Bede gives the day its conventional length of 12 hours.

5 Luke 10.1, 17.

6 Augustine, *Quaestiones evangeliorum* 2.14 (PL 35.1339).

say this, I would be afraid of being ridiculed both by those whose knowledge is certain and by those who can easily perceive that when it is night with us, the presence of the light illuminates those parts of the world through which the Sun passes between its setting and its rising. And for this reason, throughout the entire circuit of the globe, there does not fail to be day in one place and night in another for the whole 24 hours.[7] And shortly thereafter, recalling the opinion of Ecclesiastes concerning the meaning of the Sun, he says, *Therefore when the Sun is in the south, it is day with us; but when it traverses the northern part*[8] *of its orbit, it is night for us. Nonetheless, it does not happen that there is no day in that other region where the Sun is present – unless perchance your heart favours poetic conceits, so that you believe that the Sun sinks into the sea,* /285/ *and rises thence in splendour from another quarter at dawn. Notwithstanding, were this so, the abyss would be lit up by the very presence of the Sun, for it can illuminate the waters, while it cannot be extinguished by them. But to suggest this, is absurd.*[9]

Before the Sun was created, what now comes to pass by means of the Sun was accomplished by the circling of the primeval light, lighting up in its circuit on the first and second days the waters of the abyss which covered the Earth, and on the third day, the empty air. But with respect to the text, *In the beginning God created the heavens and the Earth; and the Earth was without form and void, and darkness covered the face of the deep,*[10] some of the Fathers think that a formless confusion of heaven and Earth and water and all the elements is denoted, so that what is meant is not water or Earth or heaven but, if I may use such a phrase, a single seedbed-material [*seminaria ... materies*] of

7 Augustine, *De Genesi ad litteram* 1.10, ed. I. Zycha, CSEL 28.1 (1894):15.6–16.

8 Augustine's references to the "southern" and "northern" part of the Sun's orbit are obscure, for he is referring here to the daily, not the annual course of the Sun. The idea seems to be that for us in the northern hemisphere, the Sun is always visible to the south of the zenith. Since its orbit is circular, but tilted to the south, the Sun when passing through "that other region" at night, must follow a path which is tilted to the north of our nadir. This would make sense if the earth was visualized as a disk, encompassed by the tilted ring of the Sun's daily orbit. This ring shifts to the south in the winter, and north in the summer, but is always tilted at a southerly angle; therefore its night-time segment must always be tilted in the opposite direction. If this explanation is correct, it would suggest that even educated men in antiquity did not always clearly grasp the implications of the sphericity of the earth, even when they professed it.

9 Augustine, *De Genesi ad litteram* 1.10 (15.22–16.7).

them all.[11] Therefore, not finding a place within the universe capable of receiving the first light, they teach by a loftier scrutiny that everything which they read concerning the first seven days is of necessity otherwise than the custom of our age holds. /286/ But it is much easier to understand if, following the traditions of equally catholic Fathers, the circle of the upper heaven is understood to have been designated by the name "heaven", and by the name "earth" [*terrae*, i.e. the substance or element], the Earth [*tellus*, i.e. the planet] itself, enclosed in the confines which belong to it now, save that it had brought forth no growing plant or living animal, and by "abyss", an infinite expanse of water washing the whole Earth, in the midst of which, it is recorded, the firmament of heaven was afterwards made. Thus St Clement, the disciple of the Apostles and third bishop of the Roman Church after Peter, writes as follows in the sixth book of his histories: *In the beginning God made heaven and Earth. The Earth was invisible and disordered, and darkness was upon the face of the deep, and the spirit of God hovered over the waters. Like the Creator's own hand, this Spirit of God, at God's command, separated light from darkness, and then fashioned this visible heaven after [fashioning] the other invisible one, in order to make the upper regions into a home for the angels, and the lower a home for men. For your sake, therefore, the water which was on the face of the Earth retreated at God's command, that the Earth might bear fruit for you.*[12] Ambrose, bishop of Milan, concurs with this in the second book of the *Hexaemeron.*[13] But so does Basil, bishop of Caesarea in Cappadocia, who in the fourth book of a work bearing the same title says, *Let the waters draw together and the dry land appear. He draws aside the veil so that what was hitherto unseen might become manifest.*[14] Jerome as well, that erudite translator of the sacred record, /287/ in explaining the Prophet's statement (wherein he says, *You who said in your heart, "I will ascend to heaven; I will exalt my throne above the stars of God"*),[15] calls to mind the upper heavens, writing thus: *He [Lucifer] said this either before he fell from heaven or after he fell. If he were still in heaven, how could he say, "I will ascend to heaven?" But because we*

read "To the Lord belongs the heaven of heaven",[16] although he
[Lucifer] was in heaven, that is, in the firmament, he desired to ascend
into the heaven where the Lord's throne is, not out of humility, but out of
pride. But if he spoke these boastful words after he fell from heaven, we
should be aware that he does not lie quiet now that he has been cast
down, but still promises himself great things: not to be amongst the
stars, but to be above the stars of God.[17]

We who follow in the footsteps of the Fathers may conclude that
when God said *Let there be light*,[18] then the darkness which covered the
deep vanished, and light, emerging from the east in the midst of the
waves, clothed the whole surface of the Earth in the expanse of its splendour, reaching in a single instant to the northern, southern and western
shores. Gradually setting as the interval of the day was completed, it
passed in its orbit beneath the lower parts of the earth and, preceded by
the dawn, completed the second and third day in the same sequence. It
differed from solar light only in that it lacked heat, and because the stars
did not yet exist, it left those nights still blackened by the primal darkness. Should anyone think it incredible that the paths of the waters
should be capable of containing light, let him observe the work of
sailors, who make the depths of the sea visible by pouring on oil. Then
let him understand that God, the Creator of all things, is much more
capable of illuminating water, no matter how deep, by the breath of His
mouth, /**288**/ particularly since the waters are believed to have been
much more limpid at that time, before the Creator forced them together
so that dry land could appear.[19]

Now the Lord also defined the day according to common parlance in
the statement I cited above, saying, *Are there not twelve hours in a day?*
But Moses described it in its proper terms, saying, *And the evening and
the morning were the first day.*[20] The Hebrews, Chaldeans and Persians,
following the order of primal Creation, deduce that the course of the
day is from dawn to dawn, that is, they add the time of darkness onto
that of the light. On the other hand, the Egyptians prefer to calculate
their days from sunset to sunset, the Romans from midnight to midnight,

20 Genesis 1.5.

the Umbrians and Athenians from midday to midday.[21] Divine
authority, which in Genesis decreed that days should be calculated from
dawn to dawn, /**289**/ ordained in the Gospel that the period of a whole
day should begin at sunset and end at sunset.[22] For He who at the beginning of the world first called light "day" and darkness "night", at the
end of the ages first illumined the night with the glory of His Resurrection, and thus consecrated the day by showing Himself to His disciples;
He confirmed them more fully in the faith of His Resurrection by eating
with them the following evening and offering Himself to their touch,
and also by bestowing the gift of the Holy Spirit.[23] And because after
the creation of light, *the evening and the morning were the first day*[24] but
now *at the end of the sabbath it began to dawn towards the first day of the
week*,[25] this transformation of time means that we also, who once were
borne away from the light of Paradise to this vale of tears, shall very
soon be transported from the darkness of sin to heavenly joy. Nor is
there any other way, save by placing the night before the day, of
explaining how *the Lord* spent *three days and three nights in the heart of
the Earth, after the pattern of Jonah*. There is a *synecdoche* here:[26] if you
take that *part of the Day* of Preparation *on which He was buried, together
with the previous night, as a day and a night (that is, a whole day), then
the* whole *sabbath, night and day, and the night of Sunday together with
the dawn of that day* which it inaugurated, *as a part standing for the
whole, you will* indeed *have three days and three nights*.[27] It might well

21 Cf. Macrobius, *Saturnalia* 1.3.2–4, ed. J. Willis (Leipzig: Teubner, 1970):9.24–
10.8; Isidore, *Etym.* 5.30.4 and *DNR* 1.2 (173.7–11); Pliny, *HN* 2.79.188 (1.318–320). Where
Bede got his information about the Jewish day is unknown: it is not found in any of the
sources named above. The Irish *De divisionibus temporum* says that the Chaldeans began
their day at dawn, but that the Hebrew day ran from the sixth hour to the sixth hour
"because they did not compute the night" (PL 90.656B). At the end of this chapter, Bede
acknowledges a problem: the day of Creation may have begun at dawn, but the Mosaic
ceremonial law prescribes that feasts begin at sunset.
22 Cf. Matthew 28.1, Mark 16.1.
23 Cf. John 20.28, 21.12–13.
24 Genesis 1.5.
25 Matthew 28.1.
26 A synecdoche is a figure of speech in which a part stands for the whole: cf.
Bede's *De schematibus et tropis* 2.2.7 (*BOD* 156.85 *sqq.*).
27 Cassian, *De incarnatione Domini contra Nestorium* 6.23, ed. M. Petschenig,
CCSL 17 (1886):350.17–28; cf. Matthew 12.38–40. Cf. Augustine, *Quaestiones evangeliorum* 1.7 (1325) and *De doctrina christiana* 3.35.50–51 (111.15–22). Bede, *Explanatio Apocalypsis*, praef. (PL 93. 132A–B) quotes the *regulae* of Tyconius, as found in *De doctrina*

be asked why the people of Israel who, following the tradition of Moses always preserved the order of the day from dawn to dawn, should have begun all their feast days, as we do today, at sundown, and finished them at sundown; as their Lawgiver says, *From even to even you shall celebrate your sabbath.*[28]

6. THE WORLD'S FIRST DAY

Some have claimed that the first day of the world was the 8th kalends of April [25 March], and others that it was the 12th [21 March]. In both cases they adduce the same argument, namely, the equinox. For it seems reasonable that because God in the beginning divided light and darkness into two equal parts, we should believe that the beginning of the world took place specifically at the point of the equinox.[29] They conduct their inquiry well; /291/ but they do not think through their statement consistently. They would act much more knowledgeably were they to assign to the equinox not the first day, in which light was created, but the fourth, in which the luminaries were made. For He decreed the beginning of time at that point when, upon creating the luminaries, He said, *Let them be for signs, and for seasons, and for days and for years.*[30] During the preceding three days, as everyone can see, light and darkness weighed equally in the balance, for since the stars were not yet made, there was no measurement of hours. Not until the fourth morning did the Sun, rising from the midpoint of the east, with the hours running through their lines by the shadow,[31] inaugurate the equinox, which has been maintained every

---

christiana; this is probably his immediate source for this notion that the *triduum paschale* is a synecdoche.

    28 Leviticus 23.32.

    29 I.e. the time of year when the day is divided into equal portions of light and darkness. See ps.-Anatolius 1 (317).

    30 Genesis 1.14.

    31 There could be no equinox before the fourth day of Creation, because before the Sun and stars were made, there was no way to determine whether the length of the day was or was not equal to the length of the night, since there were no celestial markers of the passage of time in either. Bede imagines that on the day the Sun was made, God also made a natural sundial, since the shadows cast by the Sun (by striking what obstacle, we may wonder!) marked the passage of the hours. This is an interesting indication of how deeply Bede's concept of time was bound up with the notion of the measurable; cf. the etymology at the beginning of ch. 2. Time without the reckoning of time was inconceivable.

year. That is, [the Sun], when it first rose above the Earth, was positioned in that part of the sky which philosophers call the fourth degree of Aries.[32] It was destined to return there when a year had passed by, after 365 days and six hours. This addition amounts to a quarter of a day, so that the vernal equinox itself sometimes occurs at dawn, sometimes at noon, sometimes at sunset and sometimes at midnight.[33]

The Moon, on the other hand, was full at sunset, for the Creator, Who is justice itself, would never make something in an imperfect state. It appeared, together with the glittering stars, in the mid-point of the east, and stood in the fourth degree of Libra where the autumn equinox is fixed,[34] and by its rising, it sanctified the beginning of Easter.[35] For the only Paschal rule to observe is that the spring equinox be completed, with a full Moon following.[36] But if the full Moon precedes the equinox by a single day, it is considered to be the Moon of the last month, and not of the first. For it is fitting that just as /**292**/ the Sun at that point in time first assumed power over the day, and then the Moon and stars power over the night, so now, to connote the joy of our redemption, day should first equal night in length, and then the full Moon should suffuse [the night] with light. This is for the sake of a certain symbolism, because the created Sun which lights up all the stars signifies the *true* and eternal *light which lighteth every man that cometh into the world*,[37] while the Moon and stars, which shine, not with their own light (as they say), but with an adventitious light borrowed from the Sun,[38] suggest the body of the Church as a whole, and each individual saint. These, capable of being illumined but not of illuminating, know how to accept the gift of heavenly grace but not how to give it. And in the celebration of the supreme solemnity, it was necessary that Christ precede the

32 Bede may have drawn this idea from the authentic text of Anatolius 2, as cited in Rufinus' trans. of Eusebius, *HE* 7.32.14 (723.15), which says that the Sun at equinox has reached *quartam . . . diem*. However, Bede interprets this text in a very different sense in *Letter to Wicthed* 7. The MS used by Bede apparently read *quadram . . . in ea die*.

33 Cf. *Letter to Wicthed*; *On Times* 10 (592.4–7).

34 Cf. *Letter to Wicthed* 5.

35 Cf. Anatolius 2, as cited in Rufinus' trans. of Eusebius, *HE* 7.32.18 (725.14).

36 Cf. *Prologus Theophili* 2–3 (ed. Krusch, *Studien I*, 223–224); Proterius, *Ep. ad Leonem* 8 (ed. Krusch, *Studien I*, 276–277).

37 John 1.9.

38 Following Isidore, *DNR* 22 (255) and *Etym.* 3.43–49, Bede seems to have thought that the stars, like the Moon and planets, shone with reflected light from the Sun; cf. *The Nature of Things* 11, 18. But see ch. 7, n. 54 below.

Church, which cannot shine save through Him. For if anyone were to argue that the full Moon can come before the equinox, he would be stating either that Holy Church existed in its perfection before the Saviour came in the flesh, or that one of the faithful, before the bestowing of His grace, can have something of the supernal light.[39]

Observing the Paschal season is not meaningless, for it is fitting that through it the world's salvation both be symbolized, and actually come to pass. Thus it was arranged by the divine plan that its beginning was fixed not on the first day, in which light was created, nor the second, in which the firmament was made, nor the third, in which the dry land appeared, nor even on the beginning of the fourth, when the equinoctial Sun went forth like *a bridegroom from his chamber*,[40] but /**293**/ only at sunset when the Moon rose. That hour was awaited which would mark the illumination of the Church which was to come once and for all in Christ. In the rite of the heavenly sacrifice neither water nor wine alone should be offered, lest the oblation correspond only to God or only to man. Rather, we mingle the blood of the grape, squeezed from the wine-press of the Cross, with the water of our fragility, and likewise mix the grain of the wheat, ground on the millstone of the Passion, with water, so that *cleaving to the Lord*, as the Apostle says, we may be able *to be one with Him in spirit*.[41] Even so, in observing the season of this same sacrifice, let us not pay attention exclusively to the course of the Sun, as if we believed indeed in God, but that He is far above our care (like those who say *Thick clouds are a covering to him, that he seeth not; and he walketh in the circuit of heaven*);[42] nor let us seek only to locate the full rising of the Moon, as if, like the Pelagians, we sought to be blessed without supernatural grace. Rather, following him who said *God shall prevent me with his mercy*,[43] let us in our Paschal celebration dedicate

39 I.e. he would be guilty of Pelagianism, that is, the heretical belief that man, by nature, has the capacity to live without sin, and does not require divine grace to do so. This is the computistical heresy of Victorius; cf. *The Reckoning of Time*, ch. 51. For the analogy of the Moon and the Church, see Isidore, *DNR* 78.6 (243–245) and Ambrose, *Hexaemeron* 4.8.32 (137.27–138.22).

40 Psalm 19.5 (18.6).

41 I Corinthians 6.16–17. This classic interpretation of the meaning of the min-gling of wine and water at the Eucharist comes from Cyprian, *Ep. 63 ad Caecilium*, ed. W. Hartel, CSEL 3.2 (1871):701–717; cf. Joseph-André Jungmann, *Missarum Sollemnia* (Paris: Aubier, 1952):2.312.

42 Job 22.14.

43 Psalm 59.10 (58.11).

the equinoctial rising of the Sun (that is, of Him who overcomes the barriers of darkness), with the full Moon of our devotion following in His footsteps.[44] On the day of His own Resurrection, He who *came not to do away with the Law but to fulfil it*[45] fulfilled this requirement of the Mosaic Passover. As God gives us means, we shall explain each of these matters in its own place in what follows.

Here it suffices to point out that the equinox is to be observed on the 12th kalends of April [21 March] /294/ and the first day of the world three days before, that is, on the 15th kalends of April [18 March], and I suppose the ancients wished to place the beginning of the zodiacal circle here to indicate the pre-eminence of this day. Yet neither the month nor the year (whose starting-point [the equinox] came to be) of the Greeks, Romans or Egyptians began on this day. For astronomy [*astrologia*] arose, not with these nations – although the Greeks make this boast – but rather with the ancient Chaldeans, from whom the Patriarch Abraham learned to know God in the revolution of the firmament and stars, as Josephus relates. Later, as he understood this science more accurately, [Abraham] introduced it to the Egyptian people when he was exiled amongst them.[46] For in the book of holy Job, who lived not long after Abraham, we find the word "Mazaroth",[47] that is, the signs of the zodiac. Therefore, according to the division of the zodiac, the Sun enters the sign of Aries on the 15th kalends of April [18 March], [the day] when light was created. But according to the sequence of its initial creation, the beginning of [the Sun's] orbit, and simultaneously the beginning of all time, is the 12th kalends of April [21 March]. Bishop Anatolius of Laodicea plainly teaches this when writing about the equinox: *In that day, /295/ he says, the Sun is found not only to have entered the first part [i.e. sign], but also to have*

---

44 Cf. ps.-Anatolius 1 (317).

45 Matthew 5.17.

46 Isidore, *Etym.* 3.25 or *De computo dialogus* (650A): Josephus is named as source in both these works. The notion that Abraham was the father of astronomy is also found in the ps.-Clementine *Recognitiones* 1.32 (1226A) (a text certainly known to Bede), and became a commonplace of anonymous *computus* treatises (see examples cited by Jones, *BOT* 339).

47 Job 38.32; cf. Augustine, *Adnotationes in Iob*, ed. I. Zycha, CSEL 28.2 (1895):610.22–611.12 (who, however, does not identify "Mazaroth" with the signs of the zodiac); and Virgilius Maro Grammaticus, *Epitoma* 4, ed. J. Huemer (Leipzig: Teubner, 1886):22.8–13.

already the quadras,[48] *that is, the first of twelve parts [of the zodiac].*
But this first small part out of twelve is the vernal equinox, and is itself
the beginning of the months, the inauguration of the [zodiacal] circle
and the completion of the course of the stars which are called planets
(that is, wanderers), and the end of the twelfth small part is also the
completion of the whole circuit.[49] Here he has understood perfectly
correctly and explained in a very elegant manner that as far as nature
is concerned the beginning of the zodiac is nowhere else but in the
spring equinox, and that the 12 signs, which he calls "small parts",
have their beginning and their termination there. [Both] the quarter-
day which we call the "leap-year day", and the "leap of the Moon"[50]
start and finish there, and there the great year, that is, the course of the
planets has its appointed beginning and end.

7. NIGHT

Night [nox] is so called because it detracts from [noceat] human affairs
or vision,[51] or else because thieves and robbers find occasion therein to
injure [*nocendi*] others. Night is the absence of the Sun, when it is
concealed by the Earth's shadow from the time it sets until the time it
rises again. The poet describes its nature accordingly: /**296**/

night sank into the ocean,
wrapping Earth and pole in a mighty shadow.[52]

48 This phrase "also to have already the *quadras*" is somewhat obscure. Bede,
both in what follows and in the *Letter to Wicthed*, interprets it to mean that the calculation
of the Sun's quarter-day, i.e. its leap-year increment, begins at this point. This may not have
been what Anatolius meant: see n. 32 above. The Laon-Metz glossator thinks that it means
that the equinox occurs when the Sun is in the fourth degree of Aries, i.e. 21 March (294 *ad*
lineam 97; 298 *ad lineam* 98).

49 Ps.-Anatolius 2 (318), or the authentic text as cited in Rufinus' trans. of Eusebius, *HE* 7.32.15 (723.15); cf. *Letter to Wicthed*.

50 On the "leap of the Moon" (*saltus lunae*), see ch. 42.

51 Isidore, *Etym.* 5.31.1, cf. *De divisionibus temporum* 11 (658C), Cassiodorus, *Ex-*
positio psalmorum 1.2, ed. M. Adraien, CCSL 97 (1958):232–233. Jones speculates that
Bede may have found this etymology amongst the philological notes in an Irish *computus*
manuscript, akin to the "Bobbio *computus*" (PL 129.1308A–B). It also surfaces in the
"Merovingian *computus* of 727" (55) and elsewhere (e.g. Vatican MS Reg. lat. 123, fol.
10r): *BOT* 339.

52 The quotation from is Vergil, *Aeneid* 2.250–251, but the immediate source is
Isidore *Etym.* 5.31 (see note 51 above). Cf. Isidore, *DNR* 2.1 (181), and Bede, *On Times* 3.

And Solomon in Holy Scripture said *who feeds amongst the lilies until the day breaks and the shadows give way*,[53] alluding to the departure of night, in an altogether elegant turn of phrase, as the giving way of the shadows.

However, because the interposition of the Earth's mass blocks the Sun's splendour from us according to the location of the regions through which his path passes, that shadow, which is the very essence of night [*quae noctis natura est*], is projected so far upwards that it appears to reach to the stars. Appropriately, [Solomon] signified that by the opposite change, that is, the rising of the light, *the shadows give way*, that is, night is suppressed and driven down. Philosophers say that this shadow of night extends upwards to the frontier between air and ether, and that the Moon, the lowest of planets, is occasionally touched and obscured by the shadow as it comes together into a point like a pyramid. No other star undergoes an eclipse, that is, the loss of its light, in this fashion, because the sunlight, diffused everywhere around the confines of the Earth, shines without impediment on those [stars] which are at a great distance from the Earth. Therefore /**297**/ [the Sun] makes the tracts of ether which are beyond the Moon to be always full of daylight, either by his own brightness or by that which beams from the stars.[54] If, on a dark night, you are positioned at a distance from some blazing torches, you see some of the surrounding area suffused with their light, although the darkness of night is all about, and all you can see are the separate flames of the torches themselves. By the same token, they say that the empty space which is purest and closest to heaven is always lit up by the light of the stars, scattered everywhere. But to our vision, impeded as it is by the thicker air, the stars themselves appear to be shining lights, while the brightness which they radiate is not obvious.

But they say that when the Moon is full and seeks its lowest point,[55] it

---

53 Song of Solomon 2.16–17.

54 "However, because the interposition ... from the stars": cf. Pliny, *HN* 2.8.51 (1.200); Ambrose, *Hexaemeron* 4.3.11 (117.26–118.11); Isidore, *DNR* 28 (277–279). "Philosophers say ... loss of light, in this fashion ...": cf. Isidore, *DNR* 21.1 (251); *Etym.* 3.59. Note that Bede retracts his previous statement that the stars shine only with borrowed light; this is not in any of the sources listed above.

55 *Lunam uero aiunt cum infimas sui circuli apsidas plena petierit . . .* (297.33–34). Normally Bede uses the term *apsida* to signify a planet's apogee: cf. *The Nature of Things* 14, an excerpt from Pliny, *HN* 2.13.63–4 (1.210–212). Here, however, he apparently uses the term to mean the maximum distance of the Moon from the plane of the ecliptic. The path

sometimes is obscured by a visible shadow until, having removed itself a little bit from the centre [i.e. plane] of the Earth, it is again exposed to the Sun. So that this does not happen at every full Moon, [the Moon's] orbit runs at variable altitudes through the width of the zodiac, which is 12 degrees [*partes*] wide.[56] For three things must occur together to make a shadow: light, a body, /**298**/ and a place on which the shadow is cast. And where the light is equal [in magnitude] to the body, a shadow of constant [diameter] is thrown; where the light is smaller than the body, the shadow increases indefinitely; where the light is greater than the body, the shadow gradually diminishes and dies away.

They maintain that the Sun is larger than the Earth, though it seems small because it is so far away. Hence, because the shadow of night gradually decreases, it fades out before it reaches the ether.[57] Blessed Ambrose calls to mind this shadow or night in the sixth book of the *Hexaemeron* in the following words: *But he* – that is, Moses – *did not think it necessary to discuss how far Earth's shadow extends into the air when the Sun retreats from us and takes away the day, illuminating the lower pole, or how the Moon, climbing into the region of [Earth's] shadow, is eclipsed.*[58]

The shadow of night was given to mortals for the body's repose lest mankind perish because of unendingly immoderate exertion at its work. It was given as well to certain animals who cannot bear the light of the Sun, and likewise to those beasts who fear the presence of human beings, in order that they may have an opportunity to go about and seek their food. As the Psalmist sings in praise of God: *The Sun knew its*

---

of the Moon is inclined at about 5 degrees and 9 minutes to the ecliptic, or path of the Sun. It therefore intersects the ecliptic twice during its monthly course; once as it is ascending towards the northern side of the ecliptic (its "ascending node"), and once as it is descending towards the southern side (its "descending node"). The highest and lowest points of its ascent and descent are its *apsidae*, at least in this context. Lunar and solar eclipses occur only at the nodes, which are where the paths of the Sun and Moon cross, and they can only take place when the Moon is full or new at one of the nodes. Bede implies that lunar eclipses occur only at the descending node, but in fact they can occur when the Moon is full at either node.

56 "But they say that when the Moon ... 12 degrees wide": cf. Pliny, *HN* 2.16.66 (1.212–214).

57 Cf. Pliny, *HN* 2.7.51 (1.200); see ch. 27 below.

58 Ambrose, *Hexaemeron* 6.2.8 (209.1–4).

setting; You set out the shadows and made the night, in which all the beasts of the wood go forth, and so on.[59]

A wonderful forethought on the part of the Creator has so balanced matters [*temperavit*] that where [the climate] is colder because of the distance of the Sun, there the night is longer, in order that labour might be shortened and limbs kept warm.[60] For winter [nights] are longer than summer [nights] everywhere throughout the world, and much longer in Scythia than in Africa. Likewise, the day shines much longer in summer in Scythia than in Africa, /**299**/ for were the day which scorches Libya not briefer in proportion as it is hotter, it would surely have consumed the whole country long ago.[61]

There are, in fact, *seven parts of the night: dusk, eventide, the first part of the night, dead of night, cock-crow, early morning, and daybreak. Dusk [crepusculum] is uncertain light, for we say that something doubtful is "murky" [creperum], that is, between light and darkness. Eventide* [*vesperum*] is when the *star* of the same name appears, *of which* the poet says:

> *The evening star [vesper] would close the gates of Olympus against the day . . .*

The first part of night [conticinium] is when everything is hushed [*conti-cescunt*], that is, *silent. Dead of night [intempestum] is midnight, when, in the deep sleep of peace, there is no* time [*tempus*] for activity *for any creature. Cock-crow* is when the *rooster* lifts up its song. *Early morning [matutinum] is between the departure of darkness and the arrival of dawn. Daybreak is when a bit of light first appears. Dawn* lasts until sunrise.[62]

59 "The shadow of night . . . and so on": cf. Isidore, *DNR* 17.3 (235), which makes an analogous argument about the alternation of the seasons, quoting ps.-Clement, *Recognitiones* 8.45 (1393A). The quotation from the Psalms is Psalm 104.19–20 (103.19–20).

60 This idea is expressed in Macrobius, *Commentarium in Somnium Scipionis* 2.7, ed. J. Willis (Stuttgart & Leipzig: Teubner, 1994):161–162, which Bede apparently did not know directly.

61 Isidore, *Etym.* 5.31.

62 Isidore, *Etym.* 5.31.4–14, from which the quotation from Vergil, *Aeneid* 1.374 is also derived. Cf. Bede, *The Nature of Things* 2.

8. THE WEEK

Hebdomada [the week] is a Greek word which takes its name from the number seven. Human custom restricts [the definition] to [a period of] seven days, but according to the authority of sacred Scripture there are many notable kinds [of weeks], all of which, if I am not mistaken, point to a single end: that is, they urge us to hope for endless peace in the grace of the Holy Spirit when all good works are accomplished. Thus the first /**300**/ week of all, unique of its kind, from which the others derive their form, is honoured by divine action, because the Lord, completing the adornment of the universe in six days, rested from His labours on the seventh.[63] We should note here that the number six is perfect, not because the Lord perfected the works of the universe in [six days], but because (as St Augustine says) the Lord, who was capable of creating everything in an instant, deigned to do His work within this number, because it is a perfect number, in order that He might demonstrate the perfection of His achievement through this number, which is the first [number] to be formed of its factors, that is, of a sixth, a third and a half – namely one, two and three, and which together make six.[64]

The second week was ordained to be observed by men according to the pattern of this divine week. The Lord said, *Six days shalt thou labour and do all thou hast to do; but on the seventh day, the sabbath of the Lord thy God, thou shalt do no work. For in six days the Lord made heaven and earth, the sea and all that is therein, and he rested on the seventh day.*[65] In olden times, the people of God enumerated the week in this way: *the first [day] of the sabbath, or [day] one of the sabbath* or sabbaths; *the second of the sabbath; the third of the sabbath; the fourth of the sabbath; the fifth of the sabbath; the sixth of the sabbath*; the seventh of the sabbath, or, the *sabbath*.[66] Not that all days could be sabbaths or days of rest, but that the first [day], the second, third and so on should be counted in their sequence from the day of rest, which took precedence because of its name and observance. Of course the Gentiles,

---

63 Cf. Genesis 2.1–2.
64 Cf. Augustine, *DCD* 11.30 (350.1–12).
65 Exodus 20.9–11.
66 Bede's source is the Irish *computus* tract beginning *De numero igitur . . .* found in MS Bodleian Library, Bodley 309 fol. 62 (the "Sirmond manuscript") and transcribed from Bern 417 fols. 52v–53v by Jones, *BOT* Appendix, 395. See Introduction, section 5, "The technical literature of *computus*".

when they learned to observe the week from the people of Israel,[67] subsequently twisted it around to the praise of their own gods, dedicating the first day to the Sun, /**301**/ the second to the Moon, the third to Mars, the fourth to Mercury, the fifth to Jupiter, the sixth to Venus, and the seventh to Saturn, consecrating their days to the same monstrosities to whom they had consecrated the planets, although in a different order.[68] For *they thought* that they *received spirit from the Sun, body* from the *Moon*, *ardour* from *Mars*, *wisdom* and language from *Mercury*, *moderation* from *Jupiter*, *pleasure* from *Venus*, and *slowness* from *Saturn*.[69] I believe that this came to pass because the Sun, positioned in the midst of the planets, seems to give heat to the whole universe and, as it were, to vivify it, as spirit does. Ecclesiastes bears witness to this, saying of it, *The spirit goes gyring in a gyre, and returns in its circle*.[70] The Moon, by supplying moisture, nourishes growth in all bodies. The star Mars, because it is very close to the Sun, is fiery in colour and likewise in nature. By its perpetual circling close to the Sun, Mercury was deemed to radiate the inexhaustible light of wisdom, so to speak. Jupiter is tempered on the one hand by the cold of Saturn, and on the other by the heat of Mars. In the loveliness of her light, which she receives from her proximity to the Sun, Venus attracts by her mien those who behold her.

67 Cf. ch. 2 above.

68 Jones thinks that Bede is referring here to different ancient notions of the order of the planets, e.g. the "Egyptian" and "Chaldean" arrangements discussed by Macrobius, *Comm. in Somnium Scipionis* 1.19 (73–78), or to the Hipparchan model of the perihelial orbit of Mercury and Venus, recorded in Chalcidius' commentary on the *Timaeus* of Plato 110 (ed. J.H. Waszink (London: Warburg Institute, 1972):157.6–21), in Martianus Capella, *De nuptiis Philologiae et Mercurii* 8.880–882 (ed. J. Willis (Leipzig: Teubner, 1983):333–335), and extracted in medieval astronomical compilations: see Jones, *BOT* 341. Cf. Bruce Eastwood, " 'The Chaster Path of Venus' (*orbis Veneris castior*) in the Astronomy of Martianus Capella", *Archives internationales d'histoire des sciences* 32 (1982):145–158, and "Notes on the Planetary Configuration in Aberystwyth N.L.W. MS 735C, f.4v", *The National Library of Wales Journal* 22 (1981):129–140. It seems to me, however, that Bede's reference to the "different order" refers to the difference between the physical arrangement of the planets in the heavens, and the order of their corresponding weekdays in the pagan calendar. Note that Bede goes on to attempt to rationalize this non-astronomical order by a kind of astrological anthropology derived from Isidore.

69 Isidore, *DNR* 3.4 (185.22–28).

70 Ecclesiastes 1.6. Bede's version of this passage (*Gyrans gyrando uadit spiritus* . . .) is not that of the Vulgate, but rather that found in Jerome, *In Ezechielem* 1.1, ed. F. Glorie, CCSL 75 (1964):20.481, and Isidore, *DNR* 27.2 (277.10), which is probably Bede's source; cf. Jones, *BOT* 341.

In that he is the highest [i.e. "farthest out" from Earth], Saturn travels more slowly than the other planets, for he finishes his course through the zodiac in 30 years. After him comes Jupiter in 12 years, third comes Mars in 2 years, fourth comes the Sun in 365¼ days; under the Sun comes Venus, /302/ called both Lucifer and Vesper, in *348 days, never departing from the Sun by more than 46 degrees. The star of Mercury is close to her, its orbit shorter by 9 days. It shines sometimes before sunrise and sometimes after sunset, and is never more than 22 degrees away [from the Sun].*[71] *Finally, the Moon completes the zodiac in 27 days and 8 hours.*[72]

This, then was the foolishness of the Gentiles, buttressed by faulty reasoning, for they sought to dedicate the first day of the week to the Sun by right, so to speak, because he is the greatest of the luminaries, and the second day to the Moon because it is the second luminary. Thereafter, following due sequence, they assigned the first star from the Sun to the third day, the first from the Moon to the fourth day, the second from the Sun to the fifth day, the second from the Moon to the sixth, and the third from the Sun to the seventh.[73] Pope Silvester was the first to instruct the clergy to have *feriae* [weekdays].[74] [Clergy] are never permitted to involve themselves with worldly service or business, /303/ for they are devoted to God alone; as the Psalmist says, *Be still and know that I am God;*[75] and likewise the Apostle, *No one who fights for God entangles himself in worldly business.*[76] Now [Silvester] called the first day, on

---

71 "The Moon, by supplying ... from the Sun": cf. Pliny, *HN* 2.6.32–40 (1.188–192), the last phrases being a direct quotation; cf. Isidore, *Etym.* 3.66.

72 Pliny, *HN* 2.6.44 (194).

73 The conventional ancient explanation of the order of the planetary weekdays was an astrological one, but Bede, if he knew about this at all, was silent on the subject: cf. Jones, *BOT* 342; Eviatar Zerubavel, *The Seven Day Circle: The History and Meaning of the Week* (New York: Free Press, 1985):12–20.

74 *De divisionibus temporum* 10 (658A). Bede and this Irish *computus* tract are the earliest texts to ascribe this peculiar Christian style of naming weekdays to Silvester, but there is much evidence for its promotion by the Fathers, e.g. Augustine, *Enarrationes in Psalmos* 93.3 (ed. E. Dekkers and J. Fraipont, CCSL 39 (1961):1302.5–1303.33), Caesarius of Arles, *Sermo* 193.4 (ed. G. Morin, CCSL 104 (1963):785–786), Martin of Braga, *De correctione rusticorum* (ed. Claude W. Barlow, *Opera omnia* (New Haven: Yale University Press, 1950):189). The word *feria* changed its meaning almost completely from classical to medieval usage. In the former, it meant "holiday", and in the latter, "ordinary day": see Jones, *BOT* 342.

75 Psalm 46.10 (45.11).

76 II Timothy 2.4.

which light was created and also Christ's Resurrection celebrated, the Lord's Day. That this name was given to it in the earliest days of the Church is testified by John, who says in the Apocalypse, *I was in the spirit on the Lord's Day.*[77] To this, [Silvester] added the "second *feria*", "third *feria*", "fourth", "fifth" and "sixth" on his own authority, and retained the sabbath from the ancient Scriptures. He did not dread the rods of those grammarians who decree that, like kalends, nones and ides, so also the weekdays should be designated strictly by a plural number.[78]

A third kind of week occurs in the celebration of Pentecost; it is completed in 7 times 7 weeks, plus one, which is 50 days.[79] On this day Moses, climbing the fiery mountain, received the Law from heaven, and Christ sent the grace of the Holy Spirit from heaven in tongues of fire. The fourth week was that of the seventh month, which was almost entirely taken up by special ceremonies. Amongst these the Day of Atonement was of particular importance. Only on this day of the year did the High Priest, leaving the people outside, enter the Holy of Holies after the year's fruits of grain, wine and oil had first been collected in order.[80] This signifies that Jesus the great High Priest, having fulfilled the dispensation of the flesh through His own blood, prepares to enter the gates of the heavenly kingdom, in order that He might now appear before the face of God on our behalf who, still standing outside, remain at the ready and attentive for His coming.[81] /**304**/ It should be noted here that just as certain unclean persons are commanded by the Law to be purified on the first, third and seventh day, so the first, third and seventh month are dedicated to particular ceremonies.

The fifth week is that of the seventh year, in which the entire people, by order of the Law, rested from agricultural labour. The Lord said, *Six years shall you sow your field, and on the seventh you shall cease.*[82] The

---

77 Apocalypse 1.10; cf. *De divisionibus temporum* 10 (658B).

78 I.e. just as Roman dates are designated by a number, followed by the possessive plural of the marker-day and month (e.g. *VIII kalendarum aprilium*), so (according to captious grammarians) the Christian weekdays ought to be designated as *secunda sabbatarum*, rather than *secunda sabbati*.

79 Exodus 19.18. On Bede's patristic sources for this exegesis, see Jones, *BOT* 342–343.

80 Leviticus 16.29–34.

81 Cf. Hebrews 9, *passim*.

82 Leviticus 25.3.

sixth *week* is that of the *Jubilee year*, the year *of remission, which is encompassed in seven weeks of years, that is, in 49 years.*[83] When these were completed – that is, at the beginning of the 50th year – *trumpets* rang out loud and clear, and *everyone recovered their former possessions according to the Law.*[84]

9. THE SEVENTY PROPHETIC WEEKS

The seventh kind of week is that employed by the Prophet Daniel, comprising (after the manner of the Law) seven years in each week, but abbreviating these years by a new calculation; that is, /305/ he fixed each [year] at twelve lunar months, and did not include in the second or third years (as tradition decrees) the embolismic months which normally accumulate from the eleven days of the epact of every year. Rather, he counted a whole year whenever the sum of 12 [lunar months] was reached.[85] He did this, not because he begrudged the knowledge of the truth to those who sought it, but in order to exercise the inquirers' intelligence, as is the manner of prophecy. At any rate, he preferred to hide his pearls away from his sons, so that they might seek them out with fruitful effort, rather than have them scattered abroad to be trodden underfoot by swine in contemptuous disdain.[86]

Now in order to explain this more clearly, let us look at the precise words of the angel to the Prophet. He says: *Seventy weeks are diminished upon thy people and upon thy holy city, to finish the transgression, and to make an end of sins and to wipe out iniquity and bring in everlasting righteousness, and to fulfil the vision and prophecy, and to anoint the most holy.*[87] There is no question but that these words refer to the Incarnation of Christ, *who took away the sins of the world,*[88] *fulfilled the Law and the Prophets*[89] and *was anointed with the oil of gladness above his fellows,*[90]

83　Isidore, *DNR* 6.4 (195.31–32).

84　Isidore, *Etym.* 5.27.3 or *DNR* 6.4 (33–34); cf. Leviticus 25.8–31.

85　Daniel's "year", in short, is twelve ordinary lunar months, or 354 days. Normally, one or two such ordinary years are followed by an embolismic year of 13 lunar months. This 13th month is formed by cumulating the 11 days by which the lunar year falls short of the solar year, over a period of three years: see Introduction, section 3.

86　Cf. Matthew 7.6.

87　Daniel 9.24.

88　John 1.29.

89　Matthew 5.17.

90　Psalm 45.7 (44.8).

and that the seventy weeks /**306**/ multiplied by seven years work out to 490 years. But note that he claims that these weeks are not simply "observed" or "calculated" but "diminished"; that is, he covertly impresses upon the reader that he should understand that the years indicated are shorter than usual.[91] *Know therefore and understand that from the rebuilding of Jerusalem unto the Christ, the Prince, shall be seven weeks and sixty-two weeks; in a brief space of time the street and the wall shall be built again.*[92] From Ezra's account we know that Nehemiah, when he was cup-bearer to King Artaxerxes, in the twentieth year of his reign, in the month of Nisan, obtained [the king's] permission to restore the walls of Jerusalem. The Temple had been constructed long before this by permission of Cyrus.[93] As it is said, he completed this work in a very short space of time; so hard-pressed was he by the neighbouring peoples, that it is written that each builder, girt about the loins with a sword, fought with one hand and repaired the wall with the other.[94]

Calculate therefore seventy "weeks" from this time until Christ the Prince, that is, /**307**/ 490 years of twelve lunar months, which make 475 solar years. Now the Persians ruled 116 years from the aforementioned twentieth year of King Artaxerxes until the death of Darius. After that the Macedonians ruled 300 years until the downfall of Cleopatra. Then the Romans held the monarchy 59 years until the seventeenth year of Tiberius Caesar. Together these make, as we said, 475 years, and contain 25 nineteen-year cycles: 19 times 25 is 475. Because seven embolismic [months] accrue in each [nineteen-year] cycle, multiply 25 by 7 and it makes 175, which are the embolismic months in 475 years. If you wish to know how many lunar years these make, divide 175 by 12: 12 times 14 is 168, and therefore it makes 14 years with seven months remaining.[95] Add these to the abovementioned 475, and together they make 489. Add the seven extra months, and the part of the 18th year of the Emperor Tiberius in which the Lord suffered, and you will find that from the prescribed time up to His Passion there are seven "diminished weeks", that is, 490 lunar years.

At the time of His baptism, when He was anointed as the Most Holy

91 Augustine, *Ep.* 199.7.19–20 (259–261).
92 Daniel 9.25.
93 Cf. Nehemiah 1.11–2.1. Cf. ch. 66, *s.a.* 3468 and 3529.
94 Cf. *ibid.* 4.17–18.
95 Cf. ch. 45, below.

One, the Holy *Spirit /308/ descending upon* Him *like a dove*,[96] not only were the 7 and 62 weeks completed, but a part of the seventieth week had begun. *And after threescore and two weeks shall Messiah be cut off, and there will be none among the people to refuse him*.[97] Christ was killed, not immediately after the 62 "weeks", but at the end of the seventieth "week". Therefore, as far as I can see, [the angel] separated this week from the others because there was more to be related concerning it. For in it Christ was crucified, and refused by the faithless people, not only during His Passion, but in fact continuously, from the time when he began to be preached by John.

Now what follows – *And the people that shall come with the prince shall destroy the city and the sanctuary, and the end thereof shall be devastation and after the end of the war a devastation shall be determined*[98] – does not pertain to the seventy weeks, for it was predicted that these weeks would extend up until the time of Christ. But Scripture, having foretold His coming and His Passion, shows what would happen afterwards to the people who refused to receive Him. It calls Titus "the prince who shall come", who in the fortieth year of the Lord's Passion, together with the Roman people, so destroyed the city and sanctuary that there did not remain one stone standing upon another.[99]

After this foretaste by anticipation, [the angel] goes back to explain what will happen in the week he left out. *And he shall confirm the covenant with many for one week*[100] – that is, in that final week when either John the Baptist or the Lord or the Apostles by their preaching shall convert many to the faith. *And in the midst of the week the sacrifice and the oblations shall cease*.[101] The middle of this week was the fifteenth year of Tiberius Caesar when, commencing with the baptism of Christ, /309/ purification by sacrifices gradually began to be disdained by the faithful. Likewise what follows – *and in the temple shall be the abomination of desolation and the desolation shall continue until the consummation and the end*[102] – refers to the time following this, and the truth of

---

96 Matthew 3.16, Mark 1.10, Luke 3.22, John 1.32.
97 Daniel 9.26.
98 *Ibid.*
99 Cf. Mark 13.2; Luke 21.6; Matthew 24.2.
100 Daniel 9.27.
101 *Ibid.*
102 Daniel 9.27.

this prophecy is attested both by the account of men of old and by the events of our own time.

Therefore, pondering the whole of the Prophet's testimony, we have explained it as much as our wit can devise, for we realized that this was unknown to many readers, and that they were much exercised by this special type of week. Those who think the Hebrews used this type of year are mistaken, for otherwise the whole sequence of the Old Testament would be disordered [*uacillat*], nor would the age of anything be understood to be as great as it is written, but would be reduced to the course of the Moon.[103] Indeed, we read that the ancient Greeks, thinking that the year was 354 days long, according to the course of the Moon, always intercalated 90 days, *distributed into three months* of 30 days, in the eighth year. These *90 days result when the quarter-day, plus the 11 days* of the epacts, *are multiplied by eight*.[104] But the Jews were never accustomed to insert a thirteenth lunar month (which we call embolismic) except in the second or third year, as the famous rule for calculating the fourteenth Moon of Easter plainly proves.[105]

It should be noted that [Julius] Africanus thinks that the sequence of weeks, which we, following Eusebius' chronicle, have brought down to the seventeenth or eighteenth year of Tiberius Caesar – the year in which we believe the Lord suffered – is complete in the fifteenth year of that emperor. /**310**/ Beginning where we do, he thinks that by the 15th year of the said emperor – the year in which *he* thinks that Christ suffered – there had been 115 years of Persian rule, 300 of Macedonian rule, and 60 of Roman. The careful reader should choose [the version] he thinks preferable.[106]

10. THE WEEK OF THE WORLD-AGES[107]

The eighth kind of week, unvarying and unique in that it does not come back again to its beginning, is composed of the unstable Ages of this

---

103 Cf. *Letter to Plegwin* 11.

104 Macrobius, *Saturnalia* 1.13.8–10 (62.23–63.10).

105 Cf. Paschasinus of Lilybaeum, *Epistola ad Leonem papam* 2, ed. Krusch, *Studien I*, 248–249.

106 Cf. Jerome, *In Danielem* 3.9.24, ed. F. Glorie, CCSL (1964):867.190–195. But note that Jerome's quotation from Africanus accords 230 years to the Persians; cf. Eusebius, *Demonstratio euangeliorum* 7.2 (trans. Jerome, PG 10.81).

107 On the main sources for Bede's conception of the Six World-Ages, see Commentary on ch. 66.

world, and follows in all respects the pattern of the first week. For on the first day, light was created, and in the First Age man was placed in the beauties of paradise. When light was divided from darkness, evening was made, and evil came about when the sons of God were separated from [their] seed. Not long after, when the giants were born, the whole earth was corrupted; at that point the Creator, regretting that He had made man, determined to destroy the world by a flood.

On the second day, the firmament was suspended in the midst of the waters; in the Second Age, the Ark was borne up in the midst of the waters, at once carried aloft on the fountain of the deep, and deluged by the cataracts of heaven. The evening of this day was when the sons of Adam, making their way from the east, conspired to build the tower of vainglory; they were punished by the division of languages, and scattered from one another.

On the third day, when the waters were gathered together, dry land appeared, adorned with woods and grasses; and in the Third Age, when the nations were rooted in the worship of demons, Abraham the patriarch, leaving his people and his homeland, was made fruitful in the seed of the saints. The evening came when /311/ the Hebrew people, beset by wicked men, acted against the will of God and demanded a king of their own, who, as soon as he was set upon the throne, butchered the priests of the Lord and the Prophets. Afterwards, he and all his clan perished by the sword of foreigners.

On the fourth day the heaven was adorned with lights; in the Fourth Age this people, renowned for its heavenly faith and glorying in the rule of David and Solomon, won fame throughout the whole world for the splendour of its most holy Temple. But it too had its evening: because sin increased, that realm was shattered by the Chaldeans, the Temple demolished and the whole nation carried off to Babylon.

On the fifth day the fishes and birds came forth from the water; the former remained in their native waves, while the latter flew over air and land. In the Fifth Age the people of Israel multiplied in Chaldea; some of them sought out Jerusalem, fledged in the plumage of heavenly yearnings, and others abode amongst the rivers of Babylon, lacking all powers of flight. The evening arrived when, with the Saviour's advent imminent, the Jewish people was made tributary to the Romans because of the magnitude of their wicked deeds, and moreover was oppressed by foreign-born kings.

On the sixth day the earth was filled with its living creatures, and the first man was made in the image of God; subsequently, woman was fashioned from a rib taken from his side while he slept. In the Sixth Age, as the Prophets foretold, the Son of God appeared, who would recreate man in the image of God.[108] As he slept upon the Cross, water and blood flowed from his side, from which he would consecrate the Church to himself. The evening of this Age, darker than all the others, will come in Antichrist's persecution.[109] /312/

On the seventh day, when His labours were ended, God rested; blessing that day, he bade it be called "sabbath", and we do not read that it had an evening. In the Seventh Age the souls of the righteous, when the excellent labours of their lives are finished, will rest forever in another life which will never be blemished by any sorrow, but rather will culminate in the greater glory of the Resurrection. For man, this Age began when the first martyr Abel, his body having been laid to rest in the earth, entered in spirit into the sabbath of perpetual rest. It will be complete when the saints, having received their bodies [back], shall possess a double measure in their land,[110] and everlasting joy shall be theirs. This is the Eighth Age,[111] concerning which the sixth Psalm was written.[112] I believe that in the Six Ages of this world we ought to pray for the Seventh or Eighth Age of the world. Because the just shall receive joy, but the wicked punishment, the Psalm concerning this Age begins, proceeds and ends in great fear: *O Lord, rebuke me not in thine anger*, etc.

11. THE MONTHS

The months [*menses*] take their name *from the measure* [*mensura*], by which each of them *is measured*.[113] But it is more correct [to say that they take their name] *from the Moon, which in the Greek language* is

---

108 Cf. John 19.34.

109 Cf. ch. 69, below.

110 Isaiah 61.7.

111 Cf. ch. 71, below.

112 Bede is referring here to the rubric of Psalm 6: *In finem in carminibus pro octava Psalmus David.* Cf. *In Gen.* 1.1.2 (32.995–33.1002); Augustine, *Enarrationes in Psalmos* 6.1–2, ed. E. Dekkers and J. Fraipont, CCSL 38 (1956):27.1–28.48. Cf. ch. 71, below, and Jones, "Some Introductory Remarks", 196–197.

113 As with the etymology of *tempus* in ch. 2, this etymology is not derived from Isidore, but is also found in the non-Bedan "Merovingian *computus* of 727", 5 (54). An untraced common source is likely.

called *"mene"*,[114] *for the Greeks call the months "mene"*. Jerome says that amongst the Hebrews as well, the Moon, which they call *iare*, gives its name to the month.[115] /313/ Hence Jesus son of Sirach, who undoubtedly wrote in Hebrew, says, in speaking of the Moon, that *the month is called after her name*.[116]

The ancients customarily calculated their months not by the course of the Sun, but by that of the Moon. Hence whenever Holy Scripture (whether [speaking of a time] under the Law or before the Law) indicates a day of the month on which something was said or done, it signifies nothing other than the age of the Moon. The Hebrews, to whom the oracles of God are entrusted, have never ceased to observe the months after the ancient custom of their fathers. They call the first month of the new fruits, which is dedicated to the rites of Passover, Nisan. Because of the fluctuating course of the Moon it sometimes falls in March, sometimes in April, and occasionally encroaches on some days of May. But it is properly assigned to April, and always either begins, ends or is totally included within it, provided that the rule we discussed above[117] is observed: when the 15th day of the Moon follows the equinox, it constitutes the first month of the new year, and when it comes before the equinox, it forms the last month of the preceding year, and so on in sequence. By a similar calculation, their second month, Iyyar, corresponds to May; their third, Sivan, to June; their fourth, Tammuz, to July; their fifth, Av, to August; their sixth, Elul, to September; their seventh, Tishri, to October, which they call the new year, because then the harvest is gathered and special festivities held; the eighth, Marheshvan, /314/ to November; the ninth, Kislev, to December; the tenth, Tevet, to January; the eleventh, Shevat, to February; and the twelfth, Adar, to March.[118] Because the lunar orbit comprises $29\frac{1}{2}$ days,

114 Isidore, *Etym.* 5.33.1.

115 *De divisionibus temporum* 14 (659C), including the quotation from Jerome, *In Amos* 2.4.7.8–9, ed. M. Adriaen, CCSL 76 (1969):264.275. Cf. *On Times* 5.

116 Ecclesiasticus 43.8.

117 In ch. 6. Since the full Moon of Easter cannot come before the equinox on March 21, even the earliest legitimate Paschal lunation will end in the month of April. The calendar month in which a lunation ends is the month to which that lunation is assigned.

118 Cf. Polemius Silvius, *Laterculus*, ed. Theodore Mommsen, *Corpus Inscriptionum Latinarum* 1, 2nd ed. (Berlin: Georg Remer, 1893):335–357. The *Laterculus* is an annotated Julian calendar; each month contains the names of the parallel Hebrew, Egyptian, Athenian and Greek months. For a cognate list of Hebrew month names, cf. Eucherius, *Instructionum ad Salonium libri duo* 2.7, ed. C. Wotke, CSEL 31 (1894):153.10–19.

these months alternate between 30 and 29 days, and in the second or third year they intercalate an extra month made up from the days of the 11 epacts from each year.

Hence I have some misgivings about how our predecessors calculated the day on which the Law was given – which is the third day of the third month – as being the fiftieth day after the slaying of the [Paschal] lamb. They count 17 days left over from the first month (because the first 13 had elapsed before Passover), 30 in the second month, and 3 in the third. Together these make 50 days, but it is agreed that two lunar months finish, not in 60, but in 59 days. Therefore if, /315/ having calculated the Paschal month at 30 days, one retains the 17 days of its course which follow Passover, the second month ought to end not in 30, but in 29 days. By this calculation no more than 49 days will be found at the end of the specified time, unless one considers that the whole is construed from the part by synecdoche, a very common rule in Holy Scripture.[119]

But however these things were done or reckoned, it is obvious that the Hebrews were accustomed to observe their months according to the course of the Moon. When in the book of Genesis Noah is said to have gone into the Ark with his folk on the 17th day of the second month, and to have left it on the 27th day of the same month after the Flood, the only correct way of understanding this is that a whole solar year, that is, 365 days, is signified.[120] For the Moon, which this year, for instance,[121] is in its 17th day on the nones of May [7 May] will next year be 27 days old on the day before the nones of May. /316/

Note well that those who say that the month ought to be defined, or was *defined by the ancients*, as *the length of time in which the Moon traverses the zodiacal circle*, make a serious mistake.[122] As more pains-

---

119 By "predecessors" Bede is referring to Augustine, *Ep.* 55.30, CSEL 34 (1898):204.12–205.4; *Quaestiones in Heptateuchum* 70, ed. J. Fraipont, CCSL 33 (1958):102.1123–1130. Cf. ch. 8, above. The passage where the date of the giving of the Law is specified is Exodus 19.10–11. On the frequency of synecdoche in Scripture, see also ch. 5, above.

120 Since Noah's months are lunar, and he left the Ark one lunar year and 11 days after he entered it, he therefore spent 354 + 11 days in the Ark, or one solar year of 365 days. Cf. Bede, *In Gen.* 28.15–18 (126.1926–127.1946). Bede's source seems to be Ambrose, *De Noe et Arca* 17.60, ed. C. Schenkl, CSEL 32.1 (1897):456.1–4.

121 This passage must have been composed in 722, when the calculated Moon would have been 17 days old on May 7.

122 Isidore, *DNR* 4.1 (185.5–6), who does not actually endorse this error, but certainly perpetuated it; cf. Pliny, *HN* 2.6.44–45 (1.194–196).

taking inspection of nature has taught, the Moon plainly completes the zodiac in 27 days and 8 hours, but its proper course is 29 days and 12 hours,[123] setting aside the calculation of the "leap of the Moon".[124] Therefore it is more accurate to define a lunar month as the circuit and reintegration of the lunar light from new Moon to new Moon. The solar month is the passage of the Sun through a twelfth part of the zodiac, that is, the circle of the signs, which is accomplished in 30 days and $10\frac{1}{2}$ hours. It is longer than a lunar month by $22\frac{1}{2}$ hours, and from this the $11\frac{1}{4}$ days of the epacts annually accrue, for 12 times 22 makes 264, which it is easy to see are the number of hours in 11 days, because 11 times 24 makes the same total. Then, 12 times $\frac{1}{2}$ makes 6, which is the number of hours in a quarter-day in each year. Now if the Moon [has] 12 months, which (as we said) are, when cumulated, /317/ shorter by $11\frac{1}{4}$ days than [12] solar months, nonetheless it illuminates the path of the zodiac 13 times in the course of these [12 solar] months.[125]

But in reckoning both months, custom or authority or at least convenience of calculation prevails over nature. For not only are lunar months reckoned as 30 and 29 days as the necessity of calculation demands, but also most people intercalate the extra Moon, which according to the logic of nature ought to be intercalated at the end of the year, in any place they please. And what is worse, the computists are in such disagreement with each other that frequently on one and the same day one of them will say that the Moon should be reckoned as in its 14th day, and another, that it is in its 15th, and another, in its 16th, /318/ and each nation measures the solar months throughout the year by a very divergent rule and in a different order. Therefore, because the

123 The siderial lunar month of approximately $27\frac{1}{3}$ days is the length of time necessary for the Moon to depart from and return to the same point in relation to the backdrop of the constellations. The synodic month (what Bede calls its "proper course") is the period from new Moon to new Moon, $29\frac{1}{2}$ days. This is longer than the siderial month, because by the time the Moon returns to the point where it was last in conjuction with the Sun, the Sun will have moved into the next sign of the zodiac. It will take the Moon about two days to catch up with the Sun and achieve conjunction again.

124 On the *saltus*, see below, ch. 42.

125 12 synodic months of $29\frac{1}{2}$ days is less than a solar year by 11 days, plus the $\frac{1}{4}$ day for the leap-year increment; 12 siderial months of $27\frac{1}{3}$ days is $327\frac{2}{3}$ days, or more than 37 days shorter than the solar year. That means that the Moon can go once more around the zodiac before the end of the solar year, for a total of 13 zodiacal revolutions, with $7\frac{2}{3}$ days left over. Cf. chs. 16–18 below, and Commentary. Bede's sources are Isidore, *DNR* 1.5; Dionysius Exiguus, *Epistola ad Petronium*, ed. Krusch, *Studien II*, 65.28–66.4.

course of the Moon is swift, and lest an error in calculation be occasioned by that swiftness, the Egyptians were the first to begin to calculate their months according to the course of the Sun, whose slower motion could more easily be apprehended.[126] Starting in the autumn, they set up these months to comprise 30 days each. The first month, Thoth, began on the 4th kalends of September [29 August]; the second, Phaophi, on the 4th kalends of October [28 September]; the third, Hathyr, on the 5th kalends of November [29 October]; the fourth, Choiac, on the 5th kalends of December [27 November]; the fifth, Tybi, on the 6th kalends of January [27 December]; the sixth, Mecheir, on the 7th kalends of February [26 January]; the seventh, Phamenoth, on the 5th kalends of March [25 February]; the eighth, Pharmouthi, on the 6th kalends of April [27 March]; the ninth, Pachons, on the 6th kalends of May [26 April]; the tenth, Payni, on the 7th kalends of June [26 May]; the eleventh, Epieph, on the 7th kalends of July [25 June], and the twelfth, Mesore, on the 8th kalends of August [25 July].[127] This [last month] ends on the 10th kalends of September [23 August], and they call the remaining five days *epagomena* – "intercalated" or "added". Every fourth year, they add a sixth day to these [five], made up from the quarter-days. Hence it happens that their first year following the leap year /**319**/ begins on the 3rd kalends of September [30 August], but the other years on the 4th kalends of that month [29 August]. Their leap years end on the 4th kalends [29 August], and the other years on the 5th [28 August].[128] Now, the discrepancies which we say arise in leap years from the time of the intercalated day, in computing the age of the Moon or in the festivities of those days, cannot be brought into harmony with the course of our year until the time when we make the intercalation, which is on the 6th kalends of March [24 February]; rather, the same calendar date on the same day of the Moon, for example the seventh, which we calculate as falling on a Monday, they calculate as falling on a Tuesday, and so on for the rest.[129]

---

126 Isidore, *Etym.* 5.33.2.

127 Cf. Polemius Silvius, *Laterculus.*

128 Cf. Isidore, *DNR* 4.7 (191); Macrobius, *Saturnalia* 1.15.1 (69.12–32).

129 I.e. the Egyptian and Julian calendars will be out of phase by one day between September (when the Egyptians insert their leap day) and the end of February, when the Romans do so.

12. THE ROMAN MONTHS

In the debate of Chorus and Praetextatus,[130] we read of the reason why the Romans have months of varying length. *Romulus, when he had ordered the state of his realm with shrewd but rustic intelligence, took the beginning of the month /320/ from the day when it happened that the new Moon was seen. Because it did not always appear on the same day,[131] but for a variety of reasons was sometimes seen earlier and sometimes later, it happened that when it occurred later, more days were alloted to the previous month, and when it appeared earlier, fewer. But [Romulus] gave to each month a fixed number of days corresponding to the first instance,[132] [which it retained] in perpetuity. Thus it happened that some have thirty, and some thirty-one days.[133]*

[Romulus] dedicated *the first month to Mars*, whose son he wished to be taken for, because it was thought that in this month Juno gave birth to Mars in Phrygia.[134] *This month* was therefore deemed to be *the first of the year; a more likely [explanation]* is that the seventh, eighth, ninth and tenth months preserve from of old a name derived from a number [counted] from [March].[135] *The second month they called April, "opening" [aperilem] so to speak,* because then the heavens, the earth and *the sea* are opened up *to sailors,* farmers and those who observe the hours [*horoscopis*],[136] when *the clouds, hailstorms* and tempests of

130 This text is a modified excerpt from Macrobius, *Saturnalia* 1.12–15, found in Bede's "Sirmond" *computus* anthology, Oxford, Bodleian Library Bodley 309, fols. 101r–105v, and also in the related codices Vatican Reg. lat. 586, fols. 117v–125r, and Tours 334, fol. 41v: Jones, "The 'Lost' Sirmond Manuscript", 217, and *BOT* 348. Bede also used it in *De temporibus* 6. Citations here are made to the original text of Macrobius, but it should be noted that Bede knew only this excerpt, and was unaware of the identity of the author.

131 I.e. at regular intervals.

132 I.e. corresponding to the lunations as they happen to have occurred in the first year of his lunar calendar.

133 Macrobius, *Saturnalia* 1.15.5–6 (70.8–18).

134 Cf. *ibid.* 1.12.5 (54.20–22). This and subsequent explanations for the Roman month names are also found in Isidore, *DNR* 4.2–4 (187.7–189.31), *Etym.* 5.32.3–11, and Augustine, *Contra Faustum* 18.5, ed. J. Zycha, CSEL 25.1 (1891):494.5–17.

135 Isidore, *DNR* 4.2 (187.8–10).

136 *Horoscopus* properly designates the point on the eastern horizon where a constellation rises, and by extension, a horoscope in the astrological sense. This term is not found in Macrobius or Isidore (it does not even appear as an interpolation in the copy of the *Chorus and Praetextatus* found in the "Sirmond manuscript") but it seems that Bede – or his source – thought that a *horoscopus* was a person, like the farmer or sailor, who would

winter are driven away. *Trees* and herbs *also begin to open up into bud, and every living thing to open in bringing forth its young.*[137] [Romulus] *established May*[138] *as the third* and June as the fourth month *in commemoration of the "elders" [maiorum] and "juniors" into which he had divided the populace, so that one part might protect the republic by arms, and the other by counsel. Others contend that Maia, the mother of Mercury, gave her name to May, and substantiate this largely from the fact that in this month all the merchants sacrifice to Maia,* /**321**/ *and likewise to Mercury.* The month of *June* is *either* named *after a segment of the populace, as we said, or as Cingius*[139] *thinks, was previously called "Iunonius" by the Latins. This appellation for the month lingered for a long time amongst the men of old, but from "Iunonius" it was called "Iunius" after some of the letters fell out. The shrine of Juno Moneta was dedicated on the first of June.* The *month of July* took the name *"Quintilis" because of its number*, which it preserved *even after two months were inserted before March. But afterwards* it was called *"Julius" in honour of the dictator Julius Caesar by a law brought in by the consul Mark Antony, son of Mark, because Julius was born in that month, on the fourth ides of Quintilis.* The month of *August was formerly called "Sextilis" until this honour was bestowed upon Augustus by decree of the Senate*, because on the very first day of that month he defeated Antony and Cleopatra and established his rule over the Roman people. The *months of September*, October, November and December *retain their original titles*, signifying by their name the number of the month from spring, that is, from March, or else because the rains[140] arrive in these [months].

---

benefit from a clear sky. If Bede interpreted the word to mean "someone who investigates the hours", he could simply be suggesting that the heavens are clear enough in April to permit the telling of the hours by sundial during the day, and by the stars at night.

137 Cf. Macrobius, *Saturnalia* 1.12.14 (56.11–20).

138 The explanations of the month-names from "[Romulus] established May ..." up to "The former have their nones on the fifth day of the month, the latter on the seventh", are quasi-verbatim (with gaps) from Macrobius, *Saturnalia* 1.12.16–37 (56.24–61.3). The sentence about the months September to December is derived from Isidore, *Etym.* 5.33.11.

139 A Roman antiquary; his works on the history of the Roman calendar, now lost, are known to us only through Macrobius' references.

140 *Imbres*, from *imber, -bris*, c.f. Sept*ember, -bris*, etc. This etymology is intruded into the Macrobius text from Isidore, *Etym.* 5.33.11. Cf. Cassiodorus, *Variae* 1.35, ed. A.J. Fridh, CCSL 96 (1973):40.11–13: *quod ab ipsis quoque mensibus datur intelligi, quando ex numero imbrium futurorum competenter nomina susceperunt* ("... which may be under-

This was the measure ordained by Romulus for the year: it established a year of ten months, having three hundred and four days. He arranged the months so that four of them – March, May, Quintilis and October – would have *31* days; the other *six would have 30 days.*[141] *The former have their nones on the fifth day of the month, the latter on the /322/ seventh. In the case of those which have their nones on the seventh, the kalends are counted back from the ides to the seventeenth day, while for those which have their nones on the fifth, the beginning of the kalends goes back to the eighteenth day.*[142]

But since this number [i.e. 31 or 30 days] corresponded neither to the course of the Sun nor to the reckoning of the Moon, it sometimes happened that cold occurred in the summer months of the year, and contrariwise, heat in the winter months. Whenever this happened, a number of days not ascribed to any month was added to that season of the year until the position of the heavens was seen to be appropriate for that particular month.[143] *But Numa, who succeeded [Romulus], added fifty days, so that the year was extended to 354 days, which he believed comprised twelve revolutions of the Moon. To these days added by himself, Numa also added another six subtracted from those months which had 30 days – one day from each month – and made 56 days. He divided them equally into two new months, and the first of the two he called January. He wished it to be the first of the year, the month of the two-faced god, as it were, looking backwards at the end of the bygone year and forward to the beginning of the one to come.*[144] Some think that January got its name because it is *the threshold and door [ianua] of the year.*[145] *The second he called after Februus,* that is, Pluto, *who was believed to rule over purificatory sacrifice. /323/ In that month, in which [Numa] ordained that justice be done to the gods of the underworld, the city was obliged to make purificatory sacrifices.*[146]

But the Christian religion altered this custom of purificatory sacrifice

---

stood from the very months themselves, when they fittingly derive their names from the rains which are in the offing"). Cf. Jones, *BOT* 348.

141 Macrobius, *Saturnalia* 1.12.38 (61.4–8).

142 *Ibid.* 1.12.4 (54.16–19).

143 *Ibid.* 1.12.39 (61.8–14).

144 *Ibid.* 1.13.1–3 (61.15–62.4).

145 Isidore, *Etym.* 5.33.3.

146 Cf. *Saturnalia* 1.13.3 (61.28–62.4); Isidore, *Etym.* 5.33.4.

for the better, when in that same month, on the feast of St Mary, the whole populace with the priests and ministers goes on procession through the churches and the city neighbourhoods, all singing devout hymns, and carrying in their hands burning candles given them by the bishop. As this good custom grew, it provided a model for the conduct of other feasts of the blessed Mother and perpetual Virgin as well, not in the five-year lustration of a worldly empire, but in the everlasting memory of the heavenly kingdom where, according to the parable of the wise virgins, all the elect shall go out to meet the Bridegroom, their King, with the lamps of their good deeds alight, and then shall enter into the heavenly city with Him.[147]

Shortly afterwards Numa, in honour of the odd number, added a day, which he gave to January, so that both in the year and in the individual months, the odd number is preserved; the exception is February, as if both diminution and an even number were fitting for the infernal regions.[148]
Thus the Romans, following this arrangement of Pompilius, calculated their year according to the course of the Moon like the Greeks, and so they were obliged to institute an intercalated month after the fashion /324/ of the Greeks. Now the Greeks, when they observed that they had carelessly ordained their year at 354 days – for it was evident from the course of the Sun, which completes its course through the zodiac in $365\frac{1}{4}$ days, that $11\frac{1}{4}$

---

147 Jones (*BOT* 348–349) claims that this very early description of the Feast of the Purification is original to Bede. A sermon ascribed, but probably falsely, to Ildefonsus, (*Sermo* 10, PL 96.277), replicates Bede's statement closely, but is possibly copied from *The Reckoning of Time*. Ps.-Eligius, *Homelia* 2 (PL 87.597–603) is also described as a post-Bedan text. But see I. Deug-Su, "La festa della purificazione in occidente (secoli IV–VIII)", *Studi medievali* ser. 3, 15 (1974):143–216, who accepts the Idelfonsus and Eligius texts as genuine, but credits Bede with the earliest description of the Candlemas procession; see esp. pp. 146–149, 147 n. 10 for parallels in Bede's sermons. and 168–182 for parallels in the sermon on the Octave of Christmas and *In Lucam*. Certainly, Bede is the first writer to suggest that the Purification was instituted to supplant a pagan festivity, probably Lupercalia. Even in the 7th century, this ancient fertility rite, along with the celebration of the New Year on 1 January, remained a stubbornly ineradicable vestige of the ancient Roman civic calendar: see Michel Meslin, *La Fête des kalendes de janvier dans l'empire romain*, Collection Latomus 115 (Brussels: Latomus, 1970):ch. 3. For the parable of the wise virgins, see Matthew 25.10.

148 The Romans considered even numbers unlucky: cf. Vergil, *Eclogues* 8.75, ed. Wendell Clausen (Oxford: Clarendon Press, 1994):23, where it is stated that the gods like odd numbers (*numero deus impare gaudet*). The Laon-Metz glossator furnishes an interesting Christian variation on this theme: God particularly likes odd numbers because He himself is "odd", having no equal: p. 323 *ad lineam* 84.

days were missing from their year – decided to intercalate by a fixed rule, so that every eighth year they intercalated ninety days, which they counted as three months of thirty days. The Romans took it into their heads to imitate this method, but in vain, for they forgot the one day which we mentioned above, added by themselves to the Greek figure in honour of the odd number. For that reason, the number and order could not coincide after eight years.[149] He who wishes to know what error arose and how it was overcome may find it in the aforementioned disputation of Chorus and Praetextatus, from which we have made these excerpts.

At last *Julius Caesar, imitating the Egyptians,* instituted a year *according to the measure of the Sun,* such as has been maintained to this day, by adding *ten days to the old observations, so that the total came to 365 days, /325/ the period of time in which the Sun lights up the zodiac. And lest the quarter-day be wanting, he ordained as well that in the fourth year, the priests, who were responsible for months and days, should intercalate one day in the same month and place where men of old had intercalated the month, that is, before the fifth day from the end of the month of February; and he ordered that it be called the bissextus.*[150] *Every intercalation was assigned to the month of February, for this was the last month of the year. They did this in imitation of the Greeks,*[151] *for they inserted the extra days in the last month of their year. On one point, in fact, they differed from the Greeks, for while they intercalated after the end of the last month, the Romans intercalated, not at the end of February, but after its 23rd day, that is, when the Terminalia was over. Then they added the remaining days of February, which were five, after the intercalation. They were following, I believe, the ancient custom of their religion so that March would follow directly upon February.*[152]

13. KALENDS, NONES AND IDES

In olden times, the responsibility for observing the first appearance of the new Moon and of announcing its sighting to the royal sacrificing-priest

---

149 Macrobius, *Saturnalia* 1.13.5–11 (62.6–63.15).

150 *Ibid.* 1.14.3 (66.7–14), 6 (66.26–67.7).

151 This statement is erroneous, but Bede came by it honestly. The text of the *Disputatio Chori et Praetextati* in the "Sirmond" group of MSS contains an interpolated sentence extracted from *Saturnalia* 1.13.14 (64.1–3), describing a Roman experiment with intercalation in imitation of the Greeks. Cf. Jones, *BOT* 349.

152 Macrobius, *Saturnalia* 1.13.15 (64.6–12).

was delegated to a minor priest. /326/ Then, after a sacrifice had been offered by the priest-king and the minor priest, the priest, summoning [kalata] the people of the Capitoline to the Curia Calabra, very near the house of Romulus, announced the number of days from kalends until nones. He proclaimed "quintanas" [i.e. that five days remained] by saying "kalo" five times, and "septimanas" [i.e. that seven days remained] by repeating it seven times. The word "kalo" is Greek and means "I call", so it seemed good to name the day which was the first of those "called out" the "kalends". Hence also the same name, "Calabra", was given both to the court where they were announced and to the assembly in this court to which all the people were summoned. Thus the minor priest announced the number of days until the nones by calling [kalando], for the populace who worked out in the fields had to gather in the town on the ninth day after the new Moon to receive from the priest-king occasion for a holiday, and to inquire of the priests what ought to be done during that month. Hence some people think they were called "nones" because it was the beginning of a new [noua] observance, or because nine [nouem] days were always counted from that day until the ides.

Now it pleased them *to call the day which divided the month the "ides", for in the Etruscan language "iduare" means "to divide". Hence a "uidua" [widow] is, as it were, truly "idua", that is, truly divided, or else "uidua" means "a uiro diuisa" [i.e. divided from a man]. Some prefer to think* that "ides" *comes from the Greek word* for "face", which they call *"idea", because on that day the Moon shows its full face.*[153] Notice that when we read of the kalends in Holy Scripture, we should take it simply to mean the appearance of the new Moon, as in the Book of Numbers: *For at the kalends, /327/* that is, at the beginning of the months, *you will offer holocausts to the Lord.* For as we mentioned above,[154] the Hebrews knew no other beginning of the month save the new Moon.[155]

14. THE GREEK MONTHS

Neither the Greeks nor the Egyptians, of whom we spoke above, observed the distinction of kalends, nones and ides in their months.

---

153 Macrobius, *Saturnalia* 1.15.9–13 (70.27–71.24), 16–17 (72.7–18).

154 Ch. 11.

155 "Notice that ... new Moon." Cf. Jerome, *Ep.* 121.10, CSEL 56.1 (44.15–16); the biblical quotation is from Numbers 28.11.

Instead, they counted forward in the sequence of days from the beginning of the month until its end, simply and without error. We discussed the year and the months of the Egyptians above. The Greeks, when all the convolutions of the aforementioned calculations had been adjusted and corrected according to the circumstances, determined that the circle of the revolving year should be fixed at twelve months (many of them starting their year on the kalends of December) and with the same number of days used by the Romans in calculating their months. However, they took no heed, as we said before, of the Romans' kalends, nones and ides, but counted each and every month from the first to the last day in regularly increasing number.

Amongst the Greeks, December was called Apellaios; /328/ January, Eudymios [i.e. Audnaios]; February, Peritios; March, Dystros; April, Xanthicos; May, Artemisios; June, Daisios; July, Panemos; August, Loios; September, Gorpiaios; October, Hyperbere-taios; November, Dios.[156] Their annual numeration,[157] recently transmitted to us from Rome,[158] shows in what order they observed the year or the months. And the canons ascribed to the Apostles teach the same thing as the ancient writings when they observe that the 12th day of the month Hyperberetaios is the 4th ides of October [12 October]. From this we can infer that the two months begin together, because they both have the same twelfth day.[159] The book which the holy father Anatolius wrote about Easter supports this, when it says: *Therefore in the first year, the beginning of the first month, which is the starting-point of the nineteen-year cycle, is on the 26th day of the Egyptian month Phamenoth, on the 22nd day of the Macedonian month Dystros, and according to the Romans, on the 11th kalends of April*

156 It should be noted that Bede incorrectly calls these months "Greek", though Anatolius (see n. 160 below) correctly identifies them as Macedonian. This is because Bede's source, Polemius Silvius, first lists the Athenian months, and then designates these month-names as those employed by "the other Greeks". Cf. Polemius Silvius, and Jones, "Polemius Silvius", 52.

157 *computus eorum annalis*: Jones (*BOT* 350) thinks that the word *annalis* designates a calendar with these month-names inscribed. The word *annalis* is certainly used elsewhere as a substantive with this meaning (e.g. ch. 23) but here it is plainly an adjective modifying *computus*.

158 Polemius Silvius seems to have been a recent acquisition by the library of Wearmouth-Jarrow. Bede did not have this source when he composed *On Times*.

159 "And the canons . . . twelfth day." Cf. Dionysius Exiguus, *Codex canonum ec-clesiasticarum*, "Regulae ecclesiasticae sanctorum apostolorum" 38 (PL 67.145D).

[22 March].[160] Now in pointing out that the 22nd day of the month
Dystros is the equivalent of the 22nd day of March, /329/ Anatolius
shows plainly that both months begin together.[161] Should anyone claim
that Anatolius did not write "11th kalends", but rather "8th kalends", he
would prove thereby that [March] was not the same as the Egyptian
month Phamenoth, whose 26th day is on the 11th, and not the 8th
kalends of April. Both of them – that is, those who read "8th kalends" in
Anatolius and those who read "11th kalends" – understand this day to
refer to the 26th day of the Egyptian month. According to what we have
stated above in our description of the Egyptian year, this date is established beyond doubt as falling on the 11th kalends of April.[162]

15. THE ENGLISH MONTHS

In olden time the English people – for it did not seem fitting to me that I
should speak of other nations' observance of the year and yet be silent
about my own nation's – calculated their months according to the
course of the Moon. /330/ Hence, after the manner of the Greeks and
the Romans, [the months] take their name from the Moon, for the
Moon is called *mona* and the month *monath*.

The first month, which the Latins call January, is Giuli; February is
called Solmonath; March Hrethmonath; April, Eosturmonath; May,
Thrimilchi; June, Litha; July, also Litha; August, Weodmonath;
September, Halegmonath; October, Winterfilleth; November, Blodmonath; December, Giuli, the same name by which January is called. They
began the year on the 8th kalends of January [25 December], when we
celebrate the birth of the Lord. That very night, which we hold so
sacred, they used to call by the heathen word *Modranecht*, that is,
"mother's night", because (we suspect) of the ceremonies they enacted
all that night.

Whenever it was a common year, they gave three lunar months to
each season. When an embolismic year occurred (that is, one of 13 lunar
months) they assigned the extra month to summer, so that three months

160 Anatolius 2, as cited in Rufinus' trans. of Eusebius, *Historia ecclesiastica*
7.32.14 (723.10–14); cf. Bede, *Letter to Wicthed* 2.
    161 Cf. Bede, *Letter to Wicthed* 9.
    162 Bede is referring to ch. 11, where he showed that the Egyptian month Phame-
noth began on the 5th kalends of March (24 February). As all Egyptian months are 30 days
long, 26 Phamenoth falls on 22 March.

together bore the name "Litha"; hence they called [the embolismic] year "Thrilithi". It had four summer months, with the usual three for the other seasons. But originally, they divided the year as a whole into two seasons, summer and winter, /331/ assigning the six months in which the days are longer than the nights to summer, and the other six to winter. Hence they called the month in which the winter season began "Winterfilleth", a name made up from "winter" and "full Moon", because winter began on the full Moon of that month.

Nor is it irrelevant if we take the trouble to translate the names of the other months. The months of Giuli derive their name from the day when the Sun turns back [and begins] to increase, because one of [these months] precedes [this day] and the other follows. Solmonath can be called "month of cakes", which they offered to their gods in that month. Hrethmonath is named for their goddess Hretha, to whom they sacrificed at this time. Eosturmonath has a name which is now translated "Paschal month", and which was once called after a goddess of theirs named Eostre, in whose honour feasts were celebrated in that month. Now they designate that Paschal season by her name, calling the joys of the new rite by the time-honoured name of the old observance. Thrimilchi was so called because in that month the cattle were milked three times a day; such, at one time, was the fertility of Britain or Germany, from whence the English nation came to Britain. Litha means "gentle" or "navigable", because in both these months the calm breezes are gentle, and they were wont to sail upon the smooth sea. Weodmonath means "month of tares", /332/ for they are very plentiful then. Halegmonath means "month of sacred rites". Winterfilleth can be called by the invented composite name "winter-full". Blodmonath is "month of immolations", for then the cattle which were to be slaughtered were consecrated to their gods. Good Jesu, thanks be to thee, who hast turned us away from these vanities and given us [grace] to offer to thee the sacrifice of praise.

16. THE SIGNS OF THE TWELVE MONTHS

Each of the months has its own sign in which it receives the Sun: *April [the sign] of Aries; May, Taurus; June, Gemini; July, Cancer; August, Leo; September, Virgo; October, Libra; November, Scorpio; December, Sagittarius; January, Capricorn; February, Aquarius; March, Pisces. As one of the ancients explained* in heroic *verse*: /333/

The Phrygian Ram looks back on the kalends of April.
May marvels at the horns of Agenor the Bull.
June sees the Laconian pair travel the heavens.
At the solstice July brings in the constellation of fiery Cancer.
Burning Leo scorches the month of August with flame.
O Virgo, September swells Bacchus under your star.
In the sowing season, October balances Libra.
The diving Scorpion bids wintry November come in.
The Archer brings his sign to a close in mid-December.
The tropic of Capricorn marks off the commencement of January.
The star of stout Aquarius stands in the midst of Numa's month.
The twin fish go forth in the season of Mars.[163]

He shows that what he states explicitly about December should be understood of the others in every case, namely, that each sign ends in the
middle of its month, and begins in the middle of the previous month. I
beseech those who are knowledgeable not to consider it an imposition
to instruct those who do not know concerning the precise position [of
the signs].

The gyre of the heavens, perfectly round at every point, is bound by
the line of the zodiacal circle, like the discrete settings of twelve gems
adjacent to each other on a sort of girdle wrapped around a very large
sphere.[164] /334/ They are of such great size that they cannot rise or set

163 Ausonius, *Ecloga* 16, ed. Sextus Prete (Leipzig: Teubner, 1978):108–109. Bede
derived the eclogue and its introduction from a text entitled *De causis quibus nomina acceperunt duodecim signa*, ed. by Jones in *BOD* 665–667. This edition omits the poem proper,
but it is included in Jones' earlier transcription in *Bedae pseudepigrapha* 103. This same text
was the primary source for *The Nature of Things* 17. Its presence in the "Bobbio *computus*"
(Milan Ambrosiana H 150 inf; PL 129.1324–1325) suggests an Irish origin. On this, and
other possible sources for this chapter, see Jones, *BOT* 351–2. It should be noted that Bede
has rearranged the poem to begin in April rather than January. This deliberate displacement
away from the civic calendar and towards the natural phenomenon of the equinox may
reflect the positive attraction of Easter for the computist, but it may also be in part inspired
by a long-standing Christian polemic against the persistent pagan associations of the
January New Year: see ch. 12, n. 147.

164 Bede's image of the zodiac signs as jewels is striking. He may have been thinking of the 12 jewels in the foundations of the New Jerusalem (*Apoc.* 2.19–20), which were
closely connected with the 12 gates of the city, facing the four cardinal directions (*ibid.*
21.13). These 12 gates are also described as "pearls" (*ibid.* 21.21). However, unlike later
writers, Bede did not, in his *Explanatio Apocalypsis,* compare the foundations or gates of
the New Jerusalem to the zodiac, and there is no indication that his immediate lapidary

or move from a position in less than two hours. Thirty parts are ascribed to each sign, because of the thirty days in which the Sun illuminates each of them. *The 10½ hours left over are not added in right away,*[165] because they do not make up a full measure of 24 hours. Nonetheless, since the full solar year is finished, not just in 360 days, but with an additional 5¼ days, it is evident that this [period of 10½ hours], multiplied by twelve and making up 5¼ days, has been added to the thirty parts.[166]

Now *the first sign of Aries begins* in that part of the heavens where the Sun stands *in the middle of March.*[167] It ends in that part which the Sun has traversed in the middle of April. Therefore it indicates the location of the vernal equinox, according to some,[168] in its eighth part, but according to *the Egyptians, who are skilled in calculation beyond all others,*[169] [the equinox] is more correctly [said to be] in the fourth part.[170] The second [sign], Taurus, starting from that part of the celestial circle where the Sun is borne about in mid-April, finishes in [the part] which [the Sun] occupies in mid-May. Gemini, the third [sign, starts] in the part of the zodiac where the Sun is in mid-May. The fourth [sign], Cancer, begins in that [part] which it illumines in the middle of June. Therefore, according to some, it receives the solstice in its eighth part, /335/ but in fact, if one follows a more careful line of research, it is a few parts earlier. From this point, I take a downward course towards the

---

sources would have suggested such a link: cf. Peter Kitson, "Lapidary Traditions in Anglo-Saxon England: Part I, the Background; the Old English Lapidary", *Anglo-Saxon England* 7 (1978):9–60; "Part II, Bede's *Explanatio Apocalypsis* and Related Works", *Anglo-Saxon England* 12 (1983):73–123.

165 *De causis quibus nomina acceperunt duodecim signa,* 667.60–61.

166 This is an interesting illustration of a basic principle of calendar construction, which is that one cannot calculate with less than a whole day. The Sun actually spends 30 days and 10½ hours in each zodiac sign – as is evident from the length of the tropical year – but if one is going to mark its entry into each sign on a calendar, one will have to choose one or another day, not a part of a day. In this chapter, Bede only states that the Sun enters each zodiac sign "in the middle" of the calendar month, but his main source, *De causis quibus nomina acceperunt duodecim signa,* gives the exact dates, and we have included these in our reconstruction of Bede's solar calendar in Appendix 1. The extra five days are absorbed by the additional day in each of the seven 31-day months, minus 2 for the two days' shortfall in February. The quarter-day, of course, is cumulated into the leap-year day.

167 *De causis quibus nomina acceperunt duodecim signa* 666.49–50.

168 E.g. Isidore, *Etym.* 5.34.

169 Dionysius Exiguus, *Ep. ad Petronium* 65.22.

170 Pliny, *HN* 2.17.81 (2.224).

bottom of the zodiac,[171] that is, the circle of the signs. The fifth [sign], Leo, [begins] in the part where the Sun is in the middle of July; the sixth, Virgo, from where it is in the middle of August. The seventh [sign], Libra, begins to rise from where [the Sun] orbits in the middle of September; hence according to common opinion it furnishes a place in its eighth part for the autumn equinox,[172] but to a more unclouded judgement it occurs rather before this. Then, as the zodiac tips towards the winter region, the eighth sign, Scorpio, begins from where [the Sun is] in the middle of October; the ninth, Sagittarius, from where it is in the middle of November. The tenth [sign], Capricorn, takes its start from where the Sun is in mid-December. Hence it consecrates a mansion for the winter solstice in its eighth part, as is commonly believed; but as Egypt, *the mother of the arts*,[173] teaches, it is a few days before this. Then, when the course of the zodiac returns to higher regions, the eleventh constellation, Aquarius, [starts] from the Sun's position in the middle of January. The twelfth [sign], Pisces, beginning in that same part of February, comes to an end in the middle of March, where Aries, following after, shows the beginning of its rise. As all the signs agree in the fashion in which they are measured and in the conjunction of their zones[174] – albeit not in their visible form – the poet has said of them:

> *thus the golden Sun holds sway over the sphere,*
> *divided into fixed sections, by the twelve stars.*[175] /**336**/

The zodiac touches the Milky Way in Sagittarius and Gemini.[176] Much

171 The "bottom" of the zodiac (*ad inferiora situ zodiaci*) comprises the signs through which the Sun will pass from summer solstice to winter solstice, that is, from the point at which its daily orbit will begin to travel southwards. It represents not only the "bottom" of the year, but also the fact that Sun will be heading towards its lowest point in the sky, closest to the horizon.

172 Isidore, *Etym.* 3.71.29; Pliny, *HN* 2.17.81 (2.224).

173 Macrobius, *Saturnalia* 1.15.1 (69.13–14). Cf. Proterius, *Ep. ad Leonem* 2 (271), who refers to the Egyptian church *quae mater huiuscemodi laboris extitit*.

174 *regionum tamen suarum coniunctione*: Jones thinks that this obscure phrase refers to the meeting-point of Pisces and Aries, and cites in this regard Hyginus 3.29. But Bede clearly states that "all the signs" (and not just Pisces and Aries) "agree" in this. What is meant is not exactly clear: one possibility is that Bede means to say that all the signs are of the same width.

175 Vergil, *Georgics* 1.231–232; cf. *Letter to Wicthed* 6.

176 Cf. *The Nature of Things* 18, which draws on the same source, namely Pliny, *HN* 18.281 (5.367).

can be said about this, but it can be done to better effect by someone speaking than through the written word.

Because we discussed above[177] the order and periodicity [*tempore*] of the planets which course through the zodiac, we will merely repeat here that the Sun illumines the circuit of the zodiac in 365 days and 6 hours, the Moon in 27 days and 8 hours. The Sun slips through each of the signs in 30 days and $10\frac{1}{2}$ hours, the Moon in 2 days and $6\frac{2}{3}$ [*bisse*] hours. Should you ask what *bisse* means, I shall now repeat briefly what I stated in the beginning of this little work about measures.[178] A *bisse* is smaller than a whole hour in the same proportion as 8 is smaller than 12, 20 than 30, 10 than 15. Take away a third part, and two parts remain every time; these two parts are called the *bisse*, and the third is called *triens*. Therefore they are mistaken who say that the Moon traverses as much of the heavens in thirty days as the Sun does in 365 days, since (as we stipulated earlier) the plain truth demonstrates that the Moon is held to complete as much of the course in $27\frac{1}{3}$ days as the Sun completes in $365\frac{1}{4}$ /**337**/ days, and that the Moon covers as much territory in one of its [synodic] months as the orbit of the Sun does in 13 of its months.[179]

17. THE COURSE OF THE MOON THROUGH THE SIGNS

Every day the Moon either retreats from the Sun by four *puncti* with respect to [its location on] the preceding day when it is waxing, or approaches the Sun by four *puncti*, when waning.[180] As we observed

177 Ch. 8.

178 Ch. 4.

179 It is Isidore who is in error: *DNR* 19.1 (247.5–8), and *Etym.* 5.36.3. But so also is Ambrose, *Hexaemeron* 4.5.24 (131.17–21); cf. ch. 18 below. The Moon returns to the point in the zodiac where it was new in $27\frac{1}{3}$ days, but must "catch up" with the Sun, which has meanwhile moved into the next zodiac sign, before conjunction can take place. Hence, as Bede puts it, the zodiacal distance covered by a synodic month is the equivalent of 13 solar months.

180 As Bede explained above, the Moon circles the zodiac at a much faster rate than does the Sun. At conjunction, Moon and Sun will occupy the same point on the zodiac, but with each day it waxes, the Moon will pull ahead of the Sun at the rate of about $\frac{4}{5}$ of a zodiac sign per day, until it is full – or in other words, until it is exactly opposite the Sun on the zodiacal band. Then, as it wanes, it continues along the zodiac, but now returning towards the Sun, again at the rate of about $\frac{4}{5}$ of a sign per day.

above,[181] *every sign has ten "puncti", that is, two hours,*[182] *for five "puncti" make one hour.*[183] So if you want to know what sign the Moon is in, take whatever [day of the] Moon /**338**/ you wish – the fifth, for example. Multiply it by 4,[184] which makes 20. Divide by 10[185] (10 times 2 is 20); therefore the fifth Moon is separated from the Sun by two signs. Then again, take the eighth [day of the] Moon. Multiply by 4 and that makes 32. Divide by 10. 3 times 10 yields 30, with 2 remaining. Therefore the eighth Moon is separated from the Sun by three signs and two *puncti*. You know that two *puncti* equal six *partes*, that is, the distance covered by the Sun in its journey through the zodiac in six days. A *punctus* has three *partes*, so each sign has 10 *puncti* or 30 *partes*. Once again, take the nineteenth Moon. Multiply by 4, which makes 76. Divide by 10: 7 times 10 yields 70 with 6 remaining. Therefore, on the journey which it began away from the Sun, the Moon has parted company with it by seven signs, plus one hour, which is half a sign and a *punctus*, that is, three *partes*. And lest you suspect a faulty argument, find the point which no one can doubt is diametrically opposite [the position of] the fifteenth Moon. Multiply 15 by four, which makes 60. Divide by 10: 60 divided by 10 is 6. Now the fifteenth Moon can always be seen six signs away from the Sun, that is, half of the celestial sphere away, whether you look ahead or behind.

Therefore you see the orb of the Moon to be completely full whenever it is opposite the Sun; /**339**/ you see it low [in the sky] when the Sun is high, and high when it is low.[186] For to be sure, when the Sun is in the summer circle [i.e. tropic of Cancer], the Moon, when it is full, is in the winter circle; when it is full and the Sun has dipped down into the winter

181 Ch. 3.

182 "every sign ... hours": *De causis quibus nomina acceperunt duodecim signa,* 667.64–65. What Bede means is that each sign represents $\frac{1}{12}$ of the daily revolution of the heavenly sphere about the Earth, or in other words, 2 hours.

183 *De causis quibus nomina acceperunt duodecim signa* 667.64–65. Note that Bede is speaking of "lunar *puncti*" here; ordinarily, an hour has four *puncti*, as he points out in ch. 3.

184 I.e. the number of *puncti* by which the Moon advances towards or retreats from the Sun each day.

185 I.e. the number of *puncti* in a zodiac sign.

186 "Therefore you see the orb ... when it is low." Cf. ch. 6. What Bede means is that when the Sun is farthest to the north (i.e. "high" in the sky, or closest to zenith) at the summer solstice in Cancer, the Moon at full will be in Capricorn, "low" in the sky to the south, and vice versa. See ch. 26 and Commentary.

circle, the longest night [of the year] sends [the Moon] forth to climb up the [summer] solstitial circle. And when [the Sun] keeps to one equinox, the Moon, when she is full, keeps to the other. And the distance by which the Sun has passed the equinox or solstice which it has most recently illumined obviously corresponds to the distance by which the Moon has passed the opposite solstice or equinox.[187]

18. MORE ON THE COURSE OF THE MOON: FOR THOSE WHO ARE IGNORANT OF THE SIGNS

Anyone who is ignorant of the [zodiac] signs, but who wishes nonetheless to discover the course of the Moon, should know that every day the Sun completes one part of the zodiac – for we should understand that the "parts" of the zodiac are nothing but the daily advances of the Sun in the heavens – and that every day the Moon completes thirteen parts of the same zodiac, that is, four *puncti* and one part. Because the Sun completes one part while [the Moon] completes thirteen, the result (as we taught earlier) is that [the Moon] in its daily progress distances itself from the Sun exactly four *puncti*, that is, twelve parts.[188]

Choose, then, where you wish to calculate the [position of] Moon – say, on the kalends of January [1 January]. The first [Moon], when a night and a day has passed, /**340**/ occupies that position in the sky which the Sun does on the 13th day of the same month. When it is the second [day of the] Moon, multiply 2 by 4, which is 8. Then, so that you may proceed from *puncti* to parts, multiply 8 by 3,[189] which makes 24. The Moon will occupy that part of the sky on the 4th nones of January [2 January] in which the Sun will on the twenty-fourth day thereafter. When it is the third [day of the Moon], multiply 3 by 4, which makes 12. Divide by 10: 10 times 1 is 10, with 2 *puncti* (that is, 6 parts) left over. Therefore the third Moon occupies that part of the sky which the Sun

187 "For to be sure ... solstice or equinox." Cf. Anatolius 2, as reported in Rufinus' trans. of Eusebius, *HE* 7.32.18 (725.8–14); and Victor of Capua, *De pascha*. On the relationship of Bede's text to that of Victor of Capua and Anatolius, see Jones, *BOT* 353.

188 It will be recalled from ch. 17 that every zodiac sign contains 10 *puncti*. Since it is also divided into 30 parts (what we would call "degrees"), this means that each *punctus* contains 3 parts. Therefore 4 *puncti* plus 1 part equals 13 parts. Therefore the Moon advances each day 13 parts along the zodiac. Since the Sun advances by one part, this means that the Moon's net gain on the Sun is 4 *puncti*.

189 There are 3 parts in 1 *punctus*.

will [occupy] when a whole month and six days have passed from the third nones of January [3 January],[190] that is, at the end of the sixth day following the 3rd nones of February [3 February]. When it is the fourth [day of the Moon], multiply by 4, which makes 16. Divide by 10: 10 times 1 is 10, with 6 *puncti* remaining, that is, 18 parts. Therefore the fourth Moon will occupy that part of the heavens which the Sun will when a month and eighteen days have passed after the 2nd nones of January [4 January]. When it is the fifth [day of the Moon], multiply 5 by 4, which makes 20. Divide by 10: 2 times 10 is 20. Therefore when two months are over, that is, on the nones of March [5 March] the Sun will arrive in the part of the sky which the Moon occupies when five days old. When the eighth [day of the Moon] is on the 6th ides of January [8 January], multiply 8 by 4, which is 32. Divide by 10: 3 times 10 is 30 with two *puncti*, that is six parts, remaining. On the 6th ides of January, the Moon eight days old will occupy that part of the sky which the Sun will arrive at three months and six days thence, that is, on the sixth day after the 6th ides of April [8 April]. When the nineteenth Moon falls on the 14th kalends of February [19 January], /**341**/ multiply by 4, which makes 76. Divide by 10: 7 times 10 is 70, with 6 [*puncti*] remaining. Multiply by 3 and they make 18 [parts]. On the 14th kalends of February, the nineteenth Moon circles around that part of the heavens where the Sun will be seven months and eighteen days thence, that is on the eighteenth day after the 14th kalends of September [19 August], or the 8th ides of September [6 September].

In case you are worried about a fallacious argument, apply [this formula] to the diameter of the year: rare is the person who does not know that the Moon is in this position when it is fifteen days old.[191] Multiply 15 by 4, which makes 60. Divide by 10: 6 times 10 is 60. When six months are over, that is, when half his journey through the circuit of the year is complete, the Sun will enter that part of the heavens in which the Moon is borne about when fifteen days old. When that takes place on the fifteenth day of January, without doubt the Sun will arrive at that

190 10 *puncti* equal 30 parts, or a whole zodiac sign. Therefore when the product exceeds 10, one must divide by 10 to find the number of signs (i.e. months) distant the corresponding position of the Sun will be. The remainder, converted into parts, is the number of days in addition to these months.

191 The full Moon is in opposition to the Sun, and is therefore at the exact opposite point of the zodiacal circle, i.e. where the Sun will be six months thence ("the diameter of the year").

part on the fifteenth day of July.[192] And so that we may come to a conclusion, take the thirtieth [day of the] Moon, which will then be on the 3rd kalends of February [30 January]. Multiply by 4, which makes 120. Divide by 10: 12 times 10 is 120. The Moon when thirty days old runs in that part of the heaven where the Sun will be visible when twelve months are over, that is, when he has passed through the whole circle of the year. This is the very same part where he abides now /**342**/ and receives the Moon in conjunction.[193]

So that these matters might be more evident to everyone, even those of slower understanding, we have taken the trouble of carefully noting down, for each age of the Moon, how far distant from the Sun it is. So the Moon one day old, when its age is complete, is distant from the Sun by a space of twelve days;[194] when two days old by a space of twenty-four days; when three days old by a space of one month and six days; when four days old by a space of one month and eighteen days; when five days old by a space of two months; when six days old by a space of two months and twelve days; when seven days old by a space of two months and twenty-four days; when eight days old by a space of three months and six days; when nine days old by a space of three months and eighteen days; when ten days old by a space of four months; when eleven days old by a space of four months and twelve days; when twelve days old by a space of four months and twenty-four days; when thirteen days old by a space of five months and six days; when fourteen days old by a space of five months and eighteen days; when fifteen days old by a space of six months; when sixteen days old by a space of six months and twelve days; when seventeen days old by a space of six months and twenty-four days; when eighteen days old by a space of seven months and six days; when nineteen days old by a space of seven months and eighteen days; when twenty days old by a space of eight months; when twenty-one days old by a space of eight months and twelve days; when twenty-two days old by a space of eight months and twenty-four days; when twenty-three days old by a space of nine months and six days;

192 As Jones observes, Bede's formula is very approximate, and works on the assumption that a Julian month is always 30 days long: see *BOT* 354.

193 The Moon when 30 days old is in conjunction with the Sun, that is, in exactly the same part of the zodiac. The Sun will be in that same part of the zodiac again after one solar year.

194 I.e. the 13 days of the Moon's progress, minus the 1 day of the Sun's progress in the same direction.

when twenty-four days old by a space of nine months and eighteen days; when twenty-five days old by a space of ten months; when twenty-six days old by a space of ten months and twelve days; when twenty-seven days old /**343**/ by a space of ten months and twenty-four days; when twenty-eight days old by a space of eleven months and six days; when twenty-nine days old by a space of eleven months and eighteen days; when thirty days old by a space of twelve months.

19. MORE ON THE SAME SUBJECT: FOR THOSE WHO DO NOT KNOW HOW TO CALCULATE

Should someone rather less skilled in calculation nonetheless be curious about the course of the Moon, we have also for his sake devised a formula [*argumentum*] adapted to the capacity of his intelligence, so that he might find what he seeks. Thus we have marked out each day of the twelve-month annual circuit [in the calendar][195] by means of alphabets, so that the first and second [alphabetical] sequences each comprise twenty-seven days, while the third [sequence] has one [day] more, namely [the day] which accrues from three repetitions of the eight superfluous hours. In order that there might be enough letters for the days we wish to indicate, we have not placed them against every day, /**344**/ but against every other day. Hence it is not necessary to take the alphabet beyond the letter O.

At the beginning of the present work, we have included the Table of Regulars [*Pagina regularum*], which contains nineteen alphabets like those described above, each beginning with a different letter, for all the years of the 19-year cycle, together with the twelve names of each of the months and signs, arranged in this sequence according to the number of days in which the Moon traverses the zodiac. Along its length, the table has twenty-seven rows, with the names of the signs noted on the left and the names of the months on the right, so that someone unfamiliar with the signs can find what he wants by noting the months. Across its width it has nineteen columns; the order of the 19-year cycle is shown by the number of years noted at the top.

So if you wish to know what sign or what part of the month the Moon is in on any given day of any year, open the [calendar-]codex, and note the letter prefixed to that day. Then, turning to the Table of Regulars in

195 See Appendix 1.

which the sequence of letters is laid out, and having found the year[196] from the inscription at the top, find the letter pertaining to that day which you seek. Then, looking backwards and forwards,[197] you will discover what region, month and sign the Moon is in from its position. Reader, we will give you something to serve as an example. Suppose you are looking for the location of the Moon on the kalends of April in the sixth year of the 19-year cycle. Open the [calendar-] codex and look up that particular date. You will find written in front of it the letter E. Turn to the Table of Regulars. Look at the sixth year and find the letter E by looking through its alphabet. Now run your eye along to the margins. You will discover that the end of Gemini /**345**/ is noted on one side and the beginning of the month of June on the other.[198] And reader, whether you are learned or unschooled, you will rejoice to have tracked down what you were looking for, right before your eyes. Moreover, during that whole year, on whatever days you find the letter E inscribed, you will know that the Moon, whether waxing or waning, is occupying that part of the sky. You cannot discover by this formula whether the Moon is waning /**346**/ or waxing, in conjunction with or in opposition to the Sun; but if you want to know this, refer to the formula of the Egyptians, handed down by ancient observance.[199]

20. WHAT THE AGE OF THE MOON IS ON ANY GIVEN FIRST DAY OF THE MONTH

In the first year of the 19-year cycle, in which the epact[201] is zero, the Moon on the kalends of January is 9 days old; on the first of February, 10; on the first of March, 9; on the first of April, 10; on the first of May, 11; on the first of June, 12; on the first of July, 13; on the first of August, 14; on the first of September, 16; on the first of October, 16; on the first of November, 18; on the first of December, 18. Take these numbers as the "regulars" of each month. By adding the annual epacts to them, you

196 That is, the year of the 19-year cycle. Bede gives a formula for translating *annus Domini* into year of the 19-year cycle in ch. 47. Alternately, one could simply look it up in a Paschal table (Appendix 2).

197 I.e. to the left and to the right.

198 It is still, of course, April 1! What Bede means is that the Moon is in the zodiac sign in which the Sun will be found in June.

199 Cf. ch. 24.

201 I.e. the age of the Moon on March 22; see ch. 50 below.

PAGINA REGULARIS[200]

Sign	1	2	3	4	5	6	7	8	9	10	11	12	13	14	15	16	17	18	19	Month
Aries	A	k	e	o	i	d	n	h	c	m	g	b	l	f	A	k	e	o	i	April
Taurus	b	l	f	A	k	e	o	i	d	n	h	c	m	g	b	l	f	A	k	May
Taurus	c	m	g	b	l	f	A	k	e	o	i	d	n	h	c	m	g	b	l	
Gemini	d	n	h	c	m	g	b	l	f	A	k	e	o	i	d	n	h	c	m	June
Gemini	e	o	i	d	n	h	c	m	g	b	l	f	A	k	e	o	i	d	n	
Cancer	f	A	k	e	o	i	d	n	h	c	m	g	b	l	f	A	k	e	o	July
Cancer	g	b	l	f	A	k	e	o	i	d	n	h	c	m	g	b	l	f	A	
Cancer	h	c	m	g	b	l	f	A	k	e	o	i	d	n	h	c	m	g	b	
Leo	i	d	n	h	c	m	g	b	l	f	A	k	e	o	i	d	n	h	c	August
Leo	k	e	o	i	d	n	h	c	m	g	b	l	f	A	k	e	o	i	d	
Virgo	l	f	A	k	e	o	i	d	n	h	c	m	g	b	l	f	A	k	e	September
Virgo	m	g	b	l	f	A	k	e	o	i	d	n	h	c	m	g	b	l	f	
Libra	n	h	c	m	g	b	l	f	A	k	e	o	i	d	n	h	c	m	g	October
Libra	o	i	d	n	h	c	m	g	b	l	f	A	k	e	o	i	d	n	h	
Scorpio	A	k	e	o	i	d	n	h	c	m	g	b	l	f	A	k	e	o	i	November
Scorpio	b	l	f	A	k	e	o	i	d	n	h	c	m	g	b	l	f	A	k	
Scorpio	c	m	g	b	l	f	A	k	e	o	i	d	n	h	c	m	g	b	l	
Sagittarius	d	n	h	c	m	g	b	l	f	A	k	e	o	i	d	n	h	c	m	December
Sagittarius	e	o	i	d	n	h	c	m	g	b	l	f	A	k	e	o	i	d	n	
Capricorn	f	A	k	e	o	i	d	n	h	c	m	g	b	l	f	A	k	e	o	January
Capricorn	g	b	l	f	A	k	e	o	i	d	n	h	c	m	g	b	l	f	A	
Aquarius	h	c	m	g	b	l	f	A	k	e	o	i	d	n	h	c	m	g	b	February
Aquarius	i	d	n	h	c	m	g	b	l	f	A	k	e	o	i	d	n	h	c	
Pisces	k	e	o	i	d	n	h	c	m	g	b	l	f	A	k	e	o	i	d	March
Pisces	l	f	A	k	e	o	i	d	n	h	c	m	g	b	l	f	A	k	e	
Pisces	m	g	b	l	f	A	k	e	o	i	d	n	h	c	m	g	b	l	f	
Aries	n	h	c	m	g	b	l	f	A	k	e	o	i	d	n	h	c	m	g	

200 For explanation of the theory behind the *pagina regularis*, see Commentary. Since Bede himself did not explain why some zodiac signs have two, and some three spaces alloted to them, there are a number of variations, many seemingly arbitrary, in the way in which this table appears in the manuscripts. The version presented by Jones in his critical edition is, as he admits, an arbitrary reconstruction (*BOT* 354); but it is also erroneous, in that it does not position the "long signs" at three month intervals. The version presented here is based on the reconstruction by John Armstrong III, "The Old Welsh Computus Fragment and Bede's *Pagina regularis* [Part 1]", *Proceedings of the Harvard Celtic Colloquium* 2 (1982):219.

will discover without error what the age of the Moon is on any given kalends.

If you wish to know the age of the Moon for the kalends of January in the second year of the decennovennal cycle, take the regular [of January] – 9 – and add the epact – 11 –[202] and that makes 20: it is the twentieth Moon. If you want to know what Moon it is on the kalends of June in the third year, take the regular 12, add the epact for that year – 22 – and that makes 34. Subtract 30,[203] and 4 remain; on the kalends in question, the Moon is four days old.

Should anyone object that the order of either this or the preceding formula[204] is unsound at any point [*alicubi ordinem uacillare*], then let him teach a more accurate and handy formula for investigating questions of this kind, and we will gladly and gratefully accept it. /**347**/ We have given out the formula which we explained earlier on to several people to be copied, and have also recommended that it be prefixed to this present work.[205] Indeed, let the present formula that we set forth for finding the Moon on the kalends suffice for learning and teaching, at least at this point. For when the age of the Moon on the kalends is known, it will

202 The epact in the first year of the 19-year cycle, as Bede observes in the first sentence of this chapter, is zero: that means that the Moon on 22 March in year 1 of the cycle is a new Moon, in conjunction with the Sun. Because the lunar year is 354 days long, or 11 days shorter than the solar year, the Moon on 22 March in year 2 of the 19-year cycle will be 11 days old. Thus the epact of year 2 is 11.

203 No Moon can be more than 30 days old, so when the total exceeds this figure, 30 must be deducted.

204 *uel huius uel praecedentis argumenti*. Bede usually employs *argumentum* in the conventional way to mean "formula" (cf. Jones, *BOT* 357): his term for "table" is *pagina*. Nonetheless from what follows, it seems plain that a table is what Bede is referring to, one which he has included in the front matter of the manuscript of *The Reckoning of Time*.

205 The Laon-Metz glossator understood the *argumentum* referred to here as the *pagina regularis* from the preceding chapter (p. 346, *ad lineam* 18). Jones (*BOT* 355–356) however, claims it must have been a table which showed the ages of the Moon on the kalends of each month, i.e. a table corresponding to the formula explained in this chapter; his argument is supported by Armstrong, "The Old Welsh Computus Fragment", 193. But Bede explicitly refers to the "preceding formula" which he has "already explained" (*praecedens quod commemorauimus argumentum*), so he must be referring to something discussed before ch. 20. Curiously, Jones also claims that Bede is *refusing* to place this *argumentum* in front of the book (*BOT* 355), though the sense of the Latin suggests exactly the opposite: *et nonnullis ad transcribendum iam dedimus et in principiis huius nostri opusculi praefigendum esse censemus.* (347.19–21). Since in ch. 19, he expressly states that he has prefaced his book with the *pagina regularis*, I think that the reading of the Laon-Metz glossator is the correct one.

easily be seen what its age is on the other days of any month by reciting on one's fingers that month and its concurrents.[206] However, there are three years in the 19-year cycle when this formula cannot preserve the stability of its course, namely the eighth, the eleventh and the nineteenth. The cause of this discrepancy is the insertion, distributed throughout the year in various ways, of the embolismic [month].[207] Now in the eighth year, the Moon on the kalends of May, according to the logic of this formula, is calculated as being in its twenty-eighth day, but because of the embolism inserted into the month of March, it is actually in its twenty-seventh. Again, according to the formula the Moon ought to be in its thirtieth day on the kalends of July, but because of the addition of the day which the superfluity of the embolism introduced, it is the twenty-ninth. Again in the eleventh year, because the Moon of the embolismic [month] is kindled on the day before the nones of December [4 December], it makes the Moon on the kalends of March to be in its twenty-eighth day, while the logic of the formula teaches that the Moon then is the twenty-ninth. Again in the nineteenth year, because the Moon of the embolismic [month] begins on the 3rd nones of March [5 March], it dictates that the Moon on the kalends of May be calculated as in its twenty-eighth day, when according to the calculation of the formula /348/ it is tallied as the twenty-ninth.

Moreover, in this [nineteenth] year the computation of the "leap of the Moon", of which we shall say more later on,[208] vitiates the reliability of this formula. But if you start [to use] this formula at the month of September, after the manner of the Egyptians, whose year begins at that point, it is necessary that the Moon of July in that year have twenty-nine days and never more,[209] one day having been removed because of the "leap of the Moon". In consequence, the Moon of the kalends of August, calculated as the second according to the rule of the formula, will be assigned as the third. But if, following what we taught above, you prefer to begin the formula in January, according to the same sequence the seventh day of the Moon will fall on the kalends of December, which

206 Bede rather confusingly uses the term *concurrentes* here to mean "successive days of the month"; cf. Jones, *BOT* 357. It has a more common computistical meaning, which is explained in the next chapter.

207 Why the insertion of the embolismic month disrupts lunar reckoning is described in the Introduction, section 3.

208 See ch. 42.

209 The lunation of July is normally a "full" one of 30 days.

one would think ought to be the sixth according to the formula. This is because the Moon of the month of November loses a day and is considered complete with only twenty-nine [days] instead of the usual thirty. /349/ In truth, all this is easier to teach by oral explanation than in writing. However, do not overlook the fact that some people begin this formula from September, giving the regular 5 to September, 5 to October, 7 to November, 7 to December, and for the rest, as we give it above. Because of the authority of the Egyptians, they act in a thoroughly reasonable manner so as to imitate, in the calculation of the beginning of the year, those from whom the art of calculation took its origin. But to others it seems much more appropriate and efficient that all calculation, insofar as logical necessity presents no obstacle, should start from the beginning of their year, and continue onwards to the end of the year in a fixed and inviolable course: [this is the case] even among the Latins.

21. WHAT DAY OF THE WEEK IT IS ON THE KALENDS

They[210] provide a handy formula, similar to [the previous] one, for finding the [week]day of the kalends, so that by simply using another set of regulars, you may find the day of the week through the concurrents[211] by means of this [formula], just as you found [the age of the Moon] through the epacts in the [previous formula].

January has a [solar] regular of 2, February 5, March 5, April 1, May 3, June 6, July 1, August 4, September 7, October 2, November 5, December 7. These regulars indicate specifically /350/ what the weekday of the kalends is in the year to which seven concurrents are assigned. For the other years, you add the number of concurrents noted for that particular year[212] to the regulars of each month, and in this way you will find the day of the kalends without a mistake. Remember that when a leap year occurs, one of the concurrent days is dropped;

210 The Egyptians, presumably.

211 The concurrent is a number representing the weekday of 24 March. It is explained at greater length below, ch. 53. The concurrents function very much like the epacts – indeed, Bede also calls them "solar epacts" – in that they can be used with a monthly "regular" to determine the weekday of any date in the calendar. The "regular" is a number representing the constant difference between the weekday of 24 March, and the weekday of the first day of the month in question.

212 These can be found in the Paschal table, or by using the formula given in ch. 54.

however, you use the number which is going to be dropped in January and February, but on the first of March, you start to compute with that [number] contained in the solar cycle.[213]

So if you wish to find out, for example, the weekday of the kalends of January, you say "January 2",[214] add the concurrent days of the week which occur in the year for which you are calculating, for instance, 3: this makes 5. The kalends of January starts on the fifth weekday [Thursday]. Again, in a year which has 6 concurrents, take the 5 regulars for the month of March, add the 6 concurrents, and they make 11. Take away 7,[215] and 4 remain. The kalends of March is on the fourth weekday [Wednesday].

22. A FORMULA FOR ANY MOON OR WEEKDAY

There is an old formula devised for finding the Moon or the weekday not only for the kalends, but for any day between kalends. It is somewhat more difficult to learn, but it was passed down to us by the authority of our elders, /351/ and so ought to be handed on conscientiously to our juniors. If you wish to know how old the Moon is on this or that day, count the days from the beginning of January up to the day you want, and when you know this, add in the age of the Moon on the kalends of January.[216] Divide the total by 59, and if more than 30 remain, subtract 30.[217] What is left over is the day of the Moon you are seeking. Again, if you wish to know what the weekday is on such and such a day, calculate the days from the beginning of January up to the day in question, and when you have found out what [the total] is, add the weekday on which the kalends of January fell.[218] If it is a leap year, remember also to add

213 I.e. the "official" concurrent for that year, as found in the Paschal table. The concurrent valid in January and February will be one less. The insertion of the extra day in February pushes the remaining dates of the year ahead by one weekday.

214 The solar regular for January is 2.

215 In other words, should the sum of the concurrents and regulars exceed the number of days in a week, i.e. 7, these must be deducted, and the remainder will be the weekday of the kalends.

216 Using the formula given in ch. 20.

217 The synodic lunar month is $29\frac{1}{2}$ days, but as this is an awkward figure to compute with, the formula uses 59, the number of days in two synodic lunar months. Since no lunar month can be longer than 30 days, this figure must be deducted from the remainder if it exceeds 30.

218 This can be ascertained using the formula described in ch. 21.

the leap-year day after it has passed. Divide the total by seven and the remainder will show you what day of the week it is, in every case.[219]

This formula only works without difficulty if you are accustomed to reciting by heart the number of each of the months by kalends, nones and ides:

> January: 1 on the kalends, 5 on the nones, 13 on the ides;
> February: 32 on the kalends, 36 on the nones, 44 on the ides;
> March: 60 on the kalends, 66 on the nones, 74 on the ides;
> April: 91 on the kalends, 95 on the nones, 103 on the ides;
> May: 121 on the kalends, 127 on the nones, 135 on the ides;
> June: 152 on the kalends, 156 on the nones, 164 on the ides;
> July: 182 on the kalends, 188 on the nones, 196 on the ides;
> August: 213 on the kalends, 217 on the nones, 225 on the ides;
> September: 244 on the kalends, 248 on the nones, 256 on the ides;
> October: 274 on the kalends, 280 on the nones, 288 on the ides; /**352**/
> November: 305 on the kalends, 309 on the nones, 317 on the ides;
> December: 335 on the kalends, 339 on the nones, 347 on the ides.[220]

So for example, if you wish to know what the age of the Moon is on the kalends of May in a year in which the Moon is nine days old on the kalends of January, say "May: 121 on the kalends". Subtract the [day of the] kalends, and 120 remains; add 9, and that makes 129. Divide by 59: 59 times 2 is 118. Subtract 118 and 11 remain. The Moon is 11 days old on the kalends of May. If you want to know how old the Moon is on the 15th kalends of June [May 18], say "June: 152 on the kalends". Subtract the fifteen kalends of June and 137 remain. Add 9, which makes 146. Divide by 59: 59 times 2 is 118. Subtract 118 and 28 remain. The Moon on the 15th kalends of June is 28 days old. If you want to know what the Moon is on the 7th ides of December [December 7], say

---

219 "Again, if you wish to know ... in every case." Dionysius Exiguus, *Argumentum* 10, ed. Krusch, *Studien II*, 77.

220 The source of this table is ps.-Anatolius 10 (323–324). It also appears in later *computus* anthologies in a number of variant forms. One version begins in March rather than January (e.g. PL 90.706D, or the "Bobbio *computus*" in PL 129.1279D–1282A), and this form came to be labelled as *secundum Dionysium*, on the assumption that Dionysius, the champion of Alexandrian reckoning, would have devised his table according to the Alexandrian custom of beginning the computistical year in March. Ironically, Bede's version was baptised *secundum Victorium*, the logic being that Victorius championed Roman methods of *computus* against the Alexandrian, including beginning the year in January. Cf Jones, *BOT* 358.

"December: 347 on the ides". Subtract the seven ides, and 340 remain. Add 9, and it makes 349. Divide by 59: 59 times 5 is 295. Subtract 295 and 54 remain. Subtract 30, and 24 remain. On the day in question, it is the 24th [day of the] Moon.

In using this formula, it is helpful if the calculator commits the products of the fifty-nine-times table to memory:

> 59 times 6 is 354
> 59 times 5 is 295
> 59 times 4 is 236
> 59 times 3 is 177
> 59 times 2 is 118
> 59 times 1 is 59.[221]

By the same procedure, you find the weekday for any date you wish by adding the [week]day of the kalends of January. For example, if you want to know /353/ what the weekday will be on the 8th kalends of October [September 24] in a year in which the kalends of January is celebrated on the fifth day of the week [Thursday], say "October: 274 on the kalends". Subtract the eight [kalends of October] and 266 remain. Add the fifth weekday on which the kalends of January fell, and that makes 271. Divide by 7: 7 times 30 is 210, 7 times 8 is 56; 5 are left over. It is the fifth day of the week on the 8th kalends of October.

23. FOR THOSE WHO DO NOT KNOW HOW TO CALCULATE THE AGE OF THE MOON

Should anyone be so lazy or slow-witted as to wish to know the course of the Moon without any of the trouble of calculating, he must rely on the alphabets which he sees, distributed according to the course of the Moon, in the calendar (*in annali libello*), where there are three alphabets, comprising two lunar cycles, that is, fifty-nine days.[222] Whatever letter is assigned to the Moon for a given age, [the Moon] will always have this same [letter] at the same age throughout the entire year, unless perhaps (but this happens rarely) the computation of the embolisms changes it.[223]

---

221 Bede only gives the multiplication table up to 6, because that is the maximum number of times 59 can be divided into 365.

222 See Appendix 1.

223 As Bede explained in ch. 20, this occurs after the insertion of the embolismic month in the 8th, 11th and 19th years of the cycle.

	1	2	3	4	5	6	7	8	9	10	11	12	13	14	15	16	17	18	19
1	.m	.a	a	.i	s.	.r	.f	p.	.o	.c	m.	.l	u.	.t	.h	r.	.q	.e	o.
2	.n	.b	b	.k	t.	.s	.g	q.	.p	.d	n.	.m	.a	a	.i	s.	.r	.f	p.
3	.o	.c	c	.l	u.	.t	.h	r.	.q	.e	o.	.n	.b	b	.k	t.	.s	.g	q.
4	.p	.d	d	.m	.a	a	.i	s.	.r	.f	p.	.o	.c	c	.l	u.	.t	.h	r.
5	.q	.e	e	.n	.b	b	.k	t.	.s	.g	q.	.p	.d	d	.m	.a	a	.i	s.
6	.r	.f	f	.o	.c	c	.l	u.	.t	.h	r.	.q	.e	e	.n	.b	b	.k	t.
7	.s	.g	g	.p	.d	d	.m	.a	a	.i	s.	.r	.f	f	.o	.c	c	.l	u.
8	.t	.h	h	.q	.e	e	.n	.b	b	.k	t.	.s	.g	g	.p	.d	d	.m	.a
9	a	.i	i	.r	.f	f	.o	.c	c	.l	u.	.t	.h	h	.q	.e	e	.n	.b
10	b	.k	k	.s	.g	g	.p	.d	d	.m	.a	a	.i	i	.r	.f	f	.o	.c
11	c	.l	l	.t	.h	h	.q	.e	e	.n	.b	b	.k	k	.s	.g	g	.p	.d
12	d	.m	m	a	.i	i	.r	.f	f	.o	.c	c	.l	l	.t	.h	h	.q	.e
13	e	.n	n	b	.k	k	.s	.g	g	.p	.d	d	.m	m	a	.i	i	.r	.f
14	f	.o	o	c	.l	l	.t	.h	h	.q	.e	e	.n	n	b	.k	k	.s	.g
15	g	.p	p	d	.m	m	a	.i	i	.r	.f	f	.o	o	c	.l	l	.t	.h
16	h	.q	q	e	.n	n	b	.k	k	.s	.g	g	.p	p	d	.m	m	a	.i
17	i	.r	r	f	.o	o	c	.l	l	.t	.h	h	.q	q	e	.n	n	b	.k
18	k	.s	s	g	.p	p	d	.m	m	a	.i	i	.r	r	f	.o	o	c	.l
19	l	.t	t	h	.q	q	e	.n	n	b	.k	k	.s	s	g	.p	p	d	.m
20	m	a	u	i	.r	r	f	.o	o	c	.l	l	.t	t	h	.q	q	e	.n
21	n	b	a.	k	.s	s	g	.p	p	d	.m	m	a	u	i	.r	r	f	.o
22	o	c	b.	l	.t	t	h	.q	q	e	.n	n	b	a.	k	.s	s	g	.p
23	p	d	c.	m	a	u	i	.r	r	f	.o	o	c	b.	l	.t	t	h	.q
24	q	e	d.	n	b	a.	k	.s	s	g	.p	p	d	c.	m	a.	u	i	.r
25	r	f	e.	o	c	b.	l	.t	t	h	.q	q	e	d.	n	b.	a.	k	.s
26	s	g	f.	p	d	c.	m	a	u	i	.r	r	f	e.	o	c.	b.	l	.t
27	t	h	g.	q	e	d.	n	b	a.	k	.s	s	g	f.	p	d.	c.	m	a.
28	u	i	h.	r	f	e.	o	c	b.	l	.t	t	h	g.	q	e.	d.	n	b.
29	a.	k	i.	s	g	f.	p	d	c.	m	a.	u	i	h.	r	f.	e.	o	c.
30	b.	l	k.	t	h	g.	q	e	d.	n	b.	a.	k	i.	s	g.	f.	p	d.
1	c.	m	l.	u	i	h.	r	f	e.	o	c	b.	l	k.	t	h	g.	q	e
2	d.	n	m.	a.	k	i.	s	g	f.	p	d	c.	m	l.	u	i	h.	r	f
3	e.	o	n.	b.	l	k.	t	h	g.	q	e	d.	n	m.	a.	k	i.	s	g
4	f.	p	o.	c.	m	l.	u	i	h.	r	f	e.	o	n.	b.	l	k.	t	h
5	g.	q	p.	d.	n	m.	a.	k	i.	s	g	f.	p	o.	c.	m	l.	u	i
6	h.	r	q.	e.	o	n.	b.	l	k.	t	h	g.	q	p.	d.	n	m.	a.	k
7	i.	s	r.	f.	p	o.	c.	m	l.	u	i	h.	r	q.	e.	o	n.	b.	l
8	k.	t	s.	g.	q	p.	d.	n	m.	a.	k	i.	s	r.	f.	p	o.	c.	m
9	l.	u	t.	h.	r	q.	e.	o	n.	b.	l	k.	t	s.	g.	q	p.	d.	n
10	m.	a.	u.	i.	s	r.	f.	p	o.	c.	m	l.	u	t.	h.	r	q.	e.	o
11	n.	b.	.a	k.	t	s.	g.	q	p.	d.	n	m.	a.	u.	i.	s	r.	f.	p
12	o.	c.	.b	l.	u	t.	h.	r	q.	e.	o	n.	b.	.a	k.	t	s.	g.	q
13	p.	d.	.c	m.	a.	u.	i.	s	r.	f.	p	o.	c.	.b	l.	u	t.	h.	r
14	q.	e.	.d	n.	b.	.a	k.	t	s.	g.	q	p.	d.	.c	m.	a.	u.	i.	s
15	r.	f.	.e	o.	c.	.b	l.	u	t.	h.	r	q.	e.	.d	n.	b.	.a	k.	t
16	s.	g.	.f	p.	d.	.c	m.	a.	u.	i.	s	r.	f.	.e	o.	c.	.b	l.	u
17	t.	h.	.g	q.	e.	.d	n.	b.	.a	k.	t	s.	g.	.f	p.	d.	.c	m.	.a
18	u.	i.	.h	r.	f.	.e	o.	c.	.b	l.	u	t.	h.	.g	q.	e.	.d	n.	.b
19	.a	k.	.i	s.	g.	.f	p.	d.	.c	m.	a.	u.	i.	.h	r.	f.	.e	o.	.c
20	.b	l.	.k	t.	h.	.g	q.	e.	.d	n.	b.	.a	k.	.i	s.	g.	.f	p.	.d
21	.c	m.	.l	u.	i.	.h	r.	f.	.e	o.	c.	.b	l.	.k	t.	h.	.g	q.	.e
22	.d	n.	.m	.a	k.	.i	s.	g.	.f	p.	d.	.c	m.	.l	u.	i.	.h	r.	.f
23	.e	o.	.n	.b	l.	.k	t.	h.	.g	q.	e.	.d	n.	.m	.a	k.	.i	s.	.g
24	.f	p.	.o	.c	m.	.l	u.	i.	.h	r.	f.	.e	o.	.n	.b	l.	.k	t.	.h
25	.g	q.	.p	.d	n.	.m	.a	k.	.i	s.	g.	.f	p.	.o	.c	m.	.l	u.	.i
26	.h	r.	.q	.e	o.	.n	.b	l.	.k	t.	h.	.g	q.	.p	.d	n.	.m	.a	.k
27	.i	s.	.r	.f	p.	.o	.c	m.	.l	u.	i.	.h	r.	.q	.e	o.	.n	.b	.l
28	.k	t.	.s	.g	q.	.p	.d	n.	.m	.a	k.	.i	s.	.r	.f	p.	.o	.c	.m
29	.l	u.	.t	.h	r.	.q	.e	o.	.n	.b	l.	.k	t.	.s	.g	q.	.p	.d	n.

Table of *Litterae Punctatae* (ch. 23)

For example, in the third year of the 19-year cycle, a lunation which will have thirty days always begins with a plain letter A; /**354**/ the second [day of the Moon] is on B, the third on C, likewise plain, that is, not marked with any point. Thus each letter shows the age of its Moon according to its position. Again a lunation which will have twenty-nine days begins at L with a point following; the second [Moon] is always on the M and third on the N, marked with the same sign. Thus the Moon, recurring in sequence, encodes its age in each letter.[224] To make it easier to tell them apart, the ancients established that the first of the three alphabets /**355**/ should be unadorned, the second marked [with a point] afterwards and the third marked [with a point] in front, in order to distinguish them.[225]

24. THE NUMBER OF HOURS OF MOONLIGHT

The ancients also hand down a formula by which one can ascertain how many hours the Moon of any given age will shine. Since, they say, the first Moon shines for 4 *puncti*, this figure is added every day from the second Moon up until the full Moon, and then deducted in equal measure as the Moon wanes.[226] And so if you wish to know how many hours the second Moon shines, multiply 2 by 4, which makes 8 [*puncti*]. Divide by 5, because 5 *puncti* make one hour;[227] 5 times 1 is 5, and 3 remain. Therefore the second Moon shines for one hour and three *puncti*. Again, multiply 3 by 4, which makes 12. Divide by 5: 2 times 5 is

224 Literally, "changes its age back into each letter" (*suam cuique literae restituit aetatem*): 354.16–17.

225 Bede expects this system of lunar letters to function with a table, such as the one presented here. Indeed, his example based on the 3rd year of the 19-year cycle can only be worked with this table, and not merely from the calendar. A more skilled student could dispense with the table if he knew how to calculate the age of the Moon on the first of the month, but (a) this would make the lunar letters redundant, and (b) Bede specifies that this method is for those who cannot – or will not – do mathematics. For a discussion of this table, its history and variant versions, see Jones, *Bedae pseudepigrapha* 65–68, and Theodore von Sickel, "Die Lunarbuchstaben in der Kalendarien des Mittelalters", *Sitzungsberichte der Akademie der Wissenschaften zu Wien*, Phil.-hist. Kl. 38 (1861):159ff. It is not infrequently ascribed (falsely) to Jerome. All that can be said with certainty of its origins is that it predates Bede, for it is found in the Calendar of Willibrord (Paris, Bibliothèque nationale lat. 10837) written *ca.*703–717: see facsimile and transcription in H.A. Wilson, *The Calendar of St. Willibrord*, Henry Bradshaw Society 55 (London: Harrison, 1918). Cf. Jones, *BOT* 358.

226 Cf. ch. 17, n. 180.

227 On the difference between ordinary *puncti* and lunar *puncti*, see ch. 4 above.

10, with 2 left over. The third Moon shines for two hours and as many *puncti*. Again, when you come to the tenth Moon, multiply by 4, which makes 40. Divide by a fifth part: 5 times 8 is 40. The tenth Moon shines for eight hours.

Should you perhaps think this is faulty logic, take 15 (they say), and find out where the Moon shines all night long.[228] Multiply by 4, which makes 60. Divide by 5: 5 times 12 /356/ is 60. The fifteenth Moon traverses twelve hours, that is, the whole night. Again, if you wish to know how many hours the sixteenth Moon, the seventeenth, and the rest from then onwards shine, determine how much less than thirty each one is, and devise from this a basis for calculation. For example, if you want to know how long the twenty-fifth Moon will shine, tell how much less this is than thirty; obviously, it will be five. Multiply by 4: 4 times 5 is 20. Divide by 5: 5 times 4 is 20. Therefore the twenty-fifth Moon sheds its radiance for four hours, just like the fifth Moon.

To be sure, this formula functions smoothly at the time of the equinox, being established according to this prescribed position; however, in the very long nights of winter, or again in the very short ones of summer, it is plain that the space of twelve [equal] hours far surpasses the one, and cannot fill up the other. By this calculation, we should believe that the Moon shines for twelve hours, lest perchance we think that we ought not to understand these to be equinoctial hours, and divide each and every night into twelve equal parts, according to its length or brevity, and call these "hours". /357/

25. WHEN AND WHY THE MOON APPEARS TO BE FACING UPWARDS, FACING DOWNWARDS OR STANDING UPRIGHT

Those who have attempted to investigate the upper air say that whenever a new Moon is seen with the crescent lying flat out, it portends a stormy month, and when it is upright, a fair one.[229] Natural reason [*naturalis ratio*] shows that this is far from the case. How is this so? Is it really cred-

228 A Moon 15 days old will be full. Being in opposition to the Sun, it will rise at sunset and set at dawn, thereby illuminating the entire night. Cf. ch. 6.

229 Bede's immediate source is not known, but Jones has assembled a number of likely classical and post-classical references, *BOT* 360. Isidore, *DNR* 38, relates a number of weather superstitions about the Moon, but not this particular one. The same is true of Vegetius, *Epitoma rei militaris* 4.40–41, ed. K. Lang (Leipzig: Teubner, 1869):158.13–160.17.

ible that the position of the Moon, which remains fixed in the ether, could be altered under the influence of a change in the winds or clouds which lie beneath it, and that it should lift up its horns any higher than nature [*naturae ordo*] dictates, as if it dreaded bad weather to come, particularly when such a blast of wayward wind would not occur everywhere on earth? The rotation of the Moon's position ought to be constant with respect to its varying degree of separation from the Sun. As St Augustine says in his exegesis of Psalm 10, they say that *[the Moon] does not have its own light, but is lit up by the Sun. But when it is [in conjunction] with [the Sun], it turns towards us the part which is not lit up, and so no light can be seen. But when it begins to draw away from [the Sun], it is lit up in the part which faces the earth, beginning (as it must) with a crescent, until the fifteenth Moon is in opposition to the Sun. At that point, [the Moon] rises when the Sun sets, so that anyone /358/ who is observing the setting Sun, if he turns to the east when he can no longer see [the Sun], may see the Moon rise. Then, when it begins to approach the Sun in the opposite direction, [the Moon] turns towards us that part which is not lit up, until it is reduced to a crescent. Eventually, it does not appear at all, because the part which is lit up is the upper part, the one facing towards the heavens, and the part which the Sun cannot shine upon is facing towards the Earth.*[230]

So when the Sun gradually ascends from the southern clime to the northern regions, and daylight increases, it is necessary that the Moon which appears at that time[231] should precede the Sun towards the northern signs at a more leisurely gait. Therefore the new Moon which is seen about to set after sunset is situated, without a doubt, not beside the Sun, but above the Sun, by which it is lit up from below. [The Moon's] horns extend virtually parallel, and it seems to move along flat on its back, like a ship.[232] But after the summer solstice, when the Sun's

---

230 Augustine, *Enarrationes in Psalmos* 10.3 (76.30–42), where this is one of a number of possible explanations advanced; cf. Augustine, *Ep.* 55.6–7, CSEL 34 (175.15–177.16), where he is rather more committed to the explanation proposed here.

231 I.e. the crescent Moon on the first day of the lunation.

232 This is a rather convoluted way of saying that in the northern hemisphere, the crescent Moon at sunset at the spring equinox will be to the north of the celestial equator. Since the equator slants to the south, this means that the Moon will seem to be directly above the setting Sun. The Moon's underside, that is, the side directly opposite the horizon, will be illuminated from below by the setting Sun. The crescent will look like a boat, with the two horns pointing up. For illustration, see Commentary.

course turns back downwards towards the south, the Moon born in these months must likewise make haste towards the lower regions [of the sky]. Hence it happens that when [the Moon] is about to set in the southern regions, where the Sun has just set, without doubt it will not still appear above [the Sun], but beside it, positioned towards the south, when it first appears after sunset. Thus [the Moon's] northern flanks, facing the Sun, will seem to advance in an erect position, for with its horns turned away from the Sun, /359/ the Moon always presents its round part to [the Sun].[233]

By this logic it follows that the longer the day is, the higher the new Moon will be; and the shorter and more inclined towards the south the day is, the lower down the new Moon will be seen. Hence the opinion of the untutored came to the conclusion that when the new Moon appeared higher up and supine, it betokened whirlwinds of storm, and when erect and lower down towards the south, fair weather. For such, indeed, is the condition of the revolving year: the motion of the atmosphere is much more gentle in the six months during which the day decreases than during the other six. The same logic explains why the waning Moon in its morning rising appears now upright, and now supine. It also explains why she will often emerge during the day bent downwards [prona],[234] since the rays of the sun, of course, are striking her upper part. Therefore the alteration of the Moon, which is natural and fixed, cannot portend the condition of the month to come. But those who are inquisitive about this sort of thing will often pronounce on atmospheric conditions by the colour of [the Moon] or sun, or from the heavens themselves, the stars or the shifting shapes of clouds, or other omens.[235] Hence they claim that if the fourth moon is unblemished and its horns unblunted, it is a sign of fair weather for the rest of the month, and things of this sort.[236]

---

233 "for with its horns ... round part to him": cf. Augustine, *Ep.* 55.7 (176.20–177.7). Again, what Bede is saying is that at the autumn equinox, the crescent Moon at sunset is to the south of the celestial equator. Since in the northern hemisphere the equator slants to the south, this means that the crescent Moon will seem to be to the south and left of the setting Sun. Its northern side will be illuminated, forming a crescent standing upright on one of its horns. See Commentary.

234 I.e. with the horns pointing downwards, like an eyebrow.

235 Bede is alluding to Isidore, *DNR* 38 (299–303) and Vegetius (see n. 229 above).

236 Isidore, *DNR* 38.3 (301.27–28); Pliny, *HN* 2.48.128 (1.268).

26. WHY THE MOON, THOUGH SITUATED BENEATH THE SUN, SOMETIMES APPEARS TO BE ABOVE IT

It is no cause for marvel when we see that the Moon, as it passes through the southern signs, /360/ is much lower down and closer to the Earth than the Sun is when it tarries in its journey through these same regions. This is because the Moon travels within the frontier region separating this agitated atmosphere and the pure ether,[237] for in fact it is much lower not only than the Sun, but also than Venus and Mercury, the lowest of the stars. Hence many philosophers say that *the Sun's distance from the Moon is nineteen times the distance from the Moon to Earth. Pythagoras, a man of keen perception, deduced that it is 126,000 stadia from the Earth to the Moon, double that to the Sun, and three times that to the zodiac.*[238]

But one may with good reason ask, and wonder in puzzlement why the Moon, when it is travelling in the solstitial circle [i.e. in the Tropic of Cancer] seems to ride as far above the summer Sun as the shadows he makes are shorter. Hence it has been intimated by some that this progress of the Moon beyond the Sun in both regions of the sky, the north and the south, extends the whole width of the zodiac. Indeed in the south, the very depression of the Moon itself contributes to this. The zodiac itself is three hundred and sixty-five and one quarter degrees long about the circuit of the heavens, but twelve degrees wide. The Sun usually travels in the two middle degrees, but the Moon [travels] through them all.[239] /361/ When the Moon comes to the southern part of the zodiac, it occasionally appears lower than the winter Sun, not only because it is closer to the Earth, but because it is perhaps five or sometimes six degrees

237 Cf. Pliny, *HN* 2.7.48 (1.198).

238 *Ibid.* 2.19.83 (2.236).

239 Cf. *ibid.* 2.13.66–67 (1.212–214). Bede's statement that the Sun travels "in the two middle [degrees]" (*harum duas tantum medias sol . . . peruagere consueuit*) is, on the face of it, incorrect: the Sun's path never deviates from the centre of the ecliptic, because it defines the ecliptic. However, it seems to be an attempt to paraphrase a rather obscure statement in Pliny: *sol deinde medio fertur inter duas partes flexuoso draconum meatu* ("then the Sun travels unevenly in the middle of the zodiac between the two halves with a wavy serpentine course": Rackham's trans. 1.215). It is possible that Bede meant to say that the Sun travels in the middle, between the two parts of the zodiac, i.e. the six degrees to the north and the six degrees to the south of the ecliptic. Given his numerous statements elsewhere about the Sun's undeviating course, particularly in contrast to the Moon's, this construction would seem plausible, but the Latin does not support it.

from the line of the Sun's path through the middle [of the zodiac]. The farther it penetrates into the southern zone, the lower it appears from our point of view, who look upon it from the north. But when [the Moon] climbs into the solstitial circle, she frequently appears to be higher than the summer Sun, for although she is closer to the Earth than to the Sun, she sometimes travels beyond the limits of the Sun by five or even six degrees, and so from our point of view, standing on the Earth and looking at both stars from underneath, the further she moves towards the north, the more loftily she seems to seek the summits of heaven.

This will be demonstrated by the following example. At night-time, you go into a very large hall, or better, a church, immense in its length, breadth and height, and ablaze with countless lamps burning in honour of a martyr's feastday. Amongst these are two very large lamps of marvellous workmanship, hanging by chains from the ceiling, but the one which is nearer to you when you enter is also closer to the floor. However the hall is so vast, and the height of those distant lamps so great, that with your nightvision [*nocturno uisu*] you can make out the light and rays of flame more than you can the vessel itself which contains the fire. Now indeed, as you start to advance towards the lamps, looking straight at them,[240] and beyond them towards the ceiling or opposite walls, the lamp which is nearer appears higher to you. The closer you approach, the higher up will the one which is lower appear to you, until, by a more evident truth, you see where they are all positioned. So /362/ likewise we, situated beneath the two great luminaries of heaven, see them both at the meridian in such a way that the one which is lower, in rising further and further to the north, seems ever higher and higher, and as we train our eyes upon them and through them to the heavens, the one which, by obvious reason, is patently riding lower down appears to be higher than the other.

27. ON THE SIZE, OR ECLIPSE, OF THE SUN AND MOON

Pliny relates the following information[241] concerning the size or eclipse of the Sun and Moon in that most delightful book, the *Natural History*: *It is obvious that the Sun is obscured by the intervention of the Moon, and*

240 I.e. keeping them aligned, one behind the other.
241 Cf. Bede, *The Nature of Things* 22, "Bobbio *computus*" (PL 129.1329).

*the Moon by the interposition of the Earth, and each affects the other. The Moon takes away by its interposition the very same rays of the Sun which the Earth takes away from the Moon. When [the Moon] makes its transit, [the Earth] is darkened by sudden gloom; the star [i.e. the Moon], on the other hand, is quenched by [Earth's] shadow. The darkness is nothing but the Earth's shadow. Both eclipses occur, not every month, but at intervals; this is due to the incline of the zodiac, and the twists and turns, as they say, of the Moon's course, for the movement of these stars [Sun and Moon] is not always matched down to the finest divisions. This theory draws mortal minds into the heavens, and discloses to their contemplation from this height, as it were, the magnitude of the dimensions of the three largest things in nature. /***363***/ Were the Earth larger than the Moon, it would not be possible for the entire Sun to be erased by the Moon when it comes between [the Sun] and the Earth.*[242] *The Sun's magnitude is seen to be the largest of the three, nor would it be necessary [to demonstrate] that its size is something immense by the evidence of the eyes and the mind's induction, given that [the Sun] casts, from the mid-point of heaven, shadows of equal length from trees spread out for any number of miles.*[243] And a little later, he says: *It is certain that a solar eclipse can only take place when the Moon is new or in its first day, what they call "in conjunction"; [an eclipse] of the Moon on the other hand can only happen when the Moon is full, and always when it is as close as possible.*[244]*There are eclipses of both stars below the Earth*[245] *every year at the appointed days and hours, but even when they take place above the Earth they are not visible everywhere, sometimes because of cloud, and often because the globe of the Earth blocks out the dome of heaven. Two hundred years ago, this was discovered by the insight of Hipparchus.*[246]

But lest we seem to take up a whole chapter with the statements of even such a great pagan [as Pliny], let us also ask the doctors of the Church what they think of this. St Jerome, interpreting the passage in the Gospels which states that darkness covered the whole Earth at the

---

242 This is, of course, wrong, but no ancient author known to Bede other than Pliny discusses the size of the Moon relative to the Earth, though many compare the Moon to the Sun. Hence Bede never had occasion to question Pliny on this point.

243 Pliny, *HN* 2.7.47–2.8.50 (1.196–200).

244 I.e. to the path of the Sun.

245 I.e. in the opposite hemisphere.

246 *HN* 2.10.56–57 (1.204).

Passion, says: *Those who write against the Gospels suspect that the disciples, being ignorant men, were making sense of a solar eclipse, such as usually takes place in spring and autumn, by [ascribing it to] the Lord's Resurrection. But no one doubts that at the time of the Passover the Moon is at its fullest, though a solar eclipse never happens except at the new Moon.*[247] /**364**/

28. WHAT THE POWER OF THE MOON CAN DO

In the fourth book of the *Hexaemeron*, the blessed Bishop Ambrose describes in this wise what the power of the Moon can do: *Similar things concerning the nature of the Moon are in accord with what we have observed about her consort and brother [i.e. the Sun], since she assumes the same office as does her brother, so that she illuminates the darkness, makes seed to sprout and fruit to swell. Nonetheless, there are many things about her which are different from her brother. All day his heat dries out the moisture which the dew restores in the space of a brief night, for the Moon herself is said to abound in dew. Consequently, they say that when the night is clear and the Moon full, a greater quantity of dew bathes the lilies. Many who sleep in the open air claim that the more moonlight there is, the more dampness collects upon their head. Hence in the Canticle, Christ says to the Church, 'For my head is full of dew, and my hair with the drops of the night.'*[248] *Then again, [the Moon] waxes and wanes, so that it is small when it appears as a new Moon, but though it is reduced in size, it grows. There is a great mystery in this, for the elements respond in sympathy to her diminution, and as she grows, those things which were depleted increase in size, for example the humid brains of sea animals. For they say that as the Moon grows to fullness, oysters and many other such things are found to be larger. Those who have learned about this from their own experience claim that the same is true of the inner parts of trees.*[249] The art of the architects and their daily experience confirms these words of St Ambrose. They *teach that one should take particular note that the trees from which sailing ships are to be fashioned,* /**365**/ or any public works to be constructed, *be felled between the*

---

247 Jerome, *In Evangelium Matthei* 4.27.45 (273.1751–1756); cf. Augustine, *DCD* 3.15 (78.10–18), and his *Ep.* 199.10, ed. A. Goldbacher, CSEL 67 (1911):273.18–274.2.

248 Song of Solomon 5.2.

249 Ambrose, *Hexaemeron* 4.7.29 (134.13–135.6).

fifteenth day of the Moon and the twenty-second. For timber felled during these eight days will remain immune from rot, but wood felled on the other days will turn to dust within a year, consumed from within by the destructive action of worms. They also take care to fell timber *after the summer solstice, that is, from the month of July and August up to the kalends of January, for in these months, with the moisture dried up, the wood is dryer and therefore stronger.*[250] The *Persian selenite stone* demonstrates the effect of the Moon's power in a marvellous fashion. *Containing the image of the Moon,* it *shines* with a snowy *white* glow, and it is claimed that it *increases or diminishes day by day, following the course of that star.*[251] In the sixth book of the *Hexaemeron* Basil, the very reverend bishop of Caesarea, writes to corroborate these claims, saying: *For I think that in the generation of animals, and in everything else which the Earth produces, formation is conferred in no small degree by the changes of the Moon. When she grows old, their bodies seem weaker and empty, and when she waxes, they are whole and full; for in a hidden fashion she pours a certain humour, mixed with warmth, into their inward parts. Those who sleep under the open sky when the Moon is shining, prove that this is true when they wake up afterwards and find their head drenched with very heavy dew. Fresh meat, if it lies out under the Moon, is quickly corrupted by a damp rot. The same thing is shown by the brains of cattle, and even by the viscera of sea creatures, which are quite moist, and by the pith of trees also.* And a little later: *But the movement of the atmosphere is controlled by these very same variations, as is proved by the new Moon, which when it follows a long period of very fine weather stirs up heaps of cloud, and disturbances* /**366**/ *[of the air].* This

250 "The art of the architects . . . therefore stronger." Almost verbatim from Vegetius 4.35.6 (152.10–18). For other classical and patristic expressions of the same notion, see references in Jones, *BOT* 362.

251 Pliny, *HN* 37.67.181 (310); cf. Isidore, *Etym.* 16.4.6 and 10.7 and Augustine, *DCD* 21.5.1 (765.30–32). See below, ch. 43, where Bede, quoting the author of the *Epistola Cyrilli*, again invokes the selenite, this time as an experimental test for establishing the true Paschal computation. Bede's use of the phrase "it is claimed" here may hint at a slight scepticism on his part; likewise, the passage in ch. 43 is more a rhetorical posture than a claim to factuality (i.e. even if the rule established by the Nicene Council did not exist to justify the orthodox *computus*, the selenite stone would prove it right – but of course, the Nicene rule *does* exist . . .). It should be noted that Pliny also expresses some reservations about the selenite, though Isidore and Augustine do not. Moreover, neither Isidore nor Augustine mention the image of the Moon within the selenite stone, but Pliny does. This may indicate that Bede had access to all or part of Book 37 of *HN*.

also presages a receding current in channels, or even a reflux of the sand-
banks which are found hard by the ocean; they say that these, following
the phases of the Moon, are violently shifted to regions nearby. Currents
in both directions are also wont to change their usual gentle flow for the
remainder of this period; for when the Moon is new they cannot lie still,
but rage with fierce turmoil until the Moon appears again to bestow calm
upon the agitated waves.[252]

29. THE HARMONY OF THE MOON AND THE SEA

But more marvellous than anything else is the great fellowship that exists
between the ocean and the course of the Moon.[253] For at [the Moon's]
every rising and setting, [the ocean] sends forth the strength of his
ardour, *which the Greeks call "rheuma"*,[254] to cover the coasts far and
wide; and when it retreats, it lays them bare. The sweet streams of the
rivers it abundantly mingles together and covers over with salty waves.
When the Moon passes on, [the ocean] retreats and restores the rivers to
their natural sweetness and level, without delay. *It is as if [the ocean]*
were dragged forwards against its will *by certain exhalations of the*

---

252 Basil, *Hexaemeron* 6.10–11. The translation is not Eustathius', and Jones' apparatus suggests that it may have been by Bede. However, the consensus of most recent
scholarship on the knowledge of Greek in 8th-century England, and Bede's in particular, is
that this is not a very likely possibility. K.M. Lynch, in "The Venerable Bede's Knowledge
of Greek", *Traditio* 39 (1983):432–439, argues that Bede's knowledge of Greek increased
throughout his life. In his early commentary on the Acts of the Apostles, he was dependent
on Greek/Latin interlinear texts and frequently transcribed their errors. In the *Retractatio*,
he corrected these, and demonstrated other signs of deeper knowledge of biblical Greek.
But whether this amounted to the ability to translate texts independently is another
matter. Even the students of Theodore and Hadrian seem to have received no instruction
in Greek grammar, nor could they actually read Greek: see Michael Lapidge, "The Study of
Greek at the School of Canterbury in the Seventh Century", in *The Sacred Nectar of the*
Greeks: The Study of Greek in the West in the Early Middle Ages, ed. M.W. Herren, *King's*
College London Medieval Studies 2 (1992):164–169. See also A.C. Dionisotti, "Bede, Grammars and Greek", *Revue bénédictine* 92 (1982):111–141; J. Gribomont, "Saint Bède et ses
dictionnaires grecs", *Revue bénédictine* 89 (1979):271–280; Mary Catherine Bodden, "The
Preservation and Transmission of Greek in Early England", in *Sources of Anglo-Saxon*
Culture, ed. Paul E. Szarmach, Studies in Medieval Culture 20 (Kalamazoo: Medieval Institute Publications, 1986):53–63.
253 Cf. ps.-Isidore, *Liber de ordine creaturarum* 9.4 (148.28–30); Virgilius Maro
Grammaticus, *Epitoma* 4 (17.6–7).
254 Vegetius 4.42 (161.4); by contrast, Isidore uses the word *rheuma* only in its
medical sense of "morbid discharge": *Etym.* 4.7.11.

Moon, and when her power ceases, *it is poured back again into his proper measure.*[255]

For as we taught above,[256] the Moon /**367**/ rises and sets each day 4 *puncti* later than it rose or set the previous day; likewise, both ocean tides, be they by day or by night, morning or evening, never cease to come and go each day at a time which is later by almost the same interval. Now a *punctus* is one-fifth of an hour, and five *puncti* make an hour. Furthermore, because the Moon in two lunar months (that is, in 59 days) goes around the globe of the earth 57 times,[257] therefore the ocean tide during this same period of time surges up to its maximum twice this number of times, that is, 114, and sinks back again to its bed the same number of times. For in the course of 29 days, the Moon lights up the confines of Earth 28 times, and in the twelve hours which are added on to make up the fullness of the natural month, it circles half the globe of the Earth, so that, for example, /**368**/ the new Moon which emerged [from conjunction] last month above the Earth at noon, will this month meet up with the Sun to be kindled at midnight beneath the Earth. Through this length of time, the tides will come twice as often, and 57 times

> *the high seas swell,*
> *breaking their barriers and once again retreat unto themselves.*[258]

Because the Moon in half a month (that is, in 15 days and nights) circles about the Earth 14 times, and once besides around half the Earth, it happens that it is in the east at evening when it is full, while earlier it was in the west at evening when it was new. In this period of time, the sea ebbs and flows 29 times. Just as the Moon in 15 days (as we said) is flung back by its natural retardation towards the east from its position in the west at evening,[259] [so that while it] occupies the east today at morning,

---

255 Ambrose, *Hexaemeron* 4.7.30 (136.3–5). Cf. Isidore, *DNR* 40.1 (307.8–11).

256 Ch. 17.

257 See n. 260 below.

258 Vergil, *Georgics* 2.479–480, ed. R.A.B. Mynors (Oxford: Clarendon Press, 1990):li.

259 The Moon moves across the sky in two ways. Each day, it rises in the east and sets in the west: from the medieval perspective, this is because it is borne about by the daily east-west rotation of the outermost heavenly sphere. But the Moon also progresses eastwards around the zodiac once each month. Bede imagines the Moon "falling behind" the daily rotation of the outermost heavenly sphere, so that it rises each day a little later. The Moon that rose and set with the Sun at conjunction, rises and sets 12 hours later, and half the sky away, when it is full.

it will be in the west at morning 15 days from now, so also the ocean tide which now occurs at evening, will after 15 days be in the morning. On the other hand, the morning tide, impeded by this daily drag, will then rise at evening. And because the Moon in one year (that is, in 12 of its months, which makes 354 days) circles the globe of the Earth 12 fewer times (that is, 342 times), /**369**/ so the ocean tide within the same period washes up against the land and recoils 684 times.[260]

The sea reflects the course of the Moon not only by sharing its comings and goings, but also by a certain augmenting and diminishing of its size, so that the tide recurs not only at a later time than it did yesterday, but also to a greater or lesser extent. When the tides are increasing, they are called *malinae*, and when they are decreasing, *ledones*.[261] In alternating periods of seven or eight days, they divide up every month among themselves in their fourfold diversity of change. Sometimes by equal shares both fill up their course in $7\frac{1}{2}$ days; sometimes they will arrive earlier or later, or be more or less strong than usual when they are pushed onwards or forced back by the winds, or by the pressure of some other phenomenon or natural force, with the result that sometimes their order is upset, and the *malina* tide claims more in this month, and less in the next. Hence both directions will begin now in the evening, now in the morning. Indeed, when an evening tide occurs at the full or new Moon, it will be a *malina*, and for the next seven days this same *malina* tide will be greater and stronger than the morning tide. Similarly if a *malina* tide starts in the morning, the morning tide for days afterwards will cover /**370**/ the land with more sea. The evening tide, confined to the boundaries laid down by the morning tide, refrains from extending its course further, although in some months both tides grow at utterly different rates.

The more the stronger tide covers the shore and lands and fills up the

260 Bede imagines that the Moon every day "runs ahead" of the Sun by 4 *puncti*, or $\frac{4}{5}$ of an hour (48 minutes), so that it can meet up with the Sun at conjunction $29\frac{1}{2}$ days later (cf. ch. 17). Over 354 days, this 50-minute advance amounts to about 12 days ($50 \times 354 =$ 17,700 minutes $\div$ 60 = 295 hours $\div$ 24 = approximately 12 days). The "gain" is therefore about 1 day each month. Hence, as Bede said earlier, the Moon circles the Earth 28 times each 29 days, 57 times each 59 days, etc. From a modern heliocentric perspective, we would say that the Earth rotates 29 times, but the retardation of the Moon means that it seems to circle our planet one less time.

261 These terms are usually translated as "spring tide" and "neap tide" respectively. For further discussion, see Commentary.

rivers and straits, the more it is wont to leave these same coastlines empty and bare as it recedes. So let him who is capable, see if what Philip says is true or no: *There are those who claim and affirm that an enormous outpouring of the ocean takes place in all the streams of every region and land at one and the same time.*[262] But we who live at various places along the coastline of the British Sea know that where the tide begins to run in one place, it will start to ebb at another at the same time. Hence it appears to some that the wave, while retreating from one place, is coming back somewhere else; then leaving behind the territory where it was, it swiftly seeks again the region where it first began. Therefore at a given time a greater *malina* deserts these shores in order to be able all the more to flood other [shores] when it arrives there.

This can easily be grasped from the Moon's course. For example, positioned at about the winter sunrise or at the [summer] solstice sunset, the Moon at any given age, whether it be above the earth or beneath it, will pull at the tide here, but will repel it when she is positioned at winter sunset or summer solstice sunrise. But the Moon which signals the rise of the tide here, signals its retreat in other regions far from this quarter of the heavens. And there is more: those who live north of me on the same coastline usually receive and give back each tide sooner than I do, and those /371/ to the south much later. In every region, the Moon holds to whatever rule of fellowship with the sea it received in the beginning. Hence we have ascertained that the *malina* often begins about five days before the new or full Moon, and the *ledones* the same number of days before the half Moon. Near the two equinoxes, the tide rises higher than usual, and is much lower at the winter and summer solstices. The Moon always *exerts less force when she is in the north and at her highest point away from the Earth*[263] *than when she ventures into the south and treads closer*. Natural reason [*naturalis ratio*] convinces us that the tides flow according to the pattern of the Moon's cycle *of* 19 *years*; hence the course of the sea returns *to the beginning of its movements and to equal increments*.[264]

262 Philippus Presbyter, *Commentarii in librum Job* 38 (PL 26.752C).

263 By "Earth", Bede means "horizon"; cf. ch. 26.

264 Pliny, *HN* 2.99.215–216 (2.344). However, Pliny proposes an 8-year cycle for the tides. Bede accepted this model in *The Nature of Things* 39, but modifies it here.

30. EQUINOXES AND SOLSTICES

On the subject of the equinoxes and solstices, the opinion of many learned men, both worldly [philosophers] and Christians, is straightforward: the equinoxes are to be observed on the 8th kalends of April [25 March] and the 8th kalends of October [24 September], the solstices on the 8th kalends of July [24 June] and the 8th kalends of January [25 December].[265] Thus Pliny, who was both an orator and a philosopher, says in the second book of his /372/ *Natural History*: *Now the Sun itself has four turning-points: twice when the night is equal to the day, in spring and autumn, when he arrives at the mid-point of the Earth in the eighth degree of Aries or Libra, and twice when the proportions [of day and night] are reversed: in winter, in the eighth degree of Capricorn, when the days grow longer, and in summer, in the eighth degree of Cancer when the nights grow longer. Since an equal part of the universe is above and below the Earth at all times, it is the angle of the zodiac which is the cause of this inequality. The signs which are upright when they rise hold their light for a longer time; those which rise at an angle, pass on in a briefer period of time.*[266] Hippocrates the premier physician, writing to King Antigonus on how he should should watch over himself during the course of the year in order to prevent illness, says this: *Let us begin therefore with the solstice, that is, with the 8th kalends of January [25 December]. From this day forward up until the spring equinox, a period of 90 days, moisture [humor] increases in bodies. This season activates a man's phlegm, /* 373/ *so that people frequently catch catarrh, [suffer from] dripping of the uvula, pain in the side, weakness of vision, and ringing in the ears, and cannot smell anything. At such a season, take high-quality food, hot and seasoned with assafoetida, with pepper and mustard. Wash your head seldom. However, purge without ceasing. Indulge in wine and do not stint on sex for the first forty days. Ninety days follow from the above-mentioned day until the vernal equinox. From the above-mentioned 8th kalends of April [25 March], until the 8th ides of May [8 May], there*

---

265 Cf. Isidore, *DNR* 8 (205.1–5); *Etym.* 5.34; Bede himself held this opinion when he wrote *On Times* 7.

266 Pliny, *HN* 2.17.81 (1.224). Because of variations in the angle formed by the intersection of the ecliptic and the horizon (see Commentary on ch. 25), some zodiac signs will rise more or less "upright", i.e. perpendicular to the horizon, while others will rise at an angle to the horizon. The former will require less time to clear the horizon than the latter, and thus will "hold their light for a longer time".

are 45 days. In these days the sweet humours (that is, blood) increase in man. Eat food that is fragrant and very pungent. Again, from the 8th ides of May [8 May] until the 8th kalends of July [24 June] are 45 days. In these days the bitter bile (that is, red choler) increases. Eat sweet foods, indulge in wine, abstain from sex, and fast very little. The summer season begins on the 8th kalends of July [24 June]. At that time, the increase of red bile falls off, and black bile waxes; it is held that this occurs until the autumn equinox, that is, until the cold season. For ninety days, use all such foods as are sweet, fragrant and somewhat cold, and which act gently upon the stomach. From the autumn equinox, which is from the 8th kalends of October [24 September], until the 8th kalends of January [25 December], the bitterness of the black bile wanes and the density of the [phlegmatic] humour increases. Use all such foods as are hot and very pungent, abstain from sex and wash little. From the above-mentioned [date] /374/ until the Vergiliae [=Pleiades] set (that is, until the 6th ides of February [8 February]) there are 96 days, and then the Pleiades set. From that hour blood increases in men. It behoves one, therefore, to eat food that is very pleasant, and to indulge in wine and sex. There are 97 days in winter.[267]

This is what some of the pagans say; and very many of the Church's teachers recount things which are not dissimilar[268] to these about time, saying that our Lord was conceived and suffered on the 8th kalends of April [25 March], at the spring equinox, and that he was born at the winter solstice on the 8th kalends of January [25 December]. And again, that the Lord's blessed precursor and Baptist was conceived at the autumn equinox on the 8th kalends of October [24 September] and born at the summer solstice on the 8th kalends of July [24 June]. To this they add the explanation that it was fitting that the Creator of eternal light should be conceived and born along with the increase of temporal light, and that the herald of penance, who must decrease, should be engendered and born at a time when the light is diminishing.[269] But because, as we have learned in connection with the calculation of Easter, the judgment

---

267 Ps.-Hippocrates, *Ad Antigonum regem* 8–9, ed. E. Liechtenhan, J. Kollesch and D. Nickel, in *Marcelli de medicamentis liber*, ed. M. Niedermann, 2nd ed. Corpus medicorum latinorum 5 (Berlin: Akademie-Verlag, 1968):1.18–25.

268 For examples of such statements in the works of the Fathers, see Jones, *BOT* 366.

269 See *BOT* 366; for John the Baptist's statement "he must increase, but I must decrease", see John 3.30.

of all the men of the East (and especially of the Egyptians, who, it is agreed, were the most skilled in calculation) is in particular agreement that the spring equinox is on the 12th kalends of April [21 March], we think that the three other turning-points of the seasons ought to be observed a little before [the date] given in the popular treatises.[270]

So let us speak briefly about the spring equinox, which is the chief of the four annual changes we mentioned, as the Creation of the world indicates.[271] The rule of the Church's observance, confirmed at the Council of Nicaea, holds that Easter Day /375/ is to be sought between the 11th kalends of April [22 March] and the 7th kalends of May [25 April]. Again, the rule of catholic teaching commands that Easter is not to be celebrated before the passing of the spring equinox.[272] Therefore he who thinks that the 8th kalends of April [25 March] is the equinox is obliged either to declare that it is licit to celebrate Easter before the equinox, or to deny that it is licit to celebrate Easter before the 7th kalends of April [26 March], and also to confirm that the Pasch which our Lord kept with his disciples on the night before he suffered was either not on the 9th kalends of April [24 March] or was before the equinox.[273] For a doctrine not only of our own day but also of the Mosaic law decrees that the day of the Paschal feast cannot be celebrated before this equinox has passed. Anatolius declares: *It is plain that Philo and Josephus teach thus; thus did their elders Agathabolus, and Aristobolus of Panaeda his student, who was one of the seventy elders sent by the priests to King Ptolemy to translate the Hebrew books into the Greek tongue, and who answered many [questions] about the traditions of Moses which the king proposed and asked particularly. So these men, when they explained the questions about Exodus, said that the Pasch could not be sacrificed before the spring equinox had passed.*[274]

---

270 Bede is not interested in fixing the dates of the autumn equinox or the solstices, as they are of no computistical consequence. For a signal exception, see *Letter to Wicthed* 6.

271 See above, ch. 6.

272 Cf. Dionysius Exiguus, *Ep. ad Petronium* (65.13–27); *Prologus Cyrilli* 4 (339–340).

273 The traditional "historic" date of the Passion was 25 March (see above); therefore Christ's Last Supper could not have been a Passover meal (as the Synoptics indicate) if the equinox were on the 25th, since Passover must follow the equinox. Cf. *Letter to Wicthed* 12.

274 Anatolius 2 as recorded in Rufinus' trans. of Eusebius, *HE* 7.32.16–17 (725.1–8).

Hence it is necessary, in order to preserve the rule of truth, that we state clearly both that the Pasch is not to be sacrificed before the equinox and before the darkness has been overcome, and that this equinox is rightfully to be assigned to the 12th kalends of April [March 21], as we are instructed not only by the authority of the Fathers, but also by examining the sundial [*horologica consideratione*]. /376/ But the other three termini of the seasons are, by an analogous reasoning, to be observed the same number of days before the 8th kalends of the following [month].

31. THE VARYING LENGTH OF DAYS AND THE DIFFERENT POSITION OF THE SHADOWS

While the equinox falls on the same day, and that day is of the same length, everywhere in the world, the solstice and all the others days are of different lengths because of disparate climates, and the [disparate] shadows [cast by the Sun]. The books composed through the efforts of both gentiles and Christians, and competent observers hailing from both regions (that is, north and south) show this plainly. For St Ambrose, discussing the seasons and the path of the Sun in the fourth book of his *Hexaemeron*, says, amongst other things: *A shadow is less at midday than it is either at the beginning of the day or at the end, and this is true where we live, in the west. There are those in southern climes who have no shadow at noon on two days of the year; having the Sun directly overhead, they are illuminated all around on every side. Hence they are called "shadowless" [ascii] from the Greek [ἄσκιος]. Many claim that the Sun is so high in the sky that through the narrow mouth of a well, they can see the water glinting in the depths. In the south there are said to be people called "amfiscii"*[275] ["shadowed all about"] *because they cast a shadow on both sides.*[276] And a little later: *In this part of the world in which we live, there are people dwelling in the south who seem to cast a shadow towards the south. This is said to happen in high summer, when the Sun travels towards the north.*[277] Basil himself writes things which are similar to this in his commentary on Genesis.

But /377/ Pliny also, in a book which is non-religious, yet not to be

275 From the Greek ἀμφίσκιος.
276 Ambrose, *Hexaemeron* 4.5.23 (130.11–20).
277 *Ibid.* (131.4–7).

condemned, expounding these matters at greater length, writes as follows: *Sundials cannot be used everywhere because the Sun's shadows change every 300, or at most 500 stadia. Hence in Egypt at noon on the equinox, the shadow of the pin, which they call the gnomon, is a little more than half the length of the gnomon. In the city of Rome the shadow is $\frac{1}{19}$ shorter than the gnomon, in the town of Ancona it is longer by $\frac{1}{35}$, and in the region of Italy they call Venetia the shadow at the same hour is equal to the gnomon. Likewise they report that in the town of Syene, which is 5,000 stadia upstream from Alexandria, no shadow is cast at noon on the solstice, and that a well, constructed for the sake of this experiment, is illuminated from top to bottom. From this it would appear that the Sun is straight overhead in that region, and Onesicretus writes that this occurs at the same time in India above the river Hypasis. They say that in Berenice, the city of the Trogodytes, and 4,820 stadia away in the same region in the town of Ptolomais, which was built on the shores of the Red Sea for the first elephant hunts, this same thing occurs forty-five days before the solstice, /378/ and forty-five days after, and after those ninety days the shadows fall to the south. Again, in Meroe, an inhabited island in the Nile River and the capital of the Ethiopian people 5,000 stadia from Syene, the shadows disappear twice each year, when the Sun reaches the eighteenth degree of Taurus and fourteenth of Leo. In the midst of the Indian tribe of the Oretes there is a mountain called Maleus; near it, the shadows in summer fall to the south and those in the winter to the north; the north [celestial pole] appears there on only 15 nights. In the famous port of Patalia in India, the sun rises on the right and the shadows fall to the south. When Alexander stayed there, it was noticed that the north [celestial pole] could be seen only during the first part of the night. His guide Onesicretus wrote that in those parts of India where there are no shadows, the north [celestial pole] cannot be seen, and therefore these parts are called shadowless, and the hours are not reckoned there. And Eratosthenes relates that over the whole land of the Trogodytes the shadows fall in the wrong direction for 90 days out of the year. Thus it happens that, due to variation in the increment of daylight, the longest day in Meroe measures twelve equinoctial hours and eight parts of an hour, but in Alexandria fourteen hours, in Italy fifteen, and in Britain seventeen, where the bright nights of summer confirm what reason compels us to believe: /379/ that when the Sun is climbing closer to the topmost point of the world during the [summer] solstice, the lands which lie beneath [the Pole], because of the narrow circuit of light, have continuous day for six months, and then [continuous] night, when [the*

*Sun] has withdrawn towards the winter regions. Pytheas of Marseilles
writes that this happens on the island of Thule, which is six days by ship
from Britain.*[278]

In discussing these matters concerning the varying length of the solstitial days, Pliny leaves the length of the winter [solstitial] days in those
regions to be inferred. But he makes it equally clear what the length of
the night in both seasons is, because it is necessary that the days, be they
of whatever length, make up, together with the night, a space of 24
hours. But it should be noted that Solinus writes these things about
Thule in another vein: *In uttermost Thule, when the Sun passes at the
summer solstice through the star Cancer, there is no night, and likewise no
day at the winter solstice.*[279] Nor does Pliny contradict this in his seventh
book: *The uttermost place of all they call Thule, in which we have indicated
that there are no nights at the solstice when the Sun passes through the
sign of Cancer, and on the other hand no days at the winter [solstice].
Some are of the opinion that this happens for six months.*[280] /**380**/

32. WHY THE SAME DAYS ARE UNEQUAL IN LENGTH

The reason why the same [calendar] days are of unequal length is the
roundness of the Earth, for not without reason is it called "the orb of
the world" on the pages of Holy Scripture and of ordinary literature. It
is, in fact, a sphere set in the middle of the whole universe. It is not
merely circular like a shield [or] spread out like a wheel, but resembles
more a ball, being equally round in all directions, but not in a mass of
equal magnitude – although I would believe that the enormous distance
of mountains and valleys neither adds to it nor diminishes it any more
than a finger would a playing ball.[281]

278 Pliny, *HN* 2.74.182–2.77.187 (1.314–318). "Thus it happens that ... by ship
from Britain." Cf. Bede, *On Times* 7.

279 Solinus, *Collectanea rerum memorabilium* 22.9, ed. Th. Mommsen (Berlin:
Weidmann, 1895):101.11–102.2; cf. Isidore, *Etym.* 14.6.4.

280 Pliny, *HN* 4.16.104 (2.198). All the MSS agree that Bede refers to the *seventh*
book of Pliny, though the quotation comes from the fourth. On the possibility that Bede
used an anthology of Pliny excerpts, see Introduction, section 5. Perhaps the passage was
misidentified in this abridgement.

281 "The reason why ... playing ball." Cf. Pliny, *HN* 2.64.160 (2.294–296). The
seemingly ineradicable belief that medieval people thought the world to be flat, and considered it heretical to think otherwise, was refuted by C.W. Jones in "The Flat Earth", *Thought* 9
(1934):296–307. The weed continues to flourish nonetheless, though one hopes that it will be

The Earth being thus shaped and given to mortals as their habitation, the orbit of the Sun, which is always shining in this universe, gives daylight to one place and leaves another in night by the firm decree of God's law. And because, as Ecclesiastes says, *the Sun rises and sets and returns to its place*,[282] when it is being reborn there, it circles through the south and inclines towards the north.[283] When orbiting through eastern regions in the morning before western ones on the same latitude, it necessarily brings noon and evening in its train. Nonetheless, through the whole year it makes days as well as nights to be of equal length in both [regions]. Again, it is necessary that *the fiery Sun mount up the middle orb of heaven*[284] at one and the same point in time through the whole circle of the year for all who are placed facing one another in the same [vertical] line under the northern or the southern clime. However, it does not rise and set at the same moment or hour for both, /381/ but travelling through the southern zone in wintertime, it rises earlier and sets later for those who inhabit the southern regions of Earth than [it does] for us who, placed towards the north, with the globe of the earth blocking the way, receive its rising later, and its setting earlier. But on the other hand, in summer the Sun rises very much earlier for us who live under the same latitude, and seems to remain much longer on the point of setting [with us] than with those who dwell on Earth's southern flank.[285] They are denied an earlier sight of [the Sun] by the imposition of the Earth, and then are obliged to renounce it sooner. Therefore they have shorter days in summer than we do, but longer ones in winter.

This can be known not only from the orbit of the Sun, but also from

---

definitively eradicated by Jeffrey Burton Russell's *Inventing the Flat Earth: Columbus and Modern Historians* (New York: Praeger, 1991). However, as Marina Smyth points out (271–278) the Irish cosmographers of the 7th century were ambivalent or confused about the difference between "round" and "spherical". Belief in the spherical Earth was confirmed by increased exposure to classical and Patristic literature. Bede knew these Irish sources well, and may therefore have seen fit to hammer home the point about sphericity.

282 Ecclesiastes 1.5.

283 I.e. when the days are lengthening (the Sun is "being reborn"), the Sun is in the southern sky, but day by day edging northward.

284 I.e. cross the equator. The phrase is from Vergil, *Aeneid* 8.97, ed. R.A.B. Mynors (Oxford: Clarendon Press, 1969):285.

285 *in meridiano terrae sinu* (381.30–31). Since Bede vehemently denies the existence of the antipodes in ch. 34, one must presume that he is referring here to those who dwell further south on the Earth's flank, but still in the northern hemisphere.

the location of all the stars which take their courses beneath the different regions of the pole. Indeed, it is because of this same sphericity of the Earth that many of the most brilliant stars of the southern region are never seen by us. On the other hand, our northern stars are to a large degree concealed from them. Hence, the *Trogodyte and his Egyptian neighbour* do not see those polar stars which are straight over our heads, and which never set.[286] In fact, not only we in Britain, but even the Italians cannot see their brightest star Canopus, which once was worshipped under the name of a god. This is not because the light of the stars is withdrawn by gradually fading, and fails entirely for those at a greater distance, but because the mass of the Earth standing in the way prevents our seeing. All this can easily be proved by any very large mountain with settlements all around it.

33. IN WHAT PLACES THE SHADOWS OR DAYS ARE EQUAL

Because we demonstrated earlier on, through the statements of Pliny, the difference in the solstitial days for peoples which dwell far from one another to the north or to the south, /382/ it seems appropriate that we state these matters at greater length in the words of the same man, and also [discuss how] the days correspond for those who dwell in any part of the world that lies beneath the same east-west line.

Writing on this subject in the sixth book of the *Natural History*, he says: *The world has a number of segments, which our writers call 'circles' and the Greeks 'parallels'. The first runs from southern India through Arabia and the neighbourhood of the Red Sea. In it are contained the Gedrosi, the Carmani, the Persians, the Elamites, the Parthians, Naria, Susiana, Mesopotamia, Babylonian Seleucia, Arabia up to the rocks, Syria Coele, Pelusium, lower Egypt (called Chora), Alexandria, coastal Africa, all of the towns of Cyrene, Thapsus, Hadrumentum, Clupea, Carthage, Utica, both Hippos, Numidia, both Mauritanias, the Atlantic Ocean and the Pillars of Hercules. Under this belt of the heavens, the pin of a sundial (called a gnomon) which is seven feet long casts a shadow no more than four feet long at noon on the equinox. The longest night and the longest day have fourteen equinoctial hours; on the other hand, the shortest ones have ten. The next circle begins from western India, and*

---

286 Pliny, *HN* 2.71.178 (1.310); cf. ch. 31.

travels through the Medes and the Persians, Persepolis, the part of Persia closest to us, hither Arabia, /**383**/ Judaea, and the vicinity of the mountains of Lebanon; it includes Babylon, Idumea, Samaria, Jerusalem, Ascalon, Joppa, Caesarea, Phoenicia, Ptolemais, Sidon, Tyre, Beirut, Botris, Tripoli, Byblos, Antioch, Laodicea, Seleucia, coastal Cilicia, southern Cyprus, Crete, Lilybaeum in Sicily, and the northern parts of Africa and Numidia. At the equinox a pin 35 feet long casts a shadow of 23 feet.[287] The longest day and night are 14 equinoctial hours plus two-fifths of one hour. The third circle starts from the part of the Indies nearest the Himalaya mountains. It extends over the region of the Caspian, the hither part of Media, Cataonia, Cappadocia, Taurus, the Amanus range, Issus, the Cilician Gates, Soli, Tarsus, Cyprus, Pisidia, Pamphylia, Sida, Lycaonia, Lycia, Patara, Xanthus, Caunus, Rhodes, Cos, Halicarnassus, Cnidos, the Doric towns, Chios, Delos, the central Cyclades, Cythnos [=Gythium], Malea, Argos, Laconia, Elis, Olympia, Messana [=Messenia] in the Peloponnese, Syracuse, Catanina, central Sicily, southern Sardinia, Carteia and Cadiz. All cast a shadow on the sundial of 38 inches.[288] The longest day is $14\frac{1}{2}$ equinoctial hours plus the thirtieth part of an hour. /**384**/ Those lands lie under the fourth circle which are on the other side of the Himalayan range: southern Cappadocia, Galatia, Mysia, Sardis, Smyrna, Sipulus, Mount Imolus [=Tmolus], Lydia, Caria, Ionia, Trallis, Colophon, Ephesus, Miletus, Chios, Samos, the Icarian Sea, the northern Cyclades, Athens, Megara, Corinth, Sicyon, Achaia, Patras, the Isthmus, Epirus, northern Sicily, the eastern part of Narbonese Gaul, coastal Spain from New Carthage, and thence westward. Shadows of 16 feet correspond to a gnomon of 21 feet. The longest day has $14\frac{2}{3}$ equinoctial hours. In the fifth segment are included, beginning from the entrance of the Caspian Sea, the Bactri, Hiberia [=Georgia], Armenia, Mysia, Phrygia, the Hellespont, Troas, Tenedos, Ab⟨d⟩ios, ⟨S⟩Cepsis, Ilium, Mount Ida, Cyzicus, Lampsacus, Sinope, Amisus, Heraclea in Pontus, Paphlagonia, Lemnos, Imbros, Thasos, Cassandria, Thessaly, Macedonia, Larissa, Amphipolis, Thessalonica, Pella, Edesus, Beroea, Pharsalia, Carystum [belonging to] Euboea,

---

287 "23 feet" is the reading in all codices save Karlsruhe, Landesbibliothek Augiensis 167 (Reichenau, s. IX[1]), which reads "24 feet" (*BOD* 383). This latter is the reading of Rackham's critical edition, 2.496–497.

288 Bede's copy of Pliny seems corrupt here: Pliny says that on this parallel, a 100-inch gnomon casts a shadow 77 inches long: 6.39.214 (2.496).

Boeotia, Chalcis, Delphi, Acarnania, Aetolia, Apollonia, Brindisi, Tarentum, Thuri, Locri, Reggio, the Lucani,[289] *Naples, Puteoli, the Tuscan Sea, Corsica, the Balearic Islands and central Spain. A gnomon of 7 feet casts a shadow of 6 [feet]. The extent of the longest day is 15 equinoctial hours. The sixth band, in which the city of Rome is contained, includes the Caspian races, /385/ the Caucasus, northern Armenia, Apollonia on the Rhyndacus, Nicomedia, Nicaea, Chalcedon, Byzantium, Lysimachea, the Cheronese, the gulf of Melas, Abdera, Samothrace, Maronea, the Aenus River, Bessi⟨c⟩a, Thrace, Maedica, Paeonia, the Illyrians, Durazzo, Canosa, the borders of Apulia, Campania, Etruria, Pisa, Luna, Lucca, Genoa, Liguria, Antibes, Marseille, Narbonne, Tarragona, central Tarragonean Spain and hence through Lusitania [=Portugal]. A gnomon of 9 feet casts a shadow of 8 [feet]. The longest day is* $15\frac{1}{9}$ *hours, or as Nigidius would have it,* $15\frac{1}{5}$. *The seventh division begins on the far shore of the Caspian Sea, and extends over Callatis, the Bosphorus, the Dnieper, Stomos [=Tomos], the northern [literally: "rear"] part of Thrace, the Trivalli, the rest of Illyria, the Adriatic Sea, Aquileia, Altino, Venice, Nicetia [=Vicenza], Padua, Verona, Cremona, Ravenna, Ancona, the Picenum, the territory of the Marsi, the Peligni, and the Sabines, Umbria, Rimini, Bologna, Piacenza, Milan and everything from the Apennine range, and beyond the Alps, Gaul, Aquitaine, Vienne, the Pyrenees and Celtiberia. A sundial pin of 35 [feet] casts a shadow of 36, although in parts of the Veneto the shadow is equal to the gnomon. The longest day is* $15\frac{3}{5}$ *hours. /386/ Hitherto we have recorded the measurements of the ancients. The most assiduous of those who have followed them divide what remains of the Earth into three segments: from the Don, through Lake Meotius [=Sea of Azov] and the Sarmatians up to the Dnieper, and thence through the Dacias and part of Germany, and Gaul on the shores of the encircling Ocean, which has 16 hours [on its longest day], and a second through the Hyperboreans and Britain which has 17. The last one [goes through] Scythia from the Riphean Mountains into Thule where, as we said, there are alternating periods of continuous day and night. Also, they place two circles before the point where we began: the first through the island of Meroe and Ptolemais on the Red Sea, constructed for the elephant hunt, where the longest day is greater by* $\frac{1}{2}$ *hour than 12 hours; the second running through Syene in Egypt, which*

---

289 A people of southern Italy.

has 13 hours.[290] These excerpts from the writings of Pliny deserve to have a place in our work.

34. THE FIVE CIRCLES OF THE UNIVERSE AND THE PASSAGE OF THE STARS UNDER THE EARTH

When we discuss time, we are occasionally obliged /**387**/ to refer to the circles or zones – equinoctial, summer solstice or winter solstice – and we have decided to speak about these in greater detail.[291] For the philosophers customarily distinguish by these terms the inequalities of the seasons or the annual circuits of the Sun, so that they call that zone or region of the heavens through which the Sun circles the world around the time of the equinoxes, the "equinoctial",[292] and that through which which it circles the world at the summer solstice, the "[summer] solstitial",[293] and that [in which it is found] in winter, the "winter [solstitial] circle".[294] They are called zones or circles because they are made by the orbit of the Sun. The equinoctial, which is the middle zone, encircles the lands with an equal space above and below it, but the [summer] solstitial circle passes under the Earth by as much as the winter solstitial circle passes above it.[295] Likewise the winter [solstitial circle], which on the upper side of the Earth is circumscribed, has a circuit on the underside of the Earth of the same amplitude as the summer solstitial circle has on the top side. For undoubtedly, the Sun at the equinox passes the same amount of time at night under the Earth[296] as it does by day above the Earth. Everywhere in the

290 Pliny, *HN* 6.39.211–220 (2.494–502).

291 The circles are the equator, tropics, and Arctic and Antarctic circles. Projected onto the heavens, the two tropics constitute the boundaries of the Sun's annual journey to the southern sky in winter, and to the northern sky in summer. The Sun crosses the equator at the equinoxes. Projected onto the Earth, these five lines form five parallel "zones" or bands running around the sphere of the Earth. These form the five climates of Earth.

292 I.e. the Equator.

293 I.e. the Tropic of Cancer.

294 I.e. the Tropic of Capricorn.

295 We cannot see the entire celestial circle of the Tropic of Cancer, or of the Tropic of Capricorn projected onto the dome of heaven overhead. What Bede means is that the segment of the Tropic of Cancer which is invisible to us at our latitude is equal to the segment of the Tropic of Capricorn which is visible to us. With some exceptions noted below, Bede understands "top" and "upper" to mean "visible to us at this latitude"; "bottom" and "lower" mean "invisible to us".

296 I.e. in the opposite hemisphere.

north, [the Sun] is concealed by an intervening distance equal to that by which its course is visible when it turns to those in the south. And indeed, in winter it is borne about far and wide through the underside of the Earth[297] the same distance as it circles the upper parts of the Earth in its broad orbit at the [summer] solstice. Similarly, when travelling through the summer (that is, solstitial) circle, /**388**/ [the Sun's] disappearance by night under the northern parts of the Earth is as brief as the ascent it makes on winter days over the southern parts. For just as any given night in the revolving year has that duration which the day had six months before and will have six months after, so the Sun travels every night under the Earth the same distance as it travelled by day six months ago and will travel six months hence, revolving now by night towards the north as much as it will then do by day towards the south. When six months have run their course, all the stars travel by day through the same field of the heavens as they illumined by night six months previously, and to which they will revert in the same number of months. At night, according to their accustomed watches, they traverse [a path] beneath the Earth as wide proportionally as that above the Earth is narrow; this also makes their journey beneath all the more brief as their journey above is more prolonged. Hence it happens that during the solar year Arcturus, Orion and the Dog, but also the Milky Way and all the rest of the army of heaven circle the Earth once more often than does the Sun itself.[298]

Besides /**389**/ these three circles of the Sun, they posit two more circles, one in the north and one in the south. The northern one[299] is

297 I.e. the southern hemisphere.

298 The constellations were called the "fixed stars", because unlike the planets, they never change their position with respect to the backdrop of the dome of heaven. This dome was imagined as a hollow sphere, rotating from east to west around the earth, once each day, and carrying the stars along with it. It also carries the Sun and other planets around the earth each day. However, the planets also seem to move slowly in the opposite direction, from west to east, around the zodiac. The Moon, as we have seen, does this every $27\frac{2}{3}$ days, while the Sun makes this reverse journey once a year. The Sun thus seems to "run ahead" of the sphere of the fixed stars, much as the Moon seems to "run ahead" of the Sun (cf. ch. 29). As the Sun moves eastward day by day along the ecliptic, the dome of the stars edges a bit more above the eastern horizon night by night. In one solar year, this dome will have made a complete revolution, and the Sun will return to exactly the same position with respect to the fixed stars (i.e. it will be in the same degree of the same zodiac sign) as it was a year before. Thus, as Bede observes, the dome of the stars makes one more revolution each year around the Earth than does the Sun.

299 I.e. the Arctic circle.

always visible to us. It is not [a circle] of the Sun, but of Arcturus and the stars close to him. Because they lack proximity to the Sun, they never cease to be cold. The southern circle[300] is like it: frozen because the Sun is so far off, it is always hidden from us because the Earth blocks it out.[301] Of the two of them it is said, in praise of God, [that He] *made Arcturus and Orion and the Hyades, and the chambers of the south,*[302] and elsewhere, [that He] *stretcheth out the north over nothing.*[303] The poet also calls these to mind:

> *The heaven holds five zones of which one glows*
> *red with the blazing Sun and ever swelters in his flame.*[304]

This is the equinoctial circle, which, because the Sun shines upon it directly, or close to it on one side or the other, is scorched beyond a doubt, *stripped bare by flames and baked by steam.*[305]

> *At the extremities of the world, [two zones] extend around to right*
> *and left,*
> *locked fast in ice and black tempests.*[306]

Here he speaks of the northern and southern [circles]. All that lies beneath them, due to the absence of the Sun and the milder constellations, *is gripped by hostile stiffness and everlasting cold*, as that frozen sea bears witness which extends for a day's sailing from the island of Thule, up towards the north.[307] /**390**/

> *The two regions that lie between these*
> *were given by divine gift to frail mankind.*[308]

300 I.e. the Antarctic circle.

301 The chapter to this point is based on Pliny, *HN* 2.68.171–172 (1.304–306), 2.71.177–179 (1.310–12), Isidore, *DNR* 10 (209–212) and *Etym.* 3.44, 13.6; cf. Bede, *The Nature of Things* 9.

302 Job 9.9.

303 Job 26.7.

304 Vergil, *Georgics* 1.233–234 (xxvi); quoted in Isidore, *DNR* 10.1 (209.5), probably Bede's immediate source.

305 Pliny, *HN* 2.68.172 (1.306).

306 Vergil, *Georgics* 1.235–236 (xxvi).

307 Pliny, *HN* 2.68.172 (1.306) and 4.16.104 (2.198); cf. Solinus 22.9 (101.11–102.3).

308 Vergil, *Georgics* 1.237–238 (xxvi); quoted in Isidore, *DNR* 10.4 (213.26–27), which is probably Bede's immediate source.

Here he means the [circles of] the summer and winter solstices, which, placed near both the torrid zone and the frozen regions, are tempered by their power. Therefore both are called habitable, that is suitable for habitation, nor do they impede the advance of mortal men by excessive heat or cold.

However, it can only be proved that one of them is inhabited; no credence whatever should be given to fables about the antipodes, nor does any writer claim to have seen, heard, or read of any who have followed the winter Sun to southern climes, so that, with this [region] behind their backs, and having traversed the scorching heat of Ethiopia, they have discovered, beyond these, temperate lands habitable by humans, with heat on this side and cold on the other.[309] Therefore consider what Pliny, that most astute researcher into natural history – who does not deny that *although the Earth is shaped like a pine cone, it is nonetheless inhabited all over*[310] – says when he writes about these zones: *Only two are temperate between the one that is scorching and the others that are frozen, and these two are not accessible to each other because of the heat of the heavenly body.*[311]

Indeed, an illustration of these zones which is quite easy to grasp is furnished by people who in the icy cold of winter warm themselves at an oblong hearth. Like the middle zone, the fire itself, as well as whatever is near to it, cannot be touched because of the heat, while things placed at some distance from the flames on either side grow stiff with the ambient cold. But those which are between /**391**/ the two are tempered and appropriately positioned to be warmed, whether they who construct this hearth for themselves for light and warmth under the sky in the dark of a cold night care to stand on one side of the fire or the other. Could the fire move around in a circle like the Sun, it would undoubtedly produce five bands. Since it stands still, it makes five lines: one blazing hot in the middle, two frozen at the outsides, and two temperate ones in between these.

309 Cf. Augustine, *DCD* 16.9 (510.83–511.2); Isidore, *DNR* 10.2–4 (209–213); *Etym.* 9.2.133.

310 Pliny, *HN* 2.65.161 (1.296).

311 *Ibid.* 2.68.172 (1.306).

35. THE FOUR SEASONS, ELEMENTS AND HUMOURS

There are four seasons in the year, in which the Sun, by taking its course through the different regions of the sky, tempers the globe which lies beneath it, according to the universal solicitude of Divine Wisdom, so that by not always remaining in the same place, it does not devastate Earth's lovely vesture by its devouring heat. Rather, travelling through diverse regions by gradual stages, it preserves temperate conditions for sprouting and ripening the fruits of the Earth.[312]

The seasons [*tempora*] take their name from this temperateness;[313] or else they are rightly called *tempora* because they turn one into the other, being tempered one to another by some qualitative likeness. For winter is cold and wet, inasmuch as the Sun is quite far off;[314] spring, when [the Sun] comes back above the Earth,[315] is wet and warm; summer, when it waxes very hot, is warm and dry; autumn, when it falls to the lower regions, dry and cold. And so it happens that with each one embracing what is on either side of it, through the moderating mean, the whole is linked up to itself like a sphere.[316] **/392/** It is also said that the very elements of the universe are distinguished by these divergent qualities, and that they are knit into a company with each other, but each to each. For earth is dry and cold, water cold and wet, air wet and warm, fire warm and dry, and therefore the first is likened to autumn, the next to winter, the next to spring, and the last to summer.[317] And man himself, who is called "microcosm" by the wise, that is, "a smaller universe", has

312 Bede is revisiting a theme discussed in *The Nature of Things* 4 and *On Times* 8. There, as here, his main source is Isidore, *DNR* 7 (199–205) and *Etym.* 5.35.

313 Cf. ch. 2 above.

314 I.e. in the south.

315 I.e. it is higher in the sky for people in the northern hemisphere.

316 Bede seems to have in mind a circular diagram of the type called *syzygia elementorum*, showing the seasons "holding hands" so to speak through shared qualities; cf. Isidore, *DNR* 7 (202 *bis*). For a more elaborate version, showing the seasons, elements, and humours, see *ibid.* 11 (216 *bis*). This schema was very widely diffused, and appears in a number of contexts: see Peter Vössen, "Über die Elementen-Syzygien", in *Liber Floridus. Mittelalterliche Studien Paul Lehmann . . . gewidmet* (St Ottilien: Eos-Verlag, 1950):33–46; Ernest Wickersheimer, "Figures médico-astrologiques des IXe, Xe et XIe siècles", *Janus* 19 (1914):157–177; and L. Pressouyre, "Le cosmos platonicienne de la cathédrale d'Anagni", *Mélanges d'archéologie et d'histoire de l'École française de Rome* 78 (1966):551–593.

317 Cf. Isidore, *DNR* 11.2–3 (215.14–217.37); Ambrose, *Hexaemeron* 3.4.18 (71.16–72.14).

his body tempered in every respect by these same qualities; indeed each of its constituent humours imitates the manner of the season in which it prevails.[318] For blood, which increases in the spring, is moist and warm; red bile, which [increases in] the summer, is hot and dry; black bile, which [increases in] the autumn, is dry and cold; and phlegmatic humours, which [increase in] the winter, are cold and moist. Indeed, blood is at its most active in children, red bile in young people, melancholia (that is, gall mingled with the dregs of black blood) in the middle-aged, and phlegmatic humours dominate in the elderly. Moreover, blood makes those in whom its potency is greatest cheerful, joyous, tender-hearted, much given to laughter and speech; red bile makes people lean, even though they eat a lot, swift, bold, irritable and agile; black bile makes them stolid, solemn, set in their ways and gloomy; phlegmatic humours produce people who are slow, /**393**/ sleepy and forgetful.[319]

However, different people place the beginnings of the seasons at different times. Bishop Isidore the Spaniard said that winter begins on the 9th kalends of December [23 November], spring on the 8th kalends of March [22 February], summer on the 9th kalends of June [24 May], and autumn on the 10th kalends of September [23 August].[320] But the Greeks and Romans, whose authority on these matters, rather than that of the Spaniards, it is generally preferable to follow, deem that winter begins on the 7th ides of November [7 November], spring on the 7th ides of February [7 February], summer on the 7th ides of May [9 May], and autumn on the 7th ides of August [7 August]. Noting that summer and winter begin with the evening or morning rising and setting of the Pleiades, they place the commencement of spring and autumn when the Pleiades rise and set around the middle of the night.[321] Then again, in the finest and most authoritative books of the cosmographers, we find the same seasons clearly distinguished, with the rising of the Pleiades observed at the 7th ides of May [9 May] and their setting at the 7th ides of November [7 November]. And Pliny the Elder in the second book of his *Natural History* judges that they are to be divided in the same manner.[322]

But that man of the Church St Anatolius, in his work about Easter,

318 Cf. Isidore, *Etym.* 4.5.3.

319 Cf. Vindicianus, *Epistola ad Pentadium*, ed. V. Rose, *Theodori Prisciani Euporiston libri III* (Leipzig: Teubner, 1894):487.5–489.2.

320 Isidore, *DNR* 7.5 (203.46–51).

321 Vindicianus, *Epistola* (487.2–9).

322 Pliny, *HN* 2.47.122–125 (1.262–266).

when he argues with great subtlety about the solstices and equinoxes, and about the increments of hours and *momenta*, ends his disputation, and at the same time, his book, in this wise: /**394**/ *But you ought not to be unaware of the fact that although these four aforementioned frontiers of the seasons are close to the kalends of the following months, nevertheless each one contains within itself the mid-point of the seasons, that is, spring, summer, autumn and winter, and the beginnings of the seasons do not occur where the kalends of the months begin. But each season is to begin at such a point that the equinox divides the spring season [in half, beginning] from the first day; summer is similarly divided at the 8th kalends of July [24 June], autumn at the 8th kalends of October [24 September], and winter at the 8th kalends of January [25 December].*[323] Scripture bears witness to where the people of God placed these beginnings in the law of the seasons, which it commands in these words: *Observe the month of the new fruits, and the first of the spring season, and make the Passover unto the Lord your God.*[324] From what their most learned Bishop Proterius says, the Egyptians appear to have followed this: *One should take note that they make a great mistake who place the beginning of the first month of the lunar course on the 25th day of the month of Phamenoth, which is the 12th kalends of April [21 March], because then it appears that the beginning of spring is advanced by those who wish, with all diligence, to find it here.*[325]

Spring [ver] is so called because /**395**/ everything is verdant then, that is, it *flourishes [virescant]. Summer [aestas] [takes its name] from heat [aestu]* which in *[summer]* is bestowed for the ripening of crops; autumn, from the increasing *[autumnatione]* of the crops which *are gathered then,* while "winter" is translated by learned men as "cold"[326] and "sterility". And indeed, these names for the seasons agree with our climate. They say that *the Indies,* where *the heavens show another face, and the stars have other points of rising, has two summers and two harvests each year, with winter in between them, because of the blowing of the trade winds, and they relate that while we have winter, there there are gentle breezes and the seas are navigable.*[327] Egypt is said

323 Ps.-Anatolius 14 (327).
324 Deuteronomy 16.1.
325 Proterius, *Ep. ad Leonem* 8 (276).
326 Isidore, *Etym.* 5.35.
327 Pliny, *HN* 6.21.58 (2.380).

to have fields all aflower with herbs and orchards laden with fruit in the middle of our winter.

36. NATURAL YEARS[328]

The word "year" [*annus*] derives either from the renewal [*innouando*] of all things which pass away according to the natural order, or *from the cycle of time, for the ancients were accustomed to use "an" for "circum" – as Cato says in the Origins, "oratorum an terminum", that is "circum terminum" – and to say "ambire" for "circumire".*[329]

There is a lunar year and a solar year, a separate year for [each of] the wandering stars,[330] and one for all of the planets, which is particularly called "the great year".[331] But "lunar year" is taken in four ways. The first is when the Moon, /396/ traversing the zodiac in 27 days and 8 hours, returns to the sign from which it set out. The second is two days and four hours longer, and is customarily called a month; after 29 days and 12 hours, the now waning Moon seeks the Sun from which it departed when it was new. The third is when the Moon completes 12 of these months, that is 354 days; this is called a common year, because two of them usually fall together. The fourth, which is called by the Greek name "embolismic" (that is, "augmented over and above"), has 13 months, that is, 384 days. The Hebrews begin both these years at the commencement of the Paschal month, but amongst the Romans they have their beginning and end with the lunation that begins with the month of January.[332]

Again, the Sun's year is [complete] when it returns to the same place with respect to the fixed stars after 365 days and 6 hours, that is, a quarter of a whole day. The fraction, multiplied by four, makes it necessary to interpolate one day, which the Romans call the *bissextus*, so that the Sun is brought back into the same cycle. The fourth year of the solar cycle is bissextile, and one day longer than the other three. When it is

---

328   Bede also addresses this theme in *On Times* 9, where his main source is Isidore, *DNR* 6 (193–197).

329   Macrobius, *Saturnalia* 1.14.5 (66.24–26).

330   I.e. the planets, so called because they "wander" around the zodiac, unlike the "fixed" stars.

331   Cf. Isidore, *Etym.* 5.36.

332   This, in Bede's view, is the rationale behind the the 19-year "lunar cycle" (cf. ch. 56 below, and Commentary).

finished, the Sun returns to all the places of the zodiac at the same time of day or night as it did four years previously.

The year of the wandering stars is that in which each of them lights up the circuit of the zodiac, of which /397/ we spoke above.[333] The Great Year is when all the planets return at one and the same time to the very places where they once simultaneously were.[334] Josephus, describing the longevity of the earliest men in the first book of the *Antiquities*, speaks of this as follows: *However, nobody who compares the life of the ancients to modern life and the brief span of years which we now live, would think what is said about them false, nor would they refuse to believe that they reached that length of life simply because such length of life has not extended [to our day]. They lived for so many years because they were pious men and fashioned by God; and because there existed at that time nourishment provided for them in advance, more apt, and for longer periods of time. Therefore, because of the virtues and glorious advantages which they constantly studied, that is, astronomy and geometry, God granted them a greater span of life. Now they would scarcely have been able to learn this had they not lived to be 600, for the Great Year completes its circuit in this many years.*[335]

The Hebrews begin the civil year, that is, the solar year, from the spring equinox, the Greeks from the solstice, the Egyptians in the autumn, and the Romans in the winter.[336]

37. THE DIFFERENT YEARS OF THE ANCIENTS

So much for natural years; as for the rest, St Augustine teaches that ancient peoples of different nations erred in various ways as regards the observation of the year. /398/ In the 15th book of *The City of God* he argues against those who *think* that Holy Scripture *has computed the years* from the beginning of the world *in a different way (that is, it is believed that [the years] were so short that one of our years equals ten of theirs* [i.e. of the biblical writers]), and who say that *when one hears or reads that someone lived 900 years, it ought to be understood as 90, for ten of our years is one of theirs, and ten of ours make 100 of theirs*). He

333 Ch. 8.
334 Cf. ch. 6.
335 Josephus, *Antiquitates* 1.3.9, ed. Franz Blatt, *The Latin Josephus* (Copenhagen: Munksgaard, 1958):136.24–137.2.
336 Isidore, *DNR* 6.2 (193.10–13).

says amongst other things: *Lest it seem incredible that a year should be calculated in a different way then, they add that among many writers of history it is found that the Egyptians had a year of four months, the Acarnians of six months, and the Lavinians of 13 months. Pliny, although he records that it is related in documents that someone lived to be 152, and another person ten years longer, and that others had a life of 200 years and others of 300, or that some reached 500, others 600 and a few 800, judges that all these arise from ignorance of chronology [inscitia temporum]. 'Some indeed', he says, 'start one year in the summer, and another in the winter, others at the four seasons, so that their years are three months long, like the Arcadians.' He adds, moreover, that the Egyptians, whose short years, as we said above, were four months long, finished a year at the end of the moon. 'Therefore amongst them', he says, 'it is claimed that men live a thousand years.'*[337] /**399**/

38. THE CALCULATION OF THE LEAP-YEAR DAY[338]

I have no wish to contrive anything new on the subject of the leap-year day, but rather to incorporate into this little work what I once said in a letter responding to a friend's inquiry.[339] There, after the appropriate introduction, I added this: I said, that just as that anticipated place and time of the appearance of [the new Moon] which they call the "leap of the Moon" is produced over a period of 19 years,[340] so in the opposite way the leap-year day is produced by nothing other than a retardation of the Sun's course.[341] Some people are so clever at computation that they understand without any difficulty how many fractions of the growing leap-year day are relentlessly added on each year, or month, or even week or day. Nonetheless, they would not be able to say how this same fraction increases, or what is its cause, or how the increase is calcu-

---

337 Augustine, *DCD* 15.12 (468.48–50, 469.36–51). The passage from Pliny is from *HN* 7.48.155 (2.608–610). This same quotation appears in the *Letter to Plegwin* 11.

338 This chapter is an expanded version of *On Times* 10, and possibly inspired by the anonymous text on leap year in the "Sirmond Manuscript", MS Bodleian Library, Bodley 309, fols. 74r–76r (Jones, *BOT* 371; see Introduction, section 5). Analogous treatments of the leap year are common in *computus* anthologies, e.g. the "Bobbio *computus*", chs. 40–44 (PL 129.1295–1296).

339 Bede is referring to his *Letter to Helmwald* (Appendix 3.2).

340 See below, ch. 42.

341 See Commentary on ch. 42.

lated, or what troublesome error is engendered if the leap-year day is not intercalated in its sequence in the necessary way.[342] For the leap-year day is made up from the calculation of quarter[-days] [*quadrantes*] over four years. It is customary to call a quarter of something – for example money, time or space – a *quadrans*,[343] and therefore the fourth part of a day, which with its night is completed in 24 hours, that is, 6 hours, is usually called a *quadrans*. /**400**/

The reason why this quarter[-day] is collected over four years into a whole day and intercalated into its place, is that the Sun is held to complete its annual circuit of heaven (that is, the 12 conspicuous signs of the zodiacal circle) not in 365 days, but in 6 hours more. Hence it happens that if, for example, the Sun now enters the equinoctial part of the sky[344] when it rises in the morning, it will return to that point next year at noon, in the third year at evening, in the fourth at midnight, and in the fifth at daybreak again, that is, with the course of a whole day completed. We should therefore bear in mind that the extra day should be inserted whenever it is necessary, and that it is to be added to the total sum of the fourth year.[345] The Egyptians are accustomed to solemnly intercalate it at the end of their year, that is, on the 4th kalends of September [29 August], but the Romans do it on the 6th kalends of March [24 February], whence they call it *bissextus*.[346] For should any computist neglect to make [this intercalation], and think that all years ought to have only 365 days, he will subsequently discover that a great shortfall occurs in the course of the year; after a certain number of years have come and gone, the erring computist will be aghast to encounter spring time in the summer months, winter in the spring months, autumn in the winter months, and summer in the autumn months.

Should anyone perchance not know what we have truthfully and succinctly stated concerning the zodiac and the circuit of the

---

342 "Some people are so clever...": Jones argues (*BOT* 371) that Bede is referring to some particular text here, perhaps the one contained in the Irish-inspired "Bobbio *computus*" 87 (PL 129.1315), which works out the leap-year increment down to the level of the day!

343 See above, ch.4.

344 I.e. crosses the equator at the spring equinox.

345 "Hence it happens...": cf. Bede, *Letter to Wicthed* 3.

346 I.e. "twice six", because the 6th kalends of March is doubled in leap years, and the second day is designated *bis sextus kal. mar.* Cf. Isidore, *DNR* 6.7 (197.50–52).

heavens,[347] it has been our care to gratify him with commonplace and perhaps rather concise and straightforward reasoning, so that he who did not learn to recognize the constellations in his elementary schooling /**401**/ can at least discover what he needs to know by the lines of the sundial on the ground. He should know, then, that the Sun supplies this quarter-day to the year by brief delays and daily retardation, according to the ordering work of the Creator. Closer inspection shows that the Sun cannot be completely carried back to the same line on the sundial in 365 days. But if this year, for instance, at the [time of the] spring equinox (*which according to the Egyptians, who bear the palm as computists, ordinarily comes on the 12th kalends of April* [21 March]),[348] the Sun rises in the mid-point of the east,[349] on the same day a year later it will rise a little lower down.[350] In the third, fourth and fifth years, it will augment this diminution, according to the natural measure of its course, to such a degree that unless the bissextile day is inserted beforehand according to custom, [the Sun] will then rise at the mid-point of the east to make the equinox on the 11th kalends of April [22 March], while at the same time continuing a regular pattern of retardation in its other risings and settings throughout the whole year regardless.

39. MEASURING THE LEAP-YEAR INCREMENT

But in case you find what we said concerning the measure of the leap-year increment obscure, we say that one quarter of that same day – that is, six hours – accumulates each year: one hour every two months, /**402**/ half an hour in every natural solar month, and indeed the fourth part of an hour (that is, a *punctus*) in half a month. For we know that the Sun goes around the zodiacal circle, that is, the twelve signs of the horoscope, in

347 Jones (*BOT* 372) thinks that Bede must be referring here, not to *The Reckoning of Time*, but to another tract composed by himself. He suggests that this is the text beginning *Primo igitur anno praeparationis bissexti...* ("In the first year leading up to the leap year ...") printed in PL 90.357–361A, and edited by Jones in *BOD* 649–653. This tract matches Bede's description, in that it discusses the leap-year increment in terms of the timing of the Sun's entrance into the various zodiac signs. To students unfamiliar with the zodiac, it would have been very challenging.

348 Bede, *HE* 5.21 (524); cf. *On Times* 10 (593.14–15); Dionysius Exiguus, *Ep. ad Petronium* (65.20–22).

349 I.e. on the celestial equator.

350 I.e. towards the south.

365 days and 6 hours,[351] and covers each sign of the same zodiac, according to its natural course, in 30 days and 10½ hours, and traverses half a sign in 15 days and 5¼ hours. As for those who say that only three hours accrue to the *bissextus* every year, as if ascribing nothing to the night, we do not think that their judgement is to be accepted at all. If this were so, the whole day which accumulates would not be complete before the passage of seven years. For even the base herd [*uulgus ignobile*] know that a whole day, that is, [a day] together with its night, has 24 hours. If they do not deny that the whole [day] is filled up in four years, why do they deny that one-fourth is filled up by each of the four years?

In fact, should anyone think our position on this issue ought to be rejected, let him read the fourth book *On the Trinity* of the blessed Aurelius Augustine. In his discussion of the perfection of the number six in which the world is made, he does not neglect to mention this quarter-day – nay rather, he teaches that it was made and ordained by the Creator's omnipotent wisdom for the sake of a certain hidden meaning. He says: *For one and two and three make six, which /***403***/ number is said to be perfect because it is made up of its parts. For it has three [parts]: a sixth, a third, and a half. Nor is any other part found in it whose number can be expressed. One sixth of it is one, one third, two, and one half, three; and one and two and three total six. Holy Scripture commends its perfection to us, particularly because God finished his work in six days, and on the sixth day, man was made in the image of God. And in the sixth age of the human race, the Son of God came and was made the Son of Man, that he might reshape us into the image of God. Now one year, if one considers 12 whole months of 30 days, which manifests the orbit of the Moon (for the ancients observed a month of this length), operates according to the number six. Six produces six in the first rank of numbers, which we know arrives at ten by units; it produces 60 in the second rank, which we know arrives at 100 by tens. Therefore 60 is the number of days in a sixth part of a year.* And shortly thereafter he says: *Likewise the sixth of the second rank is multiplied by the sixth of the first rank, and six times 60 makes 360 days, which are 12 whole months. But as the orbit of the Moon marks the month for men, /***404***/ so the year is indicated by the circuit of the Sun. So there remain five and a quarter days until the Sun can complete its course and the year come to an end. For four quarter-days make one day that must be intercalated in the*

---

351 See above, ch. 16.

course of the fourth year, and which they call the bissextus, lest the order of the seasons be perturbed. But if we consider these five and a quarter days, we see that the number six works mightily within them. First, since it usually happens that the whole is computed from the part, there are now not five days but rather six, so that this quarter-day is counted as a day; then, because in these five days there is the sixth part of a month. The quarter-day itself has six hours, for the whole day with its night has 24 hours, whose fourth part, which is a quarter-day, is found to be six hours. Thus in the course of a year the number six is particularly prevalent.[352]

It was a pleasure to excerpt these words from the writings of such a great authority, so that in re-reading the views of my insignificant self on the nature of the *bissextus*, you might understand, out of the mouth of this most learned expositor, not only that the *bissextus* grows each year by six hours, but also the many-sided perfection of this number six, by which the year itself is constituted.

40. WHY IT IS INTERCALATED ON THE SIXTH KALENDS OF MARCH

The Romans saw fit to intercalate the leap-year day in the month of February because it is shorter than the others, and because it was the last month of the year. Yet they did so when the month was not yet over, for they did not wish to do this when the year had totally run its course as the Egyptians and Greeks do, /405/ lest their old rule that they should join the beginning of March to the end of February be broken. Moreover they would not [intercalate] before the 6th kalends of March [24 February], for out of devotion to the god Terminus, they had instituted the sacred Terminalia on the 23rd day of the same month, and they certainly did not dare change it to another day. For a long time they inserted many days after this [feast] had been duly celebrated. However as their expertise increased over the centuries, they inserted one day every four years.[353]

Alas, what wretches those fools were, who did not know Him whose kingdom will have no end, and who said, I am *the beginning, which I also*

---

352 Augustine, *De Trinitate* 4.4 (169.3–170.12; 171.44–172.3).

353 Up to this point, Bede is drawing on Macrobius, *Saturnalia* 1.13.13–16, 19 (63.22–64.18, 64.29–65.7). Cf. similar passage in ch. 12.

declare unto you,[354] but believed that Terminus should be honoured with divine worship, and sacrifices and rites offered to him. But much more wretched are the demented men who, having the promise of the heavenly kingdom, prefer to cling to those things which will earn them punishment in the end, being themselves on the road to destruction, rather than to hasten towards eternal things, like those destined for victory.

41. THE MOON ALSO HAS ITS QUARTER-DAY

While we are on the subject, the computist should remember that the lunation of the month of February has 29 days in the other years, but in leap year is counted as 30 [days long], whether it happens to end before or after the intercalated quarter-day. By the addition of this day, it comes to pass that the lunar year is completed in 355 days if it is common and 385 days if it is embolismic. For it is obvious that the quarter-day of which we are speaking is applicable not only to the course of the Sun, /406/ but also to the course of the Moon. If you refuse to take the Moon's quarter-day into account, and apply to February in leap year a lunar month of the same length as you were wont to do before, it will mean that the fourteenth Moon of Easter will waver, and then the course of the year will reel, and that ever-inviolable state of the 19-year cycle, being more and more perturbed, will be overthrown. Therefore, just as the calculation of the quarter-day in the fourth year (which we call bissextile) dictates, we ought to place on the same line the weekday for the 6th kalends of March, for example the third and the fourth, so that we might then remember to count in the Moon as well (for example, likewise the third and fourth). Take heed with constant zeal, that although we shall cause the Moon of the month of February to have one more day than usual, nonetheless on the first day of March, the Moon will retain the age it did before, with the sole exception of the 11th year of the 19-year cycle.[355] /407/ For we issue a

354 John 8.25.

355 The eleventh year is an exception to the rule that a lunar month "belongs to" the calendar month in which it ends. In that year the lunar month of January ends on 1 February, and the lunar month of February ends on 2 March: see Commentary on ch. 20; cf. Jones, *BOT* 375. If leap year falls on any other year of the 19-year cycle, the extra lunar day is absorbed by the extra calendar day, namely the extra 6th kalends of March, and so the lunar age of 1 March is not affected. But since in year 11, 1 March is still within the lunation of February, that lunation will still be carrying the extra leap-year day on 1 March. Therefore if year 11 is a leap year, the Moon will be 30, not 29 days old on 1 March.

particular warning that the quarter-day be attributed to the Moon, lest, being further advanced than it ought to be on the first day of March, it deflect the calculated course of the Paschal observance from the correct path.[356]

By what and how many particles of time this same lunar quarter-day accrues demands more extensive investigation, for the frequent interposition of the embolisms and also the calculation of the "leap [of the Moon]" interferes, so that it is impossible to distinguish the whole measure of the lunar course with perfect precision.

356 Bede's suggestion that the computist annotate the solar calendar as a reminder of the additional day to be added to the lunar reckoning was taken up by hundreds of medieval manuscript calendars, whose February pages bear the following admonition: "Remember in leap year to compute the Moon of February as one of 30 days [the February Moon was normally "hollow", or 29 days], so that the Moon of March will still be one of 30 days, as usual, lest the Easter reckoning waver." (*Memento quod anno bissextili luna februarii XXX dies computes ut tamen luna martii XXX dies habeat sicut semper habet. ne paschalis lunae ratio uacillet.*). For examples, see Wormald, *English Kalendars*, 45, 59, 87, 101, etc.

[III. ANOMALIES OF LUNAR RECKONING (CHS. 42–43)]

42. THE "LEAP OF THE MOON" [1]

Concerning this "leap [of the Moon]", it seems to be true that it is produced when the place and time of the appearance of the new Moon comes more quickly than is ordinarily supposed. For how could it be that over [the course of] nineteen years it becomes necessary to subtract one from the usual number of lunar days, were it not that a certain acceleration of the lunar orbit gradually brings it into effect over the whole period of the 19-year cycle? In the same way, but in the opposite sense, the course of the Sun is demonstrably affected by retardation, so that one day is added to its course over the four years of its cycle.[2]

Although it is not easy to perceive the stages [*ordo*] of [the Moon's] acceleration, and if I may put it thus, of this "advance", its measure is by no means a mystery. For it is agreed that the diminution and removal of one day /**408**/, regardless of how this takes place, is assembled over nineteen years, augmented each year by one hour, one *punctus*, and $\frac{1}{19}$ of one *punctus*. For the day has 24 hours, of which 5 remain when you distribute 19 [hours] over the same number of years of the 19-year cycle. Multiply these by four (because there are four *puncti* in one hour) and that makes 20. Give one [*punctus*] to each year, and there is one left over. Divide this by 19, and you will see, as we said, that one adds one hour, one *punctus* and $\frac{1}{19}$ of one *punctus* each year to make up the "leap of the Moon" [*saltus lunae*]. Therefore the appearance of the new Moon is not observed at one and the same point in time, or region in the heavens, but it always arrives somewhat sooner than in the previous month, whence it is that it springs over one day in the nineteenth year.[3]

Now various computists have thought that this "leap of the Moon" should be interposed at various points in the 19-year cycle cycle; that is, that a lunar month which in other years normally has thirty days should have twenty-nine.

---

1 For discussion of Bede's sources for this chapter, see Commentary.

2 Cf. ch. 38, n. 341.

3 Cf. Bede, *On Times* 12. On the implications of this gradual advance, see ch. 43.

The natural way [*ratio naturalis*] to calculate this readily, is to assign the beginning and end of all augmentations and diminutions of this type, both of the Sun and of the Moon, /**409**/ to the equinoxes wherein their beginning and end were first created: that of the Sun at the spring, and that of the Moon at the autumn equinox.[4] Therefore Anatolius, who rightly perceived that *the beginning of the months and the starting-point and end-point of the whole cycle* is *the spring equinox*,[5] did not himself place this change of the Moon at the beginning or end of his 19-year cycle, but in its fourteenth year, which is the last one of the *ogdoas*,[6] making [the epact] rise from 8 to 20 at the [spring] equinox. On the other hand, Victorius /**410**/ considered that this insertion should be made in the third year before the end of the *ogdoas*, changing the Moon [i.e. the lunar epact] on the kalends of January from 4 to 16.[7] But then the Egyptians, to whose judgement the catholic Church now gives its consent, fix the change in the first year of the 19-year cycle, making the annual lunar epact, whose locus is the 11th kalends of April [22 March],[8] to jump from 18 to zero [*in nullam*]. Hence this same year, if it is not a leap year, is finished /**411**/ in 353 days.

Dionysius seems to touch upon this in his letter, when he says: *For instance from the fifteenth Moon of the Paschal feast in the preceding year until the fourteenth in the following year there are 354 days if it is a common year and 384 if it is embolismic. If there is one day more or less, it is a manifest mistake except in the first year of the oft-mentioned 19-year cycle, which it is our concern to calculate from the fourteenth Moon of the last (that is, the nineteenth) year, up to the fourteenth Moon of this first year. Because of this, that last year, containing an epact (that is, a*

---

4 Cf. ch. 6 above, and the *Letter to Wicthed*.

5 Anatolius 2, as reported by Rufinus in his ed. of Eusebius, *HE* 7.32.15. (723.15–19); cf. *Letter to Wicthed*.

6 The *ogdoas* is the first eight years of the 19-year cycle. Since ps.-Anatolius' cycle begins six years before the 19-year cycle, the last year of the Alexandrian *ogdoas* corresponds to his year 14. This is, of course, the famous 14th-year *saltus* of the Celtic *computus*, though Bede was unaware of it. Note that in Krusch's edition of the table of ps.-Anatolius (*Studien I*, 325) the *saltus* is placed in the 13th year. Bede's concordance of ps.-Anatolius' cycle with the 19-year cycle suggests that he thought he knew when ps.-Anatolius' table began, i.e. in the 13th year of the 19-year cycle. Bede's date for Anatolius in the chronicle in *Reckoning of Time*, ch. 66 is AM 4230, or AD 279, which happens to fit this criterion: cf. Jones, *BOT* 377.

7 On the problems of Victorius' *saltus*, see Introduction, section 3.

8 Cf. ch. 50 below; cf. *On Times* 13.

lunar increment) of 18, concedes 12 days, and not 11, as is usual, to the first year. And because [the epacts] run out at the end of 30 days, the epact of zero is placed at the beginning of this cycle, but the second year takes an epact of 11. And therefore, as we said, from the fifteenth Paschal Moon of the first [year] of the cycle up to its end, there is no doubt that we find the days as established above in common and embolismic years.[9]

Divergence of just this kind compels Victorius to calculate a Moon that matches ours in age only for the first six years of the 19-year cycle, for thereafter because of the inserted alteration of the "leap", his Moon never ceases to be greater by one day than ours, up to the point where we make the insertion at the end of the cycle. Undoubtedly, it seems that this "leap" /412/ can be positioned at no better place in the year than on the 12th kalends of April [21 March], because, of course, of the origin of the creation of the stars [i.e. Sun and Moon] which we mentioned before,[10] so that the Moon of the month of March which finishes on the 29th day is thenceforward changed to a new Moon. But there are those who aver that we ought rather to do this in the Moon of the month of November,[11] so that they can begin the new year with a computation that is free from the other [year], complicating factors of this kind having been cancelled out by the ending of the previous year. This follows the example of the Egyptians, who are said to have done this in the last month of their year, which is our July. But whether you do it here, there, or anywhere else, it is always necessary, unless I am mistaken, to calculate three Moons of 29 days together.[12]

43. WHY THE MOON SOMETIMES APPEARS OLDER THAN ITS COMPUTED AGE

It should be pointed out that the calculation [*ratio*] of this lunar *saltus*, by its slowing compounding increase, causes the Moon sometimes to be visibly older than it is reckoned to be, to the point that even at sunset on

9 Dionysius Exiguus, *Ep. ad Bonifacium et Bonum* (ed. Krusch, *Studien II*, 84.5–16).

10 In ch. 6.

11 I.e. the lunar month ending on 25 November in the 19th year of the cycle.

12 The lunar months of the 19-year cycle alternate "full" and "hollow". So whether you begin the cycle in September or in January, the penultimate lunar month when the *saltus* takes place (July or November) is "hollow". Since the two months on either side are "hollow", this produced three "hollow" months of 29 days in a row.

the thirtieth day it does not appear as a thin line in the sky. And the farther the 19-year cycle has advanced, the more frequently [the Moon] will be subject to this, for the evident reason that the "leap" of which we have spoken will then be fulfilled to its maximum extent. But in discussing natural truth [*in naturalis assertione ueritatis*], which is also affirmed to be true by the Council of Nicaea, this rule is to be observed: that we recognize that the age of the Moon changes in the evening, and not [claim], as some do, that it does so at noon or mid-afternoon.[13] /413/ For to be sure, the Moon, which first rose upon the world at eventide, must always enter upon this or that phase [at eventide], completing each [phase] in 24 hours, even as the Sun, which first arose in the morning, as the book of Genesis bears witness, completes one day from morning to morning.[14] Why should lunar reckoning be calculated from the noon-tide hours, seeing that the Moon had not yet been placed in the heavens, nor gone forth over the Earth? On the contrary, none of the feast-days of the Law began and ended at noon or in the afternoon, but all did so in the evening. Or else perchance it is because sinful Adam was reproached by the Lord *in the cool of the afternoon*,[15] and thrust out from the joys of Paradise. In remembrance of that heavenly life which we changed for the tribulation of this world, the change of the Moon, which imitates our toil by its everlasting waxing and waning, ought specifically to be observed at the hour in which we began our exile, so that every day we may be reminded by the hour of the Moon's changing of that verse, *a fool changeth as the Moon*[16] while the wise man *shall live as long as the Sun*,[17] and that we may sigh more ardently for that life, supremely blessed in eternal peace, when *the light of the Moon shall be as the light of the Sun, and the light of the Sun shall be sevenfold, as the light of seven days*.[18] Indeed, because (as it is written) *from the Moon is the sign* /414/ *of the feast-day*,[19] and just as the first light of the Moon was shed upon the world at eventide, so in the Law it is compulsory that every feast-day begin in the evening and end in the evening.[20]

13 Cf. above, ch. 6.
14 Cf. *Prologus Theophili* 3 (223–225); *Prologus Cyrilli* 6 (341–342).
15 Genesis 3.8.
16 Ecclesiasticus 27.11.
17 Psalm 72 (71).5.
18 Isaiah 30.26.
19 Ecclesiasticus 43.7.
20 Cf. Exodus 12.18, and ch. 5 above.

The age of a new Moon is more appropriately calculated from the evening hour than from any other time, and it will retain the age which began in the evening until the following evening. Should it befall that the Moon is lit up by the Sun shortly before evening, it must be counted as, and it must be, the first Moon as soon as the Sun has set, and it will be the first Moon immediately after sunset, because it has reached the hour when first it began to shine forth upon the Earth. But if its light appears after sunset, the Moon will not yet have seen its first day before evening, but ought rather to be counted as the thirtieth Moon. Even if it remains illuminated for 23 hours after sunset, nevertheless it should retain the age which it had at sunset until the next sunset, lest the order of primal Creation be disturbed. Since it sometimes becomes visible six or seven hours after it is [first] illuminated, it is not surprising that as a new Moon, it can appear very conspicuously in the heavens, seeing that it has traversed so many hours. For particularly when the Moon is in Aries, it often happens that on one and the same day it will be seen both in the morning [as the old Moon] and in the evening [as the new Moon], the action of illumination having been effected around noon.

Now if someone, pursuing this question more seriously, were to say that in the company of witnesses he had seen the new Moon, in the same year in which the "leap" is inserted /**415**/ (that is, in the nineteenth year of the 19-year cycle), shine forth two days before the first [calculated Moon], that is, on the 4th nones of April [2 April], and since in that same year of the aforementioned cycle the 14th Moon of Easter is marked down for the 15th kalends of May [17 April], and therefore the new Moon could not occur save on the day before the nones of April [4 April], he would demand that we explain the cause of this. At this point, our insignificance, lest it be defeated by its weakness, will run for help to the Fathers', and indeed to God's, authority. We are bolstered by the aid of the authority of the Fathers when we follow the decrees of the Council of Nicaea, which fixed the fourteenth Moons of the Paschal feast with such firm stability [*firma stabilitas*] that their 19-year cycle can never waver [*uacillare*] and never fail. There is no doubt on the part of any computist that within this cycle, the Paschal Moon of the year under discussion will be in its first day on the 2nd nones of April [4 April]. And therefore it is not proper for any of the faithful to say that it is otherwise. What then? Is it credible that no one amongst the 318 bishops who attended the Council of Nicaea saw that new Moon which we saw on the 4th nones of April [2 April]? No one amongst that company of lesser clergy which was party to their deliberations and decisions? Should we not rather understand that when they proclaimed that the Paschal Moon of that year begins on the day before the nones of April, their intention was to avoid by this means another and greater peril, namely, that if they discerned it otherwise, they would dissolve that indissoluble state of common **/416/** and embolismic years which they recognized was to be observed by the authority of the divine Law given to the Hebrews? But we also defend the lunar observance to which we hold by means of the particular evidence of divine authority. For we read in the writings of the blessed Cyril, bishop of Alexandria, that *the monk Pachomius, famous for deeds of apostolic grace and founder of the monastic communities of Egypt, sent letters to the monasteries over which he ruled, which he had received at the dictation of an angel, in order that they should not make a mistake in calculating the Paschal solemnity, and so that they might know the Moon of the first month in the common and embolismic year.*[21] In Cyril's account, we read that *had the Council of Nicaea not written down the lunar cycle of the first month, the cycle of the selenite stone in Persia, whose inner whiteness increases and decreases with the Moon of the first month, would suffice as a model for Paschal reckoning.*[22] We read in what St Paschasinus, bishop of Lilybaeum, wrote to the most blessed Pope Leo, that in the time of Pope Zosimus, although it was the last year of the 19-year cycle, and some *refused to celebrate Easter on the 10th kalends of May [22 April], celebrating instead on the 8th kalends of April [25 March] (that is, taking it for a common instead of an embolismic year),* the truth about the Paschal observance was made clear by a multiple miracle **/417/** of supernal virtue. He says: *A certain very run-down estate called Meltinas is situated high in the mountains in the midst of thick woodlands. There is a church there, exceedingly poor and ramshackle; in its baptistery, on the most holy night of Easter at the hour when baptisms are held, the font is filled of its own accord, although there is no trough or pipe, nor any water at all in the vicinity. And when the few who are there are consecrated [i.e. baptized], the water disappears even as it came, of its own accord, though there is no drain. When therefore, under Pope Zosimus of blessed memory, the water did*

---

21 *Ep. Cyrilli* 5 (346–347); cf. Gennadius, *De uiris illustribus* 7, ed. W. Herdin (Leipzig: Teubner, 1879):73.14–17, where the angel dictates a monastic, not a Paschal rule. Cf. ch. 66, *s.a.* 4113.

22 *Ep. Cyrilli* 5 (348). Cf. ch. 28 above.

not appear, those who were to be baptized were not consecrated, and they went away. But on that night which dawned onto Sunday, the 10th kalends of May [22 April], the holy font was filled up at the appropriate hour. Thus by this evident miracle it has been made clear what the error of the western lands has been.[23]

So it is plain that this question of the age of the Moon is an old one, and at one point was even carefully investigated by the zeal of the holy Pope Leo. It is this very question which has caused the long and serious controversy between the churches of the East and of the West. This also convinced Pope Hilarius, long after the Council of Nicaea, to seek, and Victorius to construct, a new Paschal cycle. In this controversy the afore-mentioned Pope Leo, keen and eager to win in his struggle with Prosper, a most learned and sagacious man, eventually was glad to be honourably won over by the unanimity of those who stood by the decrees of the Nicene Council. Hence, on this same question, I seek for myself and for my people no better way of acting or speaking that what I know he did, /**418**/ who so far excels us in knowledge, merit and authority – that is, to follow the sure judgement of the most reverend Fathers in those matters which are doubtful and unsure to us. It is not to be thought that we either understand the variation of the lunar course with greater acuity than the ancients, or that we are able more advantageously [than they] to distinguish what path it is preferable to follow in such variations.

---

23 Paschasinus of Lilybaeum, *Ep. ad Leonem papam* 2–3 (249–250). The year to which Paschasinus is referring is AD 417.

[IV. THE PASCHAL TABLE (CHS. 44–65)]

44. THE NINETEEN-YEAR CYCLE

Eusebius, bishop of Caesarea in Palestine, first devised the sequence of the 19-year cycle in order to find the fourteenth Moons of the Paschal feast and the day of Easter itself, because after this period of time a Moon of a given age will recur on the same day of the solar year.[1] Not that hitherto the Egyptian or other Churches of Christ in the East did not know how to find the course of the Moon or the day of Easter in the correct way, but because these matters, which at that time used to be determined year by year, with much difficulty, and then promulgated throughout the world – [whence] arose many queries – could [now], by being readily established in advance according to a cyclical principle, be observed in perpetuity and learned by heart without anxiety about disputes.[2]

Therefore it is claimed that from time immemorial it was delegated to the Church of Alexandria, by way of obligation, to take upon itself the task and duty of investigating /**419**/ the Paschal computation, so that when it announced [the same] to the Pontiff of the Apostolic See, the rulers of the churches might be informed through him of the date of Easter. Hence the blessed Pope Leo, when he asked the Emperor Marcian to refer to the bishop of Alexandria the conduct of a careful investigation of the Easter [cycle] of St Theophilus,[3] in which it appears that the 8th kalends of May [24 April] is assigned to Easter Day, gave his judgement that the aforementioned day had not correctly been chosen by [Theophilus]. Amongst other things, he says: *Some of the holy Fathers sought an opportunity to remove this error, delegating this entire responsibility to the bishop of Alexandria, for it seemed that skill in this calendrical computation had been handed down from of old amongst the Egyptians, and for many years the day of the aforementioned celebration would be indicated by [the bishop] to the Apostolic See, by whose writ a*

---

1 Jerome, *De viris illustribus* 61, ed. W. Herding (Leipzig: Teubner, 1879):41.28–42.3; cf. Isidore, *Etym.* 6.17.

2 Cf. Bede, *HE* 5.21 (544–546).

3 I.e. the 19-year cycle.

general announcement might reach churches in far-off lands.[4] Indeed, in order to avoid exertion of this kind, the abovementioned man [Theophilus] composed a cycle now famous far and wide, which contains the fourteenth Moons of the first month in sequence, and permits anyone easily to find Easter Day, which will be on the very next Sunday. Proterius, bishop of Alexandria, giving a most lucid explanation of this cycle at the request of the holy Pope Leo, was found worthy to be praised by him in the course of his reply: *The letters of your well-beloved self, which our brother and fellow bishop Nestorius delivered out of pious courtesy, cause me joy. For it was fitting that such letters be sent to the Apostolic See from the bishop of the Alexandrian church, which [letters] would show what the Egyptians learned by the initial instruction of St Peter through his disciple St Mark, and which it is agreed the Romans believe.*[5] **/420/**

45. EMBOLISMIC AND COMMON YEARS

This same cycle is divided into common and embolismic years,[6] which it is agreed are also observed by virtue of the authority of the ancient Hebrews. It contains twelve common years, that is, [years] of 354 days, but seven embolismic years, that is, of 384 days. Now the first and second [years] are common, the third embolismic; the fourth and fifth are common, the sixth embolismic; the seventh is common, the eighth embolismic; the ninth and tenth common, the eleventh embolismic; the twelfth and thirteenth common, the fourteenth embolismic; the fifteenth and sixteenth common, the seventeenth embolismic; the eighteenth common, the nineteenth embolismic.

Both [years], as was said before,[7] start from the beginning of the first month, which the Hebrews call Nisan; that is, it starts from the illumination of the Paschal Moon. The beginning of this month ought to be observed according to this rule: so that the fourteenth Moon of Easter never precedes the vernal equinox, but correctly appears either on the equinox itself (that is, on the 12th kalends of April [21 March]) or after it has passed. Hence it is that these beginnings of the lunar year are to be

4 St Leo the Great, *Ep.* 121 to Marcian (Krusch, *Studien I*, 258).

5 St Leo the Great, *Ep.* 129 (PL 121.1075A–B).

6 This chapter draws heavily on Dionysius Exiguus, *Ep. ad Bonifacium et Bonum*, *passim.*

7 See above, ch. 11.

looked for from the 8th ides of March [8 March] up to the nones of April [5 April]. Given that it falls on the 12th kalends of April [21 March] at the earliest and the 14th kalends of May [18 April] at the latest, /**421**/ this same fourteenth Moon makes us look for the day of the Paschal feast from the 11th kalends of April [22 March] up to the 7th kalends of May [25 April]. As Dionysius says: *That calculation of the embolisms is demonstrably logical, which is seen to supply what is lacking to the common years so that the lunar course is equal to the solar period. For although the Moon traverses the annual circuit of the Sun every month, nonetheless it does not succeed in completing it in twelve months. For this reason the Moon is observed to fall short of the reckoning of the solar year by 11 days in common years; but in embolismic years, it is seen to overtake that same [solar] year by 19 days.*[8]

Indeed the Hebrews, who recognized and observed in their law only lunar months of 30 or 29 days according to the natural course of the Moon,[9] completed the 12 months in the common years, and in the second or third year, whenever it was appropriate, added a thirteenth embolismic month at the end of the year.[10] Furthermore /**422**/ the Romans, who have months of unequal [length], do not wish to insert the embolisms into their calculation at positions chosen at random, but rather wherever in the midst of the year they could find a suitable empty interval between the kalends [of one month and the next].[11] They said that the first embolismic Moon begins on the fourth nones of December [2 December], the second on the fourth nones of September [2 September], the third on the day before the nones of March [6 March], the fourth on the day before the nones of December [4 December], the fifth on the fourth nones of November [2 November], the sixth on the 4th nones of August [2 August], and the seventh on the 3rd nones of March [5 March]. They shrewdly take every possible precaution so that whatever age the Moon might be on the kalends, that same Moon should be ascribed to the same month.[12] However, this does not always

8 Dionysius Exiguus, *Ep. ad Bonifacium et Bonum* (83.10–16).

9 Bede's statement that the Hebrews alternated full and hollow lunar months is not accurate. Jones notes that such a statement is found in the recension of the *Prologus Theophili* edited by Bouchier (472), to which Bede may have had access.

10 See above, chs. 11 and 13.

11 See Introduction, section 3.

12 "Furthermore the Romans . . . ascribed to the same month." Cf. *De ratione embolismorum*, ed. by Jones amongst the *alia opuscula coaeva* in *BOD* 685–689.

hold true because of the intervention of the Paschal Moon, whose final days not infrequently fall on the kalends of May and can also last until the second or third day after the kalends, but which is nonetheless the Moon of April and not of May, as it ought ever to be said to be.

46. THE *OGDOAS* AND THE *HENDECAS*

There is another division of the aforementioned [19-year] cycle by which the years are separated into the *ogdoas* and the *hendecas*. This happens in the third and the sixteenth year [of the lunar cycle], because /**423**/ in the eighth and eleventh year [of the 19-year cycle],[13] the Paschal Moon reaches the furthest limits of its nativity,[14] and in an anomalous manner a single common year precedes the embolismic year in both cases. Alternately, [this came to pass] because the ancients saw that eight solar years equals the number of days in the same number of lunar years.[15] Later on, more learned doctors proved that this is impossible unless eleven years were added. The entire cycle of the Moon is distributed through the *ogdoas* and the *hendecas*, and demonstrates both observations. For at one time the Greeks, whom the ancient Romans are also said to have imitated, when they had computed eight common lunar years of equal length, intercalated three embolismic months. For should you wish to calculate 8 times $11\frac{1}{4}$, you will make 90 days, that is, three months. In fact, there are now people who think that 8 solar years equal 8 lunar years, and similarly that 11 solar years make up the same number of lunar years, with an equal number of days.[16]

At this point it should be stated, first of all, without any contradiction, that if 8 years of both planets correspond, then this must always be the case, /**424**/ and it is impossible for 11 years [of both planets to correspond]; but if 11 years of both planets begin and end together, they will always do so, nor will there be any time when the correspondence of 8 years can be restored.[17] In order to hit upon this number, it should be observed that 8 solar years, excluding the leap-year days, make 2,920

13 On the difference between the lunar cycle and the 19-year cycle, see below, ch. 56.

14 I.e. the new Moon of Easter occurs at its latest possible calendar date.

15 On the ancient eight-year cycle or *octaëteris*, see Introduction, section 3.

16 "The entire cycle of the Moon … equal number of days." Cf. Macrobius, *Saturnalia* 1.13.8–10 (62.23–63.9)

17 I.e. one cannot have both an 8-year and an 11-year luni-solar cycle.

days, for 8 times 365 equals 2,920. But note how many days that number of lunar years has – 8 times 354 is 2,832 – and then add the 90 days of the three embolismic months, and these make 2,922. But the two days lacking in the *ogdoas* of the Sun can expect to be supplied. Let us look at the *hendecas* of both planets, and see if perhaps there the Sun can summon the help of the leap-year day, for both periods of time [*ogdoas* and *hendecas*] must be bound by the same rule [concerning leap years]. Eleven times 365 makes 4,015. Likewise 11 times 354 makes 3,894. Add 120 days for the embolismic months, and it makes 4,014. Take away the *saltus*, and there remains 4,013. See, then, whether here the *hendecas* can make use of the assistance /**425**/ of the leap-year day, so that by adding two or three days, it will suffice to compensate the lunar *hendecas*. Certainly it does not need the help of the quarter-day in order to equal the lunar *hendecas*, for it already exceeds it by two days; rather, it replaces what is missing in the *ogdoas* by these excess days. So it is plain that in harmonizing their periods, there is nothing which prejudices the leap-year days, for however many occur in the 19-year cycle, they benefit both planets equally, as we have said.[18]

Now to gut the bowels of this question! In the first year of the Paschal cycle the epact is zero, because the Moon finishes its course on the 11th kalends of April [22 March]. Again in the ninth year of the cycle, that is, after 2,922 days, on the 9th kalends of April [24 March] the Moon is 30 days old, which the sequence of the epacts (which are then 28) plainly shows. What good was it /**426**/ to add two days of the Sun's course – that is, the 10th and 9th kalends of April [25 and 24 March] – in order to complete the lunar *ogdoas*, if two leap-year days could have completed them? Therefore the two days by which the *ogdoas* of the Sun lags behind the *ogdoas* of the Moon are not supplied by the intercalation of the leap-year days, but by the two days in which the Sun's *hendecas* exceeds the Moon's. Nor does the *saltus lunae* harm it, which indeed some place in the *ogdoas*,[19] but which we have stated should be placed in the *hendecas*. But wherever you position it, the question we posed is solved at the same point. The 19-year cycle has 228 solar months and 235 lunar months, making 6,935 days, excluding leap-year days: 19 times 12 is 228, plus 7 is 235; again 19 times 365 is 6,935, and 19 times 354 is 6,726, plus 210 days for the 7 embolismic months, making 6,936.

18 See above, ch. 41.
19 I.e. Victorius.

Subtract one day for the lunar *saltus*, and you will see that in one and the same number of days, the 19-year cycle comprises the course of the Sun and of the Moon. Again, the *ogdoas* has 96 solar months and 99 lunar months; the *hendecas* /**427**/ has 132 solar months and 136 lunar months. The number of days is discussed above.

47. THE YEARS OF THE LORD'S INCARNATION[20]

In the first column of the 19-year cycle is affixed the sequence [of years] which the Greek computists designated as beginning from the time of the Emperor Diocletian. But *Dionysius*, a venerable *abbot of Rome*, and *one endued* with no small knowledge *of both languages (that is, of Greek and Latin), refused*, as he himself attests, in composing his *Paschal cycles, to link them to the memory of an impious persecutor, but chose instead to designate the years from the Incarnation of our Lord Jesus Christ, so that the source of our hope might be the more evident to us, and that the cause of man's restoration, that is, our Redeemer's Passion, might be more clearly manifest.*[21] Placing the 532nd year of our Lord's Incarnation at the beginning of his first cycle, he plainly taught that the second year of his cycle was the same as that when the mystery of the same most holy /**428**/ Incarnation began.[22] For the lunar cycle is one of 19 years, but the solar cycle is completed in 28 years; multiplied together, they produce a total of 532 years. When at last it expires, the whole lunisolar sequence, going back to its beginning, revolves along the same track, producing the same years of the 19-year and lunar cycles, the same lunar epacts and also solar concurrents, the same 14th [Paschal] Moon, the same date for Easter Day, and the same Moon for that day, in sequence. Even Victorius proved this plainly by drawing up a cycle for that number of years, albeit he established Easter according to another doctrine.

Therefore, because the 532nd year from the Lord's Incarnation is complete in the second year of the first cycle composed by Dionysius, it

20 Main source for this chapter is Dionysius Exiguus, *Ep. ad Petronium* (Krusch, *Studien II*, 63–68).

21 The prologue is by the continuator of Dionysius' tables, commonly identified as Felix Ghyllitanus (ed. Krusch, *Studien II*, 87).

22 That is, the cycle begins in BC 1. Dionysius' cycle began in 532, so the previous cycle ended in 531, which means that it began in BC 1. Therefore the second year of the cycle corresponds to AD 1, the year of Christ's birth.

is without doubt the one in which He deigned to become incarnate, as far as the revolutions of the stars are concerned. For this second year of the 19-year cycle is the 18th year of the lunar cycle, having 11 epacts and 5 concurrent days of the week, and a fourteenth Paschal Moon on the 8th kalends of April [25 March]. All these things were exactly the same then, and had there been an Easter falling upon a Sunday at that time, after the Church's present custom, that day would have come in the way noted here on the 6th kalends of April [27 March], and the Moon would have been /**429**/ 16 days old. Therefore Dionysius himself in his Easter formulae demonstrates implicitly what we have said: in order to find *the year of the 19-year cycle, take the years of the Lord*, and before we *divide them by 19*, he commands us to *add one*,[23] showing that when He became incarnate, one year of the 19-year cycle was already complete. Again, in order to know *what is the year of the lunar cycle*, he tells us to *take the years of the Lord, and always subtract two*, and then *to divide by 19*, so that the calculator, with the two years deducted which then remained, and dividing the rest by 19, might arrive at what should be the remainder.[24] Then to discover *what the lunar epacts* are, *take* in due manner *the years of the Lord*, divide, and then *multiply*; again, he bids us divide without adding or subtracting anything, because the epacts, beginning in the second year of the 19-year cycle, never require that anything be added or subtracted from the year of the Lord in order to discover their [present] status, because they began with [the year of the Lord's Incarnation].[25] Again, *if we wish to know the solar increment, that is, the concurrents of the weekdays*, he bids us *take the years of the Lord*, and *having added one fourth*, /**430**/ he teaches us always to add *four* regulars to these, and then finally *to divide by 7*.[26] For the fact is that the concurrents were 5 in the year in which our Lord was born, and in order to arrive at a fixed sequence of computation, the calculator is obliged to add the four that went before.

These things being as they are, diligently seek out the year of the Lord's Passion; nor is the way of seeking it unknown, provided that the computation does not err at any point. If I am not mistaken, the faith of the Church holds that the Lord lived in the flesh, up until the time of the

---

23 Dionysius Exiguus, *Argumenta titulorum paschalium* 5 (76.15–18).
24 *Ibid.* 6 (76.19–22).
25 *Ibid.* 3 (75.9–12).
26 *Ibid.* 4 (76.4–10).

Passion, a little more than 33 years, /**431**/ for he was baptized at the age of 30, as the evangelist Luke attests,[27] and preached for three and a half years after his baptism, as John teaches not only by reminding us of the time of the returning Paschal season in his Gospel,[28] but also in his Apocalypse,[29] and as Daniel also indicates in his visions.[30] That the holy Roman apostolic Church holds this belief is testified by the legends that it customarily inscribes, year by year, upon its [Paschal] candles. Recalling the time of the Lord's Passion to the memory of the people, it records a number of years which is always 33 less than what Dionysius laid down [as having passed] since the Incarnation. Hence, in the 701st year of His Incarnation (according to Dionysius), in the 14th indiction, our brothers who were then in Rome related that they saw, and copied down, the following inscription from the candles in St Mary's: "From the Passion of our Lord there are 668 years".[31] Therefore, since the Paschal cycle revolves in 532 years, as we said above, add 33 to these, or rather 34, so that you may arrive at the very year in which the Lord suffered, and it will make 566. This is the [year corresponding to the] year of the Lord's Passion and Resurrection from the dead. For just as the 533rd year corresponds to the first, so the 566th year corresponds to the 34th year of the luni-solar cycle. And so, with the cycles of Dionysius open before you, should you find in the 566th year from the Incarnation of the Lord that the 14th moon falls on Friday the 8th kalends of April [25 March], /**432**/ and Easter on Sunday the 6th kalends of April [27 March], then give thanks to God, for He has granted that you find what you were looking for, just as He promised! For no catholic may doubt that the Lord mounted the Cross on Friday, on the 15th day of the Moon, and rose from the dead on the first day of the week, that is, on

27 Luke 3.23.

28 John 13.1.

29 Apocalypse 12.6 and 14; cf. 11.3.

30 Daniel 7.25, 12.7.

31 This visit took place during the abbacy of Ceolfrid, when monks were sent to Rome to petition Pope Sergius for letters of privilege for Wearmouth-Jarrow; cf. Bede, *Historia abbatum* 2.15, ed. C. Plummer, *Baedae opera historica* (Oxford: Clarendon Press, 1896):1.380; anon. *Historia abbatum* 20, in *ibid.* 1.394–5. The 14th indiction begins in September 700 and ends in August 701: Ginzel 3.148 *sq.* Since it is evidently a Paschal candle which is being described, this event must have taken place in the spring of 701. Bede probably assigned the indictional date on the basis of the date on the papal letters, which would therefore have been issued before September 701. For an alternative explanation, see Jones, *BOT* 381.

Sunday, lest he appear not to believe in the Law which stipulates that the Paschal lamb be sacrificed on the 14th day of the first month in the evening, nor in the Gospel which asserts that the Lord, taken captive by the Jews on that same evening and crucified in the morning, it being Friday, and buried, rose on the first day of the week.

The opinion of many doctors of the Church that He was crucified on the 8th kalends of April [25 March] and rose on the 6th kalends of the same [27 April] is widely agreed upon as common knowledge. However, Theophilus of Caesarea, an ancient doctor, and one who lived close to apostolic times, in the synodical letter which he composed together with the other bishops of Palestine against those who celebrate with the Jews on the 14th day of the Moon, speaks as follows: *And is it not impious that such a great sacramental mystery as our Lord's Passion should be celebrated outside the terminus? For our Lord suffered on the 11th kalends of April [22 March], on which night he was handed over by the Jews, and he rose on the 8th kalends of April [25 March]. How can the three days be placed outside the terminus? It was decided in that /***433***/ synod that Easter ought to be observed from the 11th kalends of April [22 March] to the 11th kalends of May [21 April].*[32] And earlier on in the same book it is written: *For the inhabitants of Gaul always celebrate Easter on whatever day is the 8th kalends of April [25 March], when Christ's Resurrection is said to have taken place.*[33] But if you are looking for such a year, and you are unable to find it in that place you thought [it would be], blame the carelessness of the chronographers, or better yet, your own slowness, being very wary lest in defending the text of the chronicles you do not appear boldly to impugn the testimony of the Law and the Gospel by saying that our Lord and Saviour underwent the most holy mystery of the Cross in either the 15th or the 16th year of the Emperor Tiberius Caesar, or in either the 29th or 30th year of his age, when the Gospels plainly indicate that the forerunner of our Lord began to preach in the 15th year of Tiberius, and that this same [forerunner] subsequently baptized (amongst others) Jesus, *who was then at about the beginning of his 30th year.*[34]

---

32 (Spurious) *Acts of the Synod of Caesarea* 3–4 (ed. Krusch, *Studien I*, 310).

33 *Ibid.* 1 (307).

34 Luke 3.23.

48. INDICTIONS

The second row [*ordo*] of the 19-year cycle contains the indictions, which always retrace their own footsteps in a cycle of 15 years. We have learned that these [indictions] were established by the ancient diligence of the Romans in order to guard against error which might perhaps arise concerning chronology. /**434**/ When, for example, a certain emperor lays down his life or his kingdom in the middle of the year, it might come about that one historian would assign that year to the time of that king, because he reigned during part of it, while another historian would consider that it ought rather to be assigned to his successor, in that he also ruled during part of it. Lest an error be introduced into the chronology through discrepancy of this type, [the Romans] decreed the indictions, by which scribes and even the common people might readily preserve the integral flow of time. They wished as well that there might be 15 [indictions] for the sake of ease of calculation, so that by a simple number, easy to multiply, the status of past time might more advanta-geously be reduced to memory. However, some think that because formerly, in Republican times, the city of Rome used to be purified after the census was finished in the fifth year, indictions were created to indi-cate the third purification and census. Indictions begin on the 8th kalends of October [24 September] and end on the same day.

49. FORMULA FOR FINDING THE INDICTION

By means of this formula, you will find what the indiction is for any year you wish to compute. *Take the years from the Incarnation* /**435**/ *of our Lord*, however many they may be: for instance, the present year 725. *Always add 3*, because according to Dionysius our Lord was born in the fourth indiction; *this makes* 728. *Divide this by 15*: 15 times 40 is 600, 15 times 8 are 120, and there are 8 left over. It is *the* eighth *indiction. If nothing is left over, it is the 15th indiction.*[35]

50. LUNAR EPACTS

The third column [*linea*] of the aforementioned cycle contains the lunar epacts, which every year habitually exceed the course of the Sun by 11

35 Dionysius Exiguus, *Argumenta titulorum paschalium* 2 (75). Note that this formula, as well as those in chs. 52 and 54, indicates that the year of composition of *The Reckoning of Time* was 725.

days. Hence [these days] are called by the Greek word "epacts", that is, "additions", for as we said, they are accumulated by the addition of 11 days every year; or else they are appropriately called "epacts", that is "additions", because in order to discover the age of the Moon on the kalends [of the] twelve [months], they are added together throughout the year, as we taught [above].[36]

In fact, each day throughout the entire circle of the revolving year has its 11 lunar "additions". For example, if as I write today the Moon is five days old, on this same day one year from now the Moon /**436**/ will be 16 days old; after two years, 27 days old; after three years, 8 days old; nor does it revert to what it is now until the cycle of 19 years is finished. But the epacts noted in the 19-year cycle specifically stand for the age of the Moon on the 11th kalends of April [22 March], the beginning of the Paschal feast. The [epacts] always observe this rule in connection with [the Moon's] course: whenever they are less than the number 16, they announce the Paschal lunation, but whenever they are more, they direct us to look for Easter in the next lunation. This is because the fullness of the Paschal Moon ought not to precede the equinox, but rather should follow it, as was ordained in the beginning of Creation when the Sun first rose at the beginning of the day, holding to the spring equinox, and the Moon therefore rose at the beginning of the night, occupying the sector of the autumn equinox.[37]

Hence it is plain that they make a great mistake who decide that the

---

36 *uel certe quia ad inueniendas quotae sint lunas kalendarum XII per totum adiciuntur annum, ut supra docuimus . . .* (435.5–7). Bede's reference to a previous discussion clearly shows that he is referring to the formula explained in ch. 20 for finding the age of the Moon on the kalends of each of the twelve months of the year – a formula which uses annual epacts and monthly lunar regulars. Some medieval MSS corrected the figure XII to XI, thus transforming it into a date: the 11th kalends [of April], or 22 March, i.e. the day from which the epacts are counted, and which Bede names a few lines later. Curiously, it also confused Jones (*BOT* 385), who likewise thought that Bede was referring to a date, but who accepted the reading *kalendarum XII*, i.e. 21 March. In Jones' view, this date reflects Bede's idea that the beginning of the tropical year is the starting-point of all astronomical calculation. Bede does allude to this "natural calculation" from time to time in *The Reckoning of Time*, and it is a major theme in the *Letter to Wicthed*, where it helps to explain why ps.-Anatolius places the equinox on 22 March. But Bede deliberately left the awkward issue of Anatolius' equinox out of *The Reckoning of Time* as it would have only served to confuse his students. Moreover, Jones' theory ignores the clause *ut supra docuimus*.

37 Cf. ch. 6 and ch. 43.

inception of the Paschal lunation should be sought starting from the 3rd nones of March [5 March], because in fact the Moon which appears then will show a full Moon before the equinox.[38] Hence it is not fit for the Paschal solemnity, in that, as we said, /437/ it is necessary that first the Sun and then the Moon rise in the position in which they were originally created, so that [the Sun] wins victory over the length of night by crossing the equinox, and [the Moon], though she may be smaller, illuminates the whole length [of night] by her fullness. Virtually no one doubts that this refers to the sacrament of Christ and his Church. We referred to some of these matters at the beginning of this work:[39] let us go back over a few of them now. Because (as they say) the Moon and stars do not shine with their own light, but take their light from the Sun, so also the Church and all the saints possess the good by which they live not by the merit of their own virtue but by the grace of the liberal Giver. And just as we maintain our fortitude not by the strength of our own willpower, but by His acceptance of us, whose mercy goes before us, nor are we capable of thinking anything on our own, as if from our own selves, but our capability comes from God, even so in the season wherein we celebrate the great events of our redemption, the perfection of solar splendour which illuminates ought to go before the lunar [splendour] which is illuminated.

51. HOW CERTAIN PEOPLE ERR CONCERNING THE BEGINNING OF THE FIRST MONTH[40]

But see whether the error of those who think otherwise seems to be confirmed and supported by the truth, at least in the eyes of those who make wicked laws of this sort, and who in writing, write what is not just. Victorius, who composed their cycles, writes in the prologue of this same work, amongst other things, as follows: *But the Latins, on the other hand, considered that [the period from] the 3rd nones of March [5 March] to the 4th nones of April [2 April] – that is, 28 days – ought to be observed [as the limits of the beginning of the Paschal lunation], /438/ so*

38 The major culprit here is Victorius of Aquitaine, *Prologus ad Hilarium archidiaconum* 4 (ed. Krusch, *Studien II*, 19.12–14).

39 Cf. ch. 6.

40 On the possible sources for this chapter, and particularly the text beginning *Constans igitur . . .* incorporated into the Byrhtferth glosses to *The Reckoning of Time* (PL 90.499C–500C and 713–714), see Jones, *BOT* 386.

*that the beginning of the Paschal month falls on whatever day [within this
period] there is a first Moon. If the 14th Moon of this [lunation] comes
on a Friday, the following Sunday – that is, the 16th Moon – is assigned
without any ambiguity to the Paschal celebration. But if it happens that
the full Moon is on Saturday and the 15th Moon is found on the following
Sunday, they said that Easter ought to be transferred to the next Sunday,
that is, the 22nd Moon, with a week having been passed over, lest they fix
a Moon which is less than 16 days old for the celebration of the mystery of
that Sunday. Nor will they at any time accept a Moon that is more than
22 days old, but they would rather that the day of the Paschal feast be
extended to the 22nd day than ever to begin it on a Sunday before the 14th
Moon. Furthermore, they assert that the 14th Moon of this month should
be counted from the 15th kalends of April [18 March] until the 16th
kalends of May* [16 April].[41] And again in the same prologue he says:
*But when it happens that the 27th Moon falls on a Saturday, and in parti-
cular, on the kalends of January without a leap-year day, Your Holiness
should know that Easter is not to be found save either on the 13th kalends
of April [20 March] according to the Latins, which is never to be celebrated even if the Moon agrees, or on the 8th kalends of May [24 April]
/439/ according to the Egyptians, which is sometimes observed.*[42]

Let us see, therefore, in what manner Victorius commends his
"Latins". He says that they prefer to think that a new Moon which
begins on the 3rd nones of March [5 March] is the start of the first month,
and that Easter Sunday ought to be celebrated on the 16th Moon. And
again he says that Easter is never to be celebrated on the 13th kalends of
April [20 March], as the Latins do, even if the Moon agrees, but rather
on the 8th kalends of May [24 April] as the Egyptians do. But he agrees
that a Moon which was new on the 3rd nones of March [5 March] will be
16 days old on the 13th kalends of April [20 March]. Now I ask you, holy
brother Victorius, if a new Moon on the 3rd nones of March [5 March]
marks the beginning of the first month, why do we not celebrate Easter in
this month, but instead put it off to the next month – when the Law
commands all who are capable of doing so to celebrate the Pasch in the
first month, and that only those on a journey, or who are unclean, should

41 Victorius of Aquitaine, *Prologus* 4 (19.12–20.4). Most MSS of Victorius read
XVII kal. mai. in the last sentence, but all MSS of Bede used in Jones' ed. save Melk Stiftsbi-
bliothek 370 read *XVI*.
    42 *Ibid.* 12 (26.15–19).

celebrate in the second month?[43] Why do you add, "without a leap-year day"? Perhaps because when the kalends of January [1 January] is a Saturday, the 13th kalends of April [20 March] will fall on a Sunday if it is not a leap year, but the same date will fall on a Monday if the leap-year day intervenes, and you wish to teach that when Monday the 13th kalends of April [20 March] is the 16th Moon, then Easter ought rightly to be celebrated on the following Sunday, 7th kalends of April [26 March] at the 22nd Moon? Indeed, even though on that same day – the 13th kalends /**440**/ of April [20 March] – the Moon is 16 days old, one cannot celebrate Easter then because the equinox has not yet passed. But neither [can one do so] on the following Sunday, because a Moon which is too old (that is, 23 days) falls on that day, and this necessity compels us to defer the solemnities of Easter until the second month. A wonderful instructor in computation you are! Your chief doctrine is that a new Moon on the 3rd nones of March [5 March] begins the first month; then, on the other hand, vanquished by reason itself, you are forced to admit that unless the leap-year day intervenes, it is not at all the new Moon on that day, but rather the one which will be kindled 29 days later which fits the Paschal feast. But if there is a leap-year day, then it becomes the first month of the following year, which without [the leap-year day] would be the last [month] of the preceding [year]. And what is this that you say: "Even if the Moon agrees, Easter is absolutely never to be celebrated on the 13th kalends of April [20 March]"? But how can the Moon agree with the Paschal solemnities at a time when the Paschal solemnities are never to be celebrated? And if Easter is never to be celebrated on the 13th kalends of April [20 March] even if the 16th Moon arrives, then patently the Latins are wrong when they claim that the Paschal lunation begins on the 3rd nones of March [5 March], since it is not permitted to celebrate Easter on the 16th or 17th day of this [lunation] because the equinox has not passed. Having spurned their observance, you rightly urge, along with the Egyptians, that Easter in that year ought to be postponed to the 8th kalends of May [24 April]. If you think that what the Egyptians teach ought to be observed instead, why do you not opt for their science on all points? In fact, you choose not to do this, but to take a middle course between the two, and to dispense the calculation of the Latins, which you despise, to your own people to be taught and followed, rather than that of the Egyptians which you prefer. We have learned to follow the Egyptian

43 Numbers 9.10–11.

practice of computation in all things, which in your own judgement and by universal agreement of the Church seems to be truer. That is, the beginning of the first month is to be sought from the 8th ides of March [8 March] to the nones of April [5 April], the 14th Moon of Easter from the 12th kalends of April [21 March] until the 14th kalends of May [18 April], Easter Sunday from /441/ the 11th kalends of April [22 March] to the 7th kalends of May [25 April], and this from the 15th up to the 21st Moon.

Lest the friends of Victorius accuse us of attacking him wantonly, let them read the book of that most learned and holy man Victor, bishop of Capua, concerning the Easter which it was thought should be celebrated on the 15th kalends of May [17 April] in the 13th indiction, in the ninth year of the proconsul Basilius, and they will discover how highly thought of their teacher is by the prudent and catholic doctors of the Church. His book begins as follows: *Although the venerable solemnity of Easter had been carefully sought by us with a view to the day upon which it should most properly arrive in this present year, the 13th indiction, and albeit we declared according to the constitutions of the venerable Fathers that the Resurrection of our Lord should beyond a doubt be celebrated on the the 8th kalends of May [24 April], our response was viewed by some as quite unreasonable because a certain Victorius fixed the Sunday of the Resurrection otherwise in the Paschal cycle which he published, although he also designated the [Sunday] which we declare ought to be celebrated.*[44] And later in the work: *But now, order demands that I throw open to plain view the errors in the cycles which Victorius published, seeing that he does not know how to determine the proper day of Easter, so that as he was shown to have failed in this manner in the past, he may lack authority in the present and the future, and that he may renounce all occasion for evil persuasion.*[45]

52. FORMULA FOR FINDING THE NUMBER OF THE LUNAR EPACTS

If you want to know what the epacts are for any given year, take the number of years of the Lord, for example in the present eighth indiction,

---

44 An example of Victorius' policy of double dates for Easter in dubious cases: see Introduction, section 3.

45 This passage from Victor of Capua is not among the fragments printed by Pitra.

725. Divide by 19: 19 times 30 is 570 and 19 times 8 is 152, with 3 left over. Multiply these by 11, which makes 33. Subtract 30, and the remainder is 3. The epact, that is, the lunar increment, is 3.[46] /**442**/

53. SOLAR EPACTS

In the fourth column [*tramite*] of the 19-year cycle the solar epacts are designated, that is, the concurrent days of the week, which for three years always increase by the addition of one, but during leap year by the addition of two, up to the number seven. Their cycle runs for four times seven, or 28 years, because assuredly it cannot be completed until the leap-year day, which recurs in the fourth year, has fallen on every day of the week, that is, Sunday, Friday, Wednesday, Monday, Saturday, Thursday, Tuesday; it runs through them in this order.

Although every day of the year has its concurrents, those concurrents which are particularly attached to the [19-year] cycle are those which designate the weekday of the 9th kalends of April [24 March]. This is so that, being placed quite close to the beginning of the Paschal festivity, they may very readily display the day of the epacts, or what day of the week the fourteenth Moon [falls on], and thereby smooth the way to the discovery of Easter Sunday. For every year, the same day of the concurrents will fall on the 2nd kalends of April [31 March], the 7th ides of April [7 April], the 18th kalends of May [14 April] and the 11th kalends of May [21 April]: it is helpful to the computist to memorize these.

The course of their cycle is such that whatever the concurrents are in leap year, they are the same five years earlier and six years later. Whatever they are in the first year after the leap year, they will be the same as those which occurred eleven years before, and those which will recur in six years. Those for the second year after the leap year are the same as those which occurred six years previously, and which will recur after eleven years. Those in the third year after leap year are the same as those which took place six years before, and which will revert after five years. And the regular pattern of this division applies to all the days of the revolving year.

Take careful note of the fact that because of this solar cycle, which runs for 28 years, it is necessary that 28 19-year cycles be completed before the identical sequence for the observation of Easter is repeated in

46 Cf. Dionysius Exiguus, *Argumenta titulorum paschalium* 3 (75).

all respects, so that every year of this [solar] cycle becomes the first year of a 19-year cycle, and similarly, that each year of the 19-year cycle follows as the first of this [solar cycle]. Thus the entire series of the Easter observance will be finished in no less that 532 years.[47] /**443**/

54. FORMULA FOR FINDING THE NUMBER OF THE SOLAR EPACTS, AND WHEN LEAP YEAR WILL FALL

Since the sequence of the [solar] epacts, that is, of the solar increment, coincides and is closely connected with the leap year, learn by means of this formula what is the state of both simultaneously.

If you *want to know when the leap year will be, take the year of the Lord [725], divide it by 4, and if there is no remainder, it is leap year; if there is a remainder of 1, 2, or 3*, it is 1, 2 or 3 years away from leap year.[48] For instance, $4 \times 100 = 400$, $4 \times 80 = 320$, $4 \times 1 = 4$, with 1 left over, because it is the first year after the leap year. *If you wish to know the solar increment, that is, the concurrent days of the week, take the number of years from the Lord's Incarnation, for instance* 725 (which is the 8th *indiction*), *and add to this one-fourth of these years, that is, at present*, 181, *which together* [with 725] *makes* 906. *Add four to this, which makes* 910. *Divide by 7*: $7 \times 100 = 700$, $7 \times 30 = 210$, with no remainder. *Therefore the solar epact or concurrent days of the week are 7.*[49]

55. THE CYCLE OF BOTH EPACTS, AND HOW TO CALCULATE THEM USING THE FINGERS

Because anyone who has memorized the cycle of the Sun and Moon grasps very readily the day of Easter and the recurrence of the other dates, assuming he knows how to multiply and divide these cycles by 19 and 28, every computist ought to remember that the same solar concurrents will follow in 30 years as will come next year, and in 60 years from now as will come in four years, and in 120 years from now as will come in eight years, and in 150 years from now as in ten years, and the others in like manner. And if you want to know the number of the concurrent

---

47 "Take careful note . . . 532 years." Cf. cognate text in London, British Library Cotton Caligula A.XV (s. VIII²) fols. 110r–117v.

48 Dionysius Exiguus, *Argumenta titulorum paschalium* 8 (76.27–77.1).

49 *Ibid.* 4 (76.4–10).

for a year occurring at any interval, find out how many times 30 goes into that number [of years], and with this 30th part doubled, you will know the harmony /**444**/ of the revolving years. For instance, 10 times 30 is 300, and hence the same concurrents occur in 300 years from now as will occur 20 years from now. If there is a remainder, add it in. In a similar manner, the order of the concurrents can be projected into the past.

Likewise, because the lunar cycle is 19 years long, the epact of the present year will occur 20 years from now; those for next year, 40 years from now; those of the third year, 60 years from now; those of the fourth year, 80 years from now; those of the fifth year, 100 years from now; those of the sixth year, 120 years from now; those of the seventh year, 140 years from now; those of the eighth year, 160 years from now; those of the ninth year, 180 years from now; those of the tenth year, 200 years from now, and so on for the rest. This should be understood to apply to the fourteenth Moon of Easter and everything else that is included in the 19-year cycle. The diligent calculator will in this manner be able to observe the recurrence of sequential dates that are still more distant from him, and by this means of these, he will consistently remember to bring even dates in the past into accord.

But it seems worth recalling that some people, in order to simplify calculation, have transferred both cycles, the lunar and the solar, onto the joints of the fingers. Because the human hand has 19 joints if we include the tips of the fingers, by applying each year to one of these joints, they begin the lunar cycle on the inside of the left hand, at the base of the thumb, and they finish it on the inside, at the tip of the little finger. Again, because the two hands together, if you do not count in the fingertips, have 28 joints, they assign the years [of the solar cycle] to these, beginning at the little finger of the left hand and ending on the thumb of the right. They do not do this, as in the lunar cycle, by proceeding through the fingers in numerical order, but because of the calculation of the quarter-day [i.e. the leap-year increment], they mark out each four-year period by traversing the four fingers, so that the six joints of the little fingers contain all the leap years. Again, the six joints of the fingers next to the little fingers comprise all the years that follow immediately after the six leap years. By the same token, those of the second finger [from the little finger] embrace the second years, and those of the third finger the third years. On the other hand, the seventh leap year with the three years following it are appropriated to the four joints of the thumbs. Whether the computist wishes to avail himself of this or of some other system, the hands can nonetheless readily hold the cycles of both planets. But many aspects of this discipline, just as of the other arts, are better conveyed by the utterance of a living voice than by the labour of an inscribing pen. /**445**/

56. THE LUNAR CYCLE

The lunar cycle is contained in the fifth column [*regione*] of the 19-year cycle. It begins in the fourth year of that cycle, and ends in the third, and is linked to the month of January, and is Roman in origin. For just as every year of the 19-year cycle begins and ends with the Paschal [lunar] month because of the observation [prescribed in] the law of the Jews, so likewise this [lunar year], as instituted by the Romans, begins with the Moon of the month of January and ends there. What applies to the former [i.e. to the 19-year cycle] also applies to this one: the first and second years are common, the third is embolismic; the fourth and fifth are common, the sixth is embolismic; the seventh is common, the eighth is embolismic. Also the *hendecas* of the lunar cycle, just like that of the 19-year cycle, has seven common years and four embolismic ones. The common years have 12 lunar months and 354 days, but the embolismic ones have 13 lunar months and 384 days, minus one from the 17th year of this cycle, which is the first year of the 19-year cycle, from which one day is subtracted because of the "leap of the Moon". In order to make this plainer, let us look at the course of each year in order, and do for January what Dionysius has done for the Paschal month.

In lunar year 1 (19-year cycle 4) running from the kalends of January [1 January] to the 13th kalends of January [20 December], there are 354 days, because it is common.

In lunar year 2 (19-year cycle 5) running from 12th kalends of January [21 December] to the 5th ides of December [9 December], there are 354 days, because it is common.

In lunar year 3 (19-year cycle 6) running from the 4th ides of December [10 December] to the 5th kalends of January [28 December], there are 384 days, because it is embolismic.

In lunar year 4 (19-year cycle 7) running from the 4th kalends of January [29 December] to the 16th kalends of January [17 December], there are 354 days, because it is common.

In lunar year 5 (19-year cycle 8) running from the 15th kalends of January [18 December] to the 8th ides of December [6 December], there are 354 days, because it is common.

In lunar year 6 (19-year cycle 9) running from the 7th ides of December [7 December] to the 8th kalends of January [25 December], there are 384 days, because it is embolismic.

In lunar year 7 (19-year cycle 10) running from the 7th kalends of January [26 December] to the 19th kalends of January [14 December], there are 354 days, because it is common.

In lunar year 8 (19-year cycle 11) running from the 18th kalends of January [15 December] to the 4th nones of January [2 January], there are 384 days, because it is embolismic.

In lunar year 9 (19-year cycle 12) running from the 3rd nones of January [3 January] /**446**/ to the 11th kalends of January [22 December], there are 354 days, because it is common.

In lunar year 10 (19-year cycle 13) running from the 10th kalends of January [23 December] to the 3rd ides of December [11 December], there are 354 days, because it is common.

In lunar year 11 (19-year cycle 14) running from the 2nd ides of December [12 December] to the 3rd kalends of January [30 December], there are 384 days, because it is embolismic.

In lunar year 12 (19-year cycle 15) running from the 2nd kalends of January [31 December] to the 14th kalends of January [19 December], there are 354 days, because it is common.

In lunar year 13 (19-year cycle 16) running from the 13th kalends of January [20 December] to the 6th ides of December [8 December], there are 354 days, because it is common.

In lunar year 14 (19-year cycle 17) running from the 5th ides of December [9 December] to the 6th kalends of January [27 December], there are 384 days, because it is embolismic.

In lunar year 15 (19-year cycle 18) running from the 5th kalends of January [28 December] to the 17th kalends of January [16 December], there are 354 days, because it is common.

In lunar year 16 (19-year cycle 19) running from the 16th kalends of January [17 December] up to the nones of December [5 December], there are 354 days, because it is common.

In lunar year 17 (19-year cycle 1) running from the nones of December [5 December] up to the 10th kalends of January [23 December], there are 384 days, because it is embolismic.

In lunar year 18 (19-year cycle 2) running from the 9th kalends of January [24 December] up to the 2nd ides of December [12 December], there are 354 days, because it is common.

In lunar year 19 (19-year cycle 3) running from the ides of December [13 December] up to the 2nd kalends of January [31 December], there are 384 days, because it is embolismic.[50]

The computation of the 17th lunar year begins on the same day on which the previous year ended, and not on the day following, as do the others, for this reason: so that one day will not seem to be lacking in that year, because of the so-called "leap" of the Moon. Dionysius, surveying the 19-year cycle in like fashion, taught the same thing: he brings the final year to a close on the same day on which the first year is to begin.[51]

57. FORMULA BASED ON THE LUNAR CYCLE FOR FINDING THE AGE OF THE MOON ON 1 JANUARY[52]

Anyone who wants to can make a formula from the lunar cycle in order to find the age of the Moon on the kalends of January. Take any year of the lunar cycle, for instance, 5. Multiply by 11, and that makes 55. Always add the regular [for January],[53] and that makes 56. Divide by 30, and the remainder is 26. In the 5th year of the lunar cycle, the Moon is 26 days old on the kalends of January. Again, take 8. Multiply by 11, and it makes 88. Add one regular, and divide by /**447**/ 30, and 29 will be left over. In the 8th year of the lunar cycle, the Moon will be 29 days old

50 Bede's model here is the analogous list of years in the 19-year cycle in Dionysius Exiguus, *Epistola ad Bonifacium et Bonum* (85–86).

51 Cf. ch. 42.

52 This chapter is based on (ps.-)Dionysius Exiguus, *Argumenta* 13 (78.9–15).

53 The lunar regular for January in this case is 1, because the lunar cycle begins in a year when the Moon is one day old on 1 January. Compare the lunar regulars for the 19-year cycle, explained in ch. 20 and Commentary.

on the above-mentioned kalends. Only remember to add 2 regulars in the 17th, 18th and 19th years of this cycle, not 1 as in other years, and you will find the Moon on the kalends without any mistake.

58. A FORMULA TO FIND WHAT YEAR OF THE LUNAR CYCLE OR OF THE NINETEEN-YEAR CYCLE IT IS

If you wish to know the year of the lunar cycle, take the year of our Lord, for instance 725, and always subtract two:[54] there remain 723. Divide these by 19, and there is a remainder of 1; it is the first year of the cycle. But whenever there is no remainder, it is the 19th year.[55] And because the 19-year cycle keeps the same course as the lunar cycle, but a bit behind it, if you wish to know what year of this cycle it is, take the year of the Lord, for instance 725, and always add 1;[56] this makes 726. Divide by 19, and there is a remainder of 4. It is the fourth year of the 19-year cycle. If there is nothing left over, it is the last year.[57]

59. THE FOURTEENTH MOON OF EASTER

The sixth section [*locus*] of the aforementioned cycle contains the four-teenth Moons of the first month, which indicate without any ambiguity the day of Easter Sunday in every year; for the Sunday which comes after the fourteenth Moon is the Paschal day of the Lord's Resurrection. This fourteenth Moon manifests its progress over the Earth at sundown on the equinox,[58] that is, the 12th kalends of April [21 March] at the earliest, and at the latest 29 days later, that is, on the 14th kalends of May [18 April]. It is agreed that according to the time appointed by the Law [*legali tempore*] the course of the Paschal observance is contained within these limits throughout the nineteen years [of the cycle]. And even were it possible for this same fourteenth Moon to fall on Saturday every year, nothing would displace the time of our Paschal observance from its lawful [time]. For [we], sacrificing according to the precept of the Law always on the fourteenth day of the Moon of the first month at

---

54 Because the lunar cycle runs 3 years ahead of the 19-year cycle, but the 19-year cycle begins in BC 1 (see following *argumentum*), only 2 is deducted from the AD year.

55 Dionysius Exiguus, *Argumenta titulorum paschalium* 6 (76.19–22).

56 Because the 19-year cycle begins in BC 1: see ch. 46 and Commentary.

57 Dionysius Exiguus, *Argumenta titulorum paschalium* 5 (76.15–18).

58 Cf. chs. 6, 43.

sunset, and eating the flesh of the immaculate lamb, and sprinkling its blood upon our doorposts to repel /**448**/ the destroyer (this is baptism), and celebrating the solemnities of the Paschal mass, would triumph over the spiritual Egypt. And at break of day on the fifteenth day of the Moon of that month, we would enter upon the first day of Unleavened Bread, and we would complete the seven appointed days of that festivity with due veneration from the morning of the fifteenth day to the evening of the twenty-first day of that first month, that is from Easter Sunday until the Sunday of the octave of Easter.

But because the same day of the Moon falls on different weekdays, it therefore comes to pass (given that we are taught to restrict the beginning of Easter to a Sunday because of the Resurrection of our Redeemer) that sometimes our festivity takes its beginning on the seventh day after the start of the [feast of] Unleavened Bread [prescribed by the] Law. However, it never happens that our Paschal solemnity does not fall on a day of the Pasch [appointed by the] Law, but it often falls on all of them within this [period]. On the other hand, those[59] who think that Easter Sunday ought to be celebrated from the sixteenth to the twenty-second Moon labour under a double burden, because they hold to the beginning of the Pasch prescribed by the Law, and because it often happens that they follow none of the days which are laid down by the Law for the observance of the Pasch. For they reject outright from their celebration both the evening of the fourteenth day, when it was decreed that the Pasch was to begin, and the morning of the fifteenth day, when the appointed solemnity of the seven days of Unleavened Bread is to commence. And on top of it all, to exacerbate this sin, they command that the twenty-second day – which is never once mentioned by Moses in all his Paschal law – be sanctified.

There are those[60] who swerve from the way of truth in the other direction, but with no less error, even though Scripture commands them to travel by the royal road, and not to turn aside from it either to the left or to the right. By decreeing that Easter Sunday should be observed from the fourteenth day of the Moon until the twentieth, these people often pre-empt the beginning of the Pasch [decreed by the] Law, when they divert what is ordained for the fourteenth day of the Moon to the thirteenth day. And what is ordained concerning the twenty-first day, conse-

59 I.e. Victorius.
60 I.e. the Celtic computists.

crating it as holy and illustrious, they utterly disdain, as if this [day] did
not belong to Eastertide at all.

Along with other defenders of catholic faith and action, Theophilus,
bishop of the Church of Alexandria, defeated these people by plain
reasoning. Writing /**449**/ to the Emperor Theodosius the Elder, he said:
*But it sometimes happens that some folk fall into error concerning the
fourteenth Moon of the first month, if this Moon should fall on a Sunday.
When this happens, the fast would have to end on Saturday, when the thirteenth Moon is shown to arrive, and [thus] we start to do what is contrary
to the Law. Hence it is appropriate to issue a vigorous warning that whenever the fourteenth Moon falls on a Sunday, we should rather postpone
Easter until the following week, and this for two reasons: first, lest when
the thirteenth Moon is found on a Saturday we end our fast [then], which
does not make sense, nor does the Law prescribe it, particularly since the
light of that Moon is still seen to be imperfect in its own sphere; and then,
lest we be obliged to fast on Sunday, as this constitutes the fourteenth
Moon, thereby doing something unseemly and illicit, for this is the
custom of the Manichean sect. Therefore, since we ought not to fast when
the fourteenth Moon falls on Sunday, nor is it reasonable to end our fast
if the thirteenth Moon occurs on a Saturday, we assert that [Easter] ought
to be put off until the following week, as is evident from what is immediately
above, but nonetheless without perpetrating any prevarication concerning
the calculation of Easter from this delay. For just as the number 10
includes 9, so also whenever the fourteenth Moon falls on a Sunday,
because fasting is not permitted on that day, it is necessary to defer
Easter to the following week. But no harm is done to Easter because of
this, because the rest of the days which follow are contained [within it].*[61]

60. A FORMULA FOR FINDING IT

The skilful computist ought to memorize the fourteenth Moons of the
first month, just as he memorizes the annual lunar epacts. But if anyone
wishes to find it by means of a formula, let him look at what the epacts
are of the year /**450**/ he wishes to compute, and if they are 14 or 15, let
him know that the fourteenth Moon will come on the 11th or 12th
kalends of April [22 or 21 March]. For as has often been mentioned, the
11th kalends of this month is the proper date of all the epacts. If the

61 *Prologus Theophili* 3 (223–225).

epacts are less [than 14], let him count upwards day by day until he reaches the number 14, and let him not doubt that he has there the fourteenth Moon of Easter. But if the epacts are more than 15, let him count upwards day by day up to 30, which is the end of this month. Then, beginning at the first Moon and arriving in sequence at the fourteenth, he will find the day duly ordained for the Paschal rites. Note also that if it is a common year, the fourteenth Moon will recur eleven days earlier than in the previous year, and if it is an embolismic year, nine days later. There is, however, one exception: the first year of the 19-year cycle, in which, because of the "leap of the Moon", it customarily precedes its course of the previous year by twelve days.

61. EASTER SUNDAY

Easter Sunday is contained under the seventh heading [*titulo*] of the 19-year cycle. It originates in the Sunday on which our Lord rose from the dead. For although in the Old Testament the Paschal season is ordained to be observed according to three formulaic criteria [*argumentorum indiciis*] – namely, it has to be celebrated after the equinox, in the first month, and in the third week of this [month], that is, from sundown of the fourteenth Moon, which is the beginning of the fifteenth Moon, up to the evening of ending of the twenty-first Moon[62] – a fourth rule for its observance has been imposed upon us from the time of our Lord's Resurrection. Hence when we have seen the fourteenth moon of the first month rise in the evening (the equinox having been passed), we should not immediately jump up to celebrate Easter. Rather, waiting for that Lord's Day on which He himself deigned to make that Pasch – that is, the "passing over" from death to life, from corruption to incorruption,[63] from punishment to glory, by his rising – let us celebrate upon that day the solemnities befitting Easter.

Should anyone /**451**/ object that the lawgiver enjoined the remembrance not of the equinox but rather only of the first month and of the third week within that month,[64] let him know that although [Moses] does not mention the equinox explicitly, nevertheless by the very fact that he orders the Pasch to be kept at the full Moon of the first month, he proves

---

62 Cf. Exodus 12.1–3 and 17–19.
63 Cf. I Corinthians 15.42–43.
64 The supporters of Victorius could have used such an argument.

by thorough-going logic that the equinox has passed. For it is agreed beyond any doubt that the Moon of the first month is the first one to show its full orb after the equinox has passed. So on any occasion when we have a Sunday after the fifteenth Moon, our Easter season in no way disagrees with what is prescribed in the Law, although we honour the solemnities of this same Easter with other kinds of sacraments. For whenever Sunday occurs on the second or third or fourth or fifth or sixth or seventh day from this [fifteenth Moon], we do not thus dismiss the Law and the prophets, but rather fulfil the sacraments of Gospel grace. For as Theophilus (the venerable bishop of Alexandria I mentioned earlier) writes: *For indeed, our Saviour was betrayed on the fourteenth day of the Moon, the fifth after the sabbath; and He was crucified on the fifteenth day of the Moon. On the third day He rose again, that is, on the seventeenth day of the Moon, which at that time seems to have fallen on a Sunday, as we learn by examining the Gospels. Therefore we have the assurance that we can rightly celebrate Easter, even though a delay has befallen because of an intervening necessity, so that whether the fourteenth Moon of the first month falls on the sabbath, or on other days before the sabbath on the following week, we may celebrate Easter without any hesitation. But if it falls on a Sunday, let us always put it off until the following week, as has often been said,* for the reasons we have mentioned above.[65] *These things, therefore, have been demonstrated and made plain. But there is also this to consider: in cases of necessity the Law often commands those who cannot celebrate the Pasch in the first month because of lack of time to do it in the second month.*[66] *But it is better for those who find themselves in necessity to follow what is higher than what is lower, for the lower is contained in the higher, but the higher is not included in a lower number. To repeat what we said earlier,* /**452**/ *the number ten contains nine within it, but nine cannot contain ten. If the Law commands us to pass over to the second month if we cannot celebrate the Pasch in the first month because of some necessity, I do not understand why we ought not reasonably to delay Easter until the following week if the fourteenth Moon occurs on a Sunday, given that it was in the first month and on the fifteenth day of the Moon that our Saviour was crucified, and given also that it was the seventeenth day when He rose again after the three days.*[67]

65 Cf. ch. 59.
66 Numbers 9.10.
67 *Prologus Theophili* 4 (225–226).

However, when the Sunday of Christ's Resurrection took place is variously related.[68] As we have mentioned above, some assert that it took place on the 8th kalends of April [25 March], but others on the 6th and some on the 5th kalends of the same month [27 or 28 March].[69] Here it should be noted that if the Lord's Resurrection took place on the 8th kalends [25 March], as the older [authorities] write, it would then assuredly have been the fifth year of the 19-year cycle, having 7 as its concurrent and the fourteenth day of the Moon, as usual, on the 11th kalends of April [22 March].[70] But if the Lord rose on the 6th kalends of April [27 March], it would have been the 13th year of the aforementioned cycle, having 5 as its concurrent and the fourteenth Moon, as usual, on the 9th kalends of April [24 March].[71] On the other hand, if the Resurrection of Christ was celebrated on the 5th kalends of the month [28 March], it being the second year of the 19-year cycle, then the year had four concurrents and the fourteenth Moon as usual was on the 8th kalends of April [25 March].[72] All the mysteries of the 17th day of the Moon, wherein on a Sunday the first deeds of His most holy Resurrection took place, are set forth in unambiguous sequence. Only [let us be] extremely careful lest, by affirming as do some that this was accomplished on the 16th Moon,[73] we incur not only an inevitable loss to our computation, but a very great danger to the catholic faith.[74]

62. THE MOON OF THAT DAY

The Moons of Easter Sunday are shown in the final column [*meta*] of this oft-mentioned cycle, enclosed, because of variations in the occurrence of

68 Cf. ch. 47.

69 Cf. *Cologne Prologue* (Krusch, *Studien I*, 230) (25 March); *Carthaginian computus of 455*, 7 (*ibid.* 289) and Martin of Braga, *Tractatus paschalis* 4 (ed. Barlow, 272.57) (27 March); Victorius of Aquitaine, *Prologus* 9 (25.7–8) (28 March). Dionysius Exiguus' Easter for AD 34 is 28 March, but the age of the Moon (21) does not match the Gospel data.

70 I.e. AD 42.

71 I.e. AD 12. Both of Jones' editions of *De temporum ratione* read *XI kalendarum* here, but the older editions (e.g. PL 90.513C) as well as the manuscripts I have consulted (e.g. Oxford, St John's College 17 fol. 101r) all read *IX kalendarum*, which is evidently correct.

72 AD 191 is the earliest year following the birth of Christ which fulfils these criteria.

73 The reference is to Victorius.

74 A similar warning is issued in the *Cologne Prologue* (230).

that Sunday, /**453**/ within a span of seven days, that is, from the fifteenth to the twenty-first [Moon]. To be sure, these days are fixed in advance by a statement frequently [found] in the Law. There the Lord says: *In the first month, on the fourteenth day of the month, you shall eat unleavened bread until the twenty-first day at sundown. For seven days, no leaven shall be found within your houses.*[75] Now the rule governing this first month, and the seven days of Unleavened Bread within it, is such that the fifteenth Moon that follows after the equinox is to be understood as belonging to the first month. And this [rule] will put forward whatever Sunday it finds within the seven days up to the twenty-first Moon as suitable for the joys of the Paschal feast. I repeat this frequently because there were certain people who, observing Easter Sunday, for instance, on the 7th kalends of April [26 March], with the Moon 20 days old, would say: "And what error in establishing the date of Easter can you show us, when no computist [*calculator*] ever forebade that Easter be celebrated on the 7th kalends of April? Everyone agrees that the equinox is passed, and no one has denied that the twenty-first Moon is appropriate for Easter Sunday."[76] To these people we reply that Easter Sunday may indeed fall on the 7th kalends of April when the Moon is in accord, and on the twentieth day of the Moon when an opportune date is favourable. But because the Moon which on the 7th kalends of April is twenty days old was full before the equinox, it is not permitted to celebrate Easter Sunday on the 7th kalends of April when the Moon is twenty days old. The Sunday following the full Moon which falls on or after the equinox will give the lawful Easter.

The same mistake gets the same response when these computists decree that Easter should take place on the 11th kalends of April [22 March] with a Moon 16 days old, on the 10th kalends of April [23 March] with a Moon 17 days old, on the 9th kalends of April [24 March] with a Moon 18 days old, on the 8th kalends of April [25 March] with a Moon 19 days old, on the 7th kalends of April [26 March] with a Moon 20 days old, on the 6th kalends of April [27 March] with a Moon 21 days old, or on the 5th kalends of April [28 March] with a Moon 22 days old. They often do this in a year when they have a Moon 25 days old on the kalends of January. When they claim that the Moon on this

75 Exodus 12.18–19.

76 Their argument, and Victorius', is that the equinox is the terminus of Easter, not the terminus of the Easter full Moon: see Introduction, section 3.

date is 27 days old, then they deviate quite far from the truth indeed, because undoubtedly they count the Moon which is new on the 3rd nones of March [5 March] as belonging to the first month, which Moon is manifestly full three days before the equinox.[77] And because /454/ it is not valid to celebrate Easter Sunday on the fifteenth, sixteenth and seventeenth days [of the Moon], since these [days] fell on the 14th, 13th and 12th kalends of April [19, 20, 21 March], assuredly none of its remaining days can be suitable for the celebration of the Lord's Resurrection, even though they follow the equinox.

63. THE DIFFERENCE BETWEEN THE PASCH AND THE FEAST OF UNLEAVENED BREAD

Because we have been surveying some matters concerning the observance of Easter, it is proper to point out here that according to the text of the Law, the celebration of the Pasch is different from the celebration of Unleavened Bread.[78] The single day of the Pasch, that is, the "passing over", is the fourteenth day of the first month, when it is commanded to sacrifice the lamb at sundown. And then, in the night which followed, the Lord passed over, striking the firstborn of the Egyptians and exempting the houses of the children of Israel which had been marked with the blood of the lamb.[79] But the seven days which follow, that is from the fifteenth to the twenty-first day of that same month, are properly called [the days] of Unleavened Bread.[80] For it is written in Exodus, where it is commanded that the lamb be slain on the fourteenth day of the first month at sundown: *And you shall eat in haste, for it is the Pasch, that is, the Lord's Passover.* And *on that night I shall pass through the land of Egypt and strike every firstborn in the land of Egypt.*[81] And shortly thereafter: *And when your sons shall say to you, what is this rite?*

---

77 The epacts in Victorius' table are keyed to 1 January. In years when the epact is 25, a new lunation will begin on 7 March, and the Moon will be 16 days old on 22 March, the earliest calendar date of Easter. But because the 14th Moon will fall on 20 March, before the equinox, in Bede's eyes this is not a valid Easter. The problem is even worse in years with a Victorian epact of 27, for then the lunation will begin on 5 March, and the Moon will be 14 days old on 18 March. This is the situation discussed in ch. 51.

78 Cf. Exodus 12.11.

79 Bede, *In Gen.* 1.19 (19.561–20.582); Exodus 12.3, 12, 16.

80 Exodus 12.18.

81 Exodus 12.11–12.

you shall say, It is the sacrifice of the Lord's Passover, when He passed over the houses of the children of Israel in Egypt, striking the Egyptians and exempting our houses.[82] And again in Leviticus: *The Lord's Passover is in the first month, on the fourteenth day of the month at sunset. And on the fifteenth day of the month, it is the Lord's Feast of Unleavened Bread. For seven days you shall eat unleavened bread. The first day shall be for you most solemn and holy; on it, you shall do no servile work. But you shall offer a sacrifice in fire to the Lord for seven days.*[83] And lest anyone object that we have understood the words of the Law otherwise than truth would have it, let him see what Josephus, a man most learned in the legal writings and a priest, has to say about this. In the book of the *Antiquities*, he writes in this wise: the lamb is sacrificed *on the fourteenth Moon* /**455**/ *of the first month; and on the fifteenth day there follows the Feast of Unleavened Bread, which is celebrated for seven days. And on the second day of Unleavened Bread, which is the sixteenth day of the Moon, they offer the first-fruits which they have gathered.*[84]

It is the custom of the Church even now to imitate, not ignobly, the tradition of the sacred laws, observing principally the single night of the Lord's Passover, that is, of His Resurrection from the dead, by which He deigned to save the faithful in triumphing over the impious, and in the shedding of whose blood – the blood of that immaculate Lamb – His people have been washed clean of every sin in the fountain of regeneration. And then we add on another seven days as a suitable festival in memory of the Lord's Resurrection. Indeed, because this day of the Pasch is also commanded to be free of leaven, the Gospel text also counts it as the first day of Unleavened Bread, saying: *And on the first day of Unleavened Bread, when they sacrifice the Passover, His disciples said to Him, Where do you wish that we should go and make ready for you that you may eat the Passover?*[85] He calls that day – the fifteenth day of the first month in which the seven days of Unleavened Bread begin – by the name of [Passover] because of its proximity to Passover, when he says: *And they did not enter the praetorium, lest they be polluted and not be able to eat the Passover.*[86] The Gospel text does not thereby contradict

82 Exodus 12.26.
83 Leviticus 23.5–7.
84 Josephus, *Antiquitates* 3.10.5 (ed. Blatt, 250.16–17, 20–22).
85 Mark 14.12.
86 John 18.28.

the Law, but [it said this] because it was concerned to inculcate in us more vividly, by this association of words, the mystery which befitted it.

But without getting into a more subtle discussion, we can understand that each of us enacts the mystical celebrations of Easter on the day of our baptism, in that we escape spiritual annihilation through the sign of the precious blood, and pass over from spiritual darkness. But in the whole span of life as it advances, which we pass through in this pilgrimage, we should celebrate the seven days of the Unleavened Bread, in which, as the Apostle teaches, we ought to *keep the feast, not with the leaven of malice and wickedness, but with the unleavened bread of sincerity and truth.*[87] And because in baptism we seek to pass over, as it were, from the power of Satan to the portion alloted to the saints, it is necessary to cling to sincerity and truth. Likewise for the whole time of our pilgrimage, which unfolds in the sevenfold number of the days[88] /456/ we are enjoined to pass over to better things by daily progress, just as we are known to eat unleavened bread in the Passover season, and in the days of Unleavened Bread to make a spiritual Passover.

64. THE ALLEGORICAL INTERPRETATION OF EASTER

The time when Easter is ordained to take place is, like the Paschal celebrations as a whole, redolent with sacred mystery. In the first place, we are careful to wait until after the equinox to celebrate the Lord's Passover, according to the decree of the Law, so that the feast-day on which the Mediator between God and man, having destroyed the power of darkness, opened the way of light for the world, might show its inner [significance] by means of the order of time. And what the light of eternal beatitude promises us is most fully celebrated when the light of the Sun, progressing according to its yearly increase, wins its first victory over the shadow of night. So we pay heed to the first month of the year, which is also called the month of first-fruits, in which we celebrate Easter. For this is the month in which the world was fashioned and the first man installed in Paradise. Through the mystery of this feast we hope that we shall recover our primal robe and return once more to that first realm of supernal joy from which we departed into a far-off

---

87 I Corinthians 5.8.
88 For the comparison of seven days of the week to seven ages of human life and seven Ages of the world, see chs. 10 and 66.

land. Of the glory of this kingdom the blessed Apostle Peter writes: *Nevertheless we, according to his promise, look for new heavens and a new Earth, wherein dwelleth righteousness.*[89] And also John in his Apocalypse: *Then said he that sat upon the throne, Lo, I make all things new.*[90]

Then again, we observe Easter on the third week of the [first month], which harmonizes in a most fitting fashion with the joys of our Lord's Resurrection. For His most holy Resurrection took place on the third day and in the Third Age of the world – that is, when by the advent of divine grace the entire dispensation in His flesh, which was consummated through the glory of the Resurrection, was manifested to the world.[91] For He deigned to illumine the First Age of the world by natural law through the patriarchs, the middle Age by the written Law through the Prophets, and the final Age by the gift of the spirit [*charismate spiritali*] through His own self.

But the revolution of the Moon also offers us a most lovely /**457**/ image of the heavenly sacrament, for the Moon, made in a round shape, takes its light from the Sun (as we said above)[92] and thus is always lit up on the half of its sphere which is turned towards the Sun, but always dark on the other half. From the first Moon to the fifteenth, its light increases [on the side] toward the Earth, but it is eclipsed [on the side] toward the heavens; but from the fifteenth to the new Moon, the increase of its light gradually reverts from the earthward to the heavenward side. Without doubt, its revolution rightly signifies the mystery of our Easter rejoicing, in which we are taught to turn all the splendour of our mind away from visible delights and transitory partialities and to fix upon the light of heavenly grace alone, through contemplation. Or if it pleases us to interpret both aspects of its revolution in a positive sense, we can understand the light of the Moon, increasing before human view, to signify the grace of the virtues by which our Lord, appearing in the flesh, illumined the world.[93] Concerning these virtues it is said: *And Jesus increased in wisdom and stature and in favour with God and men.*[94] Re-augmenting

---

89 II Peter 3.13.

90 Apocalypse 21.5.

91 Cf. Augustine, *Ep.* 55.5, CSEL 34 (1895):174.20–175.6, and Victor of Capua, *De pascha* 11 (300). Bede is alluding here to the tripartite scheme of the World-Ages: before the Law, under the Law, under grace.

92 Ch. 6.

93 Augustine, *Ep.* 55.8 (178.9–20).

94 Luke 2.52.

towards the heavenward side, the Moon denotes the glory of His Resurrection and Ascension, a glory which in itself has arrived at perfection, but which does not cease to increase in the soul of the faithful by discrete increments of its light until the end of time. Our Lord, rising from the dead, first showed Himself to one or two, but then to more, sometimes seven, sometimes eleven, sometimes twelve, sometimes to more than fifty brethren, and in the end to all the disciples together. To those who saw that He was about to ascend into heaven, He commanded that they be *witnesses* to His dispensation *in Jerusalem and in all Judaea and Samaria, even to the ends of the earth.*[95] And indeed the Moon, as it increases in our sight, gradually recedes from the Sun, and when it increases on the side of the heavens, it returns in the direction of the Sun by equal stages. For this is what He Himself said: *I went forth from the Father, and came into the world; now I leave the world again and go to the Father.*[96] And the Psalm says of Him: *His going forth is from the end of the heaven, and his circuit unto the ends of it.*[97] Therefore the Moon, through the increase of its light which unfolds before our sight as it moves away from the Sun, signifies the doctrine and virtues of our Lord and Saviour /**458**/ while He was in the flesh up to the time of His Passion. However, when in returning to the Sun it gradually regains the face of heaven invisible to us, [the Moon] manifests the miracle of His Resurrection and subsequent glory, and so the Moon is deemed suitable for the joys of the Easter offering from the fifteenth day onwards.

To these indicators of the Paschal season, derived from the observation of the Law, we, the heirs of the New Testament, have added Sunday, which Scripture called "one" or "the first of the sabbath". Not without good reason: excellent by reason of the creation of the first light, and eminent because of the triumph of the Lord's Resurrection, it also remains ever desirable to us because of our own resurrection. Moreover, the seven days of the Moon, namely from the fifteenth to the twenty-first, through which this Sunday runs in natural order, openly proclaim the universality of the Church, which throughout all the world is redeemed by the Paschal mysteries. For Scripture often denotes universality by the number seven.[98] Thus what the prophet says – *Seven*

95 Acts 1.8.
96 John 16.28.
97 Psalm 19.6 (18.7).
98 Augustine, *Ep.* 55.5 (175.2–6).

times a day shall I praise thee[99] – can best be understood as *His praise is always in my mouth.*[100] John bears witness that it particularly symbolizes the complete perfection of the catholic Church when, writing to the seven Churches of Asia, he unfolds the mysteries of the universal Church throughout the world.[101] Hence in all the exhortations which he writes to each of the seven [Churches], he takes pains to interweave this phrase: *Let him who has ears hear what the Spirit says to the Churches,*[102] proving plainly that what the Spirit says to each one individually, it says to all the Churches.

No less does the Easter season offer us a moral meaning; in the word "Passover", that we may daily make a spiritual passage over from vice to virtue; and also in the "month of first-fruits", in which the mature fruit by its arrival proclaims that the old is finished, that we may by *driving out the old man and his deeds,*[103] *be renewed in the spirit of our mind and put on the new man, created according to God in justice, holiness and truth.*[104] Invigorated, as it were, by the variety of diverse virtues, and shaded by their leaves as by the shadow of a pleasant tree, /**459**/ let us sprout forth at the full Moon like happy and fruitful fields of corn, that we may be separated from the darkness of sin as we bear the perfect splendour of faith and understanding, and that when that same lunar light turns back to the heavenward side (which commences at the fifteenth Moon), we may be made humble in all things in the same measure as we are great, each one saying with the Apostle, *By the grace of God I am what I am.*[105] Truly, because the grace of this supernal gift was openly poured out in the Third Age of the world, by the most lovely logic of symbolism, the light of the Moon in its third week, which hitherto had increased towards the Earth, now begins to increase towards the heavens.[106] At Easter we are enjoined to observe this in a lovely manner, so that we, never forgetting the grace we have received, may remember to give thanks to Him who gives the gift, by obedience in every step of our spiritual passage. Again the Moon, increasing towards

99 Psalm 119.164 (118.164).
100 Psalm 34.1 (33.2).
101 Cf. Bede, *Explanatio Apocalypsis* 1.4 (134C).
102 Apocalypse 2.3.
103 Colossians 3.9.
104 Ephesians 4.24.
105 I Corinthians 15.10.
106 Cf. Augustine, *Ep.* 55.5 (see n. 91 above).

men, shows us the symbol of the active life, and when turned back towards the heavens, of the speculative life. Or else it shows us in this phase love of our neighbour, and in that phase love of our Maker. Or else the increase of its light, turned in this direction, admonishes us to practice good works before others; turned in the other direction, it admonishes us to do the same good works solely in view of a heavenly reward. In this direction, it is so that our *light may so shine before men that they may see* our *good works*; in the other direction so that *they might glorify* our *Father who is in heaven.*[107]

By the first day of the week, which is the proper solemnity of the New Testament, we are taught to bear patiently every adversity, and even death itself, in this present time for Christ's sake, and in the hope of our future resurrection in Christ. For we hear from the Apostle: *For if His Spirit who raised Jesus from the dead lives in us, He who raised Jesus from the dead will give life to our mortal bodies as well, by the indwelling of His Spirit in us.*[108] Because the gift of this Spirit is sevenfold, it can, not inappropriately, be understood by the number of the seven lunar days through which this aforementioned first day of the week – that is, Sunday – circulates. But if anyone wishes to know more about the mystery of the Paschal season, let him read the letter of St Augustine to Januarius on the calculation of Easter.[109] /**460**/

65. THE GREAT PASCHAL CYCLE

When the lunar and solar cycles are multiplied together, a Great Paschal Cycle is completed in 532 years. For whether you multiply 19 by 28, or 28 by 19, it makes 532. Hence this cycle has 28 19-year lunar cycles and 19 solar [cycles] (which are completed in 28 years); 19 times 7 leap-year days, that is, 133; 28 times 228 solar months, that is 6,384; 28 times 235 lunar months, that is, 6,580;[110] 28 times 6,935 days (leap-year days excluded), that is, 194,180; with the leap-year days added, 194,313. When it has completed this total through the sequence of months and days, it immediately returns upon itself, and recommences everything pertaining to the course of the Sun and Moon in exactly the same fashion as it

107 Matthew 5.16.
108 Romans 8.11.
109 I.e. Augustine, *Ep.* 55.
110 Jones' ed. reads "vdlxxx" here, but this is evidently a typographical error. The medieval codices furnish the correct figure: cf. Oxford, St John's College 17, fol. 103rb.

happened before. But the years of our Lord's Incarnation continue to increase in their particular column, and the indictions in the rank in which they are conveyed [i.e. the column in which they are listed] do not affect in any way the course of the stars [i.e. Sun and Moon], and hence do not affect the order of the calculation of Easter. To make this clearer, it has pleased us to present this cycle, fully worked out, at the front of this present work.[111] It takes its beginning from the 532nd year of the Lord's Incarnation, where Dionysius began his first cycle,[112] and extends up to the year 1063 of this most holy Incarnation. Thus whoever reads them can, with unerring gaze, not only look forward to the present and future, but can also look back at each and every date of Easter in the past; and in order to clarify an ancient text, he can clearly identify all the years, since it sometimes is doubtful when and of what sort they were. /**463**/

---

111 For a transcription of Bede's 532-year great cycle, see Appendix 2.

112 It should be pointed out that the fact that Dionysius began his cycle in AD 532 has nothing to do with the 532-year Great Paschal Cycle. To begin with, Dionysius did not know about this cycle; secondly, he seems to have taken over his chronology from accepted Roman sources, without much critical attention (cf. Introduction, section 3). Bede seems to have known better than to make anything of this coincidence.

[V. THE WORLD-CHRONICLE (CH. 66)]

66. THE SIX AGES OF THIS WORLD

We have mentioned a few things about the Six Ages of this world, and about the Seventh and Eighth [Ages] of peace and heavenly life above, by way of comparison to the first week, in which the world was adorned.[1] Here I will discuss the same subject somewhat more extensively, comparing it to the ages of man, whom the philosophers are accustomed to call "microcosm" in Greek, that is, "smaller universe".

The First Age of this world, then, is from Adam to Noah, containing 1,656 years according to the Hebrew Truth, and 2,242 according to the Septuagint, and ten generations according to both versions. This [First Age] was wiped out in the universal Flood, just as the first age of every person is usually submerged in oblivion, for how many people can remember their infancy?

The Second Age from Noah to Abraham comprises ten generations and 292 years according to the Hebrew authority, but according to the Septuagint 272 years[2] and eleven generations. This was, so to speak, the childhood of God's people, and therefore it is discovered in a language, that is, in Hebrew, because from childhood on, when infancy [*infantia*] is over – which is so called because an infant cannot speak [*fari*] – a person begins to learn to speak.

The Third, from Abraham to David, contains fourteen generations and 942 years according to both authorities. This [Age] was like the adolescence of the people of God, because from this age on, a person can reproduce. For this reason, the evangelist Matthew takes the beginning of the generations [of Christ] from Abraham, who was established as the father of the nations when he received his altered name.

---

1 In ch. 10.

2 The correct figure is 942 years, and Bede records it as such in *On Times* (602.2) and *Letter to Plegwin* 5. Yet the MSS of *The Reckoning of Time* all read 272, despite the fact that it contradicts Bede's statement below, *s.a.* 1693, that the Septuagint's reckoning of the first two Ages is longer than the Vulgate's. A marginal gloss on fol. 103v of an early 12th-century copy, Oxford St John's College 17, corrects the number to 2072, which shows that at least one reader realized the absurdity and tried to rectify it.

The Fourth, from David up to the exile to Babylon has 473 years according to the Hebrew Truth, twelve more according to the Septuagint, and seventeen generations according to both texts. However, the evangelist Matthew /**464**/ puts these [generations] at fourteen, for the sake of a certain symbolism.[3] From this Age – youth, so to speak – the era of the kings began among the people of God, for this age in man is normally apt for governing a kingdom.

The Fifth Age – maturity, if you will – from the exile into Babylon until the coming of our Lord and Saviour in the flesh, extends for fourteen generations and 589 years. In this Age the Hebrew people were weakened by many evils, as if wearied by heavy age.

The Sixth Age, which is now in progress, is not fixed according to any sequence of generations or times, but like senility, this [Age] will come to an end in the death of the whole world.[4]

By a happy death, everyone will overcome these Ages of the world, and when they have been received into the Seventh Age of perennial sabbath, they look forward to the Eighth Age of the blessed Resurrection, in which they will reign forever with the Lord.

In the First Age of the new-made world, on its first day, God made light, which he called "day".[5] On the second [day], he balanced the firmament in the midst of the waters: these waters and the Earth, together with the upper heaven and the powers which praise the Creator therein, were created before the beginning of those six days. On the third day, when the waters that covered everything were gathered into their place, He commanded the dry land to appear. On the fourth [day] He placed the stars in the firmament of heaven; this day we now call the 12th kalends of April [21 March], inasmuch as we deduce that it was the equinox.[6] On the fifth [day] He created swimming and flying creatures. On the

---

3 Bede explains this later on, *s.a.* 3065.

4 Bede's text here amalgamates portions of Augustine, *DCD* 16.43 (550.57–84) with *De Genesi contra Manichaeos* 1.23.35–41 (PL34.190–193). For the numerous other Augustinian texts on the same theme, see Jones' *apparatus fontium* on this passage in *BOD*; Elizabeth Sears, *The Ages of Man: Medieval Interpretations of the Life Cycle* (Princeton: Princeton University Press, 1985):174, n. 5; and especially Auguste Luneau, *L'Histoire du salut chez les Pères de l'Église: la doctrine des âges du monde*, Théologie historique 2 (Paris: Beauchesne, 1964): ch. 12.

5 This summary of Creation, drawn from Genesis 1, corresponds to Isidore, *Chronica maiora* 3, ed. T. Mommsen, MGH AA 11 (1894):426. Cf. Bede, *In Gen.* 1.2.3 (lines 1093 *sq.*).

6 Cf. ch. 6 above.

sixth [day] He formed terrestrial creatures, and man himself, Adam. From his side, as he slept, He brought forth Eve, the mother of all. This day (for so it seems probable to me) is called the 10th kalends of April [23 March]. Hence, unless some stronger argument should prevail, it should rightly be believed what blessed Theophilus, together with other bishops not only of Palestine but from many other regions, wrote when he debated the matter of Easter, [namely] that the Lord was crucified on that same 10th kalends of April [23 March].[7] For it is fitting that not only on the same day of the week but on the same day of the month /465/ the second Adam, stricken by a living death for the sake of the salvation of the human race should, from the heavenly sacraments drawn from his own side, sanctify the Church to himself as a bride. For on that [same] day He created the first Adam, father of the human race, and by removing a rib from his side, he fashioned woman, with whose aid [Adam] would propagate the human race.

130

Adam at the age of 130 begat Seth, and lived a further 800 years.[8] The Septuagint in fact puts 230 years before the birth of Seth, and 700 after.[9] Seth means "resurrection", signifying the Resurrection of Christ from the dead; Abel, which means "mourning"[10] and who was killed by Cain, signifies [Christ's] death, which was brought about by the Jews.

235

Seth at the age of 105 begat Enosh and lived a further 807 years.[11] But the Seventy put 205 years before the birth of Enosh, and 707 afterwards. Enosh means "man",[12] and of him it is well said, *he began to call upon the name of the Lord.*[13] For it behoves men, mindful of their fragility, to

7 Bede is alluding here to the spurious Acts of the Synod of Caesarea (307). But see ch. 61 and Commentary.

8 Genesis 5.3.

9 Cf. Isidore, *Chronica maiora* 4 (426); Bede, *In Gen.* 2.5.3 (93.749–94.753).

10 Bede, *In Gen.* 2.4.25 (91.650–651). These two etymologies are drawn from Jerome, *Liber interpretationis hebraicorum nominum* 10 and 12, ed. P. de Lagarde, CCSL 72 (1959):60.17 and 71.12–13.

11 Genesis 5.6.

12 Bede, *In Gen.* 2.5.2 (92.716). The etymology of "Enosh" is from Jerome, *Lib. int. heb. nom.* (65.17). Bede may also have used Isidore, *Chronica maiora* 5 (426), although Isidore's etymology is different.

13 Genesis 4.26; cf. Bede, *In Gen.* 2.4.26 (91.657–660).

invoke the aid of the Creator – at any rate, it behoves those who, living in the faith of Christ, rejoice to be sons of the Resurrection.

325

Enosh at the age of 90 begat Kenan, and lived thereafter 815 years.[14] But the Septuagint puts 190 years before the birth of Kenan, and 715 afterwards.

395

Kenan at the age of 70 begat Mahalalel, and after his birth lived 840 years.[15] The Septuagint says that there were 170 years before the birth of Mahalalel, and 740 afterwards.

460

Mahalalel at the age of 65 begat Jared and lived thereafter 830 years.[16] The Septuagint puts 165 years before the birth of Jared, and 730 afterwards.

622

Jared at the age of 162 begat Enoch, and lived 800 years thereafter.[17] With regard to this generation, the two codices are nowhere at variance. We find that this Enoch wrote certain things of a divine nature, as the Apostle Jude attests.[18] But as St Augustine says: *Not /466/ in vain are these [books] not included in the canon of the Scriptures which was preserved in the temple of the Hebrew people by the diligence of a long line of priests. For these books were long ago judged to be of doubtful reliability, nor could it be discovered whether these were the things which [Enoch] himself had written. Hence those things which are circulated under his name, and contain those fables about the giants – namely that they did not have human fathers – are rightly judged by the prudent as not to be attributed to him.*[19]

14 Genesis 5.9.
15 Genesis 5.12.
16 Genesis 5.15.
17 Genesis 5.18–19.
18 Jude 14.
19 Augustine, *DCD* 15.23 (491.112–120); cf. Isidore, *Chronica maiora* 9 (427).

687

Enoch at the age of 65 begat Methuselah,[20] after whose birth [Enoch] *walked with God for 30 years.*[21] The Septuagint puts 165 years before the birth of Methuselah, and 200 afterwards. *And rightly in the seventh generation did God withdraw Enoch, whose name means 'dedication', from the company of mortal men,*[22] because the city of the elect, labouring on behalf of God in the Six Ages of this world, anticipates the splendour of dedication in the Seventh [Age] of the coming sabbath. But because the reprobate are content merely with present happiness, *Cain consecrated the city which he had founded not in the seventh generation, but in his firstborn son Enoch.*[23]

874

At the age of 187, Methuselah begat Lamech, and lived thereafter 772 years, that is, until the Flood.[24] The Septuagint puts 167 years before the birth of Lamech, and 802 thereafter. This number, as the reader will readily see, exceeds the time of the Flood by 20 years according to the Hebrew Truth, and according to [the Septuagint's] authority by 14 years. The most learned fathers, Jerome in his book *On Hebrew Questions*[25] and Augustine in his book *On the City of God*[26] have debated this notorious question at great length.

1056

At the age of 182, Lamech begat Noah, and lived thereafter 595 years.[27] The Septuagint puts 188 years before the birth of Noah, and 565 afterwards. In this generation alone, the sum total is at odds, because Lamech is found to have lived 24 years longer /**467**/ in the Hebrew codices than in those of the Septuagint.

1656

The Flood arrived in the 600th year of Noah, in the second month, on the

20 Genesis 5.21.

21 Genesis 5.22.

22 Augustine, *DCD* 15.19 (481.3–5).

23 *Ibid.* 15.17 (480.47–52).

24 Genesis 5.25; cf. Bede, *Letter to Plegwin* 10.

25 Jerome, *Hebraicae quaestiones in Geneseos* 5.25, ed. P. de Lagarde, CCSL 72 (1959):8.19–9.17.

26 Cf. Augustine, *DCD* 15.11, *passim.* Cf. *Letter to Plegwin* 10.

27 Genesis 5.28–30.

17th day of the month.[28] Indeed, should anyone chastise us for putting forth novel questions concerning the difference of years between the Hebrew authority and that of the Seventy Translators, let him read the works of the Fathers named above, and he will see that this discrepancy is already quite well known. Because St Augustine had very astutely inquired into the origin of this gap, he stated in the 13th chapter of the work cited above, among other things: *One might plausibly suggest therefore that when these things began to be transcribed for the first time from the library of Ptolemy, something of this ilk might have been in one codex. But once it had been transcribed, what could have been a scribe's error would have been disseminated far and wide. It is not far-fetched to suppose that this is the case with respect to the question of Methuselah's life.* And a little later on: *I would certainly not be justified in doubting that when a discrepancy is found between the two texts, and both cannot be true records of historical fact, one should place greater reliance in the language from which the translators made a version in another tongue.*[29]

In the Second Age of the world, on the first day of this [Age], being the 17th [day] of the second month, Noah went forth from the ark,[30] in which a few people, that is, eight souls, were saved from the water. Calling this to mind in his epistle, the blessed Apostle Peter took pains to explain it on the spot in a marvellous way, when he added, *Baptism, which corresponds to this, now saves you, not as a removal of dirt from the body but as an appeal to God for a clean conscience, through the Resurrection of Jesus Christ, who is at the right hand of God.*[31] He teaches that baptism is symbolized by the water of the Flood, the Church and its faithful people by the ark and those it contained and the mystery of the Lord's Resurrection in the number eight of the souls.[32]

1658

Shem, at the age of 100, begat Arpachshad two years after the Flood.[33] Jerome writes that the Chaldeans took their origin from Arpachshad.[34]

---

28 Genesis 5.31. 7.11
29 Augustine, *DCD* 15.13 (471.17–22, 472.89–93); cf. *Letter to Plegwin* 10.
30 Genesis 8.4. 14 (8.4 for grounding on Ararat)
31 I Peter 3.21–22. Cf. Bede, *In Epistolas VII catholicas*, ed. David Hurst, CCSL 121 (1983):249.238–250.269.
32 Cf. Augustine, *Contra Faustum* 12.14.15 (343.23–345.14).
33 Genesis 11.10.
34 Jerome, *Hebraicae quaestiones in Geneseos* 10.22 (14.7).

/**468**/ After the birth of Arpachshad Shem lived another 500 years, that is until fifty years after the birth of Jacob.

1693

At the age of 35, Arpachshad begat Shelah.[35] Here the Seventy Translators insert one generation more than does the Hebrew Truth, saying that when Arpachshad was 135 years old he begot Cainan, who himself begat Shelah when he was 135 years old. The evangelist Luke seems to follow their translation in this instance.[36] Indeed the Greek chronographers, although they criticize this sequence of generations, and remove the one generation of Cainan, nonetheless did not take care to call into question the number of years in the generations according to [the Septuagint's] authority. Following their own authority, they gave to this age a sum total of years which is 130 less than that of the Seventy Translators, but 650 more than that of the Hebrew Truth, that is, 942.[37] But Arpachshad lived 303 years after the birth of Shelah; nonetheless the Septuagint ascribes to him 430 years after the birth of Cainan, and to Cainan 438 [years] after the birth of Shelah.

1723

Shelah at the age of 30 begat Heber, and lived thereafter for 403 years.[38] The Septuagint puts 130 years before the birth of Heber, and 330 after. The name and tribe of the Hebrews take their origin from this Heber.

1757

When Heber was 34 years old, he begat Peleg, and lived thereafter 430 years.[39] Peleg means "division" and his parents gave him this name because at the time of his birth the Earth was divided by the confusion of languages. The rhetor Arnobius recalls this division in his commentary on Psalm 104: *To Shem, Noah's firstborn, was given the portion from Persia and Bactria all the way to India and to the Rinocoruras. This expanse of land comprised 28 barbaric languages. Within these languages, the peoples form 48 nations, not of diverse languages but, as I*

---

35 Genesis 11.12.

36 Luke 3.36.

37 Bede is referring to Eusebius here: cf. *Letter to Plegwin* 6. This passage is also reproduced verbatim in Bede, *In Gen.* 3.11.12 (163.769–164.783).

38 Genesis 11.14.

39 Genesis 11.16.

*said, of diverse nations. /**469**/ For example, although Latin is a single
language, there are under this single language diverse nations of Bruttii,
Lucani, Apuli, Calabri, Picentes, Tusci and others of their ilk, if I may
say so. Ham the second son [received] from the Rinocoruras to Gadira,
containing languages of Punic speech in the region of the Garamantes,
Latin in the northern part, barbarian in the southern region of the Ethio-
pians and the Egyptians, and of various speech in the barbarous interior:
22 languages in 394 nations. But Japeth, the third [son took] from Media
to Gadira and northwards. Japeth had the Tigris River, which divides
Media and Babylonia: 200 nations, in 23 languages of different speech.
Altogether these made 72 languages, and 1,000 nations of the generations,
located in this order throughout the threefold world. Japeth, as I said, had
the Tigris River, which divides Media and Babylonia, while Shem had the
Euphrates and Ham the Geon, which is called the Nile.*[40]

1787

Peleg at the age of 30 begat Reu, and lived thereafter 209 years.[41] The
Septuagint puts 130 years before the birth of Reu, 108 thereafter.

*At this time, temples were first constructed, and certain rulers of the
nations were adored as gods.*[42]

1819

At the age of 32, Reu begat Serug, and lived thereafter 207 years.[43] The
Septuagint puts 132 years before the birth of Serug, 207 afterwards.

*The kingdom of the Scythians, where Tanaus was the first ruler, is said
to have begun.*[44]

1849

Serug at the age of 30 begat Nahor, and lived thereafter 200 years.[45] The
Septuagint puts 120 years before the birth of Nahor, 200 after.

*It is said that the Egyptians inaugurated their empire, Vizoues being
the first to reign over them.*[46]

40 Arnobius, *In Psalmos* 104, ed. Klaus-D. Daur, CCSL 25 (1990):159.60–160.80.
41 Genesis 11.18.
42 Isidore, *Chronica maiora* 24 (430).
43 Genesis 11.20.
44 Isidore, *Chronica maiora* 26 (430).
45 Genesis 11.22.
46 Isidore, *Chronica maiora* 28 (430).

1878

At the age of 29, Nahor begat Terah, and lived thereafter 119 years.[47] The Septuagint puts 70 years before the birth of Terah, and 129 after.

The kingdom of the Assyrians and Sycinians begins. Belus was the first to rule over the former, and Aegialius over the latter.[48]

1948

Terah at the age of 70 begat Abraham, and lived thereafter 135 years.[49] The Second Age of the world extends to this point. Having passed its whole sequence in review, St Augustine in *The City of God*, Book 16, ch. 10, concludes as follows: *From the Flood to Abraham, then, there were 1072 years according to the conventional text, that is, [the text] of the Seventy Translators. But far fewer years are found in the Hebrew codices. They can give no reason for this, or only a very tortuous one.*[50]

2023

The Third Age of the world began with the birth of the Patriarch Abraham, who was 75 years old when he left his native country at God's command and went to the land of Canaan, receiving the promise that a saviour would be born of his seed, in whom all the nations would be blessed, and that he himself would become a great nation.[51] One of these promises was spiritual, the other carnal.

At this time, Ninus and Semiramis ruled over the Assyrians.[52]

2034

Abraham at the age of 86 begat Ishmael, from whom the Ishmaelites [are descended].[53] Ishmael begat twelve princes and lived 137 years.[54]

2048

This same Abraham at the age of 100 begat Isaac,[55] who is the first and

47 Genesis 11.24.
48 Isidore, *Chronica maiora* 30 (431).
49 Genesis 11.26.
50 *DCD* 16.10 (512.51–55).
51 Genesis 12.4; 13.15.
52 Jerome, *Chronicon*, ed. Rudolf Helm, *Eusebius Werke* 7.1, Griechischen christlichen Schriftsteller 24 (Leipzig: J.C. Hinrichs, 1913):20a.1–6, 17–18.
53 Genesis 16.16.
54 Genesis 25.16–17.
55 Genesis 21.5.

only person in the entire Old Testament who is said to have been circumcised on the eighth day, which is a privilege given to the son of the promise, and not without profound mystery.

2108

Isaac at age 60 begat Esau and Jacob, patriarchs of the Idumean and Israelite peoples.[56] After their birth he lived 120 years.

At this time, Inachus first reigned over the Argives /471/ for 50 years, whose daughter Io the Egyptians worship under the altered name "Isis".[57]

2238

At the age of 130 Jacob went down into Egypt together with seventy souls.[58]

At this time, Memphis was constructed in Egypt by Ape, king of the Argives.[59] *Sparta also was founded by Spartus, son of Phoroneus king of the Argives.*[60]

2453

Now the sojourning of the children of Israel, who dwelt in Egypt, was 430 years, and at the end of the 430 years, even on the selfsame day, all of the hosts of the Lord went out from the land of Egypt, as the book of Exodus testifies.[61] However, the chronographers calculate the sum of these years from the 75th year following the birth of Abraham, when he entered the Promised Land, following the version of the Seventy Translators which says: *The sojourning of the children of Israel, in which they and their fathers dwelt in Egypt and in Canaan 430 years*. The Hebrew Truth itself shows that this [computation] must of necessity be followed, for it tells that Kohath, son of Levi, who, it claims, was born in the land of Canaan,[62] lived for 132 years, and his son Amram the father of Moses for 137 years,[63] and Moses himself was 80 years old at the time of the departure from Egypt, and obviously the sum of these years cannot

---

56 Genesis 25.26.
57 Jerome, *Chron.* 27b.10–15.
58 *Ibid.* 33a.4.
59 *Ibid.* 32b.22–23.
60 *Ibid.* 33b.18–20.
61 Exodus 12.40–41; Jerome, *Chron.* 23b.15–25.
62 Genesis 46.11.
63 Exodus 6.18–20.

amount to 430. The Apostle also agrees with [the Septuagint] translation when he says, *Now to Abraham and his seed were the promises made. He saith not 'And to seeds', as [if referring to] many, but as of one, 'And to thy seed', which is Christ. And this I say, that the covenant confirmed before of God, the law, which was 430 years after, cannot annul, to make what was promised to the fathers of no effect.*[64]

2493

Moses ruled over the people of Israel in the desert for 40 years after their departure from Egypt.[65] In the first year he built a tabernacle to the Lord[66] and completing the work in seven months, he erected it on the first day of the first month of the second year. *Up to this point, /472/* as Eusebius observes, *the five books of Moses contain the deeds of 3,730 years, according to the translation of the Seventy Elders.*[67] In his first book against Apion the grammarian, Josephus records the number of years which the Hebrew Truth contains: *Nor are there among us innumerable books, all disagreeing with one another, but only 22 books, which contain the sequence of all the ages,* and which are justly believed to be divinely inspired. *Of these, five are by Moses, containing the laws of life and the lineage of human succession down to the death of Moses himself, which they reveal to contain a little less than 3,000 years.*[68]

2519

Joshua ruled over the people of Israel for 26 years, as Josephus teaches,[69] for Scripture is silent on the number of years of his leadership. Later on, we will discuss why Eusebius puts it at 27 years in his Chronicle.[70] In the first year of the leadership of Joshua, in the first month, on the tenth day of the month, the people entered the Promised Land through the exposed channel of the Jordan.[71] This year, as we find in the chronicles of the aforementioned Eusebius, *was the beginning of the 51st Jubilee*

64 Galatians 3.16–17.
65 Numbers 14.34.
66 Exodus 20 *sqq.*
67 Jerome, *Chron.* 46a.1–6.
68 Josephus, *Contra Apionem* 1.38–39, *versio latina* ed. C. Boysen, CSEL 37 (1898):11.7–13; cf. *Letter to Plegwin* 8.
69 Josephus, *Antiqu.* 5.1.29, ed. Blatt (320.22).
70 Jerome, *Chron.* 46a.1–3. Bede discusses the reason for this discrepancy *s.a.* 2790, below.
71 Joshua 4.19.

according to the Hebrews,[72] that is, 2,500 years had been completed since the beginning of the world, with each cycle of the Jubilee being assigned 50 years. But in fact our research has not been able to confirm that this is the sum total of this time. It is agreed that up to the Flood there were 1,656 years; and from thence to Abraham (who was 75 years old when he received the promise) 292 years; 430 years of the promise, and 40 years of Moses' leadership. Added together, this makes, not 2,500 years, but seven less, that is 2,493, as we said above.

2559

By God's command, Othniel of the tribe of Judah was made the first judge over Israel, for 40 years.[73] At the outset of his [rule], the sons of Israel served Chushan-rishathaim, king of Mesopotamia, for eight years.[74] /**473**/

2639

Ehud, *son of Gera, son of Gemini, who used either hand like his right hand,* [ruled] for 40 years. In his first years Israel served Eglon king of Moab for eighteen years, until [Ehud] freed them by slaying Eglon.[75]

   At this time the city of Cyrene was founded in Libya.[76]

2697

The prophetess Deborah of the tribe of Ephraim, together with Barak of the tribe of Naphtali [ruled] for 40 years.[77] At the beginning of their rule Jabin, king of Canaan, who reigned in Hazor, oppressed the sons of Israel for 20 years. But when Sisera, captain of his army, was slain by Israel, he was at length humiliated and destroyed.[78]

   At this time, Miletus was founded.[79]

2719

Gideon of the tribe of Manasseh [ruled] 40 years,[80] under whom Israel

---

72 Jerome, *Chron.* 46a.4–8.
73 Jerome, *Chron.* 47a.7–9.
74 Judges 3.8.
75 Judges 3.15–30.
76 Jerome, *Chron.* 52b.18–19.
77 *Ibid.* 53a.6–8.
78 Judges 4.2, 23–24.
79 Jerome, *Chron.* 55b.3.
80 *Ibid.* 55b.5–6.

served the Midianites and the Amalekites seven years, but they were liberated by Gideon in battle.[81]

Tyre was founded 240 years before the building of the Temple in Jerusalem, according to Josephus.[82]

2722

Abimelech, son of Gideon, [ruled] for three years, and he ruled in Shechem.[83]

Hercules lays Troy waste.[84]

2745

Tola, son of Puah, uncle of Abimelech, a man of Issachar, [ruled] 23 years; *he dwelt in Shamir in Mount Ephraim.*[85]

The battle of the Lapiths and the Centaurs, who (as Palaephatus writes in his first book On Marvels) were noble horsemen of Thessaly.[86]

In Troy, *Priam ruled after Laomedontes.*[87]

2767

Jair of the tribe of Manasseh [ruled] 22 years.[88]

Hercules establishes the Olympic Games, from which 430 years are reckoned up to the first Olympiad.[89]

2773

Jephthah the Gileadite [ruled] six years. The Philistines and /**474**/ Ammonites oppressed Israel, and of these the Ammonites were defeated by Jephthah.[90] *He says in the Book of Judges that from the time of Moses up to his own time there were reckoned 300 years.*[91]

81 Judges 6.3, 14; Jerome, *Chron.* 55a.5–6, 10.
82 Jerome, *Chron.* 55a.21–23.
83 *Ibid.* 57a.7–8.
84 *Ibid.* 57b.6–7.
85 Judges 10.1–2.
86 Jerome, *Chron.* 57b.14–17.
87 *Ibid.* 57b.19–20.
88 Judges 10.3–5; Jerome, *Chron.* 58a.17.
89 Jerome, *Chron.* 59b.1–3.
90 Judges 11; Jerome, *Chron.* 59a.22–23.
91 Judges 11.26; Jerome, *Chron.* 60a.2–4.

2780

Ibzan of Bethlehem [ruled] seven years.[92]

Agamemnon ruled over Mycenae for 35 years, and in the 15th year of his [reign] Troy was captured.[93]

2790

Elon the Zebulonite ruled for ten years. This man, and these ten years, are not in the Septuagint.[94] In order to make up for their omission, Eusebius records more years for Joshua son of Nun, Samuel and Saul (whose years Scripture does not declare) than can be read in Josephus, in order that he might arrive at the sum of 480 years, which Scripture declares [elapsed] between the departure of Israel from Egypt and the construction of the Temple.

2798

Abdon of the tribe of Ephraim [ruled] eight years.[95]

In his third year, Troy was captured, 375 years *after the first year of Cecrops, who first ruled over Attica.*[96] *835 years [had passed] since the 43rd year of the reign of Ninus over the Assyrians.*[97]

After the death of Abdon, Israel served the Philistines for 40 years.[98]

2818

Samson of the tribe of Dan [ruled] 20 years.[99] Up to this point, the Book of Judges has recorded a [period of] time comprising 299 years, and twelve judges.

Aeneas ruled over the Latins, who were later called Romans, for three (or as some would have it, eight) years following the capture of Troy.[100] *After him, Ascanius [ruled] 39 years.*[101] *Prior to Aeneas, Janus,*

92 Judges 12.8–9.
93 Jerome, *Chron.* 59b.24–25.
94 Judges 12.11; Jerome, *Chron.* 60a.17–20.
95 Judges 12.14; Jerome, *Chron.* 60a.18.
96 Jerome, *Chron.* 61b.2–10.
97 *Ibid.* 61a.5–6.
98 Judges 13.1.
99 Jerome, *Chron.* 62a.12.
100 *Ibid.* 62b.1–8.
101 *Ibid.* 62b.14–15.

Saturnus, Picus, Faunus, and Latinus ruled in Italy for about 150 years.[102]
Ascanius, the son of Aeneas, built the city of Alba.[103]

2858

Eli was priest for 40 years. In the books of the Hebrews we find 40 years, in the Septuagint 20.[104] /**475**/

The sons of Hector retake Troy, after the descendants of Antenor have been expelled; Helenus gave them refuge.[105]

Silvius, the third son of Aeneas, reigned over the Latins for 29 years. Because he was born after the death of his father and raised in the country, he took the names "Silvius" and "Posthumus", whence all the kings of Alba are called "Silvius".[106]

The kings of Sicyon died out, who from Aegealus to Zeuxippus had reigned 962 years. After them the priests of Carnus were established.[107]

2870

Samuel [ruled] twelve years, as Josephus teaches, but in Holy Scripture it is not evident how long he held pre-eminence.[108] From this point on, the time of the Prophets begins.

Aeneas Silvius, the fourth [king] of the Latins, [ruled] 31 years.[109]

2890

Saul, the first king of the Hebrews, ruled 20 years.[110] We have recorded the duration of his reign from the *Antiquities* of Josephus, because it is not contained in the canonical Scripture.

Eurystheus first ruled in Lacedaemonia for 42 years, and Alethis first ruled in Corinth for 35 years.[111]

The Fourth Age of the world begins not only with the inception of the rule of the tribe of Judah, but also with a renewal of the promise of Chris-

---

102 *Ibid.* 62b.9–12.
103 *Ibid.* 62b.19–20.
104 *Ibid.* 63a.13–14, 14–17.
105 *Ibid.* 63b.17–19.
106 *Ibid.* 64b.9–13.
107 *Ibid.* 64a.15–26.
108 Josephus, *Antiqu.* 6.13 (Cologne: ex aedibus Eucharii Cervicorni, 1534): fol. 63v.
109 Jerome, *Chron.* 66b.1–2.
110 Josephus, *Antiqu.* 6 fin. (fol. 66r).
111 Jerome, *Chron.* 66b.9–11.

tian rule once given to the patriarchs, the Lord swearing *in truth unto David, of the fruit of his body would He set upon his throne.*[112]

2930

David, the first king from the tribe of Judah, [ruled] 40 years.[113]
   Latinus Silvius, fifth [king] of the Latins, [ruled] 50 years.[114]
   Ephesus is founded by Andronicus.[115]
   Carthage is founded, as some would have it, by Carcedon of Tyre, but others [say] by his daughter Dido, 143 years after the Trojan War.[116]

2979

Solomon, son of David, [ruled] 40 years.[117] In the fourth year of his reign, and in the second month, *he began to build a Temple for the Lord* /**476**/ *in Jerusalem, 480 years after the departure of Israel from Egypt, as is also the testimony of the Book of Kings.*[118] [The Temple] was finished in seven years, and dedicated in the seventh month of the eighth year, as a symbol of the totality of time in which the Church of Christ, which is made perfect in the future [age], is built up in this world.[119] *Alba Silvius, son of Aeneas Silvius, sixth [king] of the Latins*, ruled *39 years.*[120] The Queen of Sheba came to hear the wisdom of Solomon.[121]

2987

Rehoboam son of Solomon [ruled] 17 years.[122] Jeroboam of the tribe of Ephraim separated ten tribes from the house of David and from the Lord.[123] He is a symbol of the heretics, who dissociate their followers from Christ and His Church. In [Rehoboam's] fifth year, *Shishak king of Egypt* came to Jerusalem and *looted the Temple.*[124]

---

112 Psalm 132 (131).11.
113 Jerome, *Chron.* 67a.17–19.
114 *Ibid.* 67b.13–15.
115 *Ibid.* 69b.13–14.
116 *Ibid.* 69b.19–23.
117 *Ibid.* 69a.22–23.
118 *Ibid.* 70a.1–6, 70b.1–5; I Kings 6.31. Cf. Bede, *De tem.* 1 (157.393–158.451).
119 I Kings 6.38. Cf. Bede, *De tem.* 2 (196.189–197.234).
120 Jerome, *Chron.* 70b.14–15.
121 I Kings 10.1.
122 Jerome, *Chron.* 72a.5.
123 I Kings 12.20.
124 II Chronicles 12.2; Jerome, *Chron.* 72a.18–20.

Aegyptus Silvius, son of the previous king of Alba, [ruled] 24 years [as] seventh [king] of the Latins.[125]

Samos is founded and Smyrna is expanded into a city.[126]

2990

Abijah son of Rehoboam [ruled] *three years.*[127] He overcame Jeroboam when [Jeroboam] attacked him, and 15,000 of [Jeroboam's] soldiers were killed, because [the people of Judah] *relied in the Lord.*[128]

3031

Asa son of Abijah [ruled] *41 years.*[129] Ben-Haddad king of Damascus in Syria advanced against Israel and smote all the land of Naphtali.[130]

Capys Silvius, son of Aegyptus the previous king, [ruled] 28 years [as] eighth king of the Latins.[131]

Asa destroyed the idols, purified the Temple, and overthrew Zara the Ethiopian, together with his army, when he went out against him. Omri king of Israel *bought the hills of Samaria from Shemer for two talents of silver and constructed [Samaria].*[132] Hiel the Bethelite restored Jericho.[133]

3056

Jehoshaphat son of Asa [ruled] 25 years.[134] Elijah the Tishbite withheld the rain for three and a half years because of the sins of Ahab and the people of Israel.[135] Amongst other mighty deeds, he anointed Elisha son of Shaphat of Abelmeholah to be a prophet in his place.[136]

Carpentus Silvius, son of the previous king Capys, ruled as ninth [king] of the Latins for 13 years.[137] *After him, his son Tyberinus Silvius*

---

125 Jerome, *Chron.* 72b.18–19.
126 *Ibid.* 72b.24–26.
127 I Kings 15.2; Jerome, *Chron.* 73a.8.
128 II Chronicles 13.17–18.
129 II Chronicles 16.13; Jerome, *Chron.* 73a.15.
130 II Kings 13.3; II Chronicles 16.4.
131 Jerome, *Chron.* 74b.7–8.
132 I Kings 16.24.
133 I Kings 16.34.
134 Jerome, *Chron.* 75a.26.
135 I Kings 17.1.
136 I Kings 19.16.
137 Jerome, *Chron.* 75b.21–23.

[ruled] 8 years, after whom the river Tiber is named, which was called Albula before.[138] *After him, his son Agrippa Silvius [ruled] 40 years.*[139]

Jehoshaphat did what was right before the Lord.[140]

3064

Jehoram son of Jehoshaphat [ruled] *8 years.*[141] Elijah is carried off into heaven (as it were) in a fiery chariot, and the bereaved Elisha, the heir of his prophetic power, cleanses the waters of Jericho in his first miracle.[142] *In the days* of Jehoram, *Edom seceded lest it be subject to Judah, and set up a king for itself.*[143] Jehoram *walked in the ways of the house of Ahab, for the daughter of Ahab was his wife.*[144]

3065

Azariah son of Jehoram [ruled] *one year.*[145] *Jehonadab son of Rechab attained renown.*[146] The evangelist Matthew excluded Azariah from the genealogy [of Christ], together with his son Joash and grandson Amaziah, because of the enormity of their crimes, and because neither father nor son had anything good about them.

3071

Athaliah mother of Azariah [ruled] 6 years.[147] She, seeing her son Azariah killed by Jehu king of Israel, slew all the royal line of the house of Jehoram save Joash, son of Azariah, whom Jehoshabeath sister of Azariah and wife of Jehoiada stole away from the midst of the kings' sons, when they were slain.[148] In the Septuagint, Athaliah is said to have reigned 7 years.

3111

Joash the son of Azariah [ruled] *40 years.*[149] This [king], who started off

138 *Ibid.* 76b.21–24.
139 *Ibid.* 77b.10–11.
140 I Kings 22.
141 II Kings 8.17; Jerome, *Chron.* 77a.23.
142 II Kings 2.
143 II Kings 8.20.
144 II Kings 8.18.
145 II Kings 8.24–25; Jerome, *Chron.* 78a.9.
146 II Kings 10.15; Jerome, *Chron.* 78a.14–16.
147 II Chronicles 22.12.
148 II Chronicles 22.10–11.
149 II Kings 12.1; Jerome, *Chron.* 78a.24.

well and ended very badly, restored the Temple at the beginning of his reign; /**478**/ at the end, among other outrages, he ordered Zechariah, son of Jehoiada, who was once his guardian, and who made him king, to be stoned between the Temple and the altar.[150] The Lord in the Gospel gives a distinctive epithet to the son of Zechariah, that is, "blessed of the Lord", for the sake of his merits.[151]

Aremulus Silvius, son of the previous king Agrippa, ruled for 19 years as twelfth [king] of the Latins. He established the stronghold of the Albi in the hills where Rome now stands. His son was Julius, great grandfather of Julius Proculus, who, migrating to Rome with Romulus, founded the Julian clan.[152]

3140

Amaziah son of Joash [ruled] *29 years.*[153] Elisha the prophet died and was buried in Samaria.[154] *Hazael king of Syria oppressed Israel.*[155]

Aventinus Silvius, elder son of the previous king Aremulus, ruled as thirteenth king of the Latins for 37 years. He died and was buried on the hill which is now a part of the city, and bestowed the name of "eternal" on the spot.[156]

3192

Azariah, also called Uzziah, son of Amaziah, [ruled] *52 years.*[157]

Thonos Concoleros, who in Greek is called Sardanapalus, 36th [king] of the Assyrians, *built Tharsus and Anchiale, and when he was defeated in battle by Arbaces the Mede, he burnt himself to death in a fire.*[158] *History relates that there were kings of Assyria up to this point in time, and that altogether [they ruled] 1,197 years. All the years of the kingdom of Assyria from the first year of Ninus total 1240 years.*[159]

Procas Silvius, son of Aventinus the previous king, ruled as fourteenth

150 II Chronicles 24.21.
151 The allusion is to John the Baptist, whose father was named Zechariah (Luke 1.5 *sqq.*).
152 Jerome, *Chron.* 79b.22–80b.5.
153 II Kings 14.2; Jerome, *Chron.* 81a.6.
154 II Kings 13.20.
155 II Kings 13.22.
156 Jerome, *Chron.* 81b.1–7.
157 II Kings 15.1–2; Jerome, *Chron.* 82a.25–26.
158 *Ibid.* 82a.1–3, 19–24.
159 *Ibid.* 83a.7–10.

[king] of the Latins for 23 years.[160] *After him Amulius Silvius, the fifteenth [king], ruled 43 years.*[161]

Arbaces the Mede, having destroyed the empire of the Assyrians, transferred sovereignty to the Medes, where he was the first to rule, for 28 years.[162]

The kingdom of the Macedonians was established, whose first king, Caranus, ruled 28 years.[163] *The kings of the Lacedaemonians died out,*[164] and those of the Lydians began.*[165] */479/*

3208

Jotham son of Uzziah [ruled] *16 years.*[166]

Olympias was first founded by the Elisians 405 years after the capture of Troy.[167]

Mars and Ilia beget Romulus and Remus.[168]

Amongst other good deeds, Jotham *built the uppermost gate of the house of the Lord,*[169] which in the Acts of the Apostles is called "Beautiful".[170] All the gates of the Temple were [built] into the ground except the Beautiful Gate, which was suspended, and which the Jews called the gate of Jotham.

3224

Ahaz son of Jotham [ruled] *16 years.*[171]

Tiglath-Pileser king of Assyria came up and slew Rezin, king of Syria, and carried off the inhabitants of Damascus to Cyrene.[172]

Rome was founded on the Palatine hill on the 11th kalends of May [21 April] by the twins Romulus and Remus, sons of Rhea Silvia, who was the daughter of Numitor, brother of King Amulius. *She was a Vestal*

---

160 *Ibid.* 83b.10–11.
161 *Ibid.* 84b.20–21.
162 *Ibid.* 83a.13–15, 11–12.
163 *Ibid.* 83b.21–23.
164 *Ibid.* 86b.2–3.
165 *Ibid.* 85b.24–25.
166 II Kings 15.33; Jerome, *Chron.* 86a.26.
167 Jerome, *Chron.* 86a.7–8.
168 *Ibid.* 87b.1–2.
169 II Kings 15.35.
170 Acts 3.2.
171 Jerome, *Chron.* 87a.25; cf. 86a.25.
172 II Kings 16.2, 5–9.

Virgin, but was ravished.[173] *In the third year after the foundation of the City, the Sabine women were carried off at the games of the Consualia.*[174] *Remus is killed with a rustic shovel by Fabius, Romulus' general.*[175]

3253

Hezekiah son of Ahaz [reigned] *29 years.*[176] In his sixth year Shalmaneser king of Assyria, having captured Samaria, carried off Israel to Assyria, whose kingdom from the first year of Rehoboam had stood for 360 years.[177]

After the death of Romulus, who reigned for 38 years, the senators governed the republic for five days, and in this way one year was completed.[178] *After them, Numa Pompilius [ruled] 41 years; he built the Capitol from its foundations.*[179] /**480**/

3308

Manasseh, son of Hezekiah, [ruled] *55 years.*[180] On account of his crimes, he was carried off, chained and shackled, to Babylon, but was restored to his kingdom because of his penitence and prayers.[181]

Tullius Hostilius, third [king] of the Romans, ruled 32 years. He was the first king of the Romans to use purple and the fasces,[182] *and he expanded the City by annexing the Caelian hill.*[183]

3310

Amon, son of Manasseh, [ruled] *two years.*[184] In the Hebrew Truth it is read that he ruled for two years, and in the Septuagint, for 20.[185]

173 Jerome, *Chron.* 88a.5–6; Eutropius, *Breviarium ab urbe condita* 1.1., ed. C. Santini (Leipzig: Teubner, 1979):3.1–5.
174 Jerome, *Chron.* 88a.11–12.
175 *Ibid.* 88a.7–8.
176 II Chronicles 29.1; Jerome, *Chron.* 89a.12–13.
177 II Kings 18.9. The Bible says that these events took place in Shalmaneser's fourth year.
178 Jerome, *Chron.* 91a.5–8.
179 *Ibid.* 91b.9–10.
180 *Ibid.* 91a.11–12.
181 II Chronicles 33.11–13.
182 Jerome, *Chron.* 93a.25–26.
183 *Ibid.* 94a.16–17.
184 II Kings 21.19; Jerome, *Chron.* 95a.3.
185 Jerome, *Chron.* 95a.4–6.

Istrus, a city in Pontus, was founded.[186]
Amon is slain by his servants.[187]

3341

Josiah son of Amon [ruled] 31 years.[188] Having purified Judaea and Jeru-salem, he restored the Temple, and after casting out the impurities of idolatry, he made a very celebrated Passover unto the Lord in the 18th year of his reign. When he went forth against Necho, king of the Egyptians, he was slain in the field of Megiddo,[189] *which is now called Maximianopolis.*[190]

Ancus Marcius, Numa's grandson by his daughter, [ruled as] fourth [king] of the Romans for 23 years. He added the Aventine hill and the Janiculum to the City, and founded Ostia on the sea coast 16 miles from the City.[191] *After him Tarquinius Priscus ruled 37 years. He built the Circus at Rome, augmented the number of senators, instituted the Roman games, built walls and sewers, and constructed the Capitol.*[192]

In the Hebrew [version of the Bible] Josiah is read to have reigned for 31 years, and in the Septuagint for 32.[193] But Eusebius adds another year, taken from his [reign], between his reign and that of Jehoiakim, on account of the two [periods] of three months in which Jehoahaz and Jehoiakin ruled.[194] In fact, Jeremiah demonstrates what the [Hebrew] Truth would hold: he testifies that he prophesied for 33 years from the thirteenth year of Josiah to the fourth year of Jehoiakim. /**481**/

Nebuchadnezzar began to reign in the fourth year of Jehoiakim, and in the nineteenth year of his reign, Jerusalem was destroyed.[195]

3352

Jehoiakim son of Joash [ruled] 11 years.[196] After Josiah, Jehoahaz his son ruled for three months.[197] Necho defeated him and carried him off into

186 *Ibid.* 95b.4.
187 II Chronicles 33.24.
188 II Kings 22.1; II Chronicles 34.1.
189 II Chronicles 35.
190 Jerome, *In Zachariam* 3.12.11–14, ed. M. Adraien, CCSL 76A (1970):869.335.
191 Jerome, *Chron.* 96a.5–6, 97a.9–13.
192 *Ibid.* 97a.15–16, 99a.11–14, 97b.14–15.
193 II Chronicles 34.1; Jerome, *Chron.* 96a.24.
194 Jerome, *Chron.* 98a.1–4.
195 Jeremiah 25.1 *sqq.*
196 II Kings 23.36; Jerome, *Chron.* 98a.4.
197 II Kings 24.8.

Egypt, and set up Jehoiakim as king. In his third year, Nebuchadnezzar captured Jerusalem, and took many prisoners, among whom were Daniel, Ananias, Azarias and Misael, and carried off part of the Temple vessels to Babylon.[198] Scripture computes the reign of Nebuchadnezzar from the fourth year of Jehoiakim, because from this point on, he began to rule not only over the Chaldeans and the Jews, but also over the Assyrians, Egyptians, Moabites and countless other peoples. *Jehoiakin, also called Jeconiah*, son of Jehoiakim [reigned] for three months and eleven days.[199] When Jerusalem was surrounded by the Chaldeans, he went out to the king of Babylon together with his mother, and he was taken off to Babylon with his people in the eighth year of the reign of Nebuchadnezzar.[200]

3363

Zedechiah, also called Mattaniah, son of Josiah, [ruled] 11 years.[201] In his eleventh year, and in the nineteenth year of the king of Babylon, Judaea was carried away captive into Babylon and the Temple of the Lord was burnt, 430 years after its foundation.[202] The Jews who were left behind took refuge in Egypt.[203] Five years later, Egypt was smitten by the Chaldeans, and these people too were exiled to Babylon.

The Fifth Age of the world began after the extinction of the kingdom of Judah, which lasted, in accordance with the prophecy of Jeremiah, for 70 years.[204]

3377

In the fourteenth year after the city was attacked, which was the 25th year of the deportation of King Jehoiakin, with whom Ezekiel was taken captive, this same Ezekiel, borne off in the visions of God to the land of Israel, saw the restoration of the city, and of the Temple and its ceremonies.[205]

Servius reigned 34 years as twelfth [king] of the Romans. He added

---

198 II Kings 25.1; Jerome, *Chron.* 98a.11–16.
199 II Chronicles 36.9; Jerome, *Chron.* 99a.3–4.
200 II Kings 24.10–12.
201 Jerome, *Chron.* 98a.26–99a.1.
202 II Kings 25.2–12.
203 II Kings 25.26.
204 Jeremiah 25.11–12.
205 Ezekiel 40.1 *sqq.*

*three /**482**/ hills to the City – the Quirinal, Esquiline and Viminal – laid moats around the walls, and instituted the first census of the Roman people.*[206]

3389

In the 26th year after the overthrow of Jerusalem, *which is the 37th year of the deportation of King Jehoiakin, Evilmerodach, king of Babylon, in the year that he began to reign did lift up the head of Jehoiakin king of Judah out of prison, and set his throne above the throne of the kings that were with him in Babylon.*[207] Jeremiah had in mind this time, but also the future, when he wrote as follows: *Behold, I will send and take all the families of the north, saith the Lord, and Nebuchadnezzar the king of Babylon my servant, and will bring them against this land, and against the inhabitants thereof, and against all these nations round about, and will utterly destroy them, and make them an astonishment, and an hissing, and perpetual desolations and these nations shall serve the king of Babylon seventy years, and when the seventy years are accomplished, I will visit upon the king of Babylon and upon that nation (says the Lord) their iniquity, and upon the land of the Chaldeans, and will make it perpetual desolations.*[208] And writing elsewhere to the exiles which Nebuchadnezzar carried off to Babylon with King Jeconiah, he says: *After seventy years be accomplished at Babylon I will visit you, and perform my good work toward you, in causing you to return to this place, says the Lord.*[209] Again, the words of the same man called to mind the days gone by as follows: *And them that escaped from the sword carried he away into Babylon; where they were servants to him and his sons until the reign of the kingdom of Persia: to fulfil the word of the Lord by the mouth of Jeremiah, until the land had enjoyed her sabbaths: for as long as she lay desolate she kept sabbath, to fulfil threescore and ten years. Now in the first year of Cyrus king of Persia, that the word of the Lord spoken by the mouth of Jeremiah might be accomplished, the Lord stirred up the spirit of Cyrus king of Persia,* and so on.[210] It is plain by these words that after Judea had been laid waste, the Chaldeans did not send in settlers, as the Assyrians [had done] /**483**/ in Samaria, but left the land deserted until

---

206 Jerome, *Chron.* 101a.7–8, 10–14.
207 II Kings 25.27–28.
208 Jeremiah 25.9, 11–12.
209 Jeremiah 29.10.
210 II Chronicles 36.20–22.

the Jews themselves could return there after 70 years. Josephus agrees
with this; he writes in his book of *Antiquities* that the Temple, and Jerusalem, and all of Judea remained desolate for 70 years. On the other
hand, in enumerating the kings of Babylon (if indeed he wrote this, and
a faulty codex is not in error) [Josephus] calculates almost 100 years
from the overthrow of Jerusalem to the overthrow of the kingdom of
the Chaldeans. For he writes that after Nebuchadnezzar, who according
to Holy Scripture lived 25 years after the overthrow of Jerusalem, his
son Evilmerodach ruled 18 years. After him, his son Hegesar ruled 40
years, who was succeeded for nine months by his son Labosordach.
Upon his death, the kingdom passed to Belshazzar, who is called
Naboan: *when he had reigned 17 years, Babylon was captured by Cyrus
the Persian and Darius the Mede.* And there follows: *Darius, son of
Astiagis, who together with Cyrus his kinsman destroyed the realm of the
Babylonians, was 62 years of age when Babylon was invaded. However,
the Greeks call him by another name. Summoning Daniel the prophet, he
carried him off to Media and bestowed every honour upon him.*[211] Daniel
himself calls to mind this Darius: *In the first year of Darius the son of
Ahasuerus, of the seed of the Medes, who was made king over the realm of
the Chaldeans, I Daniel understood by books the number of the years,
whereof the word of the Lord came to Jeremiah the prophet, that seventy
years would be accomplished in the desolation of Jerusalem.*[212] Eusebius
in the Chronicle computes 30 years from the overthrow of Jerusalem to
the beginning [of the reign] of Cyrus, king of the Persians, but Julius Africanus [computes] 70 [years].[213] Furthermore Jerome, in his commentary
on the Prophet Daniel says this: *Concerning the 70th year, in which Jeremiah said that the captivity of the Jewish people would be loosed, and of
which Zechariah also speaks in the beginning of his book, the Hebrews
recount a legend like this. Belshazzar, thinking that God's assurance was
void and his promise false, turned it into a cause for merriment* /**484**/ *and
made a great feast, insulting the hope of the Jews and the vessels of the
Temple of God. But vengeance was executed immediately.*[214]

---

211 Josephus, *Antiqu.* 10.13 (Cologne 1534 ed., fol. 109v–110v; quotation is on fol.
110v).
212 Daniel 9.1–2.
213 Jerome, *Chron.* 100a.20–102a.12.
214 Jerome, *In Dan.* 2.5.2 (821.30–37).

3423

Cyrus, first [king] of the Persians, reigned 30 years.[215] In order that the word of the Lord from the mouth of Jeremiah might be fulfilled, in the first year of his reign he *loosed the captivity of the Hebrews. He caused about 50,000 people to return to Judea*, and restored to them 5,400 gold and silver vessels from the Temple of the Lord.[216] Gathering in Jerusalem, they built an altar in the seventh month, and from the first day of the same month they began to offer burnt sacrifices unto the Lord. In the second year after their arrival, in the second month, they laid the foundation for the Temple, 72 years after its destruction according to Africanus, but 32 years according to the Chronicle of Eusebius.[217] But because of the obstruction of the Samaritans the work was interrupted until the second year of Darius.[218] Even in the reigns of Ahasuerus and Artaxerxes [the Samaritans] wrote accusations against the Jews, and they wrote again to Artaxerxes that Jerusalem should not be rebuilt.[219]

Tarquin, seventh [king] of the Romans, ruled 35 years.[220] *He was expelled from his kingdom on account of his son Tarquin the Younger, who violated Lucretia.*[221]

3431

Cambyses, son of Cyrus [ruled] *eight years.*[222] When he had *conquered Egypt, he suppressed* its *rites and temples, for he utterly loathed its religion.*[223] *He constructed Babylon in Egypt.*[224] *They say that the Hebrews call him a second Nebuchadnezzar. The history of Judith was written in his reign.*[225]

3432

The Magi brothers rule seven months.[226]

215 Jerome, *Chron.* 102a.19–20.
216 I Esdras 2.13; Jerome, *Chron.* 102a.18–20.
217 Jerome, *Chron.* 102a.22.
218 Ezra 4.24.
219 Ezra 4.6–16.
220 Jerome, *Chron.* 104a.11.
221 Jerome, *Chron.* 104a.5–8; Eutropius 1.8. (4.25–5.2).
222 Jerome, *Chron.* 104a.11.
223 Orosius, *Historiarum adversus paganos libri quinque* 2.8.2, ed. C. Zangemeister, CSEL 5 (1882):99.17–100.2.
224 Josephus, *Antiqu.* 2.15.1, ed. Blatt (212.4–7).
225 Jerome, *Chron.* 104a.12–15.
226 *Ibid.* 104a.23–24.

Joshua [was] the high priest and Zerubbabel the prince of the people.[227] *The Prophets Haggai, Zechariah and Malachi were eminent.*[228]
Pythagoras the natural philosopher was considered famous.[229]

3468

Darius [ruled] 36 years.[230] We find in the Chronicle /**485**/ of Eusebius that two Magi brothers ruled between Darius and Cambyses.[231] In fact Jerome in his commentary on Daniel writes: *After Cambyses, Smerdes the Magus ruled. He took Panthaptes the daughter of Cambyses as his wife. After he had been killed by seven Magi, and Darius had taken over the kingdom in his place, Panthaptes married Darius and had a son by him, Xerxes.*[232]

In the second year of Darius, the seventy years of the captivity of Jerusalem were accomplished, according to Eusebius,[233] based on the testimony of Zechariah the prophet, to whom in the second year of Darius an angel said: *O Lord of hosts, how long wilt thou not have mercy on Jerusalem and on the cities of Judah, against which thou hast been indignant? Now is the seventieth year.*[234] Again in the fourth year of King Darius the same prophet says: *When ye fasted and mourned those seventy years, did ye not fast unto me?*[235] In the sixth year of Darius the building of the Temple was completed, on the third day of the month of Adar,[236] which is 46 years after its foundations were laid under Cyrus. Hence in the Gospel *the Jews say, Forty and six years was this temple in building.*[237] For they began to build in the second year of Darius, in the sixth month, on the 24th day, and they finished, as was said, in the sixth year, in the twelfth month, on the third day. From which it appears that

227 *Ibid.* 104a.16–18.
228 *Ibid.* 104a.21–22.
229 *Ibid.* 104b.12–13.
230 *Ibid.* 104a.25–26.
231 *Ibid.* 104a.23–24.
232 Jerome, *In Dan.* 3.11.2 (898.834–837).
233 Jerome, *Chron.* 105b.5–22.
234 Zechariah 1.12.
235 Zechariah 7.5.
236 Ezra 6.15. Bede's attention throughout this chapter to dating events concerning the Temple might be compared to his *Hom.* 2.24 (363.187–364.228), which recounts, with great chronological precision, the history of the three Jewish Temples from the time of Solomon to the time of Christ.
237 John 2.20.

since the work on the Temple had been carried out to a considerable degree beforehand, the 70 years should be computed from its destruction until permission was granted for its full restoration.

After the kings, who had ruled for 243 years, *had been expelled from the city, Rome held sway barely up to the fifteenth milestone.*[238] *After the kings were driven away, consuls were first instituted at Rome with Brutus, then the tribunes of the people and the dictators. The consuls ruled the republic for almost 464 years, up to Julius Caesar, who first seized exclusive power in the 183rd Olympiad.*[239]

3488

Xerxes, son of Darius, [ruled] 20 years.[240] He *captured Egypt,* which Darius had abandoned.[241] **/486/** *It is related* that he set off to fight *the Greeks* with *70,000 soldiers from his kingdoms and 300 auxiliaries, 1,200 prowed ships and 300 transport vessels.* Nevertheless he was defeated, and fled back to his homeland.[242]

Herodotus, writer of histories, [lived].[243]
Zeuxis the painter won recognition.[244]

3489

Artabanus [ruled] seven months.[245]
Socrates was born.[246]

3529

Artaxerxes, also called "Long-Hands" (that is, μακρόχειρ) [ruled] *40 years.*[247]

In his seventh year, on the first day of the first month, Ezra the priest and scribe of the Law of God *went up from Babylon* with the king's letters,[248] *and on the first day of the fifth month came to Jerusalem* with

238 Jerome, *Chron.* 106.13–15, 19.
239 *Ibid.* 106.20–26.
240 *Ibid.* 108.23.
241 *Ibid.* 108.25.
242 Orosius 2.9.1–2, 2.10 (102.10–103.1, 105.3–107.15).
243 Jerome, *Chron.* 110.2.
244 *Ibid.* 110.6.
245 *Ibid.* 110.15–16.
246 *Ibid.* 110.12.
247 *Ibid.* 110.18–21.
248 Ezra 7.9.

1700 men, and among other deeds of zeal, he chastised the sons of the exile because of their foreign wives.[249] In the twentieth year of the same [king], Nehemiah the cupbearer, arriving from the fortress of Susa, rebuilt the wall of Jerusalem[250] in 52 days, and was leader of the nation for twelve years.

Up to this point Holy Scripture follows a chronological sequence. The deeds of the Jews *which follow hereafter are recorded in the Books of the Maccabees, and in the writings of Josephus and Africanus, who survey the history of the world from the beginning up to Roman times.*[251] Africanus calls to mind this time in the fifth book of his chronicle: *The work remained unfinished until Nehemiah and the 20th year of Artaxerxes, at which time 115 years of Persian rule had passed. It was the 185th year following the capture of Jerusalem. Then Artaxerxes first commanded that the walls of Jerusalem should be erected, and Nehemiah presided over the work, and the street [platea] was built and the walls [were raised] around. And if you wish to count from that time, you will be able to find seventy weeks of years up to Christ.*[252]

Xerxes ruled two months.[253] *After him Sogdianus [ruled] seven months.*[254]

Plato was born.[255]

The physician Hippocrates was reputed famous.[256]

3548

Darius surnamed Nothus [ruled] 19 years.[257] *Egypt broke away from the Persians.*[258] When the Jews returned from captivity, it was priests and not kings who ruled over them up to the time of Aristobolus, who began to assume the royal title together with the priestly dignity.[259]

---

249 Ezra 10.10–11.
250 Jerome, *Chron.* 112.24–113.3.
251 *Ibid.* 113.7–13.
252 Jerome, *In Dan.* 3.9.24, lines 173–181. The "seventy weeks of years" allude to 70 × 7 or 490 years; cf. ch. 9.
253 Jerome, *Chron.* 115.15–16.
254 *Ibid.* 115.18.
255 *Ibid.* 115.13.
256 *Ibid.* 114.8–10.
257 *Ibid.* 115.19–20.
258 *Ibid.* 116.5–6.
259 *Ibid.* 148.6–10; Jerome, *In Dan.* 3.9.24 (873.318–320); Augustine, *DCD* 18.36 (632.12–13), and 18.45 (642.45–47).

3588

Artaxerxes, also known as Mnemon, son of Darius and Parysatidis, [ruled] 40 years.[260] *The story of Esther was written under this king. This is he who is called Ahasuerus by the Hebrews and Artaxerses by the Seventy Translators.*[261]

The Athenians, who hitherto had used sixteen letters, began to use twenty-four.[262]

The celebrated Carthaginian war [took place].[263] *The Gauls [called] Senones under their leader Brennus captured Rome except for the Capitol, and* devastated *the burnt [city] for six months.*[264] *Military tribunes start to be proconsuls.*[265]

Aristotle is Plato's disciple, and continues as such for eighteen years.[266]

3614

Artaxerxes, who is called Ochus, [reigned] 26 years.[267] He *annexed Egypt to his empire, and drove King Nectanebo into Ethiopia; with [Nechtanebo] the kingdom of the Egyptians came to an end.*[268]

The orator Demonsthenes is renowned in the opinion of all.[269]

The Romans defeat the Gauls.[270]

Plato died; Speusippus was head of the Academy after him.[271]

3618

Arses son of Ochus [ruled] four years.[272] **/488/** *Jaddus, high priest of the Jews, was pre-eminent.*[273] *His brother Manasses built the temple on Mount Gerizim.*[274]

260 Jerome, *Chron.* 116.23–117.2.
261 *Ibid.* 117.1–8.
262 *Ibid.* 117.23–26.
263 *Ibid.* 118.12–13.
264 *Ibid.* 118.16–18; Orosius 2.19.5.13 (130.5–132.11).
265 Jerome, *Chron.* 118.24–26.
266 *Ibid.* 120.17–18.
267 *Ibid.* 120.19–21.
268 *Ibid.* 121.18–21.
269 *Ibid.* 121.26–122.1.
270 *Ibid.* 122.2.
271 *Ibid.* 122.5–7.
272 *Ibid.* 122.17–18.
273 *Ibid.* 122.16–18.
274 *Ibid.* 123.1–3.

Speusippus died, and was succeeded by Xenocrates.[275]

In the fourth year of Ochus, Alexander, son of Philip and Olympias, being 20 years of age, *began to reign over the Macedonians.*[276]

3624

Darius son of Arsamus [ruled] six years.[277]

Alexander waged war successfully against the Illyrians and the Thracians, overthrew Thebes, and took up arms against the Persians. After defeating their royal generals at the river Granicus, he captured the city of Sardis.[278] *When he had captured Tyre, he invaded Judea. There he was received with favour, and made sacrifices to God, and bestowed many honours upon* Jaddus *priest of the Temple, after he had dismissed Andromachus as guardian of the [holy] places.*[279] In the seventh year of his reign he built Alexandria in Egypt.[280] Immediately thereafter *he captured Babylon and slew Darius, with whom the Persian empire*, which had endured 231 years, *died out.*[281]

At this time, the Romans subjugated the Latins.[282]

3629

Alexander reigned for five years after the death of Darius, and seven years before.[283] *Alexander captured the Hyrcanians and the Mardi. Returning to Ammon, he built Paraetonium.*[284] He reached the Indian Ocean more by victories than by battles.[285] Returning to Babylon, he died from a poisoned draught in the thirty-second year of his life and the twelfth of his reign.[286] *After [Alexander], the empire was handed on to a number of men. Ptolemy son of Lagus held Egypt. Philip called Arideus, the brother of Alexander [took] Macedon. Seleucus Nicanor reigned over Syria, Babylon, and all the kingdoms of the East, and*

275 *Ibid.* 122.19–20.
276 *Ibid.* 122.17–18.
277 *Ibid.* 123.2–3.
278 *Ibid.* 123.4–10.
279 *Ibid.* 123.16–20.
280 *Ibid.* 123.25–26.
281 *Ibid.* 124.1–3.
282 *Ibid.* 123.26–27.
283 *Ibid.* 124; cf. I Maccabees 1.7.
284 Jerome, *Chron.* 124.13–15.
285 Orosius 3.19 (179.14 *sqq.*).
286 Jerome, *Chron.* 124.25–26; *In Dan.* 3.11.3, 4a (899.860–861).

Antigonus over Asia.[287] These men are designated by Daniel as the four horns of the goat which smote the ram.[288] /**489**/

3669

Ptolemy son of Lagus first ruled over Egypt for 40 years.[289]

Appius Claudius Caecus was held in eminence in Rome. He put the Claudian aqueduct in place, and paved the Via Appia.[290]

Ptolemy brought Jerusalem and Judea under his sway by fraud, and carried off many captives into Egypt.[291]

Onias the son of Jaddus, high priest of the Jews, was pre-eminent.[292]

In the thirteenth year of Ptolemy, Seleucus Nicanor began to rule over Syria, Babylon and the upper regions.[293] *The Hebrew history of the Maccabees reckons the kingdom of the Greeks from this point;*[294] *the Edessans also compute their chronology from this [point].*[295] *Seleucis built the cities of Seleucia, Laodicea, Antioch, Apamia, Edessa, Beroea and Pella.*[296] *Simon, son of Onias, a most devout and pious high priest of the Jews, was pre-eminent.*[297] *After him, his brother Eleazar received the office of ministry in the Temple; his son Onias was passed over because he was only a little boy.*[298] *Seleucus deported the Jews into the cities he had built, granting them citizenship rights and municipal government on a par with the Greeks.*[299]

3707

Ptolemy Philadelphus ruled 38 years.[300] *Sostratus of Cnidus built the lighthouse of Alexandria.*[301] *Ptolemy allowed the Jews who were in Egypt*

287 Jerome, *In Dan.* 2.8.5b–9a (854.835–839).
288 Daniel 8.8 *sqq.*
289 Jerome, *Chron.* 125.1–5.
290 *Ibid.* 125.6–8.
291 *Ibid.* 125.17–19.
292 *Ibid.* 125.20–21.
293 *Ibid.* 126.16–20.
294 *Ibid.* 126.16–18.
295 *Ibid.* 126.24–25.
296 *Ibid.* 127.8–10.
297 *Ibid.* 127.16–18.
298 *Ibid.* 128.21–24.
299 *Ibid.* 128.15–19.
300 *Ibid.* 129.7–9.
301 *Ibid.* 129.10–11.

to go free, and gave Jerusalem to the High Priest Eleazar, along with many vessels as gifts for the Temple. He sought out the Seventy Translators, who turned Holy Scripture into the Greek language.[302] *Aratus gains recognition.*[303] *After Eleazar, his uncle Manasses takes over the priesthood of the Jews.*[304] *It is said that the power of this Ptolemy was so great* /**490**/ *that he outstripped his father Ptolemy, for the histories recount that he had 200,000 foot soldiers, 20,000 cavalry, 2,000 chariots, 4,000 elephants (which he was the first to import from Ethiopia)*, etc.[305]

3733

Ptolemy Evergetes, brother of the previous king, [ruled] *26 years.*[306] The Egyptians called him Evergetes because *when he had captured Syria and Cilicia and then all of Asia, he restored to [the Egyptian] gods, out of the innumerable pieces of silver and precious vessels which he captured, those [pieces] which Cambyses had carried off into Persia after the capture of Egypt.*[307]

Onias, son of Simon the Just, High Priest of the Jews, was pre-eminent.[308] *His son Simon enjoyed no less glory; under him, Jesus son of Sirach wrote the Book of Wisdom, which they call Panareton, and even mentions Simon in it.*[309]

3750

Ptolemy Philopator son of Evergetes [ruled] *17 years.*[310] *Antiochus king of Syria annexed Judea after the defeat of Philopator.*[311] *Onias, son of Simon, high priest of the Jews, is pre-eminent, to whom King Arius of the Lacedaemonians sent ambassadors.*[312]

---

302 *Ibid.* 129.15–23.
303 *Ibid.* 130.18.
304 *Ibid.* 131.22–24.
305 Jerome, *In Dan.* 3.11.5b (901.912–902.916).
306 Jerome, *Chron.* 132.8–10. Ptolemy III Evergetes was actually the son of Ptolemy II Philadelphus.
307 Jerome, *In Dan.* 3.11.7–9 (904.975–905.982).
308 Jerome, *Chron.* 132.10–13.
309 *Ibid.* 133.16–22.
310 *Ibid.* 134.9–12.
311 *Ibid.* 135.4–5.
312 *Ibid.* 135.8–11.

3774

Ptolemy Epiphanes son of Philopator [ruled] *24 years.*[313]

The second Book of the Maccabees contains the events which took place amongst the Jews at this time.[314] *Onias the priest fled into Egypt, taking many of the Jews with him. He was honorably received by Ptolemy, and given the region called Heliopolis. With the king's permission, he constructed a temple in Egypt similar to the Temple of the Jews. This temple endured for 250 years, until the reign of Vespasian.*[315] *Because of the favourable circumstances [created by] Onias the priest, a huge number of Jews took refuge in Egypt at this time, and Cyrene was filled with their multitude.*[316] /**491**/ The reason why Onias and the others went to Egypt was that *during the conflict between Antiochus and the generals of Ptolemy, the parties in Judea were divided into opposing camps, some favouring Antiochus and others Ptolemy.*[317]

3809

Ptolemy Philometor ruled 35 years.[318]

Aristobolus the Jew, a Peripatetic philosopher, gained recognition; he wrote commentaries on [the books of] Moses for Ptolemy Philopator.[319]

Antiochus Epiphanes, who ruled after Seleucus Philopator, reigned *eleven years* in Syria. *He offended against the law of the Jews*, filling everything with the obscenities of idols, *and set up a statue of Olympian Jupiter in the Temple.* But *he also built a shrine to Jupiter Peregrinus in Samaria on the top of Mount Gerizim; the Samaritans themselves asked him to do this. However Mattathias* the priest *vindicated the ancestral laws, taking up arms against the generals of Antiochus.*[320] After his death, his son Judas Maccabeus took on the leadership of the Jews, in the 146th year of the Greek kingdom, in the 20th year of Ptolemy, in the 155th Olympiad.[321] As soon as *he had expelled the generals of Antiochus from Judaea, he purified the Temple of images and after three years restored to*

313   *Ibid.* 135.14–16.
314   *Ibid.* 137.7–9.
315   Jerome, *In Dan.* 3.11.14b (908.1054–1060).
316   *Ibid.* 3.11.14b (908.1062–909.1065).
317   *Ibid.* 3.11.14b (908.1051–1054).
318   Jerome, *Chron.* 138.18–19.
319   *Ibid.* 139.1–6.
320   *Ibid.* 139.22–23; 140.3–12.
321   *Ibid.* 141.10–11.

his fellow countrymen their ancestral laws.[322] After the withdrawal of Onias the priest into Egypt (of which we spoke above) and the death of Alchymus, who though unworthy attempted after the flight of Onias to usurp his pontificate, *the priesthood was conferred upon Maccabeus with the assent of all the Jews.*[323] After his death his brother Jonathan, having been chosen by lot, ministered for nineteen years with great zeal.[324]

3838

Ptolemy Evergetes [II ruled] 29 years.[325]

Jonathan, leader and high priest of the Jews, made an alliance with the Romans and Spartans.[326] After he had been killed by Tryphon, *his brother Simon was elevated to the priesthood,* /**492**/ in the seventh year of the reign of Evergetes.[327] After exercising this office with great energy for eight years, he relinquished it to his son John.[328] This man *waged war against the Hyrcani, and received the name "Hyrcanus". Petitioning the Romans for the right of alliance, he was enrolled amongst the allies by a decree of the Senate.*[329] He levelled Samaria (which in our time is called Sebaste) after having captured it by siege alone. Afterwards Herod restored it, and wanted to name it Sebaste in honour of Augustus.[330]

3855

Ptolemy Physcon, also called Soter, [ruled] seventeen years.[331]

Cicero was born at Arpinum; his mother was Helvia, his father being of a family of equestrian rank from the territory of the Vulsci.[332]

Aristobolus succeeded Hyrcanus in the priesthood, which [Hyrcanus] held for 28 years, *for one year. He was the first to wear the diadem as king as well as priest, 484 years after the Babylonian captivity.*[333] *After him*

322 *Ibid.* 141.3–8.
323 *Ibid.* 141.23–25.
324 *Ibid.* 142.5–8.
325 *Ibid.* 144.3–5.
326 *Ibid.* 144.6–8.
327 *Ibid.* 144.13–15.
328 *Ibid.* 145.18.
329 *Ibid.* 146.17–21.
330 *Ibid.* 146.24–147.4.
331 *Ibid.* 147.7–9.
332 *Ibid.* 148.1–3.
333 *Ibid.* 148.5–10.

Janaeus called Alexander [reigned] for 27 years; he also exercised the office of priest, and ruled his fellow-citizens most cruelly.[334]

3865
Ptolemy, also called Alexander, [ruled] ten years.[335]

In the eighth year of his reign, *Syria came under Roman sway* when Philip was captured by Gabinus.[336] *Expelled from the kingdom by his mother Cleopatra, Ptolemy Physcon withdrew to Cyprus.*[337]

3873
Ptolemy, who had been ejected by his mother for eight years, returned from exile and gained control of the kingdom, for the citizens had expelled Alexander (who preceded him) for killing his mother.[338]

Sulla plunders the Athenians.[339]

3903
Ptolemy Dionysius [ruled] 30 years.[340]

From the fifth year of [Ptolemy's reign], *Alexandra, wife* of the high priest *Alexander, /***493***/ ruled the Jews for nine years* following his death. *From* this time forward, *confusion and diverse calamities weighed heavily upon the Jews.*[341] After her death, *Aristobolus and Hyrcanus* her *sons fought with one another for the rulership, and this furnished an opportunity for the Romans to invade Judea. Thus Pompey, advancing on Jerusalem, captured the city. Opening up the Temple, he penetrated all the way to the Holy of Holies. He carried off the defeated Aristobolus with him, and confirmed Hyrcanus as priest. Then he made Antipater, son of Herod of Ascalon, procurator of Palestine. And Hyrcanus* remained priest for *34 years.*[342]

334 *Ibid.* 148.11–14.
335 *Ibid.* 149.6–8.
336 *Ibid.* 150.4.
337 *Ibid.* 149.9–11.
338 *Ibid.* 150.14–18. There is some confusion in Bede's account here. The Ptolemy referred to in this passage is Ptolemy IX Physcon, i.e. the king whose reign is recorded in the preceding entry, and whose reign was interrupted by that of Ptolemy X Alexander I. The Ptolemy Alexander who was assassinated by the citizens of Alexandria after killing his own mother is actually Ptolemy XI Alexander II, who succeeded Ptolemy IX.
339 *Ibid.* 151.3.
340 *Ibid.* 151.23–24.
341 *Ibid.* 152.8–10, 11–14.
342 *Ibid.* 153.13–23.

Vergilius Maro was born in a country district called Andes, near Mantua, in the consulate of Pompey and Crassus.[343]

Pompey, after the capture of Jerusalem, makes the Jews tributaries.[344]

Vergil pursues his studies in Cremona.[345]

Caesar takes Germany and Gaul and Britain as well, where before the name of Roman was not even known. After taking hostages, he made [these peoples] tributaries.[346]

3925

Cleopatra, sister of Ptolemy, [ruled] *22 years.*[347]

When civil war broke out between Caesar and Pompey, Pompey fled to Alexandria, and there he died, struck down by Ptolemy himself, from whom he had hoped for assistance. Subsequently Caesar came to Alexandria, and again Ptolemy wished to prepare an ambush. War was declared against him, and he was defeated and perished in the Nile. Caesar, now in possession of Alexandria, gave the kingdom to Cleopatra, with whom he had had illicit sexual relations.[348] In the third year of her reign, [Caesar] *first obtained sole rule over the Romans; the princes of the Romans are called "Caesars" after him.*[349] *Cleopatra enters the City with a royal entourage.*[350]

3910

Four years and six months after he began to reign, Caesar, because of his arrogance, was killed, stabbed in the Curia by sixty or more Roman senators and knights who had taken an oath against him.[351] **/494/**

343 *Ibid.* 153.1–3.
344 *Ibid.* 154.7–8.
345 *Ibid.* 154.21.
346 *Ibid.* 155.19; Eutropius 6.17 (38.19–23, 38.26–39.2).
347 Jerome, *Chron.* 155.21. It should be noted that the three entries which follow are out of chronological sequence. After the end of the Jewish kingdom, Bede's chronology follows, first, the line of the Ptolemaic rulers of Egypt, and thereafter the succession of the Roman emperors. Bede interrupts the sequence at this point in order to dovetail the two chronologies by recording (a) the death of Julius Caesar (which occurred after the death of the last Ptolemy, Cleopatra) and (b) the birth of Christ (which occurred before the death of Augustus, the first Caesar). The Roman sequence resumes in 3979, with the death of Tiberius Caesar.
348 Cf. Eutropius 6.19, 21 (39.15–25, 40.7–15).
349 Jerome, *Chron.* 156.1–4.
350 *Ibid.* 156.20–21.
351 Eutropius 6.25 (41.13–22); Jerome, *Chron.* 156.5.

Cassius, having captured Judea, sacked the Temple.[352]

3966

Octavius Caesar Augustus, second [emperor] of the Romans, ruled for 56 years and six months. The kings of the Romans are called "Augusti" after him.[353] Of his [reign], 15 years were during the lifetime of Cleopatra, and 41 after.

In the eleventh year of Augustus the rule of the priests in Judea came to an end. Herod, who did not belong to their [line], but rather was the son of Antipater of Ascalon and his mother Cypros of Arabia, received the government from the hands of the Romans; he held it for 36 years.[354] This man, lest it be proven that he was base born and not of Jewish stock, burnt all the books inscribed with the pedigrees of the Jewish people and preserved in the Temple, so that he could be believed to belong to [the nobility], for want of proof [to the contrary].[355] Moreover, in order to mix his own progeny with their royal race, he cast off Dosis, a woman of Jerusalem whom he had privately married and by whom he had his son Antipater, and took to himself Mariamne, the daughter of Alexander and granddaughter of Aristobolus the brother of Hyrcanus, who before him had been king of the Jews.[356] She bore him five sons, two of which – Alexander and Aristobolus – he himself killed in Samaria.[357] And very shortly thereafter he slew their mother by a similar act of villany, though he loved her more than anything else.[358] Of these [sons], Aristobolus had a son Herod by Berenice. We read in the Acts of the Apostles that he was struck down by an angel.[359]

After the third war broke out between Augustus and Antony, Antony, who held Asia and the East, repudiated the sister of Augustus and took Cleopatra as his wife. After their defeat, Antony and Cleopatra killed themselves.[360] It is from this point on that some reckon *the first year of*

---

352 Jerome, *Chron.* 157.3.

353 *Ibid.* 157.18–20.

354 *Ibid.* 160.1–9.

355 Cf. Rufinus' trans. of Eusebius, *HE* 1.7.13 (61.2–13).

356 Cf. Hegesippus, *Hegesippi qui dicitur historiae libri quinque* 1.36.2, ed. V. Ussani, CSEL 66 (1932):70.8–25.

357 Cf. *ibid.* 1.42.1 (101.11–13).

358 Cf. *ibid.* 1.37.6 (77.6–28); 1.42.2 (101.14).

359 Acts 12.23.

360 Eutropius 7.3; Jerome, *Chron.* 162.12–15, 21–22.

the monarchy of Augustus.[361] **/495/** Hitherto *the Lagids, as they were called, had ruled in Egypt for 295 years.*[362]

3952

In the forty-second year of Caesar Augustus, and the twenty-seventh [year] after the death of Cleopatra and Antony, when Egypt was turned into a [Roman] province, in the third year of the one hundred and ninety-third Olympiad,[363] *in the seven hundred and fifty-second from the foundation of the City* [of Rome],[364] that is to say *the year in which* the movements of all the peoples throughout the world were held in check, and *by God's decree Caesar established genuine and unshakeable peace*,[365] Jesus Christ, the Son of God, hallowed the Sixth Age of the world by His coming.

In the forty-seventh year of Augustus, *Herod died in a suitably horrid way from a disease [which produced] water under the skin and worms which teemed throughout his whole body.*[366] Augustus set up [Herod's] his son Archelaus in his place, and he ruled for nine years, up till the end of the reign of Augustus.[367] *Then the Jews*, who could stand no more, *denounced his ferocity before Augustus, and he was exiled to the city of Vienne in Gaul.*[368] To reduce the power and subdue the insolence of the Jewish kingdom, his four brothers *Herod, Antipater, Lysias, and Philip* were made *tetrarchs* in his place.[369] Of these Philip and Herod (previously called Antipas) had already been appointed as tetrarchs during the lifetime of Archelaus.[370]

---

361 Jerome, *Chron.* 163.7–9.

362 *Ibid.* 163.1–3.

363 Rufinus' trans. of Eusebius, *HE* 1.5.2 (45.24–27); cf. Jerome, *Chron.* 169.9–10.

364 Orosius 6.22.1 (426.17–427.2) and 7.3.1 (437.18).

365 *Ibid.* 6.22.1, 5 (426.17–427.2, 428.3–5).

366 Jerome, *Chron.* 170.1–3. Cf. Rufinus' trans. of Eusebius, *HE* 1.8.5 (65.18–22); Josephus, *Antiqu.* 17.9 (Cologne 1534, fol. 181r). The phrase *scatentibus . . . uermis* which Bede quotes from Jerome echoes Acts 12.23, where it describes the death of Herod Agrippa.

367 Jerome, *Chron.* 170. 9–10, 6–7.

368 Josephus, *Antiqu.* 17.17–19 (Cologne 1534, fol. 185v–187v); Rufinus 1.9.1 (71.18–72.2).

369 Jerome, *Chron.* 170.9–10.

370 Hegesippus 2.3.2 (135.23–136.2).

3979

Tiberius the stepson of Augustus, that is to say the son of the latter's wife Livia born of a previous marriage, ruled for 23 years.[371]

In the twelfth year [of his reign] *Pilate* was sent by him as *procurator of Judea.*[372]

Herod the tetrarch, who ruled over the Jews for 24 years,[373] *founded Tiberias* in honour of Tiberius and *Livias* in honour of his mother Livia.[374] **/496/**

3981

In the fifteenth year of [the reign of] Tiberius, the Lord, after [receiving] the baptism preached by John, proclaimed the kingdom of heaven to the world. As Eusebius indicates in his chronicle, this happened in the 4,000th year from the beginning of the world, according to the Hebrew system of years, which holds that the sixteenth year of Tiberius was *the eighty-first Jubilee, according to the Hebrews.*[375] Why our computation calculates nineteen years less may easily be discovered by reading the previous parts of this book.[376] According to that chronicle which Eusebius composed from both versions [of the Bible], as it seemed good to him, the years [from Creation] are 5,228.[377]

3984

In the eighteenth year of the reign of Tiberius the Lord redeemed the world by His Passion,[378] and the Apostles, preparing to preach throughout the regions of Judea, ordained James the brother of the Lord to be bishop of Jerusalem. They also ordained seven deacons,[379] and after the stoning of Stephen the Church was scattered throughout the regions of Judea and Samaria.[380]

Agrippa surnamed Herod, *the son of Aristobolus* the son of king Herod,[381] *accuser of Herod the tetrarch, having been brought to Rome,*

371 Jerome, *Chron.* 171.17; Hegesippus 2.3.2.
372 Jerome, *Chron.* 173.4–5.
373 *Ibid.* 171.11–16.
374 *Ibid.* 173.9.
375 *Ibid.* 174.7–11.
376 See above *s.a.* 2519.
377 Jerome, *Chron.* 173.18–174.5.
378 Cf. *ibid.* 174.14 *sqq.*
379 Cf. *ibid.* 175.24–26.
380 Acts 8.1.
381 Cf. Rufinus' trans. of Eusebius, *HE* 2.1.8 (107.4–7).

was thrown into prison by Tiberius. He made many friends there, in particular Gaius [Caligula] the son of Germanicus.[382]

3993
Gaius, called Caligula, ruled for 3 years, 10 months, and 8 days.[383]

He *released* his friend Herod *Agrippa* from *chains* and *made him king of Judea,*[384] and he remained king for seven years,[385] that is, until the fourth year of Claudius. Then, when Herod Agrippa had been struck down by an angel,[386] his son Agrippa succeeded to the kingdom.[387] He continued in office for 26 years, until the Jews were overthrown.[388] Herod the tetrarch, who also sought the friendship of Gaius, went to Rome under compulsion from Herodias; but being accused by Agrippa, he lost his post as tetrarch. Fleeing to Spain with Herodias, he died destitute there.

*Pilate, who had pronounced the verdict of condemnation on Christ, /**497**/ was beset with such anguish inflicted by Gaius that he committed suicide.*[389]

Gaius, counting himself one of the gods, profaned the holy places of Judea with the filth of idols.[390]

Matthew, preaching *in Judea, wrote his Gospel.*[391]

4007
Claudius [ruled for] 13 years, 7 months, and 28 days.[392]

The Apostle Peter, after he had first established the Church in Antioch, went *to Rome,*[393] and there held the episcopal throne for 25 years, until the last year of the reign of Nero.

Mark, having been sent to Egypt by Peter, preached the Gospel he had written in Rome there.[394]

382 Jerome, *Chron.* 176.21–23 and Hegesippus 2.3.4 (136.16–20).
383 Jerome, *Chron.* 177.11–13; Eutropius 7.12 (45.10).
384 Jerome, *Chron.* 177.14–15.
385 *Ibid.* 179.16–17.
386 Acts 12.23.
387 Hegesippus 2.5.6 (141.15–16).
388 Jerome, *Chron.* 179.17–19.
389 Orosius 7.5.8 (445.18–446.4).
390 Jerome, *Chron.* 177.17.
391 Isidore 242.
392 Jerome, *Chron.* 179.1–2.
393 *Ibid.* 179.7–9.
394 *Ibid.* 179.12–13.

In the fourth year of Claudius there was a very severe famine, which is recorded by Luke.[395]

In the same year [Claudius] went to Britain, *which no one before Julius Caesar or since his time had dared* to invade. *Most of the island surrendered without battle or bloodshed within a very few days. He also added the Orkney islands to the empire, and he returned to Rome in the sixth month after he had set forth.*[396]

In the ninth year of his reign the Jews were expelled from Rome for disturbing the public peace, as Luke reports.[397]

The following year a great famine overcame Rome.[398]

4021

Nero [reigned for] 13 years, 7 months, and 28 days.[399]

In his second year *Festus succeeded Felix* as procurator of Judea;[400] Paul *was sent in chains to Rome* by this man. *He was held there under house arrest for two years*, and thereafter was released to preach, while Nero *burst forth in those great crimes that histories relate of him.*[401]

James the brother of the Lord, after having *ruled the Church of Jerusalem* for 30 *years*, was stoned to death by the Jews in the ninth year of Nero; they took their revenge on him because they had been prevented from killing Paul.[402] /498/

Albinus succeeded Festus in the governorship of Judea, and Florus [succeeded] Albinus.[403] Because *they could not tolerate* his wantonness, greed and other enormities, the Jews *rebelled* against the Romans. *Vespasian, Master of the Soldiers, was sent out* against them, and *captured many of the cities of Judea.*[404]

On top of all his other crimes, Nero now persecuted the Christians; of their leaders in Rome *he killed Peter by crucifixion and Paul by the sword.*[405]

395 *Ibid.* 179.20–22; Acts 11.28.

396 Orosius 7.6.9–10 (447.6–450.4); cf. Bede, *HE* 1.3 (22–24).

397 Orosius 7.6.15 (451.7–8).

398 *Ibid.* 7.6.17 (452.1–4).

399 Jerome, *Chron.* 181.24–25.

400 *Ibid.* 182.4.

401 Jerome, *De viris illustribus* 5 (10.29–11.4).

402 Jerome, *Chron.* 182.25–183.1; Jerome, *De viris illustribus* 2 (9.12–18); Rufinus' trans. of Eusebius, *HE* 2.23.16–18 (171.11–24).

403 Jerome, *Chron.* 182.24 and 183.23.

404 *Ibid.* 185.11–15.

405 *Ibid.* 185.6–10; Orosius 7.7.10 (454.15–455.1).

This [emperor] undertook nothing of a military nature, and nearly lost Britain, for in his reign two of the finest towns there were captured and sacked.[406]

4031

Vespasian [ruled for] 9 years, 11 months, and 22 days.[407]

He *had been declared emperor by his army in Judea, and leaving the war in the hands of his son Titus, he journeyed to Rome by way of Alexandria.*[408] In the second year [of his reign], Titus overthrew the kingdom of Judea and destroyed the Temple by himself,[409] 1,089 years after it was first built. This war was finished in four years: two during the lifetime of Nero, and two more after him.

Amongst other great deeds which Vespasian performed while he was *still in private life, he was sent by Claudius to Germany, and thence to Britain, and he fought the enemy thirty-two times. He added to the empire two powerful peoples, twenty towns, and the Isle of Wight, near Britain.*[410]

He erected the Colossus [of Rome], one hundred and seven feet high.[411]

4033

Titus [ruled for] 2 years and 2 months, a man so admirable in all forms of virtue that he could be called the love and delight of humankind.[412]

He *built the amphitheatre in Rome, and killed five thousand wild beasts during its dedication.*[413]

4049

Domitian the younger brother of Titus [reigned for] 16 years and 5 months.[414]

---

406 Eutropius 7.14 (46.35).
407 Jerome, *Chron.* 186.20–21.
408 *Ibid.* 186.22–25.
409 *Ibid.* 187.1–26.
410 Eutropius 7.19 (47.23–28); cf. Bede, *HE* 1.3 (22–24).
411 Jerome, *Chron.* 188.9–10. This Colossus, a gigantic statue of the Sun God, gave its name to the Colosseum: see *s.a.* 4033 and 4145.
412 *Ibid.* 188.2–3; Eutropius 7.21 (48.20–22).
413 Jerome, *Chron.* 189.14–16. This is the Flavian amphitheatre, popularly known as the Colosseum.
414 *Ibid.* 189.24–26.

After Nero, he was the *second [emperor] who persecuted Christians. Under* his [rule] /**499**/ *the Apostle John was exiled to the island of Patmos*,[415] and *Flavia Domitilla, the niece of the consul Flavius Clemens by his sister*, was exiled *to the island of Pontia* for testifying to the faith.[416]

It is said that this same John was placed in a vat of boiling oil, but he remained as immune to pain as he had always been to corruption of the flesh.[417]

4050

Nerva [ruled for] one year, 4 months, and 8 days.[418]

In his first edict, he recalled the exiles, and so the Apostle John, freed by this general pardon, returned to Ephesus.[419] Because he saw that during his absence the faith of the Church had been assailed by heretics, he immediately made it firm by describing the eternity of the Word of God in his Gospel.[420]

4069

Trajan [reigned for] 19 years, 6 months, and 15 days.[421]

In the sixty-eighth year after the Passion of the Lord and in the ninety-eighth year of his life, the Apostle John died a peaceful death at Ephesus.[422]

When Trajan began to persecute the Christians, Bishop Simeon *of Jerusalem* (also known as *Simon*), *the son of Cleophas, was crucified.*[423]

Ignatius bishop of Antioch was also brought to Rome and thrown to the wild animals.[424]

Alexander, the bishop of the city of Rome, *was* also *crowned with martyrdom*, and he was buried *where he was beheaded, on the Via Nomentana at the seventh milestone* from the City.[425]

415 *Ibid.* 192.1–4.
416 *Ibid.* 192.16–18.
417 Jerome, *Adversus Jovinianum* 1.26 (247C).
418 Jerome, *Chron.* 192.25–26; Eutropius 8.1 (50.1–11).
419 Jerome, *Chron.* 193.3–8.
420 Cf. Jerome, *De viris illustribus* 9 (14.24–15.16).
421 Jerome, *Chron.* 193.19–22; Eutropius 8.5 (51.15–17).
422 Jerome, *De viris illustribus* 9 (16.3–4).
423 Jerome, *Chron.* 194.19–22.
424 *Ibid.* 194.24–26.
425 *Liber pontificalis* 7, ed. L. Duchesne (Paris: E. de Boccard, 1955):1.127.

Pliny [the Younger], the orator and famous historian from Como, lived at this time. Many products of his genius survive.[426]

The Pantheon at Rome, which Domitian built, *was burnt by lightning;*[427] it took this name from the fact that it was to be the dwelling-place of all the gods.[428] /**500**/

The Jews, instigating sedition in various parts of the world, were struck down by the slaughter they deserved.[429]

Trajan extended the borders of the Roman empire far and wide, which since Augustus had been defended rather than nobly expanded.[430]

4090

Hadrian the son of Trajan's female cousin [ruled for] 21 years.[431]

Because he had been instructed by reading the books written about the Christian religion by Quadratus the disciple of the Apostles and by Aristides the Athenian, a man full of faith and wisdom, and by the legate Serenus Granius, he ordered in a letter that the Christians should not be condemned without being accused of crimes in a proper fashion.[432]

He utterly defeated the Jews, who had rebelled a second time, *in a final slaughter, and also deprived them of permission to enter Jerusalem, whose walls he had restored to their finest state; and he ordered [the city] to be called Aelia after himself.*[433]

He himself, *being most learned in both languages [Latin and Greek], constructed a library of marvellous workmanship in Athens.*[434]

Mark was the first gentile to be made bishop of Jerusalem,[435] when the Jewish bishops came to an end. There were fifteen of these,[436] extending for 107 years from the Passion of the Lord.

426   Jerome, *Chron.* 195.21–23. Bede conflates the Elder and Younger Pliny here.
427   *Ibid.* 195.26.
428   Cf. *ibid.* 191.8.
429   *Ibid.* 196.12–26.
430   Eutropius 8.2 (50.16–17).
431   Jerome, *Chron.* 197.9–12.
432   Orosius 7.13.2 (467.18–468.6).
433   *Ibid.* 7.13.4–5 (468.9–469.3).
434   Jerome, *Chron.* 197.21–22; 200.18–19.
435   *Ibid.* 201.15–17.
436   Cf. *ibid.* 203.25.

4113

Antoninus, surnamed Pius, with his sons [Marcus] Aurelius and Lucius [Aurelius Commodus], [ruled for] 22 years and 3 months.[437]

The philosopher Justin [Martyr] gave Antoninus the book that he had written in favour of the Christian religion, and caused him to look kindly upon the Christians.[438] Not long after, when Crescens the Cynic [philosopher] stirred up persecution, [Justin] *shed his blood for Christ.*[439]

In the time of Bishop Pius of Rome, *Hermas wrote a book* called *The Shepherd*, in which is *contained an angel's command that Easter should be celebrated on Sunday.*[440]

When he arrived in Rome, Polycarp chastised many people who had recently been corrupted by the teachings of Valentinus and Cerdo for disgracing themselves with heresy.[441]

4132

Marcus Antoninus Verus [i.e. Marcus Aurelius] [ruled] with his brother Lucius Aurelius Commodus for 19 years and one month.[442]

From the outset they administered the empire with equal authority, although hitherto there had [only ever] been single emperors.[443]

They made war against the Parthians with great valour and success.[444]

When the persecution began in Asia, Polycarp and Pionius were martyred; in Gaul as well, many gloriously shed their blood for Christ.[445]

Not long after, *plague,* the avenger of evil deeds, caused *extensive devastation in many provinces*, especially Italy and the city of Rome.[446]

After the death of his brother Commodus, *Antoninus [i.e. Marcus Aurelius] made his own son Commodus his colleague as emperor.*[447]

437 *Ibid.* 202.1–3.
438 Orosius 7.14.2 (469.10–12).
439 Jerome, *Chron.* 203.13–18.
440 *Liber pontificalis* 11 (1.132) No such episode appears in the *Shepherd of Hermas.* On similar early medieval legends about angelic authority for computistical rules, see ch. 43 and Commentary.
441 Rufinus' trans. of Eusebius, *HE* 4.14 (333.19–335.14).
442 Jerome, *Chron.* 204.7–9.
443 *Ibid.* 204.10–12.
444 Orosius 7.15.2 (470.1–2).
445 Jerome, *Chron.* 205.9–10.
446 *Ibid.* 205.13; Orosius 7.15.5 (471.1–2).
447 Jerome, *Chron.* 207.18–19.

Melitus of Asia, the bishop of Sardis, gave the emperor Antoninus his apology on behalf of the Christians.[448]

Lucius the king of Britain sent a *letter* to Bishop Eleutherius of Rome, asking *to become a Christian.*[449]

The celebrated bishops Apollinaris of Asia, bishop of Hieropolis, and *Dionysius bishop of Corinth, lived [then].*[450]

4145

Lucius Antoninus Commodus ruled for 13 years after the death of his father.[451] *He waged war successfully against the Germans*,[452] but *otherwise* he was given over *to all forms of wantonness and indecency*[453] and did nothing which can be compared to his father's virtue and piety.[454]

Irenaeus the famous bishop of Lyons lived [then].[455]

The Emperor Commodus ordered the head of the Colossus to be removed and a representation of his own head put up [in its place].[456] **/502/**

4146

Aelius Pertinax [ruled for] six months.[457]

He was murdered in the palace through the treason of Julian [i.e. Didius Julianus], a lawyer; seven months after [Julian] began to rule, *Severus defeated him in battle at the Milvian Bridge and slew him.*[458]

Through notices sent far and wide, Victor the thirteenth bishop of Rome *ordered that Easter* should be celebrated *on the Sunday between the fifteenth and the twenty-first day of the lunar month, as* his predecessor *Eleutherius* [had done].[459]

448 *Ibid.* 206.1–3.
449 *Liber pontificalis* 14.2 (1.136); cf. Bede, *HE* 1.4 (24).
450 Jerome, *Chron.* 206.4–6.
451 Orosius 7.16.1 (472.19–21).
452 *Ibid.* 7.16.2 (472.21–473.1).
453 Cf. Eutropius 8.15 (54.31).
454 *Ibid.* (54.28).
455 Jerome, *Chron.* 208.20–21.
456 *Ibid.* 209.16–18.
457 *Ibid.* 210.4–5.
458 *Ibid.* 210.16–19; Orosius 7.16.5–6 (473.14–20).
459 Jerome, *Chron.* 210.9–10 and *Liber pontificalis* 15 (1.137). The standard recension of the *Liber pontificalis* actually says that Victor ordered that Easter be kept between the 14th and the 21st day, and implies that this was the system espoused by Theophilus of Alexandria.

In support of this decree, *Theophilus bishop of Caesarea in Palestine together with other bishops* present at the same Council, wrote *a very useful synodical letter against those who* celebrated *Easter with the Jews on the fourteenth day of the lunar month.*[460]

4163

[Septimius] *Severus* Pertinax [ruled for] *17 years.*[461]

Clement, a priest of the church of Alexandria, and Panthenus, a Stoic philosopher, conducted a very vigorous disputation on the subject of [Christian] doctrine.[462]

Narcissus bishop of Jerusalem, Theophilus bishop of Caesarea, and Polycarp and Bachylus, bishops of the province of Asia, were notable at this time.[463]

Many persecutions were unleashed against the Christians in various provinces. Leonidas the father of Origen was amongst those who *received the crown of martyrdom.*[464]

After Clodius Albinus, who had made himself Caesar in Gaul, was killed at Lyon, Severus carried the war over *to Britain, where* in order to make the provinces more secure from barbarian incursions *he built a massive ditch and a very solid wall, closely set with towers, running for one hundred and thirty-two miles from sea to sea. He died at York.*[465]

Perpetua and Felicitas were thrown to the beasts for Christ in the arena at Carthage in Africa on the Nones of March [7 March].[466] **/503/**

4170

Antoninus surnamed Caracalla, the son of Severus, [ruled for] 7 years.[467]

Alexander bishop of Cappadocia came to Jerusalem out of longing for the holy places. *While Narcissus*, then a very old man, *was still* bishop of that city, [Alexander] was ordained bishop there, after the Lord indicated in a revelation that this should happen.[468]

---

460 Jerome, *De viris illustribus* 43 (34.2–6). Bede undoubtedly identified this "very useful synodical letter" with the spurious *Acts of the Synod of Caesarea*.

461 Jerome, *Chron.* 210.20–21.

462 *Ibid.* 211.1–4.

463 *Ibid.* 211.5–8.

464 *Ibid.* 212.7–9; Orosius 7.17.4 (474.10–12).

465 Orosius 7.17.6–7 (474.17–475.9); cf. Bede, *HE* 1.5 (24–26).

466 Prosper of Aquitaine, *Chronicon* 757, ed. Th. Mommsen, MGH AA 9 (1892):434.

467 Jerome, *Chron.* 213.4–5.

468 *Ibid.* 213.8–10; Rufinus' trans. of Eusebius, *HE* 6.11.1–2 (541.25–543.6).

Tertullian the African, son of a centurion in the service of the proconsul, was praised in the report of all the churches.[469]

4171

Macrinus [reigned for] one year.[470]

According to [Julius] Africanus, the holy man Abgar ruled at Edessa.[471] *Together with his son Diadumenian,* with whom *he had usurped the empire, Macrinus* was killed *during a military uprising at Archilaides.*[472]

4175

Marcus Aurelius Antoninus [ruled for] 4 years.[473]

In Palestine the city of Nicopolis, previously called Emmaeus, was [re]founded; Julius Africanus, a writer of chronicles, received the commission to carry this out.[474] This is the Emmaeus which the Lord deigned to sanctify by going there after his Resurrection, as Luke records.[475]

Bishop Hippolytus, the author of many works, brought his book *The Canon of History up to this point [in time].* His *discovery* of the *sixteen-year* Easter *cycle gave Eusebius the opportunity to compose a nineteen-year cycle based on it.*[476]

4188

Aurelius Alexander [Severus ruled for] 13 years.[477]

He *was singularly*[478] *dutiful towards his mother Mammaea and for this reason, beloved by all.*[479]

469 Jerome, *Chron.* 212.23–25.

470 *Ibid.* 213.23–24.

471 *Ibid.* 214.5–6.

472 Orosius 7.18.3 (476.1–4).

473 Jerome, *Chron.* 214.8–9. This is the emperor more commonly known as Elagabalus.

474 *Ibid.* 214.20–24.

475 Luke 24.13.

476 Jerome, *De viris illustribus* 61 (41.24–42.3); cf. Rufinus' trans. of Eusebius, *HE* 6.22.1 (569.15–20). See ch. 44 and Commentary.

477 Orosius 7.18.6 (476.11–14); Jerome, *Chron.* 215.1–2.

478 Jones and Mommsen read *unicae,* but I emend to *unice* to agree with Jerome's text.

479 Jerome, *Chron.* 215.25–26.

Urban the bishop of Rome *led many of the nobility* to the faith of Christ and *to martyrdom.*[480]

Origen of Alexandria was famous throughout the world.[481] For this reason, *Mammaea, the mother* of Alexander, *arranged to hear* him and *received him at Antioch with the highest honour.*[482] /**504**/

4191

Maximin [the Thracian ruled for] 3 years.[483]

He launched a *persecution* against the *bishops and clergy* of the Churches, *that is to say, against the teachers, largely because the family of Alexander, whom he had succeeded, and of his mother Mammaea, were Christian; but above all because of the priest Origen.*[484]

Pontianus and Antherus, bishops of the city of Rome, were *crowned with martyrdom* and *buried in the cemetery of Calixtus.*[485]

4197

Gordian [reigned for] 6 years.[486]

Julius *Africanus was celebrated amongst the authors of the Church.* In *the chronicles* that he wrote, he said that *he was hastening to Alexandria, attracted by the very celebrated reputation of Heraclas, whom rumour declared to be exceptionally learned in divine and in philosophical studies and in all Greek learning.*[487]

In Caesarea in Palestine, Origen initiated [two] young brothers from Pontus – Theodore (known as Gregory)[488] and Athenodorus, who later became very celebrated bishops – into divine philosophy.[489]

4204

Philip [the Arabian] together with his son Philip [ruled for] *7 years.*[490]

He was the first of all the emperors to be a Christian, and in the third

480 *Liber pontificalis* 18.2 (1.143).
481 Jerome, *Chron.* 215.22.
482 Orosius 7.18.7 (476.14–15); Jerome, *De viris illustribus* 54 (38.12–15).
483 Jerome, *Chron.* 216.8–9.
484 Orosius 7.19.2 (477.9–13).
485 *Liber pontificalis* 19–20 (1.145–147).
486 Jerome, *Chron.* 216.18–19.
487 Rufinus' translation of Eusebius, *HE* 6.31.1 (585.24–25, 587.3–9).
488 More commonly known as St Gregory the Thaumaturge: see below *s.a.* 4222.
489 Jerome, *De viris illustribus* 65 (44.2–9).
490 Jerome, *Chron.* 217.9–13.

year of his reign, the thousandth year from the founding of Rome was completed. Thus, this anniversary year of past events, the most august of all, was celebrated with magnificent games by a Christian emperor.[491]

Origen replied in eight volumes to a certain Epicurean philosopher called Celsus, who had written against our books. [Origen], to put it briefly, was so prolific in writing that Jerome remembers that he read somewhere that his books [numbered] five thousand.[492]

4205

Decius [ruled for] one year and 3 months.[493]

After he *had killed both the Philips, he launched a persecution against the Christians out of hatred of them,*[494] in which Fabian /505/ was crowned with martyrdom in the city of Rome. He left his episcopal see to Cornelius, who was also *martyred.*[495] *Alexander the bishop of Jerusalem was killed at Caesarea in Palestine and Babylas at Antioch.*[496] This *persecution,* as Dionysius bishop of Alexandria states, *did not originate in an imperial command, but, as he says, "A servant of the demons, who in our city was said to be divine, anticipated the imperial edicts by a whole year, stirring up the credulous crowds against us".*[497]

4207

Gallus with *his son Volusian [ruled for] 2 years and 4 months.*[498]

Dionysius the bishop of Alexandria recalls of his reign that *Gallus could neither see nor guard against Decius' misfortune, but tripped over the same stone. When his reign was prospering at the outset and all was as he wished it to be, he persecuted the holy men who were praying to God on high for the peace of his realm, and both peace and his prosperity took flight with them.*[499]

---

491 Orosius 7.20.2 (478.12–479.1).

492 The source of this quotation has not been traced. *Ep.* 82.7.2 (114.11–12) refers to the 6000 volumes of Origen's works, and *Ep.* 84.8.2 (130.24) says that Origen wrote "1000 or more books". Cf. Mommsen's introduction to his ed. of ch. 66 in MGH AA 13 (1898):228.

493 Jerome, *Chron.* 218.7–8.

494 *Ibid.* 218.11–13.

495 *Liber pontificalis* 21–22 (1.148–151).

496 Jerome, *Chron.* 218.16–18.

497 Rufinus' trans. of Eusebius, *HE* 6.41.1 (601.4–6).

498 Jerome, *Chron.* 218.11–12.

499 Rufinus' trans. of Eusebius, *HE* 7.1 (637.8–12).

Origen died and was buried in the city of Tyre, being not quite seventy years old.[500]

Bishop Cornelius of Rome *moved the bodies of the Apostles from the catacombs at night at the request of a certain lady called Lucina, and on the 3rd kalends of July [29 June] he installed the body of Paul on the road to Ostia where he had been beheaded, and that of Peter in the place in which he had been crucified, among the bodies of the holy bishops in the temple of Apollo on the Mons Aurelius in the Vatican palace of Nero.*[501]

4222

Valerian [ruled] with his son Gallienus [for] 15 years.[502]

While the persecution of the Christians raged, [Valerian] was suddenly captured by Shapur [I] king of the Persians. His eyes were gouged out, and he grew old there in miserable slavery.[503] *Gallienus, frightened by such a clear judgement from God*, therefore restored *peace* to us [Christians].[504] But because of his own wanton cravings, and his father's war against God,[505] he sustained much loss to the Roman realm from barbarian uprisings. *Bishop Cyprian of Carthage*, whose highly learned writings have survived, *was crowned with martyrdom in this persecution.*[506] /506/ His *deacon Pontius*, who *shared his exile up to the day of his death, left a celebrated book* about [Cyprian's] *life and passion.*[507]

Bishop Theodore of Neocaesarea in Pontus, who was also called *Gregory [the Thaumaturge]*, and of whom we wrote above,[508] shone with great glory of [miraculous] powers,[509] one of which was that he moved a mountain by his prayers that there might be sufficient space to build a church.[510]

Stephen and *Sixtus*, bishops of Rome, *suffered martyrdom.*[511]

500 *Ibid.* 7.1 (637.4–5); Jerome, *De viris illustribus* 54 (39.20–22).
501 *Liber pontificalis* 22.4 (1.150).
502 Jerome, *Chron.* 220.1–2; Rufinus' trans. of Eusebius, *HE* 7.10.1 (649.13–15).
503 Jerome, *Chron.* 220.12–14; cf. Orosius 7.22.4 (481.12–14).
504 Orosius 7.22.5 (482.2–4).
505 I.e. the persecution of the Christians.
506 Jerome, *Chron.* 220.9–11.
507 Jerome, *De viris illustribus* 68 (45.11–13).
508 *s.a.* 4197.
509 Jerome, *De viris illustribus* 65 (44.16–17). Jerome, however, states only that Theodore/Gregory worked miracles, and does not mention this specific one.
510 Rufinus' trans. of Eusebius, *HE* 7.28 (954.11–12).
511 *Liber pontificalis* 24–25 (1.154–155).

4224

Claudius [reigned for] one year and 9 months.[512]

He defeated the Goths who had been ravaging Illyria and Macedonia for fifteen years, and because of this a golden shield was erected to him in the Senate house and a golden statue of him erected on the Capitol.[513]

Marcion, a very eloquent priest of the church of Antioch, who also taught rhetoric in the same city, disputed with Paul of Samosata, bishop of Antioch,[514] who was teaching that Christ *had the same nature as man.*[515] *This dialogue still survives today, for it was taken down by notaries.*[516]

4229

Aurelian [ruled for] 5 years and 6 months.[517]

After he had launched a persecution against us, lightning struck directly in front of him, greatly terrifying bystanders, and not long after he was killed by some soldiers while *travelling between Constantinople and Heraclea, at a place on the old highway called Caenofrurium.*[518]

Eutychius bishop of Rome was *crowned with martyrdom* and buried *in the cemetery of Calixtus*; he himself *had buried three hundred and thirteen martyrs with his own hands.*[519]

4230

Tacitus [reigned for] 6 months.[520]

After [Tacitus] was killed in Pontus, Florian was emperor for 88 days and was killed at Tarsus.[521] /**507**/

Bishop Anatolius of Laodicea in Syria, an Alexandrian by birth, who was very learned in philosophical subjects, was widely renowned. The greatness of his genius can readily be appreciated from the book *that he*

---

512 Jerome, *Chron.* 221.22–23.

513 *Ibid.* 221.26–222.3; Orosius 7.23 (485.1–5).

514 Jerome, *De viris illustribus* 71 (47.7–10).

515 Cf. Rufinus' trans. of Eusebius, *HE* 7.27.2 (703.20–26). Neither Rufinus nor Jerome mention Paul's specific heresy.

516 Jerome, *De viris illustribus* 71 (47.11–12).

517 Jerome, *Chron.* 222.8–9.

518 *Ibid.* 223.7, 9–10; Orosius 7.23.6 (486.10–13); Eutropius 9.15 (60.20–26).

519 *Liber pontificalis* 28 (1.159).

520 Jerome, *Chron.* 223.12.

521 *Ibid.* 223.13–14.

wrote about Easter, and from his ten books on the principles of arith-metic.[522]

The insane heresy of the Manichees was born at this time.[523]

4236

Probus [ruled for] 6 years and 4 months.[524]

By many fierce battles he completely liberated the Gallic [provinces], which had been occupied for some time by barbarians.[525]

As we read in Eusebius' chronicle, the *second year* of his reign was *the 335th year of the Antiochenes, the 402nd of the Tyrians, the 334th of the Laodiceans, the 588th of the Edessans, the 318th of the city of Ascalon, and the beginning of the eighty-sixth Jubilee for the Hebrews*, which means their year 4250.[526]

Bishop Archelaus of Mesopotamia wrote an account in Syriac of the debate he had with a Manichee who had come from Persia; this became known to many through its Greek translation.[527]

4238

Carus [reigned] with his sons Carinus and Numerian [for] 2 years.[528]

Bishop *Gaius* of Rome, who later suffered *martyrdom* under Diocletian, was illustrious.[529]

Pierius, a priest of Alexandria under Bishop Theon taught the people very effectively. His words and various treatises, which have survived to the present, were of such refinement that he was called "the younger Origen". He was amazingly frugal and *longed for voluntary poverty. After the persecution he spent the rest of his life in Rome.*[530]

4258

Diocletian [ruled] with *Herculius Maximian for 20 years.*[531]

---

522 *Ibid.* 223.21–22; *De viris illustribus* 73 (47.24–48.2).

523 Jerome, *Chron.* 223.25–26.

524 *Ibid.* 223.16.

525 Orosius 7.24.2 (487.2–4).

526 Jerome, *Chron.* 223.23–24, 20–22. Bede explains *s.a.* 2519 and 3981 why his reckoning deviates from Eusebius' by 19 years.

527 Jerome, *De viris illustribus* 72 (47.17–20).

528 Jerome, *Chron.* 224.19–20.

529 *Liber pontificalis* 29 (1.161).

530 Jerome, *De viris illustribus* 76 (49.2–10). Jerome identifies Pierius' bishop as Thomas.

531 Jerome, *Chron.* 225.7–8, 13–14.

*Carausius, having assumed the purple, occupied the British [provinces]. Narseh king /**508**/ of the Persians declared war in the East. The "Five Nations" invaded Africa. Achilleus took control of Egypt. For these reasons Constantius and Galerius Maximianus were taken on as Caesars.*[532]

Constantius took Theodora the stepdaughter of Herculius [as his wife], and afterwards had by her the six half-brothers of Constantine. Galerius [married] Valeria the daughter of Diocletian.[533]

After ten years, Britain was recaptured by the praetorian prefect Asclepiodotus.[534]

In the seventeenth year [of their reign], Diocletian in the East and Herculius Maximian in the West ordered the churches to be destroyed, and the Christians attacked and killed.[535]

In the second year of the persecution Diocletian laid aside the purple[536] *at Nicomedia and Maximian at Milan.*[537] However, the persecution now under way did not cease to rage until the seventh year of the reign of Constantine. *In the sixteenth year of his reign, Constantius,* a man *singularly mild and unpretentious, died in Britain, at York.*[538]

This *persecution* raged with such cruelty and intensity *that in* the course of a single month it was discovered that *17,000* had suffered *martyrdom for Christ.*[539]

It even crossed the band of the Ocean, and in Britain *Alban, Aaron, and Julius,* with many other men and women, were condemned to that happy slaughter.[540]

The priest Pamphilius, a close connection of Bishop Eusebius of Caesarea, was martyred.[541] [Eusebius] wrote his life in three books.

---

532 *Ibid.* 225.18–23; cf. Bede, *HE* 1.6. "Caesar" in this case means a junior co-emperor. Senior emperors in the Diocletianic tetrarchy were called "Augusti".

533 Jerome, *Chron.* 225.26–226.3.

534 *Ibid.* 227.2–3.

535 *Ibid.* 228.5–6; Orosius 7.25.13 (491.13–15).

536 I.e. abdicated.

537 Jerome, *Chron.* 228.12–14; Orosius 7.25.14 (492.1–7).

538 Jerome, *Chron.* 228.20–21; Orosius 7.25.16 (493.1–2); cf. Bede, *HE* 1.8 (36).

539 *Liber pontificalis* 30.2 (1.162); cf. Bede, *HE* 1.6 (28).

540 Gildas, *De excidio et conquestu Britanniae* 10, ed. T. Mommsen, MGH AA 13 (1898):31.14–22. Cf. Bede, *HE* 1.7 (28–34).

541 Rufinus' trans. of Eusebius, *HE* 8.13.6 (773.10–12); Jerome, *De viris illustribus* 81 (50.26).

4259

In the third year of the persecution, *Constantine, Maximin and Severus were made Caesars* by Galerius Maximianus.[542] Of these, Maximin added crimes and debauchery to the persecution of the Christians.

During this time, *Bishop Peter of Alexandria* suffered [martyrdom] along /**509**/ with many other Egyptian bishops; *so did Lucian a priest of Antioch, a man of exceptional chastity and erudition,*[543] *and Timothy, [martyred] at Rome on the 10th kalends of July* [22 June].[544]

4290

Constantine, the son of Constantius by his concubine Helena, was made emperor in Britain, and ruled for 30 years and 10 months.[545]

In the fourth year of the persecution *Maxentius, the son of Herculius Maximianus, was made Augustus in Rome.*[546]

Licinius, the husband of Constantia the sister of Constantine, was made emperor at Carnuntum.[547]

Constantine turned from a persecutor into a Christian.

At the Council of Nicaea, in the 636th year after Alexander, on the 19th day of the Greek month Desi, which is the 10th kalends of July [22 June], in the consulship of Paulinus and Julian, "viri clarissimi", the catholic faith was laid out.[548]

In Rome, *where he was baptized, Constantine built the basilica* of St John, known as the *Constantinian* basilica.[549] He also built *a basilica dedicated to the the blessed Peter on [the site of] the temple of Apollo,*[550] and also [one] *to the blessed Paul,*[551] and he encased both their *bodies in [sarcophagi of] copper five feet thick.* He built a *basilica in the Sessorian Palace, which is known as [the basilica of the Holy Cross] of Jerusalem,*

542   Jerome, *Chron.* 228.20–21, 18–19.

543   Rufinus' trans. of Eusebius, *HE* 9.6.2–3 (813.1–12).

544   Prosper 974 (447).

545   Jerome, *Chron.* 228.26–229.1; Eutropius 10.2 (65.20–23) and 10.8 (67.21–24); cf. Bede, *HE* 1.8 (36).

546   Jerome, *Chron.* 229.1–3.

547   *Ibid.* 229.6; Eutropius 10.4 (66.13).

548   *Liber pontificalis* 34.4 (1.171); the full dating clause comes from the *Acta* of the Council of Chalcedon 2, ed. J.D. Mansi, *Sacrorum conciliorum amplissima collectio* (Paris: Hubert Welter, 1901):6.955.

549   St John Lateran.

550   St Peter's on the Vatican.

551   St Paul outside the Walls.

where he deposited some of the wood of the Lord's Cross. He built a *basi-lica dedicated to the holy martyr Agnes at his daughter's request, and a baptistery*[552] *in the same place, where his sister Constantia was baptized together with the emperor's daughter.* Again, he built a *basilica to the blessed martyr Lawrence on the Via Tiburtina, in the field of Veranus,*[553] and also one *to the blessed martyrs Peter and Marcellinus, between two laurel trees on the Via Labicana, and a mausoleum where he laid his mother [to rest] in a purple sarcophagus.* He built a *basilica in the town of Ostia to the holy Apostles Peter and Paul and John the Baptist, near the port of the city of Rome,* and a *basilica in the city of Alba to St John the Baptist,* and a *basilica in the city of Naples.*[554] /**510**/

When he restored the town of Drepanum in Bithynia in honour of the martyr Lucian, who was buried there, Constantine renamed it Helenopolis after his mother.[555]

He also wished to make the *city* in Thrace to which he had given *his own name*[556] *the seat of the* Roman *empire and the capital of all the East.*[557]

He ordered that the temples of the pagans be closed, but without killing [anyone].[558]

4314

Constantius ruled with his brothers *Constantine* and *Constans* for *24 years, 6 months, and 13 days.*[559]

Bishop James of Nisibis enjoyed renown; the city was often saved in times of crisis by his prayers.[560]

Fostered by the protection of the Emperor Constantius, the Arian heresy persecuted first Athanasius and then all bishops not of its party with exile, imprisonment, and various types of affliction.[561]

Bishop Maximin of Trier was famous. Bishop Athanasius of Alexan-

552 St Constance.

553 St Lawrence without the Walls.

554 *Liber pontificalis* 34.9 (1.172), 13 (1.174), 16 (1.176), 21–24 (1.178–181), 26 (1.182), 28 (1.183–184), 30 (1.184–185), 32 (1.186).

555 Jerome, *Chron.* 231.22–25.

556 I.e. Constantinople.

557 Orosius 7.28.27 (504.12–16); 3.13.2 (163.18–164.3).

558 *Ibid.* 7.28.28 (505.1–3).

559 Jerome, *Chron.* 234.13–15.

560 *Ibid.* 234.24–25.

561 *Ibid.* 234.16–235.4.

dria was honourably received by him when Constantius wished to punish him.[562]

The monk Anthony died in the desert at the age of 105.[563]

The relics of the Apostle Timothy were brought to Constantinople.[564]

After Constantius had visited Rome, the bones of the Apostle Andrew and Luke the evangelist were received by the citizens of Constantinople with marvellous enthusiasm.[565]

Bishop Hilary of Poitiers who had been *driven out* by the *Arians* and exiled *to Phrygia, returned to Gaul after a book [written] on his behalf was presented to Constantius in Constantinople.*[566]

4316

Julian [reigned] for 2 years and 8 months.[567]

Julian, having converted to the worship of idols, persecuted the Christians.[568]

In the city of Sebaste in Palestine, pagans *broke into the tomb of John the Baptist and scattered the bones. After these had been collected up again they burnt them and scattered them* more widely. *But by the providence of God, some monks from Jerusalem* were there. Collecting whatever they could, they brought [the bones], all *mixed* together, as an offering to their *father [i.e. abbot] Philip.* He /**511**/ immediately *sent them by his deacon Julian to the metropolitan [ad pontificem maximum]*, who *at that time* was *Athanasius* [of Alexandria], for [Philip] felt it was *beyond his powers to guard so great a treasure by his own vigilance. After he had received them, [Athanasius], inspired by a prophetic spirit, preserved the relics for a generation yet to come in a hole under the sanctuary wall in the presence of a few witnesses.*[569] This prophecy was accomplished in the reign of Theodosius by Theophilus, bishop of the same city, who dedicated a church to St John after the tomb of Serapis had been destroyed.[570]

562 *Ibid.* 236.1–4.
563 *Ibid.* 240.5–6.
564 *Ibid.* 240.15–16.
565 *Ibid.* 240.26–241.2.
566 *Ibid.* 240.11–14; 241.17–19.
567 *Ibid.* 242.10–11.
568 *Ibid.* 242.11–12.
569 Rufinus' trans. of Eusebius, *HE* 11.28 (1033.20–1034.16).
570 *Ibid.* 11.26–27 (1031.10–1032.18). For a different account of the fate of the Baptist's relics, see below *s.a.* 4410.

4317

Jovian [ruled for] 8 months.[571]

A synod was held at Antioch by Melitius and his supporters, in which, having rejected "homoousion" and "anomoio" as a compromise, they affirmed the "homoeousion" of the Macedonians.[572]

Jovian, *warned by the fall of his predecessor* Constantius, *in most courteous and respectful letters inquired of Athanasius, and received from him the form of the faith, and the manner in which the Church should be ordered. But an early death cut short* his *pious and happy reign.*[573]

4328

Valentinian [ruled] with his brother Valens for 11 years.[574]

Apollinaris bishop of Laodicea wrote a variety of works concerning religion, but later, deviating from [true] belief, he instituted the heresy that bears his name.[575]

Damasus bishop of Rome *built the basilica of St Lawrence next to the theatre, and another in the catacombs, where the holy bodies of the Apostles Peter and Paul had lain. There he adorned with verses the very marble slab*[576] *where the bodies had lain.*[577]

Having been baptized by Eudoxius, a bishop of the Arians, Valens persecuted our people.[578]

Gratian, the son of Valentinian, was made emperor at Amiens in [Valentinian's] third year.[579]

The martyrium of the Apostles at Constantinople was dedicated.[580]

---

571 Jerome, *Chron.* 243.16–17; cf. 247.23–26.

572 *Ibid.* 243.21–24.

573 Rufinus' trans. of Eusebius, *HE* 11.1 (1002.10–13).

574 Cf. Jerome, *Chron.* 244.7–8; Orosius 7.32.1 (512.1–2).

575 Jerome, *Chron.* 244.14–15; Prosper 1129 (457).

576 I emend Jones' *platoniam* to *platomam.*

577 *Liber pontificalis* 39 (1.212). The basilica "next to the theatre" is St Lawrence *a Damaso* in the Campus Martius, which Damasus built on the site of his family's house. The basilica in the catacombs is St Lawrence "without the Walls": it was actually constructed by Constantine the Great (see above, *s.a.* 4290), but remodelled during the pontificate of Damasus: see Richard Krautheimer, *Rome, Profile of a City* (Princeton: Princeton University Press, 1980):33–34.

578 Jerome, *Chron.* 245.4–5.

579 *Ibid.* 245.6–7.

580 *Ibid.* 245.24–25.

After Auxentius died late in life, Ambrose, elected bishop of Milan, converted all Italy to the true belief.[581] **/512/**

Bishop Hilary of Poitiers died.[582]

4332

Valens [ruled for] 4 years with Gratian and with Valentinian, the son of his brother Valentinian.[583]

Valens issued a law that monks should serve in the army, and he ordered that those who refused should be cudgelled to death.[584]

The people of the Huns, who had long dwelt secluded in inaccessible mountains, struck by sudden madness, burst forth against the Goths, and scattering them hither and thither *drove them from their ancient settlements. The fugitive Goths, having crossed the Danube, were received by Valens without [being made to] lay down their arms.* Soon *they were driven to rebel because of the greed of general Maximus, and having defeated Valens' army they spread abroad throughout Thrace, all the while perpetrating massacres, arson, and rapine.*[585]

4338

Gratian [ruled] with his brother Valentinian [for] 6 years.[586]

Theodosius was made emperor by Gratian, and in many terrible battles defeated the greatest of those Scythian peoples, namely *the Alans, the Huns, and the Goths.*[587]

Because they could not reach an agreement with [Theodosius], the Arians relinquished the churches which they had held by force for forty years.[588]

A synod of one hundred and fifty bishops under Bishop Damasus of Rome was held in the Augustan city[589] against [the heresy of] Macedonius.[590]

---

581 *Ibid.* 247.16–18.

582 *Ibid.* 245.13.

583 Orosius 7.33.1 (515.8–9); cf. Jerome, *Chron.* 247.23–26.

584 Jerome, *Chron.* 248.3–4.

585 Orosius 7.33.10–11 (518.2–10); Jerome, *Chron.* 248.23–25.

586 Prosper 1167 (460).

587 Marcellinus Comes, *Chronicon* 379, ed. T. Mommsen, MGH AA 11 (1894):60.17–22; Orosius 7.34.5 (522.7–523.2).

588 Cf. Marcellinus 380 (61.3–6).

589 I.e. Rome.

590 Cf. Marcellinus 381 (61.8–12).

Theodosius made his son Arcadius his colleague as emperor.[591]

In the second year of Gratian's reign, and in his and Theodosius' sixth consulship,[592] Theophilus wrote his paschal *computus*.[593]

Maximus, a man of energy and integrity, who would have been worthy to be an emperor if he had not emerged as a usurper in defiance of his oath of allegiance, was invited to Britain and there virtually made emperor by the army. He crossed into Gaul, and there waylaid the Emperor Gratian at Lyons, and killed him, and expelled his brother Valentinian from Italy.[594] The latter suffered a richly deserved penalty of exile with his mother Justina, because he had polluted himself with the Arian heresy and had tormented Ambrose, **/513/** the most eminent citadel of the catholic faith, by a treacherous seige.[595] Not until the relics of the blessed martyrs Gervasius and Protasius, revealed by God, were brought forth incorrupt did he abandon the wicked deeds he had begun.[596]

4349

Theodosius, who had *already ruled the East for 6 years while Gratian was alive*, reigned for 11 years after his death.[597]

He received Valentinian, who had been expelled from Italy, kindly. They killed *the usurper Maximus at the third milestone from Aquileia*.[598]

For [Maximus] had *despoiled Britain* of almost *all of its armed youth and large numbers of soldiers*, who had followed *in the footsteps of this tyrant* to Gaul. Seeing that they would *never return home*, some *wild peoples from across the sea – the Irish from the north-west and the Picts from the north* – came to the island thus abandoned by its troops and by its defender, and after devastating and pillaging it, they oppressed it *for many years*.[599]

591 Orosius 7.34.9 (524.2–5).

592 *Recte* fifth. This error was detected by Roger Collins: see above, p. xcix.

593 Cf. ch. 44 above. This is a unique instance of Bede using the word *computus* to denote a book or text about calendar reckoning.

594 Orosius 7.34.9–10 (524.5–11); cf. Bede, *HE* 1.9 (36); the information that Gratian was killed at Lyons is from Marcellinus 383 (61.26–27). Maximus was in fact officially recognized as emperor in the West from 383 to 388.

595 Cf. Rufinus' trans. of Eusebius, *HE* 11.15 (1020.18–1021.15).

596 Cf. Paulinus, *Vita Ambrosii* 14–15, ed. M. Pellegrino, Verba seniorum n.s. 1 (Rome, 1961).

597 Orosius 7.35.1 (525.1–4).

598 Prosper 1191 (462); Orosius 7.34.10 (524.10–13).

599 Gildas 13–14 (32.26–33.14); cf. Bede, *HE* 1.12 (40).

Jerome, the translator of sacred history, *wrote* a book *about the most illustrious men* of the Church, which he brought down *to the fourteenth year of Theodosius' reign.*[600]

4362

Arcadius, the son of Theodosius [ruled] *with his brother Honorius for 13 years.*[601]

By a divine revelation, the bodies of the holy prophets Habakkuk and Micah were discovered.[602]

The Goths entered Italy and the Vandals and Alans entered the Gallic [provinces].[603]

Bishop *Innocent* of Rome *dedicated the basilica* of the blessed martyrs *Gervasius and Protasius, [built with] the testamentary bequest of a certain illustrious lady called Vestina.*[604]

Pelagius the Briton impugned the *grace* of God.[605]

4377

Honorius [ruled] with Theodosius the Younger, the son of his brother, for 15 years.[606]

Alaric, king of the Goths, *captured Rome and burnt part of it*[607] **/514/** *on the 9th kalends of September* [24 August][608] *in the 1,164th year* after its foundation.[609] *On the sixth day after entering it, he left the pillaged city.*[610]

In the seventh year of the reign of Honorius, the priest Lucian, to whom God had revealed the place where the relics of the blessed protomartyr Stephen and of Gamaliel and Nicodemus (about whom it is possible to read in the Gospel and in the Acts of the Apostles) were buried, wrote down his revelation in Greek for the whole Church.[611]

600 Marcellinus 392 (63.5–9).

601 Isidore, *Chronica maiora* 365 (470).

602 *Ibid.* 367 (471).

603 *Ibid.* 368 (471).

604 *Liber pontificalis* 42.3 (1.220).

605 Isidore, *Chronica maiora* 374 (471); cf. Bede, *HE* 1.10 (38).

606 Isidore, *Chronica maiora* 371 (471).

607 Marcellinus 410 (70.7–8).

608 Prosper 1240 (466).

609 Orosius 7.40.1 (548.14–15); cf. Bede, *HE* 1.11 (40).

610 Marcellinus 410 (70.9–10).

611 Gennadius, *De viris illustribus* 47 (92.7–11). Lucian's *Revelatio sancti Stephani* has been edited by E. Vanderlinden in *Revue des études byzantines* 4 (1946):190–217.

The *priest Avitus, a Spaniard by birth*, translated this revelation *into Latin, and, adding a letter of his own, sent it to the West by the priest Orosius*.[612] When he arrived in the Holy Land, whither he had been *sent by Augustine to Jerome to learn about the nature of the soul*, Orosius received *relics of the blessed Stephen* and returning home, was the *first to bring them to the West*.[613]

The Britons, who could bear *the vexations* of the Irish and the Picts no longer, *sent to Rome*, and promising *their submission*, /**515**/ besought help against the enemy. Immediately, *a legion* was sent to them which struck down *the great horde* of barbarians and expelled others from *the borders* of Britain. When it was preparing to return home, [the legion] ordered the allies[614] to build *a wall across the island between the two seas in order to keep the enemy out*. As there was no master builder, this [wall] *was built without* craftsmen, and *from turf rather than stone. Shortly after* the Romans had departed, the *former* enemy came in ships and *cut down whatever stood in their way, devouring it* as if it were *ripe grain*. Their *help* having been sought *again, the Romans* swooped down and defeated the enemy, *chasing them across the seas. With* the Britons, they built [another] *wall* from sea to sea, *for fear of the enemy,* not as before out of crumbly earth, but in solid stone, between the towns which were sited there. But *along the southern* sea *coast, because they also feared* an enemy approach there, they built *towers at intervals over-looking the sea*. Then *they bade farewell* to their allies, *but they were never to return again*.[615]

Bishop *Boniface* of Rome *built an oratory in the cemetary of Saint Felicitas, and decorated* her *sepulchre and that of Saint Silvanus*.[616]

The priest Jerome died *at the age of ninety-one* on the 2nd *kalends of October* [30 September] in the twelfth year of Honorius.[617] /**516**/

4403
Theodosius the Younger, the son of Arcadius, [ruled for] 26 years.[618]

---

612 *Ibid.* 48. Bede knew the works of Avitus at first hand, and cites him in his *Retractatio in Actus Apostolorum* and his commentary on the Epistle of James.

613 Gennadius, *De viris illustribus* 40 (89.5–8).

614 I.e. the Britons.

615 Gildas 15–18 (33.14–35.7); cf. Bede, *HE* 1.12 (40–42).

616 *Liber pontificalis* 44.6 (1.227).

617 Prosper 1274 (469).

618 Isidore, *Chronica maiora* 376 (472).

Valentinian the Younger, son of Constantius, was made *emperor at Ravenna.*[619]

His *mother [Galla] Placidia was called Augusta.*[620]

The ferocious *race of the Vandals, the Alans, and the Goths, crossing from Spain into Africa,* ravaged everything with fire, sword, rapine, and *the Arian heresy.*[621] But the blessed Bishop Augustine of Hippo, the foremost doctor of the Church, lest he see the ruin of his city, went to the Lord *on the 5th kalends of September* [28 August] *in the third month of the siege, having lived for seventy-six years* and having completed *nearly forty years in holy orders and as a bishop.*[622]

At this time, the Vandals, having captured Carthage, annihilated Sicily as well. Paschasinus bishop of Lilybaeum recalled its capture in the letter which he sent to Pope Leo about the calculation of Easter.[623]

In the eighth year of Theodosius, *Palladius was ordained and sent by Pope Celestine to the Irish who believed in Christ, to be their first bishop.*[624]

When word got around that the Roman army had left Britain with no intention of returning, *the Irish and Picts came back again* and *captured the whole island from its indigenous inhabitants, from the north right up to the wall.* With the guardians of the wall defeated, captive, or in flight and [the wall] itself broken, the savage brigands without delay prowled through [the land]. In the twenty-third year of the reign of the Emperor Theodosius, a letter was sent bearing [the Britons'] tears and groans to the commander of the Romans, Aetius thrice consul, /517/ seeking help, but without success. *Meanwhile a terrible and very notable famine* attacked the fugitives, forcing some *to surrender* to the enemy, but others to fight back vigorously *from the mountains, caves, and defiles,* and to *inflict defeat* upon the enemies. The Irish *returned home, though shortly to come back again. The Picts* took hold of *the uttermost part of the island*[625] *for the first time* and inhabited it *from that time forward.*

619 Marcellinus 425 (76.22–23), 419 (74.5–7).

620 Marcellinus 424 (76.11–12).

621 Isidore, *Chronica maiora* 377 (472).

622 Possidius, *Vita Augustini* 28.4, ed. M. Pellegrino, Verba seniorum 4 (Alba: Edizioni Paolini, 1955):148.26–27, 29.1 (156.11–12 and 158.3–4); the date of Augustine's death is supplied from Prosper 1304 (473).

623 Ed. Krusch, *Studien I*, 247; see Introduction, p. l.

624 Prosper 1307 (473); cf. Bede, *HE* 1.13 (46).

625 I.e. the northern part.

The famine mentioned earlier was followed by a great abundance of crops, the abundance by indulgence and carelessness, the carelessness by a very severe epidemic, and soon the fiercer plague of a new enemy, namely the Angles. By unanimous agreement with their king Vortigern, [the Britons] had chosen to invite them in as defenders of [their] homeland; but soon they realized that those they had chosen were assailants and conquerors.[626]

Sixtus the bishop of Rome *built the basilica of Saint Mary* the mother of the Lord, which *was called by the ancients [the basilica] of Liberius.*[627]

Eudoxia the wife of the Emperor Theodosius returned from Jerusalem, bringing with her the relics of the most blessed Stephen the first martyr; these were venerated in the basilica of Saint Lawrence, where they had been placed.[628]

The brothers Bleda and Attila, kings of many peoples, depopulated Illyria and Thrace.[629]

4410
Marcian and Valentinian [ruled for] 7 years.

The people of the Angles or of the Saxons were conveyed to Britain *in three longships*. When their voyage turned out to be a success, news of them was carried back home. A stronger army set out, which, joined to the previous one, first of all drove away the enemy they were seeking.[630] Then they turned their arms on their allies, and subjugated almost the entire island by fire or the sword, from the eastern shore as far as the western one, /**518**/ on the trumped-up excuse that the Britons had given them a less than adequate stipend for their military services.[631]

John the Baptist revealed [the location of] his head, next to a former dwelling of King Herod, to two eastern monks who had come to Jerusalem to pray.[632] It was then taken to the city of Emesa in Phoenicia and venerated with appropriate honour.

The Pelagian heresy disturbed the faith of the Britons. Having sought

626 Gildas 19–22 (35.8–38.11); cf. Bede, *HE* 1.12 (42–44).

627 *Liber pontificalis* 46.3 (1.232).

628 Marcellinus 439 (80.4–8).

629 *Ibid.* 442 (80.4–8).

630 I.e. the Picts and Irish.

631 Gildas 23 (38.12–39.9); cf. Bede, *HE* 1.15 (48–52). For discussion of Bede's dating of the *Adventus Saxonum*, see Harrison, *Framework* 124–125.

632 Marcellinus 453.1–2 (84.21–85.11). Cf. above *s.a.* 4316.

help from the bishops of Gaul, they received Bishop Germanus of Auxerre, and Lupus, equally by apostolic grace bishop of Trier, as the defenders of the faith. The bishops strengthened the faith by the word of truth and at the same time by miraculous signs. Having *gathered some men* they stemmed *the campaign of the Saxons and Picts against the Britons* by divine power. With Germanus himself as their general, the brutal enemy was put to flight, not by the sound of the trumpet, but by the shout of "Alleluia" raised to the stars. [Germanus] then went to Ravenna, where after being received with the utmost reverence by Valentinian [III] and [Galla] Placidia, he migrated to Christ. His body was brought to Auxerre accompanied by a guard of honour, and by the working of miracles.[633] The patrician *Aetius, the great salvation of the Western empire and* once *the terror of King Attila,* was killed *by Valentinian*; with him *fell the Western realm, and to this day it has not had the strength to be revived.*[634]

4427

Leo [ruled for] 17 years. /**519**/

He *sent identical letters to all the bishops throughout the world in support of the Chalcedonian Tome*, asking them to write back to him *what they thought of this Tome.* He *received* concordant replies from *all of them* concerning the true Incarnation of Christ; it was as if all had been written *at one time* and at one dictation.[635]

Theodoret, bishop of the city called Cyrrhus, after its founder the Persian king Cyrus, wrote about the true Incarnation of the Lord our Saviour *against Eutyches and Bishop Dioscorus of Alexandria, who denied the human flesh in Christ.* He also wrote an *Ecclesiastical History from the point where Eusebius' work breaks off down to his own time; which is to say down to the reign of Leo, under whom he died.*[636]

Victorius, at the command of Pope Hilarius, wrote a paschal cycle of five hundred and thirty-two years.[637]

633 Constantius, *Vita Germani* 12, ed. W. Levison, MGH Script. rerum merov. 7 (1920): 259; 14–18 (260–265); 35 (276–277); 42 (281); 44–46 (281–282); cf. Bede, *HE* 1.17 (54–58), 20–21 (62–66).

634 Marcellinus 454.2 (86.6–10); cf. Bede, *HE* 1.21 (66).

635 Marcellinus 458 (87.9–17).

636 Gennadius, *De viris illustribus* 89 (108.9–25).

637 *Ibid.* 88 (107.25–108.7); cf. ch. 51 above, and Introduction, pp. l–li.

4444

Zeno ruled for 17 years.

The body of the Apostle Barnabas and the Gospel of Matthew written by his own hand were found through [Matthew's] own revelation.[638]

Odoacer king of the Goths gained control of Rome, which their kings henceforward held for a long time.[639] After the death of Theodoric the son of Triarius, another *Theodoric, surnamed Valamer*, ruled as king of the Goths, and *depopulated both Macedonia and Thessaly.*[640] Having put *various parts* of the royal city [of Constantinople] to the *torch*, he also invaded and *occupied Italy.*[641]

Huneric the Arian king of the Vandals in Africa *having driven into exile or flight more than three hundred and thirty-four catholic bishops, closed their churches and afflicted the populace with various torments.* Although he cut off countless hands and cut out innumerable tongues he still could not eliminate the expression of the catholic confession.[642]

/520/

Under the leadership of Ambrosius Aurelianus – a man of modest means, who alone of the mighty Romans had survived the slaughter [inflicted] by the Saxons *in which his parents, who had worn the purple, had been killed –* the Britons *goaded* the victors *into battle* and defeated them. And *from then on*, first one side then the other had the palm, until the intruders, being stronger, gained possession of the whole island for a long time.[643]

4472

Anastasius [ruled for] 28 years.

Trasamund the king of the Vandals closed the catholic churches and sent two hundred and twenty bishops into exile in Sardinia.[644]

Pope *Symmachus*, amongst the many church buildings which he built from the ground up or restored, *constructed a dwelling for the poor, dedicated to Saints Peter, Paul and Lawrence. And every year he supplied*

---

638 Isidore, *Chronica maiora* 388 (474).
639 Marcellinus 476.2 (91.21–26).
640 *Ibid.* 481.1–482.2 (92.14–18, 24–25).
641 *Ibid.* 487–489 (93.18–21, 25–30).
642 Cf. *ibid.* 484.2–4 (92.3–93.9).
643 Gildas 25–26 (40.1–41.14); cf. Bede, *HE* 1.16 (52–54).
644 Isidore, *Chronica maiora* 390 (474).

money and clothes to the bishops who were in exile throughout Africa or in Sardinia.[645]

Anastasius, who supported the heresy of Eutyches, harassed the catholics, and perished by *a divine thunderbolt.*[646]

4480

Justin the Elder [ruled for] 8 years.

John, bishop of the Roman church, when he came to Constantinople, was met by a great crowd at the gate called "Golden", where, in the sight of all, he restored the sight of a blind man who besought him [to do this].[647]

When he returned to Ravenna, Theodoric killed him and his companions by imprisoning them in a dungeon, out of spite that Justin, the defender of catholic piety, had received [John] with so much honour.[648] In that year, the consulship of Probus the Younger, he killed the patrician Symmachus at Ravenna, and in the following year he himself died suddenly, and was succeeded in the kingship by his grandson Athalaric.[649]

Hilderic king of the Vandals *ordered the bishops to return from exile* and /**521**/ *the churches* to be restored, after seventy-four years of heretical profanation.[650]

Abbot Benedict shone forth in the glory of his miracles, which the blessed Pope Gregory wrote down in his book of *Dialogues*.

4518

Justinian, nephew by a sister of Justin [ruled for] 38 years.

The patrician Belisarius, sent to Africa by Justinian, exterminated the Vandals.[651] *Carthage was recovered in the ninety-sixth year after its loss. The Vandals were defeated and expelled, and their king Gelimer was sent as a captive to Constantinople.*[652]

The body of St Anthony the monk, discovered by divine revelation, was

645 *Liber pontificalis* 53.10–11 (1.263).

646 *Ibid.* 54.5 (720.4).

647 Gregory, *Dialogi* 3.2.3, ed. A. de Vogüé, Sources chrétiennes 251 (Paris: Cerf, 1978):2.268.22–26.

648 *Liber pontificalis* 55.6 (1.276).

649 Marius Aventicensis, *Chronica, anno* 525, 526, ed. Th. Mommsen, MGH AA 11 (1892): 235; *Liber pontificalis* 55.6 (1.276).

650 Isidore, *Chronica maiora* 396 (475).

651 *Ibid.* 398–399 (475).

652 Marcellinus 534 (103.33–104.3).

brought to *Alexandria, and buried in the church* of the blessed *John the Baptist.*[653]

Dionysius wrote paschal tables beginning in the 532nd year from the Lord's Incarnation, which is the 248th year after Diocletian.[654]

After [came] the consulship of Lampadius and Orestes, and in this year *the Code of Justinian was promulgated to the world.*[655]

Victor bishop of Capua wrote a book about Easter, criticizing the errors of Victorius.[656]

4529

Justin the younger [ruled] for 11 years.

The patrician Narses overcame and killed Totila king of the Goths in Italy.[657]

Then, due to the malice of the Romans, on whose behalf he had laboured long against the Goths, [Narses] was accused before Justin and his wife Sophia of claiming Italy in subjection as his reward. He withdrew to Naples in Campania, and *wrote to the people of the Lombards [telling them] that they should come and possess Italy.*[658] /**522**/

John bishop of the Roman Church *completed and dedicated the church of the Apostles Philip and James,* which his predecessor Pelagius had begun.[659]

4536

Tiberius Constantine [ruled for] 7 years.

Gregory, then *apocrisiarius* in Constantinople and later pope, wrote his commentaries on Job. He also showed, in the presence of the Emperor Tiberius, that Bishop Eutychius of Constantinople had erred concerning our faith in the Resurrection, so that the emperor ordered that *the book he had written about the Resurrection be burned, refuting it with catholic arguments.* For Eutychius had taught that *our bodies* will be *impalpable in the glory of resurrection,* and will be *more subtle than winds*

---

653 Isidore, *Chronica maiora* 400 (476).
654 Counting from his accession in AD 284. Cf. ch. 47.
655 Marcellinus 531 (103.12).
656 See ch. 51 above.
657 Isidore, *Chronica maiora* 402 (476); *Liber pontificalis* 61.8 (1.299).
658 *Ibid.* 63.3 (1.305).
659 *Ibid.* 62.3 (1.303) and 63.1 (1.305).

and air; contrary to [what] the Lord [said]: *Touch me and see, because a spirit does not have flesh and bones in the way that you see I have.*[660]

The plundering people of the Lombards, accompanied by famine and plague, seized hold of all Italy, and besieged the city of Rome; at this time King Alboin was pre-eminent.[661]

4557

Maurice [ruled for] 21 years.

Herminigild, son of Leuvigild the king of the Goths, because his profession of the catholic faith was proof against all persuasion, was deprived of his kingdom by his Arian father, and was thrown into prison in chains, until he was struck on the head at the end of the night of Easter Sunday, and exchanged an earthly kingdom for the heavenly one, which he entered as a king and martyr. His brother Reccared soon after took over the kingdom, following their father, and converted to the catholic faith all of the Gothic people over whom he ruled under the instigation of Leander bishop of Seville, who had also taught Herminigild.[662] /523/

In the thirteenth year of the reign of Maurice and the thirteenth indiction, Gregory, the bishop of Rome and outstanding teacher, assembled a synod of twenty-four bishops at the tomb of the blessed Apostle Peter, to make decisions concerning the needs of the Church.[663]

He sent to Britain *Augustine, Mellitus and John, and many others, with God-fearing monks with them, to convert the English to Christ.*[664] Aethelberht was soon converted to the grace of Christ, together with the people of the Cantuarii over whom he ruled, and those of neighbouring kingdoms. [Gregory] gave him Augustine to be his bishop and teacher, as well as other holy priests to become bishops.[665] However, the people of the Angles north of the river Humber, under Kings Aelle and Aethelfrith, did not at this time hear the Word of life.[666]

660 Gregory, *Moralia in Job* 14.56.74, ed. M. Adriaen, CCSL 143A (1979):745.70.85; the biblical quotation is from Luke 24.39.

661 *Liber pontificalis* 64–65 (1.308.9); Marius Aventicensis 569 (238).

662 Gregory, *Dialogi* 3.31 (2.384–390).

663 Gregory, *Registrum* 5.57a, ed. P. Ewald and L.M. Hartman, MGH Ep. 1 (1887):362–366; cf. Bede, *HE* 2.1 (128).

664 *Liber pontificalis* 66.3 (1.312); cf. Bede, *HE* 2.1–3 (130–142).

665 Cf. Bede, *HE* 1.23 (114–116).

666 Cf. *ibid.* 2.1 (134), 2.2 (140–142).

Gregory, in the eighteenth year of Maurice and the fourth indiction, decreed in a letter to Augustine that the bishops of London and York, having received the pallium from the apostolic see, should be metropolitan bishops.[667]

4565

Phocas [ruled for] 8 years.

In his second year and the eighth indiction, Pope Gregory went to the Lord.

At the request of Pope Boniface, [Phocas] proclaimed that the see of the Roman and apostolic Church *was the head of all other churches, because the Church of Constantinople wrote that it was the first of churches.*[668]

The same emperor, [responding to] another request of Pope Boniface, ordered that a church should be constructed in the old temple called the Pantheon, with the stains of idolatry removed, and dedicated *to the blessed and ever virgin Mary and all the martyrs;*[669] so that where once the worship, not of all the gods but rather of all the demons had taken place, there should thenceforth be a memorial to all the saints. /**524**/

The Persians, waging a *terrible* war *against the empire*, seized many *Roman provinces, and* even took *Jerusalem itself*,[670] destroying churches and profaning the holy furnishings of the places of the saints and of ordinary people. They carried them off, and also stole the standard of the Lord's Cross.

4591

Heraclius [ruled for] 25 years.

Anastasius the Persian monk suffered a noble martyrdom for Christ.[671] Born in Persia, he learned the magic arts from his father when a boy, but when he heard the name of Christ from Christian captives, he soon converted to Him with all his heart. Leaving Persia, he came to

667 Cf. *ibid.* 1.29 (104–106).
668 *Liber pontificalis* 68.1 (1.316).
669 *Ibid.* 69 (1.317); cf. Bede, *HE* 2.4 (148).
670 Isidore, *Chronica maiora* 413 (479).
671 Bede revised a poorly translated Greek *vita* of Anastasius, of which what follows is a summary: cf. *HE* 5.24 (568–580); see C. Vircillo Franklin and Paul Meyvaert, "Has Bede's Version of the 'Passio S. Anastasii' come down to us in 'BHL' 408?", *Analecta Bollandiana* 100 (1982):373–400.

Chalcedon and then Hierapolis in search of Christ, and finally reached Jerusalem, where he received the grace of baptism, and entered the monastery of Abbot Anastasius, at the fourth milestone from the city. There he lived under a monastic rule [*regulariter*] for seven years, until he came for the sake of prayer to Caesarea in Palestine. Captured by the Persians, he was kept in prison by the sentence of Marzban the judge, whipped frequently and held in chains for a long time. Then he was sent back to Persia, to their king Chosroes. Over a period of time, he was almost flogged to death by them three times, and suspended by one hand for three hours a day, before he was beheaded, along with seventy others, and his martyrdom was completed. Shortly after, a man possessed by a demon was cured by putting on his tunic. He was amongst those who were survivors when the Emperor Heraclius defeated the Persians with his army and brought the Christian captives back rejoicing. The relics of the blessed martyr Anastasius were first of all venerated at his own monastery, and then after they had been brought to Rome, in the monastery of the blessed Apostle Paul, called "By the Salvian Waters". /**525**/

In the sixteenth year of Heraclius' reign and the fifteenth indiction, and more or less one hundred and eighty years after the arrival of the English in Britain, Edwin, the most excellent king of the English living across the Humber in north Britain, received with his people the Word of salvation through the preaching of Bishop Paulinus, whom the venerable Archbishop Justus had sent from Kent. This was in the eleventh year of [Edwin's] reign. He gave Paulinus the episcopal see at York.[672] As an omen of the coming of the faith and of the heavenly kingdom, the power of this king's earthly kingdom was increased, so that all the confines of Britain, wherever either the English or the Britons dwelt, were under his authority. None of the English before him [had accomplished this].[673] At this time Pope Honorius condemned in a letter the Quartodeciman error concerning the observance of Easter, which had appeared amongst the Irish. Also John, who came after [Honorius'] successor Severinus, while he was still pope-elect, wrote to them about this same problem of Easter and at the same time about the Pelagian heresy, which was reviving amongst them.[674]

672 Cf. Bede, *HE* 2.14 (186–188).

673 Cf. *ibid.* 2.9 (162).

674 Cf. *ibid.* 2.19 (198–202). Though Honorius definitely accused the Irish of Quartodecimanism, an accusation frequently repeated by Eddius Stephanus (*Vita sancti*

4593

Heracleonas [ruled] with his mother for 2 years.

Cyrus of Alexandria, with Sergius and Pyrrhus, bishop of the royal city [of Constantinople], instigators of the Acephalite heresy, taught that there was one operation of divinity and humanity in Christ and one will.[675] Of these [two], *Pyrrhus*, coming to Rome *from Africa at this time*, namely [the time] of Pope Theodore, *displayed* (as it later became evident) a *false* penitence *in the presence of* the pope and *all the clergy and people* when he presented *a declaration, signed by himself*, in which *he condemned everything that he or his predecessors had either by word or deed done against the* catholic *faith*. Hence he was graciously received by Pope Theodore, as if he really were **/526/** the bishop *of the royal city*. But when he returned home and again took up his wonted error, the aforementioned *Pope Theodore*, having called together all *priests and clergy in the church of the blessed Peter, Prince of the Apostles, condemned him under the bond of anathema.*[676]

4594

Constantine the son of Heraclius [ruled for] 6 months.

Paul the successor of Pyrrhus, tormented the catholics not only by his insane teaching, as his predecessors had done, but by open persecution, afflicting the envoys of the holy Roman Church, who had been sent to correct him, with imprisonment, exiling some and flogging others. He also *overturned and destroyed their altar, consecrated in the ancient oratory of the house of Placidia, forbidding* them to celebrate mass there.[677] Thus he, like his predecessors, was condemned *by the apostolic see by a just sentence of deposition.*[678]

Wilfridi, ed. Bertram Colgrave (Cambridge: Cambridge University Press, 1927) chs. 12, 14, 15), Bede seems to have realized that this was not the case, for in *HE*, he edits Pope Honorius' letter to excise the reference to Quartodecimanism; cf. Alan Thacker, "Bede and the Irish", 38–39. The other heresy associated with the Celtic Paschal observance was Pelagianism: see D. Ó Cróinín, "'New Heresy for Old': Pelagianism in Ireland and the Papal Letter of 640", *Speculum* 60 (1985):505–516. However, Bede seems to associate Pelagianism with celebrating Easter in a lunation which has not reached its 14th day before the equinox, which is a problem with Victorius' tables: cf. ch. 6 and ch. 51.

675 Cf. *Liber pontificalis* 76.3 (1.336–337).

676 *Ibid.* 76.3 (1.337).

677 *Ibid.* 76.1–2 (1.336).

678 *Ibid.* 76.6 (1.338).

4622

Constantine [i.e. Constans II] the son of Constantine [ruled for] 27 years.

Deceived by Paul in the way that his grandfather Heraclius had been deceived by Sergius, bishop of that same royal city, he promulgated a *Typos* against the catholic faith, in which he confessed that neither one nor two wills or operations in Christ should be specified, as if Christ could be believed neither to have willed nor to have acted.[679] In consequence Pope Martin convened a synod in Rome of one hundred and five bishops, and under anathema condemned the aforementioned Cyrus, Sergius, Pyrrhus and Paul as heretics.[680]

Following this, *the exarch Theodore, sent by the emperor*, carried off *Pope Martin from the Constantinian basilica and sent him to Constantinople*. Afterwards [Martin] was exiled to Cherson, where he ended his life, refulgent in many signs of miracles in that place right up to today.[681] The above-mentioned synod was held in the month of October, in the ninth year of the Emperor Constantine, in the eighth indiction.

When Pope Vitalian had just been consecrated, the Emperor Constantine /527/ sent *to the blessed Apostle Peter a golden Gospel book, decorated all around the edge with white gems of great size*.[682] *Some years later, in the sixth indiction, he came to Rome and laid on [St Peter's] altar a cloak made of cloth of gold*, while the entire *army* entered the church *carrying wax candles*.[683]

In the following year there was a solar eclipse, which is still remembered in our days, around the fifth hour on the 5th Nones of May [3 May].[684]

Archbishop Theodore and Abbot Hadrian, a man equally learned, were sent to Britain by Pope Vitalian, and made many of the churches of the English fertile with the fruit of ecclesiastical teaching.[685]

Constantine, after many and unprecedented raids had been made on

---

679 *Ibid.* 76.1 (1.336).
680 *Ibid.* 76.3 (1.336–337).
681 *Ibid.* 76.8 (1.338).
682 *Ibid.* 78.1 (1.343).
683 *Ibid.* 78.2–3 (1.343).
684 Cf. Bede, *HE* 3.27 (310). This eclipse actually took place on 1 May, 664: cf. Commentary on ch. 43.
685 Cf. Bede, *HE* 4.1–2 (328–336).

the provinces, *was killed in his bath in the twelfth indiction*. Not long *after this*, Pope Vitalian also sought the heavenly kingdom.[686]

4639

Constantine, the son of the previous Constantine [ruled for] 17 years.

The Saracens invaded *Sicily*, and then returned *to Alexandria, taking with them much booty*.[687]

Pope *Agatho*, at the request of *the most pious rulers Constantine, Heraclius and Tiberius*, sent envoys *to the royal city*, amongst whom was *John*, then a *deacon*, but not long after bishop of Rome, *to bring about the unification of the holy churches of God*. They were *received* very graciously by Constantine, the most reverend defender of the catholic faith, and were ordered *to lay aside* all *philosophical* debates in order to search for the true faith in *peaceful* discussion, and all the works of the Fathers that they asked for were given to them from the library of Constantinople. One hundred and fifty bishops were present there, under the presidency of the Patriarchs George of the royal city [Constantinople] and Macarius of Antioch. And they were convinced. They added that those who asserted that there was *one will and one operation /528/ in Christ* falsified very many statements of the catholic Fathers. With the conflict settled, George was corrected, and *Macarius, together with* his followers and their predecessors Cyrus, Sergius, Honorius, Pyrrhus, Paul, and Peter, were anathematized. *In [Macarius'] place Theophanius, an abbot* from *Sicily*, was made bishop of *Antioch. Such great gratitude* accompanied the envoys of the catholic peace, that John bishop of Porta, who was one of them, *publicly celebrated mass in Latin in the presence of the emperor and the patriarch in the church of Sancta Sophia, on the Sunday of the octave of Easter*.[688]

This was the *sixth* universal *synod held in Constantinople*[689] and *written up in Greek* in the time of Pope Agatho, enforced and presided over by the most pious Prince Constantine *in* his *palace*, together with *the legates of the Apostolic See and with one hundred and fifty bishops* taking part. The first universal synod was convened in Nicaea against Arius, with three hundred and eighteen bishops, in the time of Pope

686 *Liber pontificalis* 78.4 (1.344).

687 *Ibid.* 79.3 (1.346).

688 *Ibid.* 81.3–4 (1.350–351), 6 (1.351), 12 (1.353–354), 14–15 (1.354); the reference to the condemnation of Honorius is from 82.2 (1.359).

689 *Ibid.* 82.2 (1.359).

Julius and under the Emperor Constantine. The second was in Constantinople with one hundred and fifty bishops, against Macedonius and Eudoxius in the time of Pope Damasus and the Emperor Gratian, when Nectarius was ordained bishop of Constantinople. The third was held in Ephesus, with two hundred bishops present, against Nestorius, the bishop of the imperial city, under the Emperor Theodosius the Great and Pope Celestine. The fourth was in Chalcedon, with six hundred and thirty bishops under Pope Leo and in the time of the Emperor Marcian, against Eutyches the most villainous leader of the monks. The fifth was also in Constantinople in the time of Pope Vigilius and under the Emperor Justinian, against Theodore and all heretics. The sixth is the one of which we have just spoken.[690]

The holy and perpetual virgin of Christ Aethelthryth, daughter of Anna king of the English, was first married to a very eminent man, and afterwards /**529**/ to King Egfrid. After she had preserved the marriage bed uncorrupted for twelve years, she who was once a queen became a consecrated virgin by taking the holy veil. Immediately she also became a mother of virgins and the pious nurse of holy women, and received the place called Ely in order to build a monastery. Her dead body, discovered uncorrupted along with the garment in which it had been wrapped, having been buried for sixteen years, also attests to her enduring merits.[691]

4649

Justinian the Younger, the son of Constantine, [ruled for] 10 years.

He *made a ten-year peace on land and sea with the Saracens. But the province of Africa was brought under the control of the Roman empire. It had been occupied by the Saracens,*[692] and Carthage itself was captured by them and destroyed.

[Justinian], sending out his *protospatharius* Zacharias, ordered that the Roman Bishop Sergius, of blessed memory, be deported to Constantinople because he was unwilling to approve of and subscribe to the heretical synod that [Justinian] was holding in Constantinople. But the militia of the city of Ravenna and of the surrounding regions forestalled

---

690 Cf. Bede, *HE* 4.17 (386).
691 Cf. *ibid.* 4.19–20 (390–400).
692 *Liber pontificalis* 84.3 (1.366).

the wicked orders of the emperor and drove Zacharias from Rome with insults and injuries.[693]

The same Pope Sergius ordained the venerable man Willibrord, called Clement, as bishop of the Frisian people. Even now, as a pilgrim for the eternal homeland (for he is one of the people of the English from Britain) he achieves there every day innumerable daily losses for the devil and gains for the Christian faith.[694]

Because he was guilty of treason, Justinian was deprived of the glory of his kingdom, and withdrew to Pontus as an exile.

4652

Leo [ruled for] three years.

By divine revelation, Pope Sergius found *a silver chest in the sanctuary of the blessed Apostle Peter*, where it had long lain hidden *in a very dark corner*. In it was *a cross adorned with diverse precious stones*. /**530**/ *When he had removed the four metal plates by which the gems were embedded*, he discovered *inside the cross a piece of the salvific Cross of the Lord, of marvellous size*. From that time forward, each year *on the day of the Exaltation [of the Cross] this [relic] is kissed and adored by all the people in the basilica of the Saviour, known as the Constantinian [basilica]*.[695]

The most reverend Bishop Cuthbert, who from being an anchorite became bishop of the church of Lindisfarne in Britain, led from infancy to old age a life justly celebrated for its miraculous signs. When his body had remained buried but uncorrupted for eleven years, it was thereafter discovered, together with the clothes by which he was covered, as if he had just died, as we have recounted in the book of his life and miracles recently composed in prose, and some years ago in hexameter verse.[696]

4659

Tiberius [ruled for] 7 years.

A synod held at Aquileia out of ignorance of the faith *was reluctant to*

693 *Ibid.* 86.6–8 (1.372–374).
694 Cf. Bede, *HE* 5.11 (584–586).
695 *Liber pontificalis* 86.10 (1.374).
696 Cf. *HE* 4.27–32; Bede's prose life of Cuthbert is edited and translated by Bertram Colgrave, *Two Lives of St. Cuthbert. A Life by an Anonymous Monk of Lindisfarne and Bede's Prose Life* (Oxford: Clarendon Press, 1940). The verse life has been edited by W. Jaeger, *Bedas metrische Vita Sancti Cuthberti* (Leipzig: Palaestra, 1935).

accept the fifth universal council; but *instructed by the* salutary *warnings of the blessed Pope* Sergius, it consented to endorse it along with the other churches of Christ.[697]

Gisulf, the duke of the Lombards of Benevento, ravaged *Campania* with fire and sword and the [taking of] captives. Because there was no one who could resist his advance, the apostolic Pope John, who had succeeded Sergius, having sent priests and very many *gifts to him, ransomed all of the captives* and *made the enemy* return home.[698]

He was succeeded by another *John*, who amongst many notable deeds, *built the oratory of the holy Mother of God*, of the most beautiful workmanship, *inside the church of the blessed Apostle Peter.*[699]

King Aripert of the Lombards restored many farms *and estates* /**531**/ *in the Cottian Alps* which previously had belonged to the Apostolic See, but which had been stolen by the Lombards a long time since, to the same See, and he sent this deed of gift to Rome written *in letters of gold.*[700]

4665

Justinian [reigned] for a second time with his son Tiberius for 6 years.

With the help of Terbellius, the king of the Bulgars, he killed those who had expelled him, the patricians *Leo, who had usurped his place*, and also his successor *Tiberius*, who ejected [Leo] from the throne and held him a captive in Constantinople throughout his reign.[701] He sent the patriarch Callinicus to Rome with his eyes gouged out, and gave the bishopric [of Constantinople] to Cyrus, who had been an abbot in Pontus and had aided him when in exile. He ordered Pope Constantine to come to him, and received and saw him off with great honour. He even ordered [Constantine] to *celebrate mass* on Sunday, and received communion from his own hand. He prostrated himself on the ground asking [the pope] to intercede for his sins, and he *restored* all *the privileges of the Church.*[702] When he sent a large army into Pontus in order to overtake Philip[picus], whom he had exiled there, though the pope forbade it, the whole army went over to Philip[picus]' side, and made him emperor. Returning with him to Constantinople, [the army] fought

697 *Liber pontificalis* 86.15 (1.376).
698 *Ibid.* 87.2 (1.383).
699 *Ibid.* 87.1 (1.385).
700 *Ibid.* 88.3 (1.385).
701 *Ibid.* 88.4 (1.385).
702 *Ibid.* 90.6 (1.390–391).

against Justinian at the twelfth milestone from the city. With Justinian defeated and killed, Philip[picus] took over the empire.

4667

Philip[picus] [ruled for] one year and 6 months.

He ejected Cyrus from the bishopric and ordered him to go back and govern his monastery in Pontus. The emperor sent letters containing *deviant teaching* to Pope Constantine, which /**531**/ the latter, *along with the council of the Apostolic See, rejected, and for this reason* [the Pope] set up pictures in the portico *of Saint Peter* depicting the deeds *of the six holy universal synods.*[703] For Philip[picus] had ordered similar pictures in the royal city [of Constantinople] to be removed. *The Roman people ordained that they would not recognize the name of the heretical emperor or his decrees or his face on the coinage. Hence, his portraits were not carried into church, nor was his name mentioned in the solemnities of the mass.*[704]

4670

Anastasius [ruled for] 3 years.

He captured Philip[picus] and blinded him, but did not kill him.

He sent letters to Pope Constantine in Rome *via* the patrician Scholasticus, the exarch of Italy, in which he demonstrated that he supported the Catholic faith and proclaimed the sixth holy Council.[705]

Reproved by Pope Gregory, King Liutprand of the Lombards *confirmed the gift of the estates in the Cottian Alps, which King Aripert had made and he had claimed back.*[706]

In the seven hundred and sixteenth year from the Incarnation of the Lord, Egbert, a holy man of the English people and priest /**533**/ in monastic life, training himself for the celestial homeland as a pilgrim, converted through his pious preaching many provinces of the Irish to the canonical observance of the timing of Easter, from which they had long strayed.[707]

---

703 *Ibid.* 90.8 (1.391).
704 *Ibid.* 90.10 (1.392).
705 *Ibid.* 90.11 (1.392–393).
706 *Ibid.* 91.4 (1.398).
707 Cf. Bede, *HE* 5.22 (553–554).

4671

Theodosius [reigned for] one year.

Elected emperor, he defeated Anastasius in a terrible battle at the city of Nicaea, and *having administered an oath to him,* he forced him to become *a cleric* and be ordained *a priest.* In order to receive the imperial office, since he was a catholic, he soon restored to its former place *in the royal city that image, worthy of veneration, depicting the six holy synods, which had been torn down by Philip[picus].*[708]

The river Tiber overflowed its bed and caused a great exodus from the city of Rome. The waters, flowing together and exceeding their normal height by one and a half times, *descended in a broad sweep from the gate* of St Peter as far as the Milvian Bridge. [The flood] lasted for *for seven days,* /534/ until, with the citizens performing *frequent litanies,* it at last receded *on the eighth day.*[709]

At this time many of the English, both nobles and commoners, men and women, leaders and people in private life, were wont to go from Britain to Rome, inspired by divine love.[710] Amongst these was my most reverend Abbot Ceolfrid, 74 years of age, who had been a priest for 47 years and an abbot for 35. When he reached Langres, he died and was buried there in the church of the blessed twin martyrs. Amongst other gifts which he had arranged to take with him, he sent to the church of Saint Peter a complete Bible [*pandectem*], translated by the blessed Jerome from the Hebrew and Greek sources.[711]

4680

Leo [ruled for] 8 years.

The Saracens, coming to Constantinople with an immense army, besieged it for three years until, with the citizens calling on God on numerous occasions, many of [the Saracens] died of hunger, cold /535/ and pestilence, and withdrew, as if wearied of the siege. As they retreated, they started a war with the people of the Bulgars on the river Danube, and being likewise defeated by these people, they fled and sought their

---

708 *Liber pontificalis* 91.5 (1.398–399).

709 *Ibid.* 91.6 (1.399).

710 Cf. Bede, *HE* 5.7 (472).

711 Cf. Bede, *Historia abbatum* 2.15–23 (379–387); *Historia abbatum auctore anonymo* 21–22 (395–400). Judith McClure has argued for Bede's authorship of the latter: "Bede and the Life of St Ceolfrid", *Peritia* 3 (1984):71–84. The "pandect" is the famous Codex Amiatinus.

ships. When they were on the high sea, a storm suddenly blew up and many were killed when their ships were sunk or wrecked upon the shore.

Hearing that the Saracens had depopulated Sardinia and had dug up the place where the bones of holy Bishop Augustine had once been moved on account of the barbarian raids and honourably buried, Liutprand sent and paid a great price [for them], received [them] and transported them to Pavia, and reburied them there with the honour due to so great a Father.

67. THE REMAINDER OF THE SIXTH AGE

We have striven to the best of our ability to work out all these matters concerning the course of past time from the Hebrew Truth, in the conviction that it is reasonable that just as the Greeks, /536/ using the version of the Seventy Translators, compose books about time for themselves and their own people on the basis of [this translation], even so we, who drink from the pure fountain of Hebrew Truth, thanks to the industry of the holy translator Jerome, should also seek to understand the reckoning of time according to [this version]. Should anyone accuse this work of ours of being superfluous, let them, whoever they may be, accept this just response (without doing violence to brotherly love) which the aforementioned Jerome gave to those who criticized an ancient cosmography[1] – that if it displeased them, they should not read it. For the rest, I urge everyone to note the course of time as he sees it, whether from the Hebrew Truth, which is acknowledged even by the hostile Jews to have been transmitted to us by the abovementioned translator in its pure form, or from the translation of the Seventy Translators, which was either composed with less care to begin with (as many assert) or else afterwards corrupted by the Gentiles (as Augustine holds);[2] or else, indeed, from both works taken together.

Whether he indicates the length of past time to be shorter or longer, or finds it so indicated, nonetheless let him not conclude from this that the time remaining in this World-Age is longer or shorter, remembering always that the Lord said that *no one knows the* last *day and hour, not even the angels, but only the Father.*[3] No one should pay heed to those who speculate that the existence of this world was determined from the beginning at 6,000 years, and who add (lest they seem to deny the Lord's statement) that it is unclear to mortal men in what year of the

1 Jones' text reads *priscae cosmographae* (536.11), but the manuscripts I have consulted read *cosmographiae*. The reference to Jerome has not been traced.

2 Cf. *Letter to Plegwin* 10.

3 Matthew 24.36.

sixth millenium the day of Judgement will come, though its arrival is generally to be expected around the end of the sixth millenium. If you ask them where they have read that such things should be thought or believed, they immediately become vexed, and because they have nothing else to answer they say: "Have you not read in Genesis how God made the world in six days? Hence it deservedly ought to be believed that it will exist for more or less six thousand years."

What is more serious, there were some who believed, on account of the seventh day on which God rested from his works, that after six thousand years in which the saints labour in this life, they would reign with Christ in the Seventh Age after the Resurrection, in this same life [but] immortal, and in the midst of pleasures and great happiness. Giving such things a wide berth, because they are heretical and frivolous, let us understand in a plain and catholic manner those six days in which God finished adorning this world, and the seventh in which He rested from all His work, and which for this reason He consecrated with a blessing of perpetual rest. These signify, not six thousand years /**537**/ of labour and a seventh of the reign of the blessed on earth with Christ, but rather six Ages of this fleeting world in which the saints labour in this life for Christ, and the Seventh Age of rest in another life which the holy souls, released from their bodies, will possess in Christ. This sabbath of the souls is rightly believed to have begun when Christ's first martyr [i.e. Abel], slain in the flesh by his brother, was translated in spirit into eternal rest. But it will come to an end when the souls shall have received incorrupt bodies on the day of resurrection. And because none of the five Ages in the past is found to have run its course in a thousand years, but some in more, some in less, and none had the same total of years as another, it follows that this [Age] likewise, which is now running its course, will also have a duration uncertain to mortal men, but known to Him alone who commanded His servants to keep watch with loins girded and lamps alight *like men waiting for their lord, when he shall return from the marriage feast.*[4]

68. THREE OPINIONS OF THE FAITHFUL AS TO WHEN THE LORD WILL COME

To be sure, all the saints wait attentively for the hour of His coming and long for it to arrive quickly. But they behave rather dangerously if any

4 Luke 12.35–36.

of them presumes to speculate or to teach that this [hour] is near at hand or far off. Hence St Augustine, rebuffing that wicked servant who says in his heart, *My master delays his homecoming*[5] – *surely this man hates the coming of his lord, without any doubt* – gives the example of the three superior servants *who look forward with longing to the coming of their master, watching for it vigilantly and loving faithfully. One of them says, "Let us watch and pray, because the Lord is coming swiftly." The second one says, "Let is watch and pray, for this life is brief and uncertain, though our Lord delays His coming." The third says, "Let us watch and pray because this life is brief and uncertain, and we do not know the time when the Lord will come." Accordingly, if it turns out as the first one predicted, the second and third ones will rejoice with him. But if it does not come to pass, one might dread lest those who believed him be upset, and begin to think that the coming of the Lord was not delayed, but rather non-existent.* /**538**/ *Those who believe what the second one says – that the Lord will come later – will in no way be disturbed in their faith if [that servant] is proved wrong and the Lord comes quickly, but will rejoice greatly with unexpected delight. For this reason, he who says that the Lord is coming quickly speaks more in line with what is desirable, but poses a danger if he is mistaken. He who says the Lord will come later, but believes, hopes for, and longs for His coming nonetheless, even if he is mistaken about His delay, is happily mistaken. For he will have greater patience if it is so, and greater joy if it is not so. And for this reason he is heard with greater pleasure by those who desire the Lord's appearing, and he is believed with greater security. But he who confesses himself truly ignorant of the hour, longs for the former and bears with the latter and errs in neither, since he neither affirms nor denies either.*[6]

69. THE TIME OF ANTICHRIST[7]

We have two very certain indicators of the approach of the Day of Judgement, namely the conversion of the Jewish people, and the reign and persecution of Antichrist, which persecution the Church believes will last three and a half years. But lest this [persecution] come unexpectedly and

---

5 Matthew 24.48.

6 Augustine, *Ep.* 199.13.52–54, CSEL 57 (289.10–292.14).

7 This chapter is largely based on Augustine, *DCD* 20.13 and Jerome, *[De Antichristo] In Danielem* 4.

involve everyone whom it finds unprepared, [the Church believes] that Enoch and Elijah, great prophets and teachers, will come into the world before [Antichrist's] arrival, and will convert the Jewish people to the grace of faith, and will surrender to the insuperable affliction of this mighty whirlwind directed against the elect. After they have preached for three and a half years, and as the prophet Malachi predicts, one of them – Elijah – has *turned the hearts of the fathers to the children*[8] (that is, when they will have planted the faith and love of the saints of old in the mind of those who will then be persuaded), then that horrific persecution will burst into flame. First it will crown [Enoch and Elijah] with the virtue of martyrdom, and then engulfing the rest of the faithful, it will make them either glorious martyrs of Christ or condemned apostates. This is what the Apostle John seems to signify when he writes in the Apocalypse: *But do not measure the court outside the temple; leave that out, for it is given over to the nations, and they will trample over the holy city for forty-two months.*[9] That is, show that those who love external things only, but claim the title of faithful, are to be set apart from the destiny of the elect. For they themselves will turn to persecuting the Church in that final persecution of three and a half years. *And I will grant /**539**/ my two witnesses power to prophesy for one thousand two hundred and sixty days, clothed in sackcloth,*[10] that is, they will preach clothed in the confining labours of continence and affliction. And a little later on he says: *And when they have finished their testimony, the beast that ascends from the bottomless pit will make war upon them and conquer them and kill them*, and so on.[11] He records that the ministers of this same beast, that is, Antichrist, will rejoice over the slaying of these two witnesses, that is, martyrs, and will even mock them after their death.[12]

And in another place he says: *And I saw a beast rising out of the sea. And to it the dragon gave his power and great authority*,[13] that is, I saw a man of savage mind, begotten from the turbulent stock of the impious, to whom after his birth and training by the most despicable teachers in the magic arts, the Devil will add all the authority of his power, through which he will accomplish feats of magic greater than any others. [The

---

8 Malachi 4.6.
9 Apocalyse 11.2; cf. Bede, *Explanatio Apocalypsis* 11.2 (PL 93.162B).
10 Apocalypse 11.3.
11 Apocalypse 11.7.
12 Cf. Apocalypse 11.10.
13 Apocalypse 13.1–2.

Devil] presented [Antichrist] as his sole companion, and (he says) *the beast was allowed to exercise authority for forty-two months.*[14] When that son of perdition will have been struck down either by the Lord himself or by the archangel Michael, as some teach, and damned with an eternal verdict, it is not to be believed that the Day of Judgement will arrive immediately. Otherwise, the men of that age would be able to know the time of the Judgement, if it followed immediately upon the three and a half years after the beginning of Antichrist's persecution. Everyone is allowed to know that the Day of Judgement will not come before that persecution is completed; but it is granted to no one to know how long after the end of that persecution it will come. Therefore the prophet Daniel, who writes that the reign of Antichrist will be for 1,290 days, concludes thus: *Blessed is he who waits and comes to one thousand three hundred and thirty-five days.*[15] Jerome explains it in this way: *He says, blessed is he who, after Antichrist is slain, waits until the forty-fifth day* after the 1,290 days, that is, three and a half years, *when the Lord and Saviour shall return in his majesty. But why there should be 45 days of silence after the slaying of Antichrist, is known to God alone, unless perhaps we might say that the postponing of the Kingdom is a test of the patience of the saints.*[16]

70. THE DAY OF JUDGEMENT

For the day of the Lord shall come as a thief because, as He himself declares, *we know not when He will come, whether in the evening, or in the middle of the night, or /*540*/ at cock-crow, or in the morning.*[17] Then the heavens, [St Peter] says, *shall pass away with a great noise, and the elements shall melt with fervent heat.*[18] What these heavens are which shall pass away, the same Apostle Peter teaches a little earlier on, saying: *By the word of God the heavens were of old, the Earth standing out of the water and in the water, whereby the world that then was perished. But the heavens and Earth which are now, by the same word are kept in store, reserved unto fire in the Day of Judgement.*[19] Not, therefore, the

14 Apocalypse 13.5.
15 Daniel 12.12.
16 Jerome, *In Dan.* 4.12.12 (943.671–944.677).
17 Mark 13.35.
18 II Peter 3.10.
19 II Peter 3.5–7.

firmament of heaven in which the fixed stars revolve,[20] nor the ethereal heaven, that is, that great empty space between the starry heaven and our troubled atmosphere, full of pure and tranquil diurnal light, in which it is believed that the seven wandering stars roam, but this heaven of air [*aerium*],[21] the one close to the Earth, from which the birds [*aues*] of heaven that fly therein take their name,[22] which the waters of the Flood once destroyed when they overflowed the annihilated Earth[23] – this [heaven] the fire of the Last Judgement will destroy, extending as far as [this heaven] and battening upon it.[24] Not only does the statement of St Peter[25] in which he says that those heavens which the flood waters destroyed are to be destroyed by the fire of Judgement, testify that the starry heaven will not be touched by this fire, no matter how great it will be, but so also does the word of the Lord which says: *Immediately after the tribulation of those days the Sun will be darkened, and the Moon will not give its light, and the stars will fall from heaven.*[26] For the Sun could not be darkened, nor the Moon deprived of light, nor could the stars fall from heaven, if the heaven itself (that is, their place) were to perish, consumed by fire. So at that time the airy heaven will shrivel up in fire, [but the heaven] of the stars will remain undamaged.[27] In fact, the heavenly bodies will be darkened, not by being drained of their light, but by the force of a greater light at the coming of the Supreme Judge, on the supposition that they [merely] appear to be concealed, which sometimes happens to the Moon and all the stars because of the greater power of the Sun. But when there will be a new heaven and a new Earth after the Judgement – which is not one [heaven and Earth] replacing another, but these very same ones [which] will shine forth, having been renewed by fire and glorified by the power of the Resurrection – then, as Isaiah predicts: *The light of the Moon will be as the light of the Sun, and the light of the Sun will be sevenfold, as the light of seven days.*[28] John

20 Cf. Bede, *The Nature of Things* 25.

21 Cf. Bede, *In Gen.* 1.21 (21.631–22.649).

22 The source of this etymology is unknown. It is not one of those offered by Isidore, *Etym.* 12.7.3.

23 Augustine, *De Genesi ad litteram* 3.2 (63.22–64.10).

24 Augustine, *Enarrationes in Psalmos* 101(II).13 (1448.6–12).

25 II Peter 3.5–7.

26 Matthew 24.29.

27 Augustine, *DCD* 20.24 (744.15–745.71).

28 Isaiah 30.26.

states in the Apocalypse: *And I saw a new heaven and a new Earth, for the first heaven and the first Earth had passed away*, and he adds, *and there was no more sea*.[29] Whether the sea will be dried up by that mighty heat, or whether it will also be transformed into something better, is not evident. We read that there will be a new heaven and a new Earth, but nowhere do we read that there will be a new sea. /**541**/ For this statement, *And there was no more sea*, can be understood allegorically, for then this world [*saeculum*], turbulent with the life of mortal men, which is often represented in Scripture by the word "sea", will be no more.

When this critical moment of the Last Judgement will take place is sought in many places. However, it is agreed that when the resurrection of all the dead takes place *in the twinking of an eye*[30] at the descent of the Lord to judgement, the saints will be immediately caught up *to meet Him in the air*. The Apostle is understood to indicate this when he says: *For the Lord himself will descend from heaven with the cry of command, with the archangel's call, and with the sound of the trumpet of God. And the dead in Christ will rise first; then we who are alive, who are left, shall be caught up together with them in the clouds to meet the Lord in the air*.[31] But whether the reprobate as well will be lifted above the Earth to meet the Judge when He comes, or whether the deserts of the sinners will weigh so heavily that even though they have immortal bodies they will be incapable of rising on high, it would appear more likely that, with the Lord presiding in judgement, the saints would take their place on high at his right hand, and [the reprobate] lower down on his left. But if at that time that lofty and mighty fire covers the whole surface of the Earth, and if the wicked who rise from the dead cannot be caught up on high, it is understood that, as one might expect, they would await the Judge's verdict standing on earth and surrounded by fire.

But who would dare to decide in advance whether they will be burned by [this fire], who are not to be punished by it, but rather are to be condemned to eternal fire? For St Augustine in Book 20 of *The City of God* understands from the statements of the prophets that some of the elect will be purged by this [fire] from certain less serious sins.[32] And in his homilies on the Gospel, the holy Pope Gregory explains this passage

29 Apocalypse 21.1.
30 I Corinthians 15.52.
31 1 Thessalonians 4.15–16.
32 Cf. Augustine, *DCD* 20.26 (750.61–70).

of the Psalmist: *"Before Him is a devouring fire, round about Him a mightly tempest."*[33] *Tempest and fire accompany the chastisement of such great justice, for the tempest interrogates those whom the fire burns.*[34] But if the fire of the furnace which surrounded them failed to touch the mortal bodies of the three children,[35] then it is amply plain that the world-wide conflagration will not harm His perfect servants who are caught up at the sound of the trumpet to meet the Lord in the air. Indeed, in all these matters it is more expedient that one present oneself chaste before the eyes of the chastening Judge, than debate where or how that Judgement will take place.

Although the Apostle rightly states, *We will be caught up together with them in the clouds to meet the Lord in the air*, and adds, *and so we shall always /542/ be with the Lord*,[36] this should not be interpreted to mean that he says that we shall always remain with the Lord in the air in this way. For [the Lord] Himself will certainly not remain there, because in coming, He will continue onwards. Indeed, one meets someone who is approaching, not someone who is standing in place. But we will be with the Lord in this manner: namely, wheresoever we are with Him, we will be there with everlasting bodies.

71. THE SEVENTH AGE, AND THE EIGHTH AGE OF THE WORLD TO COME

And this is that Eighth Age which is ever to be loved, hoped for, sighed for by the faithful, when Christ shall convey their souls, endowed with the gift of incorruptible bodies, into the possession of the kingdom of heaven and into the contemplation of His divine majesty. He will not take from them the glory which they, released from their bodies, receive in blessed peace from the moment of their departure [from this life], but will heap upon them the even greater glory of their restored bodies. As a foreshadowing of this continuous and uninterrupted beatitude, Moses, although he said that the six days in which the world was made began at light and morning and ended at evening,[37] [in the case of] the seventh day, when God rested from His labours, mentioned only the morning,

---

33 Psalm 50.3 (49.3).
34 Gregory, *Hom. in Evang.* I, 1.6 (PL 76.1081).
35 Cf. Daniel 3.19 *sqq.*
36 I Thessalonians 4.16.
37 Genesis 2.2; cf. Bede, *In Gen.* 1.2 (32.975–34.1049).

and not the evening. Everything which He thought worthy to be commemorated concerning [the seventh day], He comprised in the light of
eternal rest and beatitude. For as we observed earlier,[38] all the Six Ages
of this world, in which the just apply themselves to good works in co-
operation with God, are so disposed by heavenly arrangement that,
while they each began with some element of joy, they ended in many
shadows of trouble and constriction. However, the repose of the souls,
which they will receive in the world to come for the sake of their good
works, will not be brought to a close by the eruption of any time of
fearful anxiety. Rather, when the time of Judgement and resurrection
comes, it will be fulfilled in the more glorious perfection of eternal bliss.

The sacrosanct time of our Lord's Passion, burial and Resurrection
can be compared to these Ages.[39] For we read in the Gospel of John that
Jesus came to Bethany six days before the Passover,[40] where Judas,
offended at the service performed by the devout woman, betrayed Him
to the chief priests. On the following day, [Jesus] entered into Jerusalem
upon an ass, with a crowd of people singing praises to God. For the next
five days He was assailed by their insidious questions, /**543**/ and finally
on the sixth day He was crucified. On the seventh day He lay in the tomb,
but on the eighth, that is, on the first day of the week, He rose from the
dead. For in the Five Ages of this world that are passed and gone, the
saints never had respite from the hatred and treachery of the reprobate.
But in the Sixth [Age], which the Lord deigned to establish in faith by His
Incarnation, redeemed from hell by His Passion, and fired with hope and
love for the heavenly kingdom by His Resurrection and Ascension, the
loftier virtue of the blessed martyrs has borne with more atrocious
conflicts of persecution at the hands of the infidel. But the more assured
they are that they suffer for Him who promised to the one who died with
Him, first as a thief and then as a confessor, *Today you shall be with me in
paradise*,[41] the more resolutely they confront these things.

This blessed repose of paradise has no end other than the beginning
of the glorified Resurrection. The evangelist Matthew wishes to indicate
this, and after writing that the Lord died on the sixth day and was

38 See above, ch. 10.
39 Cf. Bede's *Hom.* 1.23 (167.209–227) and 2.7 (225.17–226.26), where the *triduum
paschale* is compared to the Six Ages.
40 John 12.1.
41 Luke 23.43.

buried on the seventh, he begins to speak of the Resurrection thus: *At the evening of the sabbath, as it began to dawn towards the first day of the week, came Mary Magdalene and the other Mary to see the sepulchre.*[42] He says that the evening of the sabbath, in which the Lord rested in the tomb, did not darken into night, but brightened into the first day of the week. For indeed His sepulchre was not to be defiled by the corruption of the body it received, but as Isaiah says, *His grave shall be glorious*,[43] and exalted by the power of His swift Resurrection. Likewise the repose of our souls after we have put off our bodies will not be darkened by any shadows of anxiety, but in the end will be taken up and enhanced by the gifts of the true Sun and everlasting light.[44]

This is that great and unique octave for which the sixth and the eleventh Psalms were written, whose words commemorate the day as follows: *And Mattithiah and Elipheleh and the other Levites sung a song of victory [epinicion] for the octave*,[45] that is, they sung the praises of the victory of God and His judgement in the endless *ogdoad*, for *epinicion* means song of triumph and palm.[46] Or they prophesied the mystery of the Resurrection of Christ, which deservedly is called the great and unique octave. For every eighth day of this faltering age is an eighth day following a sabbath, so that it is likewise the first day of the following week. But just as the first day of the world was "first" in such a way that it did not have seven other days preceding it, of which it was the eighth, and so was "first" in a unique manner, so also the day of the future resurrection is not only uniquely great, but also eighth in a unique way, because this eighth day will so follow upon the preceding seven, /**544**/ that it will not have other days following it of which it will be the first, but it alone will abide, one and unending, in the eternal light. Hence the prophet, thirsting for the vision of that day, rightly called it "one", saying, *Better is one day in thy courts than a thousand.*[47]

42 Matthew 28.1. Cf. *Hom.* 2.7, and ch. 5 above.

43 Isaiah 11.10.

44 Cf. Augustine, *DCD* 22.30 *passim.*

45 I Chronicles 15.21.

46 Ambrose, *Explanatio super psalmos* 38.1.2, ed. M. Petschenig, CSEL 64 (1919):183.21–184.6.

47 Psalm 84.10 (83.11); cf. Augustine, *Enarrationes in Psalmos* 83.14 (1159.5–10): . . . *contemnant milia dierum, desiderent unum diem, diem sempiternum, cui non cedit hesternus, quem non urget crastinus.* ("Let them scorn these thousands of days, let them long for the one day, the everlasting day, for which no yesterday makes way, and which no tomorrow urges on.")

Therefore when we read of the octave in Scripture, we know that it can be understood symbolically of both the day and the age. For the Lord rose from the dead on the eighth day, that is after the seventh day of the sabbath; and we also shall rise again, not only after the seven fleeting days of this World-Age, but also after the aforementioned Seven Ages, at once in the Eighth Age and upon the eighth day. To be sure, the day of this life has always abided, abides, and will abide, eternal in itself. But for us it will begin when we deserve to enter into it in order to see it, where the saints, renewed in the blessed immortality of flesh and spirit, are occupied in doing what the Psalmist invokes, who sings to God in praise of His love: *Blessed are they who inhabit your house; they shall praise you, world without end.*[48] He then tells of the vision which shall delight them; for He who gave the law will give blessing, and *they will go from strength to strength, and the God of Gods will be seen in Zion.*[49] The Lord himself, who is "the way, the truth and the life"[50] declares what sort of men will be able to come thither: *Blessed are the pure in heart, for they shall see God.*[51]

And so our little book concerning the fleeting and wave-tossed course of time comes to a fitting end in eternal stability and stable eternity. And should those who read it deem it worthy, I ask that they commend me in their prayers to the Lord, and that they behave with pious zeal towards God and their neighbour, to the best of their ability, so that after temporal exertions in heavenly deeds, we may all deserve to receive the palm of heavenly reward.

48 Psalm 84.5 (83.4).
49 Psalm 84.8 (83.7).
50 John 14.6.
51 Matthew 5.8.

COMMENTARY

Preface

Bede's preface to *The Reckoning of Time*, addressed to Abbot Hwætbert (also known as Eusebius) of Wearmouth and Jarrow, is by no means a straightforward introduction to a didactic work. To begin with, its account of the reason why the book was written is somewhat ambiguous. On the one hand, Bede tells a conventional tale of a request from his brother-monks to produce a fuller and more detailed account of *computus* than *On Time* and *The Nature of Things*. The brethren were especially perplexed by the details of Paschal reckoning, and so Bede obliged with a longer treatment of "certain matters concerning the nature, course and end of time".

Most of the preface, however, is occupied, not with the Paschal dispute, but with a somewhat defensive apology for the scheme for dating biblical events employed in ch. 66, "The Six Ages of the world", also known as the "Great Chronicle". This excursus summarizes the arguments laid out at length by Bede in his *Letter to Plegwin*, where he answered charges that this same scheme, as published in the chronicle in *On Times*, was heretical, because it dates Christ's birth to *annus mundi* 3952. If each World-Age is 1,000 years long – for this is implied in II Peter 3.8, which states that for God, a thousand years is as a day – this would mean that Christ's Incarnation did not take place at the beginning of the Sixth Age. Bede counters that his chronology is based on the Vulgate translation of the Bible, which he likes to call "the Hebrew Truth", and he invokes the authority of Jerome, Augustine, Eusebius and even Josephus to justify his preference, while disclaiming any intention of denigrating the Septuagint. That Bede had this particular controversy in mind is indicated by the fact that he specifically summons his patristic authorities to witness that the Vulgate "contains a shorter span of time than is commonly conveyed in the edition of the Seventy".

But why is Bede reiterating the arguments of the *Letter to Plegwin* in *The Reckoning of Time*? and why in the preface, rather than in chapter 66? Jones asserts that Bede's chronology "challenged the orthodox",[1] but this is something of an overstatement. The position taken by Bede in

---

1 *BOT* 329.

the *Letter to Plegwin*, far from "challenging the orthodox", was entirely in line with orthodox refutations of chiliasm.[2] It is interesting that no objection to Bede's chronology is ever voiced in the glosses to the preface; all the glossators pass over his apology in silence, as if they cannot understand why he should be apologizing. Nor indeed could they know why, unless they had access to the *Letter to Plegwin*, since Bede does not recount the circumstances of the original controversy in the preface to *The Reckoning of Time*. Nonetheless, it is plain that Bede wished to ensure that the substance of the *Letter to Plegwin* was incorporated into his new book on *computus*, probably to secure its survival and diffusion. But to argue, as Jones does, that this incident underlies the entire structure and content of *The Reckoning of Time* is excessive and unnecessary.[3] On the contrary, it would seem that Bede incorporated this summary of the *Letter to Plegwin* into the preface precisely because he did not wish to make an issue of it in chapter 66.

Chapter 1:

The first four chapters of *The Reckoning of Time* are a technical preparation for the study of *computus*. They deal with ways of representing number, modes of discussing time, and divisions of time smaller than the day, which is the basic unit of calendar-reckoning. These divisions and fractions will be important later for understanding such issues as the leap year and the *saltus lunae*. *Computus* manuscripts, reflecting this fact, often reproduce one or more of these chapters as a detached excerpt or paraphrase.[4]

Chapter 1 is really about representing numbers *for the purpose of calculation*. Roman numerals are all but impossible to use for ciphered arithmetic. Although they obey an overall place system (like ours, starting with units on the right and reading to the left with tens, hundreds, thousands and so forth), the individual numbers are

---

2 This is discussed at greater length in the Commentary to ch. 66.

3 See Introduction, pp. xxx–xxxi.

4 E.g. the treatise beginning *Leva manus totum* . . . in Paris, B.N. lat. 8429A, fol. 1, or the summary of finger-reckoning in London, British Library Harley 3017, fols. 117v–118r. Chapter 1 is frequently excerpted in *computus* anthologies, e.g. Bern 110 (s. X) (with ch. 4); Bern 207 (779–797); Bern 417 (s. IX), as a marginal gloss in fols. 18v–19r. It is also found in non-computistical manuscript contexts: for examples see Jones, *BOT* 330. See also Introduction, section 6.

compounded of separate sigla, sometimes by addition to the right of a base number (e.g. the number 8 – VIII – is made by adding III to V) and sometimes by subtraction from the left (e.g. 9 – IX – which is formed by putting I before X). Moreover, the Roman composite method of notation muddles the place system and makes it impossible to use for written calculations. In the number 876 (DCCCLXXVI), for instance, the units place has two characters (VI), the tens place three (LXX) and the hundreds place four (DCCC). Trying to write the number 120 (CXX) under 876 so that the ranks align will readily demonstrate that in the Roman system, numbers cannot easily be computed by visual juxtaposition of their notated forms.

The early Middle Ages inherited two systems for coping with this deficiency: finger-calculation and Greek letter-numerals. Bede describes them both in this chapter.

Roman finger notation[5] used both hands, the left for units and tens, the right for hundreds and thousands, and it formed the individual numbers in two ways: (a) by bending the last three fingers to various degrees, and in varying combinations, and (b) by forming different figures with the thumb and index finger. As the positions are sometimes difficult to visualize from Bede's text, an illustration is provided (fig. 1), based on MS Vatican City, Urb. lat. 290 (s. XI) fol. 31r.

With these finger signs, one could form integers up to 9,999, and perform arithmetical calculations with them. The system incorporated an implicit concept of "place" in that the integers were formed on the last three fingers of the left hand (hence the technical term for the integers was *digiti*, "fingers"), the tens with the index and thumb of the left hand (hence these numbers were called *articuli*, "joints"), the hundreds with the index and thumb of the right hand, and the thousands with the last three digits of the right hand.

As Bede describes it, the hands alone are used for numbers up to 9,999, while other parts of the body are called into play to represent numbers beyond 10,000. In one of his sermons, however, Bede implies that finger-calculation is only performed up to the thousands place, so

5 The most recent treatment of Roman finger-reckoning is Burma P. Williams and Richard S. Williams, "Finger Numbers in the Greco-Roman World and the Early Middle Ages", *Isis* 86 (1995):587–608. This article includes an extensive anthology of classical and early medieval references to finger-reckoning, including a translation of this chapter of *The Reckoning of Time*, as well as a thorough survey of the older literature.

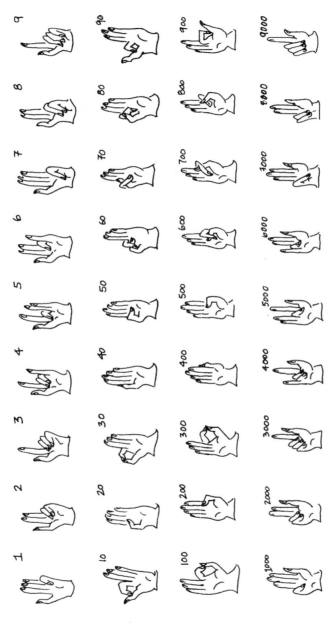

Fig. 1. Illustration of finger positions for numbers 1–9000, based on MS Vatican City, Biblioteca Apostolica Vaticana, Urb. lat. 290, fol. 31r. The perspective is that of the calculator looking down at his own hands.

the larger numbers must have been employed only for representation.[6] Exactly how calculations could have been performed using *computus digitorum* is not explained in ch. 1, but elsewhere in *The Reckoning of Time*, Bede gives us some clues. First, Bede knew how to make composite numbers: that is, he could represent numbers in two ranks on the same hand. His code for secret communication translates letters into number-values based on alphabetical position (A=1, B=2, etc.), with which one could literally cipher messages to a friend across a crowded room without anyone overhearing, by making number-signs for the coded letters upon the hands. One has to use composite numbers to do this, because there are 23 letters in the Latin alphabet, and indeed Bede's example uses composite numbers. This game, if such it is, may also have deep antique roots: in Martianus Capella's *De nuptiis Philologiae et Mercurii* (a text not known to Bede), Lady Arithmetic salutes Jupiter by making the number 717 on her fingers, which is the numerical value of the letters in his title ἡ ἀρχή, "The First".[7] The Roman counters or tesserae studied by Elisabeth Alföldi-Rosenbaum likewise depict composite numbers.[8] The number 12, for instance is made by forming 2 with the last fingers, and 10 with the thumb and forefinger.

But if Bede used composite numbers, he must also have been aware of another feature of finger-reckoning: its place-system. Each rank of numbers has a fixed field: units on the last three fingers (*digiti*) of the left hand, tens on the joints (*articuli*) of the left thumb and index finger; hundreds formed by analogy with the tens on the right hand, thousands by analogy with the units.[9] Moreover, it also has an implicit zero. If in any rank the fingers are simply held upright and relaxed, no numbers are to be read in that rank.

6 In *Hom.* 2.2, Bede is explaining the mystical significance of the 5000 who were fed with the loaves and fishes: *Millenarius autem numerus ultra quem nulla nostra conputatio succrescit plenitudinem rerum de quibus agitur indicare consueuit* ... ("The number one thousand, beyond which no calculation of ours extends, customarily indicates the fulness of the things which are under consideration ...": 168.259–264). The *conputatio* alluded to here must be finger-reckoning, since Bede could certainly count, and compute, numbers beyond 9999; cf. ch. 65, where he calculates the number of days in the 532-year Paschal cycle as 194,313.

7 7.729, ed. J. Willis (Leipzig: Teubner, 1983):261.11–13.

8 "The Finger Calculus in Antiquity and in the Middle Ages. Studies in Roman Game Counters I", *Frühmittelalterliche Studien* 5 (1971):1–9.

9 Karl Menninger, *Number Words and Number Symbols: A Cultural History of Numbers*, tr. Paul Broneer (Cambridge, Mass.: M.I.T. Press, 1969):205.

But how were calculations actually performed? No ancient source reveals the secret, nor has any convincing modern reconstruction been advanced. The most ambitious attempt at a hypothesis is that of Karl Menninger. His analysis is based on St Augustine's sermon on the miraculous draught of fishes, where Augustine says:

> Count for yourselves and figure as follows: 10 and 7 make 153, for if you count from one to seventeen and add together all the numbers – 1, 2, 3: 1 and 2 and 3 are 6; 6 and 4 and 5 are 15 – you finally arrive at 17 and carry 153 on your fingers.[10]

Menninger comments: "Thus as one counts up to 17, the fingers constantly join together the previous numbers until they finally 'carry' 153. This example is very important, for it reveals the main purpose of finger-counting: to record temporarily the intermediate sums in mental arithmetic . . .".[11] It is true that the procedure for calculation involves "constantly joining together" a sequence of numbers and then "reading" the results in the hand, but it seems probable that this involves something more than "temporarily recording the intermediate sums in mental arithmetic". Computation with the Roman system must have exploited the place-value inherent in the calculating hand, for while mentally adding together 1, 2 and 3 seems simple, how would Augustine's parishioner have performed the last operation in the exercise, namely adding 17 and 136? It is doubtful that he did it in his head. Rather, the fingers must have been used to actually tabulate the addition, rank by rank.

I would propose the following reconstruction of the process. To add 17 to 136, begin by showing 136 on one's hands. Then add 7 to the 6 already on the *digiti* of the left hand. One might know by heart that 7 and 6 make 13, but even if one did not, one had simply to count forward 7 places on one's fingers, tally-fashion. When one reaches 10 on the *digiti* (i.e. after counting forward four steps), one carries the 10 over to the *articuli*, and adds this 10 to the 30 already shown on the *articuli* by

---

10 *Apud vos numerate; sic computate. Decem et septem faciunt centum quinquaginta tres: si vero computes ab uno ad decem et septem et addas numeros omnes – unum, duo, tria: sicut unum et duo et tres faciunt sex; sex, quattuor et quinque faciunt quindecim: sic pervenes usque ad decem et septem, potans in digitis centum quinquaginta tres. Sermo* 270.7 (PL 38.1245).

11 Menninger 210; he is repeating here the judgement of Moritz Cantor, *Vorlesungen über Geschichte der Mathematik*, 3rd ed. (Leipzig: Teubner, 1907):1.829.

moving the *articuli* up one notch, raising the total on the *articuli* to 40. One then finishes the count to 7 on the *digiti*, for a total of 3. Finally, one adds the 10 of 17 to the *articuli* by moving the *articuli* forward one notch, raising the total to 50. The computation being finished, one looks at one's hands, as Augustine says, and reads the "digital display": 153. All you need to know in order to "add" is the *sequence* of the finger positions – in short, how to count on your fingers. With the fingers recording one's count, one can temporarily "forget" the first number and concentrate on keeping a running total on the fingers. It is rather like pressing "+1" over and over again on a modern electronic calculator, and then reading the result on the liquid crystal display. Subtraction can be done by reversing the addition method, and one can even multiply, provided multiplication is performed as a series of additions: i.e. 7×3 is performed as $7 + 7 + 7$. This is of course a very laborious way of multiplying, but the only alternatives were memorizing the multiplication table or consulting a multiplication table in a book.[12] If one was incapable of the one and had no access to the other, multiplication by additions was still possible by finger-reckoning. Division would be carried out by analogous procedures. For the computist, division usually meant dividing the AD – a three- or four-digit number – by 19, 28 or 15, to find the position of any given year within the lunar, solar or indictional cycle.

Elsewhere in *The Reckoning of Time*, Bede furnishes verbal descriptions of the calculating process which reveal something of how finger-computation might have functioned for medieval computists. In chapter 22, he gives the following formula for finding the age of the Moon:

> If you wish to know how old the Moon is on this or that day, count the days from the beginning of January up to the day you want, and when you know this, add in the age of the Moon on the kalends of January. Divide the total by 59, and if more than 30 remain, subtract 30. What is left over is the day of the Moon you are seeking ... So, for example, if you wish to know what the Moon is on the kalends of May in a year in which the Moon is nine days old on the kalends of January, say "May: 121 on the kalends". Subtract the [day of the] kalends, and 120 remains;

---

12 Edward Bechtel suggested that the Romans multiplied by successive addition even on the counting-board: "Finger-Counting Among the Romans in the Fourth Century", *Classical Philology* 4 (1909):31.

add 9, and that makes 129. Divide by 59: 59 times 2 is 118. Subtract 118 and 11 remain. The Moon is 11 days old on the kalends of May ... In using this formula, it is helpful if the calculator commits the products of the fifty-nine-times table to memory ...

Helpful indeed, but not strictly speaking necessary. Notice the terms in which Bede describes the process of division. First one "says" the dividend out loud because one cannot show it on the hands, so one will have to fix it in the memory. It will be recalled that in describing the finger positions, Bede likewise instructs his pupils to "say" the numbers as they form them, so that they can engrave the connection between the number and the symbol in their memory. The next step is the actual process of division, which takes place as a sequence of multiplications up to a point where the next multiplication would produce a product in excess of the dividend: Bede "divides" by *multiplying* 59 by 2, and when he realizes he cannot go any further, subtracting the product from the dividend to find the remainder. Hence he recommends memorizing the 59-times table. However, this can even be done by successive additions should one not know the 59-times table by heart. Once you have this product, you can subtract it from your memorized figure for the number of days between 1 January and your date.

If *computus digitorum* was indeed used for calculations, why did Bede not tell his students how to perform it? One answer may be that he was constrained by his sources. Finger-reckoning is nowhere expounded in ancient scientific literature, because it is a technique of practical reckoning, without relevance to mathematics as a science;[13] nonetheless, literary allusions to its practice abound.[14] The earliest Latin description is found in the Irish *computus* anthology used by Bede, the so-called "Sirmond manuscript". This *Romana computatio*[15] merely describes

13 Jones ("Bede's Place", 279, n. 29) notes that it is for this very reason that finger-reckoning and later, abacus texts appear in *computus* manuscripts. *Computus* was "vocational", not discipine-based: c.f. Introduction, section 2.

14 See anthology assembled by Williams and Williams.

15 Edited by Jones in *Bedae pseudepigrapha: Scientific Works Falsely Attributed to Bede* (Ithaca and London: Cornell University Press, 1939):106–108, and in *BOD* 671–672. For English translation, see Williams and Williams 605–606. Two other versions of this early text survive in the "Bobbio *computus*", Milan Ambrosiana D.150 inf. (Bobbio, s. IX) and in London, British Library Cotton Caligula A.XV (n. France, s. VIII), from which latter MS the text was transcribed and discussed by Alfred Cordoliani, "A propos du premier chapitre du *De temporum ratione* de Bède", *Le moyen âge*, 54 (1948):209–223. For

the positions representing the various numbers, and says nothing about operations. For this reason, Bede may have felt unsure about how to put the operations into words. Bede claims that he often found it difficult to reduce mathematical procedures to prose,[16] but his silence is also evidence that he expected that many people knew how to do finger-calculation, and how to teach others to do it. Perhaps as well, the technique of calculation was only taught to some advanced and dedicated students. At the very outset of chapter 1, Bede admits that there was some resistance to learning finger-calculation, and elsewhere in *The Reckoning of Time*, he half-heartedly provides alternative methods for students who cannot, or will not, learn to calculate. Had he subjected all his readers to a full-dress explanation of finger-reckoning in ch. 1 of *The Reckoning of Time*, few would have had the courage to go on to ch. 2.

The usefulness of the system had ultimately to be justified, not in mathematical terms, but by invoking patristic exegetical authority. For this reason, Bede paraphrases Jerome's exegesis of the Parable of the Sower. The thirty-, sixty- and hundredfold yield of the seed which falls on good ground is interpreted as three states of life – marriage, widowhood, and virginity – and each state of life is symbolized by the finger-position representing its number.[17] Having convinced his students that knowledge of finger-reckoning is not only relevant, but endorsed by patristic authority, Bede can proceed to at least indoctrinate the basic positions.

The second system described in chapter 1 is that of the Greek letter-numerals. Bede refers to this system as *computus Graecorum*, possibly

---

the numbers above 9,999, the London manuscript differs from the Milan and "Sirmond" codices, and Bede's text agrees with the London version. There is also a tabular depiction of finger-reckoning reproduced in PL 90.693. Abbé Lejay, in an important study of this chapter (*Compte rendu du quatrième congrès scientifique international des catholiques* [1897], vi sect., sc. philol. [Fribourg, 1989], pp. 129–136), believed that this form was created by Hrabanus Maurus (cf. PL 107.673–674). But as Jones points out, there are numerous manuscripts of this table dating from the early 9th century (*BOT* 329–330). All the manuscripts listed by Jones have Insular links, so this version, like the *Romana computatio* itself, may also have originated in the British Isles.

16 See Introduction, n. 40.

17 Jones, "Some Introductory Remarks on Bede's *Commentary on Genesis*," particularly his remarks on the symbolism of 30 (172) and 100 (173). This allegory certainly caught the attention of Bede's readers: see Brian Stock and Edward Synan, "A Tenth-Century Preface to Bede's *De temporum ratione*", *Manuscripta* 23 (1979):113–115.

to distinguish it from *Romana computatio*. It consists in assigning a single letter of the Greek alphabet to each digit from one to ten, then to each of the tens to 90, and of the hundreds to 900. Since this requires 27 symbols, and the Greek alphabet contains only 24 letters, the symbol called ἐπίησμον or *stigma* was used to represent 6, the archaic letter *koppa* stood for 90, and a symbol which Latin-speakers called *ennacosi* was employed for 900: this last term is actually the Greek word for 900, ἐνακόσιοι. The advantage of this system, as Bede himself declares, is that it employs a single and distinct siglum for each number within each rank. Therefore when compound numbers are written vertically, the place columns will align, which permits one to add each column and carry the excess to the next column. Curiously, however, Bede presents the Greek letter-numbers, not as a system for ciphered computation, but as a substitution-code. Like the substitution-code described in connection with finger-reckoning, the position of the letter within the alphabet determines its numerical value. Once again, the modern reader will be taken aback by what seems a rather irrelevant and frivolous application of an arithmetical technique – all the more so in this case, in that no arithmetical application is offered.

Indeed, Bede and his contemporaries would probably have encountered the Greek letter-numbers *only* as a code, in the so-called *litterae formatae*. These were official documents – in the early Middle Ages, usually papal or episcopal "passports" for travelling clergy – which were protected from forgery by a special numerical cryptogram. If the numbers of the cryptogram were resolved according to the *computus Graecorum*, they would yield the initial letters of the sender, recipient, place of origin, and date.[18] The *regula formatorum* actually ascribed the system to the 318 Fathers of the Council of Nicaea – the same 318 who, according to Dionysius, legislated the 19-year Paschal cycle.[19]

---

18 Clara Fabricius, "Die *litterae formatae* im Frühmittelalter", *Archiv für Urkundenforschung* 9 (1926):168–194.

19 Until recently, it was believed that some Irish ecclesiastical centres, notably Bangor, converted to the Dionysian reckoning as early as 610, the year when Mo-Sinnu moccu Min, abbot of Bangor, died. According to a note in an 8th-century Irish manuscript of the Gospel of Matthew, now Würzburg Universitätsbibliothek M. p. th. fol. 61, he was "the first of the Irish who learned the Computus by heart from a certain Greek". His disciple, Mo-Chuaróc later "committed this knowledge in writing, lest it should lapse from memory". Jones and others concluded that this Greek *computus* was Alexandrian reckoning, but this hypothesis has been refuted by Ó Cróinín, who was the first to transcribe the full

Strictly speaking, *computus Graecorum* has nothing to do with the calendar, or even with calculation: Bede states that it is useful for cryptography, not computation.[20] Nonetheless, Bede's inclusion of the Greek numerals in *The Reckoning of Time* did result in their being used by computists, albeit neither as a mode of calculation nor as a substitution-code. Until the early twelfth century, Greek letter-numbers were not uncommonly used in *computus* tables, in preference to, or in parallel to, Roman numerals. Why the designers of these tables preferred Greek numerals is not altogether clear. A plausible explanation is that they are more compact, and therefore easier to inscribe into the restricted and inelastic spaces of a grid, than are Roman numerals. They could also be read more quickly and easily, and mistakes in transcription were less likely.[21] But some computists, such as Abbo of Fleury (d. 1004) designed parallel versions of tables in both Greek and Roman notation.[22] This would suggest that their use was regarded as a kind of esoteric technique or quasi-secret knowledge, even when the intent was not cryptographic.

---

text of the Würzburg note. The final sentence of the note identifies the *"computus"* which Mo-sinnu learned as nothing more or less than the Greek system of letter-numerals described by Bede's *computus Graecorum*, for it closes with an explanation of the three additional symbols of *episinon* (ἐπίσημον, i.e. *stigma*), *koppa*, and *enacosse* (ἐνακόσιοι), "which the Greeks insert in amongst the letters, lest the order of the numbers be disturbed". (*Sed tamen inseruntur apud Grecos inter literas, ne turbetur ordo numerorum*): "Mo-Sinnu moccu Min and the Computus of Bangor", *Peritia* 1 (1982):290.

20 Ó Cróinín (*ibid.*) argues that these Greek letter-symbols were an adjunct to finger-computation and learned as part of that system. But there is no functional link between the two at all. Finger-computation, indeed, would be unnecessary if Greek letter-numerals were actually used for arithmetic. Moreover, tracts describing the two systems are rarely fused in the MSS: in Bede's own "Sirmond" compilation, for instance, the Greek letter-numerals are found on fol. 61v, and the *Romana computatio* on fols. 97v–98r.

21 These advantages are signalled by the Laon-Metz glossator: *Facilius enim cognoscunt Greci numerum ex suis literis quam nos ex nostris, quia unusquisque numerus apud illos habet suam propriam literam per quam intelligatur. Latini autem no intelligunt numeros nisi paucis literis isdemque etiam geminatis.* ("The Greeks learn number[s] more readily from their letters, than we do from ours. For with them, each number has its own proper letter, through which it is recognized. The Latins do not recognize numbers save by means of a very few letters, but these doubled": 272 *ad lineam* 80).

22 See tables from his *computus* reproduced by Migne amongst the works of Bede in PL 90.735–736, 741–742.

Chapter 2:

In ch. 2, Bede introduces a second propaedeutic theme: the various modes in which one can speak about time. Though the chapter as a whole follows Bede's Irish source *De divisionibus temporum* (PL 90.653 *sqq.*), the distinction between natural, customary (or conventional), and authoritative time-reckoning is original to Bede, and one which is critical to his understanding of *computus*. Some divisions of time are imposed by the natural movements of astronomical time-markers, such as the solar year of $365\frac{1}{4}$ days; others derive from convention (Bede's example of the 30-day "month" is a good one); and others are legislated by human or divine authority, such as the Olympiad or the Jubilee. Terms like "year" or "month" can therefore have different meanings, depending on the mode of time-reckoning to which they refer.

What is noteworthy in this chapter is Bede's consciousness of "nature" as a separate, autonomous, yet God-created system of laws for the natural universe. There is, so to speak, a whole "natural system" of time, separate from the system legislated by religious law. Indeed, Bede almost personifies nature when he says that "with nature as our guide" (*natura ducente*) we learn a different value for the length of the year from that which calendrical computation dictates. This distinction, already enunciated in *On Times* 1, and reiterated in *Letter to Wicthed* 7,[23] was appreciated by Bede's readers. Commenting on Bede's statement in the preface that he intends to discuss "the nature, course and end of time", the Laon-Metz glossator remarks:

> All that is contained in this book concerns time. For the status of times is contained in the natural order of hours, days, weeks, months, years and cycles, which ever return upon themselves when the natural circle is completed. For it is according to nature that the day has 24 hours, the week seven days, and so forth. Thus the day returns after 24 hours, and so forth. But the course of time, that is, the various actions of men, are contained in the chronicles which are in the second part of this book [i.e. *The Reckoning of Time* ch. 66]. And the end of time is contained

23 In the case of the *Letter to Wicthed*, the issue is the leap-year increment, which *secundum naturam* accrues at the end of the solar year (i.e. at the spring equinox) but which "human custom" inserts at arbitrary and varying dates.

in the end of this book, where [Bede] treats of the Seventh and Eighth Ages.[24]

There is a distinction to be drawn, then, between the "nature" of time, its "course" (which is human history) and its "end". Nature obeys immutable, repeatable laws, and natural time flows in predictable cycles; historical time, on the other hand, is linear, and frames the distinct, irreproducible actions of men.

The early medieval period is not generally credited with such an understanding of nature. Studies like Charles Radding's *A World Made by Men*,[25] or Giselle de Nie's *Views from a Many-Windowed Tower*[26] argue that the early Middle Ages did not conceive nature as a category apart from divine intervention, or as a system characterized by regular and repeatable laws. Their conclusions are certainly not entirely mistaken. However, as Bede's text demonstrates, a concept of autonomous and regular nature was never "lost". Indeed, it was to some degree consciously resisted by a new Christian ideology of the physical world. The main concern of this ideology was that acknowledging the autonomy of nature was an invitation to divinizing her. This threatened the idea of divine providence and omnipotence, particularly as manifested in miracles,[27] and might even suggest the validity of magic. For this reason, Bede explicitly denies the divinity of nature. As far as time goes, "nature" is what God created on the fourth day, namely the planets and stars that he made to be the signs of seasons, days and years. "It is not," he adds, "as the folly of the pagans asserts, a creating goddess, one amongst many."

---

24 *Quae omnia continentur in isto libro qui est de tempore. Status temporum autem continetur in naturali ordine horarum, dierum, ebdomadarum, mensium, annorum, ciclorum, qui in se semper peracto naturali circulo reuertitur. Natura est enim ut dies habet xxiiii horas, ebdomada vii dies, reliqua. Ergo post xxiiii horas reuertitur dies, sic et reliqua. Cursus autem, id est diuersi actus hominum, in chronicis continetur quae sunt secunda pars huius libri. Finis uero temporum in fine istius libri, ubi de vii^a et viii^a aetate tractauit, continetur.* (263, *ad lineam* 7).

25 Chapel Hill: University of North Carolina Press, 1985.

26 Amsterdam: Rodopi, 1987.

27 On the early medieval debate over "nature" in the context of medicine, see Faith Wallis, "The Experience of the Book: Manuscripts, Texts, and the Role of Epistemology in Early Medieval Medicine", in *Knowledge and the Scholarly Medical Traditions*, ed. Don G. Bates (Cambridge: Cambridge University Press, 1995):117–124, and literature cited therein.

Yet at the same time, Bede follows the lead of the Irish cosmographers of the seventh century in carving out a new role for nature. No longer the governing deity of the lower world, nature is now God's law and instruction to His creation. Natural time was "hard-wired" into the heavens on the fourth day of Creation; even the miracles of the Sun standing still on Gibeon (Joshua 10.12–14), or the shadow moving backwards up the Temple steps for King Hezekiah (II Kings 20.1–11) cannot stop time's course. Indeed, when God came to the aid of Joshua's troops, he took care to stop the Moon, as well as the Sun, precisely so as not to dislocate the rhythm of natural time.[28] Dethroned from the pantheon, deprived of a name and identity, nature resurfaces as, if anything, more powerful and unshakable than before. Indeed, Marina Smyth suggests that the very orderliness of *computus* may have reinforced the belief that the celestial world was one of absolute order.[29]

The Reckoning of Time shows how *computus* furnished a setting for a discussion (however circumscribed and cautious) of "nature" in the early Middle Ages. Though Bede's contemporaries certainly understood what he was talking about, it was the computists of a later age who reflected most deeply on this theme. Virtually every major treatise written after 1100 begins with an explanation of the distinction between "natural" and "authoritative" *computus*.[30] By then, however, with the arrival of new texts on astronomy through the Arab world, and in particular with the discovery of astrology, it had become plain that the *kind* of reckoning required for the Church's calendar was substantially different from that demanded by astronomical measurement or astrological calculations. In particular, the former dealt only in whole days; the latter, in fractions of days.

This is not to suggest that Bede shows no interest in divisions of time smaller than a day, or in fractions; indeed, the remainder of the technical propaedeutic of *The Reckoning of Time* is devoted to these subjects.

---

28 Smyth 160–165, 169; cf. Bede, *In Regum librum XXX Quaestiones* 25, ed. D. Hurst, CCSL 119 (1972):316–317.

29 Smyth 171.

30 The outstanding example is the *Computus* of Sacrobosco, but it is found in 11th- and 12th-century antecedents, such as the unpublished treatises of Gerlandus and of "Magister Cunestabulus". I am preparing an edition of Gerlandus, and Dr Jennifer Moreton one of Cunestabulus.

Chapter 3:

Chapter 3 discusses intervals of time smaller than the day, the basic unit of *computus*. These include the hour, defined in both its sundial sense and its astronomical sense. Ancient sundials were divided into twelve equal segments, so that the entire span of daylight, be it long or short, was divided into twelve hours. These were called "artifical" hours, because they were defined by the artifice of the sundial. Of course, an "hour" in winter would be shorter than an "hour" in summer. "Natural" hours, or as Bede terms them, "equinoctial" hours, are of a constant length, namely $\frac{1}{24}$ of a day. There are more equinoctial hours of sunlight in summer, fewer in winter.

Bede assumes that his reader is familiar with both kinds of hour. While monks may have used sundial hours for managing their daily round,[31] the calendars in their *computus* manuscripts would have made them aware of equinoctial hours, for one of the conventional addenda to the monthly calendar page was a note on the number of equinoctial hours of daylight and night in that month.[32] Another common feature of *computus* manuscripts, the *horologium*, showed this variation in diagrammatic form.[33] Certainly, whatever convention was used for daily life, the student of *computus* needed to know about equinoctial hours, for they were the basis of all smaller divisions of time. It is also noteworthy that in his discussion of the hours of moonlight through the course of the lunar month (ch. 24), Bede uses only equinoctial hours.[34]

The intervals of time described in this chapter include the *punctus*

---

31 But see J.D. North, "Monasticism and the First Mechanical Clocks", in *The Study of Time*, Proceedings of the Second Conference of the International Society for the Study of Time, 1973 (New York: Springer, 1973):381–398. North argues that monks were pioneers in the development of mechanical clocks precisely because they preferred both to conceive of and to manage time in terms of "equinoctial" hours.

32 Every calendar included by Francis Wormald in his *English Kalendars Before A.D. 1100* contains such an addendum stating, e.g., that night in May is 8 hours long, and day 16 hours long.

33 E.g. PL 90.953–954.

34 Max Lejbowicz argues that by using the term "equinoctial hours", Bede reveals the essentially unquantitative cast of his thinking, for "equinoctial hours" invokes a concrete experience of time on a certain day, the equinox, not an abstract unit of measurement: "Postérité", 6–7. However, the fact that Bede not only described the equinoctial hour but used it in calculations would seem to contradict this hypothesis. The "equinoctial hour" is precisely the abstraction used for computation, while the unequal hour is the lived experience of time.

(one-quarter of an hour), the *minutum* ($\frac{1}{10}$ of an hour, i.e. six minutes of our time), the *pars* ($\frac{1}{15}$ of an hour, i.e. 4 minutes of our time), the *momentum* ($\frac{1}{40}$ of an hour, or 1.5 minutes).[35] Significantly, Bede points out that these divisions are conventional rather than natural; they were invented out of the necessity to divide the hour into segments of various lengths for the purposes of calculation, and the names are arbitrary. After explaining the etymology of these terms (following Isidore), Bede reinforces his point about their conventional character by pointing out that astrologers (*mathematici*) use virtually the same terms, but with rather different meanings. For example, an astrological *punctus* is $\frac{1}{12}$ of an astrological *pars*, which is $\frac{1}{30}$ of a zodiac sign, or in other words, one day; hence an astrological *punctus* is two equinoctial hours.[36]

It should be pointed out that Bede evidently understood what ancient technical astrology purported to do, and rather ostentatiously disclaimed it. Like the notion of nature, technical astrology may not have been as unknown to the early Middle Ages as is sometimes asserted, and Bede's denunciation seems more than mere convention.[37] His

35 On the ancient background to these divisions, see P. Tannery, "Sur les subdivisions de l'heure durant l'Antiquité", *Revue archéologique* 26 (1895), rep. in his *Mémoires scientifiques*, vol. 2 (Toulouse: Privat, 1912):517–526; Grumel, *Chronologie* 164–165; on the *momentum*, see F. de Gandt, "Ébauche d'une notice sur *momentum*", *Documents pour l'histoire du vocabulaire scientifique*, no. 1 (Paris: Editions du CNRS, 1980):1–20; Robert Steele, *Compotus fratris Rogeri*, Opera hactenus inedita fratris Rogeri Baconi, fasc. 6 (Oxford: Clarendon Press, 1926):290–291.

36 Jones asserts that Bede's use of the term *punctus* is "confusing", because it means something different in this passage from its meaning in his enumeration of the computistical divisions of the day given above. But that is precisely Bede's point: these terms are conventional, not natural, and can be arbitrarily defined according to any system. As Bede states with reference to the *atomus*, it means something different to a grammarian and a computist. Bede also points out that *punctus* means $\frac{1}{5}$, not $\frac{1}{4}$ of an hour when it is the Moon which is being reckoned (cf. ch. 17), a variation which did, indeed, sow some confusion amongst computistical writers: see *BOT* 333.

37 Lejbowicz, "Postérité" 17–19, argues that Bede objects to the *mathematici* primarily because they spatialize divisions of time by assigning them to the zodiac and ecliptic, and because they claim to be able to measure infinitesimally small units. While such arguments appear in patristic literature, I do not think they are exactly Bede's. Bede after all claims that the computists (*calculatores*) make strict distinctions between *momentum*, *punctus* and *atomus*, though ordinary usage treats them as synonyms. It is the astrologers' claim to deduce the fate of the newborn from the position of the stars that draws his criticism, not the division of the zodiac *per se*. The conventional argument that technical astrology was unknown in the early Middle Ages is represented by J. Tester, *A History of Western Astrology* (New York: Ballantine, 1987):ch. 3, and McCluskey, *Astronomies* 145–

concern about astrology will re-emerge in chs. 11 and 16, when he discusses the signs of the zodiac. What is noteworthy here is that while Bede decries the pretensions of astrologers to predict fate from the stars, he does not mention the classical argument that man's free will transcends celestial determinacy. Possibly he felt nervous about the associations of this idea with Pelagianism.[38]

What is the purpose of learning these divisions? Bede does not answer this question directly, but his response is implicit. Computists have to resolve their cycles into whole days in order to establish a calendar. On the other hand, arriving at the correct number of whole days will involve calculating not only fractions of days, but even fractions of hours, as is the case with the *saltus lunae*, the calendrical correction in lunar reckoning at the end of the 19-year Easter cycle.[39] The evidence of the *computus* manuscripts suggests that teachers drilled their students in these terms.[40] Certainly, the fact that these units were studied and discussed is of the utmost importance for the history of *computus*, and perhaps even for the history of science. The critique of conventional *computus* launched in the eleventh century most notably by Gerlandus Compotista rested on the idea that the problems in lunar reckoning which were causing the Church's calendar to become embarrassingly out of phase with astronomical reality could only be detected and corrected by calculating lunar and solar periodicity at a very minute level. That Bede furnished and sanctioned the tools for such a calculus allowed Gerlandus to claim that his new *computus* was quite conventional. In fact, Gerlandus was moving towards a new definition of time-reckoning as a scientific problem, concerned not with the calculation of Easter, but with the calculation of eclipses.[41]

149; for a revisionist perspective, see Valerie I.J. Flint, *The Rise of Magic in Early Medieval Europe* (Princeton: Princeton University Press, 1991):esp. ch. 6. Astrology, however, seems entirely absent from Bede's Irish source-material: see Smyth 172.

38 Lejbowicz, "Postérité", 22.

39 Cf. *The Reckoning of Time*, ch. 42.

40 For examples, see Jones, *BOT* 331–332.

41 The content and significance of Gerlandus' treatise will be discussed more fully in my forthcoming edition. Note that Alfred Cordoliani, "Notes sur un auteur peu connu: Gerland de Besançon", *Revue du moyen âge latin* 1 (1945):411–419, and "Le comput de Gerland de Besançon", *Revue du moyen âge latin* 2 (1946):309–313, are to be used with caution, as he confuses Gerlandus the computist with Gerlandus of Besançon, a 12th-century *scholasticus*.

Chapter 4:

The final section of the technical propaedeutic concerns duodecimal fractions. Such fractions would be used in much the same context as the smaller divisions discussed above, i.e. in the calculations designed to resolve varying cycles into whole days, or in explanations of how such cycles work.[42] Bede's purpose here is essentially to clarify terminology, especially since (as he points out) young pupils tend to misuse the terms. The terminology is fairly complex, as Romans had separate words for the fraction itself, and for the remainder of the unit, e.g. *uncia* $= \frac{1}{12}$, *deunx* $= \frac{11}{12}$, *quadrans* $= \frac{1}{4}$, *dodrans* $= \frac{3}{4}$.

A unique feature of Bede's discussion is the number of synonyms he knows for different kinds of fractions, e.g. *libra*, *as* and *assis* for "pound". This suggests that he drew on a number of sources for this chapter.[43] Like ch. 1, ch. 4 frequently circulated separately in *computus* anthologies.[44] The fractions took on a new lease on life in the eleventh century, when they were revived by Abbo of Fleury (cf. Bern MS 250, fol. 11v) and Gerlandus Compotista to calculate lunar periods. An editor of *On Times*, quite possibly Abbo, actually inserted them into Bede's text (cf. British Library Egerton 3088, fol. 79r, and Cotton Tib. E. IV, fols. 134v–135r).

Noteworthy as well is Bede's concern to make the study of numbers relevant to the reading of Scripture, while at the same time turning his reading of Scripture commentaries into material for his scientific interests. The coherence of Bede's interests illustrates *doctrina christiana* in practice: reading Philip's commentary on Job, he discovers information on tides; commenting on the word *semis*, he recalls its appearance in the description of the Mosaic tabernacle.

Chapter 5:

The Julian calendar is essentially a list of days, and computists do not commonly deal with units of time smaller than a day. Nonetheless ch. 5,

---

42 Bede in fact uses these fractions rather seldom in *The Reckoning of Time*, e.g the *bisse* ($\frac{2}{3}$) in ch. 16, or the *quadrans* in ch. 38. Jones (*BOT* 334) judged this chapter to be a relic of *computi* produced for the Roman 84-year cycle, which lent itself especially to duodecimal division, especially as the *saltus* fell in the 12th year of the cycle.

43 The synonyms suggest a grammatical source, perhaps a lexicon: cf. Florence Yeldham, "Notation of Fractions in the Earlier Middle Ages", *Archeion* 8 (1927):313–329.

44 Cf. Jones, "Byrhtferth Glosses", 85–86.

"The Day", follows logically from the preceding chapters in that it begins by focusing on terminology, and particularly on the distinction between conventional and specialized usage. Both the ordinary definition of "day" as "period of daylight" and the scientific meaning of "period of 24 hours" are sanctioned by Scripture. This is not only an important issue for Bede personally, but a premiss of Christian learning, as defined by the Fathers, and especially Augustine: all knowledge, human and divine, is a unity, and apparent disparities find their resolution in revealed authority. Nonetheless, it is interesting that in ch. 3, Bede reminds his readers that Christ himself used "unequal hours", before demoting them to the rank of mere human convention; later, in ch. 24, he will imply that thinking in terms of unequal hours is vulgar, and something which computists should avoid as "unprofessional". In this chapter as well, scientific usage is promoted at the expense of the ordinary definitions.

Scripture is also the framework in which scientific curiosity and erudition comes into play. Bede provides an excellent illustration of this in his discussion of what "day" means in those passages of Genesis referring to the time before the creation of the Sun. The problem really lies in the first few verses of the Creation account. The narrative opens by stating that God "created heaven and Earth", and then goes on to state that the Earth was unformed and chaotic. Are the initial verses a summary of the entire Creation story, or do they form a temporal sequence with the statement about the Earth's chaotic form? Like Augustine in *De doctrina christiana*, Bede feels that there are a number of legitimate and orthodox ways of interpreting these passages.[45] One of them, adopted by Augustine himself in *De Genesi ad litteram*, was allegorical. Augustine's premiss is that the initial verses are recapitulatory. Since there is as yet no division between heaven and Earth on the first day, there is no "place" for the primal light to come from or go to, so it cannot have behaved like the Sun and produced a conventional "day". Bede reports this opinion, which indicates that he valued it.[46] But he seems to prefer, or think his pupils will understand better, the more straightforward approach of ps.-Clement, Basil, Ambrose and Jerome: namely that the first verses denote temporal sequence, and that the

45 *De doctrina christiana* 3.27.38 (99–100).
46 I disagree with Jones (*BOT* 335–336) who thinks that Bede is challenging Augustine here. Bede normally avoids disagreements with his authorities by passing over contentious matters in silence. That he cited Augustine at all indicates that he approved of him.

heavens, Earth and "abyss" were separated at the very outset of Creation. The angels were created before man, and therefore heaven must have already existed as their home, as well as a place from which Lucifer could be cast out. There is thus a "place" through which the primal light can move, to "make" a day.

Bede then offers his own contribution to the patristic dialogue. The primal light on the first day rose in the east, between heaven and Earth, and illuminated every corner of the Creation, even the depths of the sea. This last is not as improbable as it may seem, Bede avers, for one can see into the depths of the ocean if one pours oil on the waters to create a smooth surface; moreover, at the beginning of Creation, when the primal waters had not been compacted together into seas, they would have been more limpid, and more permeable to light. And in the end, God's power can make light shine anywhere. What is noteworthy here is how Bede combines natural and supernatural explanations, without feeling obliged to choose or rank or compare them. The primal light orbited, and finally set, just like the Sun; the only difference was that it produced no heat, and because the stars were not yet created, it left absolute darkness in its wake. The idea of the "primal light" implies that the "heavens" created on the second day revolved once each day around the Earth, as they do now, and that they shone with the light created on the first day.

The ensuing discussion of the various starting-points of the day is not merely an antiquarian exercise but relates directly to *computus*. Like all historic peoples, Christians have their own starting-point for the day, defined by Christ's Resurrection. Since this took place on the night between Holy Saturday and Easter Sunday, that night is counted into the following day, and so Easter is deemed to begin at sundown on Holy Saturday. Since the Moon also changes its age at evening, a Moon 14 days old on Saturday will by 15 days old by Saturday evening, and the following day can be celebrated as Easter. Bede may also here be silently correcting a troubling statement in the *Liber Anatolii*. This text claims that the computistical lunar day begins at half past the sixth hour, i.e. about 12:30 p.m.[47] Therefore if the Moon is 13 days old on Saturday, it will be well into its 14th day by the time Easter services begin on Saturday evening.[48] Many genuine primitive Alexandrian texts – including the authentic statements of Anatolius incorporated into the *Liber Anatolii* –

---

47 Ed. Krusch, *Studien I*, 320.
48 *Ibid.*; cf. Cordoliani, "Computistes insulaires", 12.

condoned celebrating on the 14th Moon. Bede was prepared to rationalize this,[49] but had to draw the line at celebrating on the 13th day. Hence,
perhaps, his concern to locate the change firmly in the evening.

The new Christian definition of day involves a synecdoche, or reading
of the whole in the part. Christ spent three days in the tomb if one counts
Friday evening as comprising all of Friday, and Saturday night as part
of Sunday. The manner in which Bede weaves a little lesson on
"schemes and tropes" into this discussion is a good example of how
Bede understands the unitary quality of *doctrina christiana*, and how he
borrows from exegesis to illuminate *computus*, and vice versa. But this
digression also has implications for the theology that Bede is endea-
vouring to articulate. Christ's Resurrection changed time in every
respect. It ushered in a new age of history; but it also changed the time
when the day began. At Creation, the day began at dawn; but the last
Age of Earth was inaugurated by a *mutatio temporis* that, significantly,
shifted the beginning of the day to the evening, so that time flows from
darkness into light, not from light into darkness.[50]

Chapter 6:

Chapter 6 may seem like a digression, but in fact it is integral to Bede's
defence of the Alexandrian *computus*. Its computistical significance
revolves around the equinox as terminus for the Paschal full Moon, and
the consequences for the calendar limits of Easter.[51] Bede's resolution of
this problem, interestingly, rested on a natural rather than an authoritative argument.

Behind Bede's argument lies an ancient allegory: the week of the
Passion replicates the week of Creation. Just as man was created on the
sixth day of the first week, so was he re-created in the sacrifice of the Son
of Man on Good Friday.[52] Since Passover, the time of the historic Good
Friday, falls in the spring, it follows that the world was created in the

---

49 At the Synod of Whitby (*HE* 3.25 [304]), Wilfred explains away Anatolius'
lower lunar limit of 14 by claiming that Anatolius really meant that Easter celebrations
could begin on the evening of the 14th day, when the Moon turned 15 days old. Cf. ch. 43
below.

50 This theme is discussed by Bede again in *The Reckoning of Time* 71, and also in
Hom. 2.7. This typology hangs, of course, on Bede's curious notion that the Jewish day ran
from dawn to dawn: see ch. 5.

51 See Introduction, section 3.

52 Cf. *The Reckoning of Time*, chs. 10, 66 (introduction) and 71.

spring.[53] Ps.-Anatolius and others argued that the first day of Creation was the equinox itself, but Bede asserts otherwise. Day and night can only be said to be equal when there are hours by which they can be measured. The equinox occurred on the first day when there was a Sun and stars to mark those hours; this was the fourth day of Creation, when the heavenly bodies were made. The world, therefore, was created on 18 March, but the equinox was created on 21 March. But because the Moon was also created on 21 March, and God would not have made it in any other state than full, this "first day of time" was also the very day when both the criteria for the Easter terminus were present: the equinox, and the full Moon. Since the full Moon will not rise until evening, the earliest date of Easter will be the *following* day, 22 March. Bede goes on to clinch his argument by answering the objection that the "days" before the first "day of time" (the fourth day of Creation) were indeed days in the natural sense of the term – but days without hours.

Bede's sources for this argument are difficult to trace. The closest parallel is the *De pascha computus* composed about 243, and falsely attributed to Cyprian of Carthage.[54] "Cyprian's" theological explanation convinced the Romans, who clung for some time thereafter to 18 March as the earliest possible date of 14 Nisan: this is, in fact, Victorius' definition of Nisan, which Bede will refute in ch. 51.

However, Bede's interest in demonstrating this point is also passionately theological. The correct computation of the date of Easter is, in his view, sacramental. It is the union of a sign with the reality it effects: "Observing the Paschal season is not meaningless, for it is fitting that by means of it the world's salvation both be symbolized, and actually come to pass." The equinox itself, which began with the rising of the Sun at dawn, was not sufficient; it required also the rising of the full Moon at evening. In this way, Bede elevates the criteria for Paschal *computus*, and particularly his own argument for the Alexandrian Easter limits, to the level of a theology of time. Celebrating Easter at a full Moon before the dawn of the equinoctial day, as Victorius proposed, reverses an

53 Bede would certainly have encountered this theme in Ambrose, *Hexaemeron* 1.4.13 (11 *sq.*), and Victor of Capua, *De pascha* 6 (298); cf. P. Siniscalco, "Le età del mondo in Beda", *Romanobarbarica* 3 (1978):310–320, and T. Halton, "The Coming of Spring: A Patristic Motif", *Classical Folia* 30 (1976):150–164. The motif has ancient pagan roots: cf. Vergil, *Georgics* 2.316 *sqq.*

54 See Introduction, section 3, p. xxxvii for an explanation of "Cyprian's" argument.

order inscribed at the beginning of Creation, and dissolves the symbolic relationship of grace which exists between Christ and his Church. It is, in short, a kind of calendrical Pelagianism.[55]

A theological perspective on *computus* also underpins Bede's championing of Abraham as the true father of astronomy. His immediate source for the story of how Abraham brought *astrologia* to Egypt is probably Isidore's *Etym.* 3.25, for though the Irish *De computo dialogus* also relates the tale, it is *scientia numeri*, not astronomy, that the patriarch invents. Both sources cite Josephus. The modifications which Bede introduces, however, are very telling. Isidore states that the Egyptians invented *astronomia* (astronomical theory) and the Chaldeans *astrologia* (applied astronomy), and that Abraham was the intermediary between the two. He also relates the Greek legend that Atlas invented astronomy. Bede excises the references to Atlas, and transforms Abraham from mere intermediary into a creative scientist who improved on Chaldean *astrologia*, and taught it to the Egyptians, who apparently had no knowledge of the stars before his arrival.[56]

At every step of his exposition, Bede is determined to weave natural science and theological truth into a single seamless fabric. But he is also concerned to circumscribe certain elements which might endanger this harmonious relationship, such as the zodiac. He stresses that the zodiac and its signs are conventional divisions of the Sun's annual course. "As far as nature is concerned", the zodiac is simply the track that course takes, a by-product of God's Creation, and not its governor. As if to anticipate possible objections to any reference to the zodiac whatsoever – for Bede, like Isidore, is always conscious of the spectre of judicial astrology – he adds that the zodiac was known to the Old Testament patriarchs, and is not the property of pagans or philosophers. Here once again we see how Bede's *natura*, while a system or force in its own right, always serves Her maker, and teaches His truth.[57]

---

55 Cf. *The Reckoning of Time*, ch. 59.

56 Lejbowicz, "Postérité", 2–3.

57 Cf. C.W. Jones' remark on Bede's approach to nature in his commentary on Genesis: "[Bede's] exegetical aim is not a philosophic Order, but a law which is natural because it is divine": "Some Introductory Remarks", 117.

Chapter 7:

Though this chapter on night begins with an Isidorean etymology, Bede's development of this theme has a rather Augustinian flavour: like evil, night has no existence in itself, but rather is the privation of light, the shadow cast by the Earth as the Sun moves around to its far side. This shadow is the dominant theme of the chapter, and it allows Bede to summon up a number of related themes from his scientific encyclopaedia: eclipses, the relative size of Sun and Earth, and how season and latitude contrive to produce variations in the length of the night. The result is an excellent illustration of how *computus* served as an organizing principle for a broad range of scientific knowledge in the early medieval world.

Night is a sinister time, and especially so for the computist, because there is no instrument comparable to a sundial with which one can measure it in precise and uniform units. Bede describes its divisions, not as hours (equal or unequal) but as seven *partes*, each with a distinctive name. These names evoke an image of the night as a time without light (*crepusculum*) or sound (*conticinium*), a time without time (*intempestum*). Time, as Bede observed in ch. 2, is only what can be separated out into units (*spatia*) and measured (*temperatum*). The night contrasts with the day in that its time is qualitative, not quantitative.[58]

Yet it would be well to bear in mind that for Bede and his contemporaries, day and night could only be subdivided in a very approximate and contingent way. Whether one used a sundial or the passage of the stars, one was dependent on the weather for one's success in reading the time. The very word *tempus* means both time and weather (cf. modern French *temps*), but the wayward effect of weather on time-keeping is perhaps the reason why Bede does not discuss meteorology in *The Reckoning of Time*, though it occupies an important place in *The Nature of Things*.[59] By contrast to the day, calendar-time, in which the day was the smallest unit, was independent of accidental interference; it was pure *ratio*.

This chapter also furnishes a typical example of Bede's talent for inventing concrete examples to demonstrate scientific truths. Bede's

58 It is noteworthy that Bede does not discuss the rising and setting of constellations as a form of nocturnal clock. He may not have known Gregory of Tours' *De cursu stellarum*, or if he did, may have been anxious about its possible astrological implications: cf. McCluskey, *Astronomies*, ch. 6.

59 Lejbowicz, "Postérité", 11.

sources – Ambrose, Isidore, Pliny – teach that the stars are never eclipsed, because they lie beyond the cone of the Earth's shadow, which terminates at the Moon; beyond the Moon there is perpetual light, either from the Sun or from the stars. Bede anticipates scepticism on the part of his students: if this is so, why then is the night sky black? Bede invokes a picture of distant torches, which not only shine of themselves, but illuminate some of the surrounding area. The stars evidently behave in the same way, and if we can see only a little of this diffused light, it is because of the thick atmosphere which shrouds the Earth. Incidently, an implicit assumption here is that the stars are simply natural fires, like their earthly counterparts: another unobtrusive blow against any divinization or astrological mystification of the heavens.[60]

One cannot help but be struck by Bede's casual assumption that the heavens are really filled with boundless light. The darkness which we experience at night, pierced only by the light of the stars, is nothing but an atmospheric effect of our planet. Few passages convey in a more striking manner the immense difference between the medieval view of the universe – all brightness and order, with Earth as the one dark spot – and our modern perceptions.

Chapter 8:

Because the Julian calendar was a generic or perpetual one, it did not indicate weeks. The calendar had to be customized for each year by using formulae or systems of key letters to determine the weekdays of specific calendar dates. Bede therefore reserves his discussion of this matter for his section on the month (ch. 20.), since both the formulae and the key letters function in relation to particular months.

Though it is invisible in the Julian calendar itself, the week is nonetheless of great importance for Bede and for his readers. From the very earliest days of the Church, Christians have ordered their common worship according to its seven-day rhythm. The medieval attitude to the week was not only compounded of Biblical and Jewish concepts, but was also coloured by ancient astrological and numerical symbolism. Both elements combined to transform the week into a trope for time

---

60 The Fathers were not entirely agreed as to the nature of the stars: see Alan Scott, *Origen and the Life of the Stars: A History of an Idea* (Oxford: Clarendon Press, 1991).

itself, linear in its progress, cyclical in its rhythmic return. Sunday, the first day of the week, the day on which the Lord's Resurrection was commemorated, was for Christians both a day of rest and a symbol of new beginnings, of eternity, heaven, and the consummation of all things at the end of time. It was both the first day and the eighth, a renewal of time and a consummation beyond time.[61] It is for this reason that Bede devotes three chapters to the week, chapters which underscore its status as the privileged symbol of sacred time.

Bede draws heavily on his Irish sources for the etymological and philological material in chapter 8, but in many respects he goes considerably beyond what his sources had to offer. While *De divisionibus temporum*, for instance, enumerated the ancient planetary names for the weekdays, Bede lends his own characteristic stamp to this material in two ways with which we have already become familiar. First, he takes a robustly no-nonsense approach to pagan mythology, a combination of rationalization and dismissal; secondly, he salvages from his disparaging account of the astrological names the opportunity to discuss the order of the planets and the interval each requires to traverse the zodiac, a typically Bedan "encyclopaedic" digression.

Bede also elaborates a theme found in Irish *computi*, and ultimately derived from a hint in Isidore's *De natura rerum* 3. This is the notion that the "week" is not a univocal category; rather there are eight (clearly not an insignificant number) different kinds of weeks. The message is that whereas human authority restricts the meaning of time, the divine authority of Scripture expands and diversifies that meaning.[62] These Scriptural weeks include:

---

61 Cf. *The Reckoning of Time*, ch. 71. On the Christian symbolism of the week, see F.H. Colson, *The Week: An Essay on the Origin and Development of the Seven-Day Cycle* (Cambridge: Cambridge University Press, 1926); Charles Pietri, "Le temps de la semaine à Rome et dans l'Italie chrétienne (IV–VIe s.)", in *Le temps chrétien de la fin de l'Antiquité au Moyen Age IIIe–XIIIe siècles*, Colloques internationaux du Centre national de la recherche scientifique 604 (Paris: Éditions de CNRS, 1984):63–79.

62 Paolo Siniscalco, "Le età del mondo in Beda", 313 *sqq*. Bede may be taking his cue here from Augustine's *De Genesi ad litteram* 2.13.29. Here Augustine, the enemy of astrology, is concerned to demote the heavenly bodies to mere *signs* of the passage of time. But Augustine goes on to point out that even the divisions of time are relative and conventional: a year, for example, could mean a solar siderial year, or the period of any other planet's passage throught the zodiac. In short, "year" could mean whatever Scripture wanted it to mean: cf. Eileen Reeves, "Augustine and Galileo on Reading the Heavens", *Journal of the History of Ideas* 52 (1991):563–579, esp. 567–573.

1. the divine week of Creation

2. the conventional 7-day week

3. Pentecost or the "feast of weeks", i.e. the 50 days after Easter ($7 \times 7 + 1$ days)

4. the Jewish seventh month, during which the feast of the Atonement was celebrated

5. The sabbath of the land in the seventh year

6. the Jubilee, the annual analogy of Pentecost ($7 \times 7 + 1$ years)

7. the "prophetic week" of the Old Testament, looking forward to the coming of Christ (to be discussed in ch. 9)

8. the week of the World-Ages, which like Sunday, both encompasses all time and reaches beyond time to eternity (to be discussed in ch. 10).

Like Russian dolls, these weeks of ever-increasing dimensions nest one inside the other, and closely resemble on another. They are, or course, all formed of sevens, but they also reflect a consistent pattern of religious themes of fulfilment, rest, and restoration.[63] Here again, Bede fills out the skeletal allusions of his immediate sources with the flesh and blood of patristic theology.

Chapter 9:

Of particular interest to medieval readers was chapter 9, on the "seventy prophetic weeks" of the book of Daniel, in which Bede unravels the dark number imagery of the prophecies using *computus* as his guide.[64] His purpose in so doing is not immediately evident, but it is essentially chronological. The "seventy weeks" provided the only firm dating framework for the events of the Fifth World-Age, from the Babylonian Captivity to the birth of Christ,[65] and Bede has much to say about his

63 These themes, as applied to the 7×7 weeks of Pentecost, are expounded in Bede's *Hom.* 2.16 (154, 159–160).

64 On Bede's sources, see Jones, *BOT* 341–342. Bede's immediate source is Jerome's *In Danielem*, and through Jerome, the chronographical writings of Julius Africanus (*ca.* 220). Bede repeats the substance of this section in *The Reckoning of Time*, ch. 66 (*s.a.* 3529) and in *In Ezram et Nehemiam* 3, ed. D. Hurst, CCSL 119A (1969):342.132–343.157. It should be noted that Bede wrote a commentary, now lost, on the Book of Daniel: cf. *HE* 5.25.

65 Croke, "The Origins of the Christian World Chronicle", 121. One of the earliest computists, Hippolytus of Rome, was also deeply interested in this question, for he sought

own particular reading of the chronological data in ch. 66. Here he simply notes that since the "seventy weeks" of Daniel 9.24 must evidently refer to years, and since Daniel says (in the Vulgate translation) that these week are "diminished", Bede argues that they are lunar, not solar years, since 12 lunar months are shorter than a solar year by 11 days. From this basic premiss, Bede produces calculations that demonstrate that the Prophet accurately anticipated the Passion of Christ. Though not an original argument – Bede found it in a quotation from Julius Africanus embedded in Jerome's commentary on Daniel – it seems to have impressed medieval readers, for ch. 9 was sometimes copied as a separate extract in non-computistical manuscripts,[66] and was eventually incorporated into the *Glossa ordinaria*.

Chapter 10:

Bede also knits *computus* and exegesis together in his brief allusion to the World-Ages,[67] a subject which he will develop further in ch. 66. Here he emphasizes the analogy between the events of history in each of the six World-Ages, and the events of the day of Creation (in the "primal week") to which each corresponds. This picks up on a theme implicitly enunciated in ch. 6, where the sixth day of Creation is compared to the Friday of Christ's Passion, for the Sixth Age is inaugurated by the coming of Christ. Though utterly traditional,[68] this allegory reveals Bede's holistic approach to sacred erudition, and to the place of *computus* within that learning. Scripture justifies and enobles *computus* by endorsing the sacred character of time itself; *computus*, in return, joins the philological and encyclopaedic sciences which Augustine, in *De doctrina christiana*, approved as worthy handmaids to the Christian scholar's sublime task of unfolding the meaning of the Bible.

---

to test the accuracy of his 112-year Paschal table by comparing it to the recorded dates of Passovers in Biblical times. Hence the establishment of an accurate chronology for the post-exilic period was important: see Marcel Richard, "Comput et chronologie chez saint Hippolyte [part 1]", *Mélanges de sciences religieuses* 7 (1950):237–268, esp. section 2; *ibid.* [part 2], *Mélanges de sciences religieuses* 8 (1951):19–50, esp. 25 *sqq.*

66 E.g. in Paris, Bibl. nat. lat. 579 (s. XII), in Cambridge, Trinity College B.3.5 (233) (Christ Church, Canterbury, s. XII), both theological miscellanies.

67 For discussion of this chapter, see Siniscalco 299–303.

68 See Siniscalco 316–318 and Landes 143, nn. 20–22 for references to patristic tradition in general; cf. Tristram 22–24, on Augustine.

Chapter 11:

The largest block of chapters in the first part of *The Reckoning of Time* (11–29) concerns the month. This is because the solar calendar itself, whose structure and contents underpin this section, is articulated as twelve distinctively titled units called "months". Within these months, individual days are named in numerical sequence. The months of the Julian calendar are not natural months in either the lunar or the solar sense: that is, they represent neither lunations, nor divisions of the zodiac. Instead, they form an artificial grid against which the true months of the Sun and Moon must be plotted for any given year. The pattern of weekdays is also independent of the division of the year into months. For this reason, formulae are necessary to transform the "generic" month of the calendar page into *this* month, with its distinctive pattern of weekdays and lunar phases. Therefore, introducing the months of the Julian calendar meant explaining the entire system of lunar reckoning, at least in so far as this involved locating the Moon in the heavens and determining its phase. Understanding this lunar reckoning raised further issues, such as the number of hours of moonlight associated with each phase (ch. 24); why the tilt or angle of the waxing or waning Moon varies (ch. 25); why the Moon, though closer to Earth than the Sun, sometimes rides higher in the sky (ch. 26); how eclipses occur (ch. 27); the effect of the lunar cycle on life on Earth (ch. 28), and the mechanism of the tides (ch. 29).

The calendar itself, as a graphic document, is very present in this section. Every month of the Roman calendar is unique, not only in name, but also with respect to length, to the position of its internal divisions (kalends, nones and ides), and to the location of the point where the Sun enters each zodiac sign. The latter is especially important, as the zodiac signs provide a point of reference for locating and determining the age of the Moon on any given day. While Bede would prefer that his students learn to "customize" the calendar through mathematical formulae, he also describes systems of key-letters attached to the individual months, which accomplish the same thing, and invites his students to consult them *in annali* (in the calendar).

But the relationship between the calendar and *The Reckoning of Time* also went in the reverse direction. Medieval calendars came to incorporate onto their pages much of the information discussed by Bede in these chapters, such as his concordance of the Roman months with those of the Jews, Egyptians, Macedonians and Anglo-Saxons, or his enumeration of the lunar and ferial regulars. The relation of textbook to reference document was a constantly evolving and complex one, and in some respects defines the ambivalent status of *computus* itself, as neither exactly a *disciplina* conveyed through authoritative texts, nor yet purely a technique.

Chapters 11–15 constitute a subsection on the structure of the month. This section weaves together two rather complex themes: the relation of calendar months to natural months (lunar or solar), and the concordance of the Roman months with those of other ancient peoples. The second theme serves two purposes: it illustrates the first, in that the Jewish calendar serves as an example of a purely lunar calendar, and the Egyptian of a purely solar one; and it also enables Bede to prove an important point concerning the claims put forward by the Celtic computists about the patristic authority of their system.

The Roman calendar might be solar, but the month, for Bede, was grounded in the natural phenomenon of the lunation, as his etymology makes plain. The ancient Jewish calendar, pre-eminent both historically and theologically, was lunar. Just as in the case of the week, Bede puts biblical categories of time into the foreground to demonstrate both their "natural" truth, and their role as types and exemplars of all other systems.

The names of the Hebrew months, like those of the Egyptian and "Greek" months which Bede will discuss later, are drawn from the prologue to the calendar of Polemius Silvius,[69] a fourth-century ecclesiastical adaptation of the Roman civic calendar. This prologue was known to Bede in the form of an anonymous treatise *De anno*.[70] Polemius Silvius did not attempt a concordance of these various national monthly systems, but Bede did. In chapter 11, he argues that Nisan, the first Jewish month, should be considered the equivalent of April, because it "either begins, ends, or is totally included within it" – or at least will do so if the computist is following the "orthodox" rule and designating Nisan as the first lunar month with a full Moon after the equinox. Moving Nisan to April was an innovation on Bede's part, in that Nisan

---

69 For a full discussion of Bede's use of this document, see C.W. Jones, "Polemius Silvius, Bede, and the Names of the Months", *Speculum* 9 (1934):50–56.

70 J.F. Mountford, "De mensium nominibus", *Journal of Hellenic Studies* 43 (1923):102–116.

had traditionally been associated with March (and was so by Polemius Silvius); but it was adopted almost universally thereafter.[71]

The lunar character of the Jewish month did, however, pose some difficulty to exegetes unaccustomed to reckoning time in this way. Bede takes the opportunity to prove once again the value of *computus* to Christian learning, by unravelling problems associated with the timing of the Hebrew Pentecost, and the sojourn of Noah in the Ark. It is not always clear, Bede says, that a lunar month is a lunation, or synodic month, and not the period of time required for the Moon to return to the same position in the zodiac (a siderial lunar month). The synodic month is $29\frac{1}{2}$ days, but the inconvenience of dealing with half-days led to distributing the lunations into alternating months of 29 and 30 days. This distibution, combined with the variable times chosen for intercalating additional lunar months in order to bring the lunar and solar years back into phase (cf. *The Reckoning of Time*, ch. 43), meant that no two calculators could agree on what the exact age of the Moon on any particular day was. When it comes to *computus*, says Bede, "custom or authority or at least convenience of calculation prevails over nature".

It is worth our while to pause over this sentence, and try to capture Bede's tone of voice. On the one hand, he is exasperated that human arbitrariness is allowed to override logic: the embolisms should be inserted at the end of the lunar year, not in the middle, where they will upset the lunar reckoning. On the other hand, he is aware that the artifice of the calculated Moon is an unavoidable necessity: calendars deal in days, so the Moon's course must be reduced to a whole number of solar days, even in defiance of appearances. How far can time be schematized before the calendar ceases to be an instrument of order, and becomes instead the victim of human wilfulness? Of course, Bede does not pose the question in quite this way, let alone answer it, but it lies close beneath the surface of his text, and informs his concept of *ratio* – a creative yet orderly dialogue between God's Creation and man's divinely-inspired quest for order.

Faced with the confusion engendered by luni-solar calendars and the chaos of embolisms, the Egyptians jettisoned lunar reckoning completely, and adopted a solar year, divided into equal, if arbitrary, months of 30 days, with 5 – or in leap years, 6 – extra days at the end. Bede now

71 Jones, "Polemius Silvius", 52–53.

pauses to establish a precise concordance of the Egyptian months to the Roman calendar months – an exercise of considerable significance to the broader argument underlying *The Reckoning of Time*.

Chapter 12:

The first thing which Bede's students would have remarked in examining their *annale* is that the Roman months are neither natural months, like the Hebrew ones, nor uniform, like the Egyptian. Bede's chapter on the Roman months sets out to explain why.

Bede's main point is that the eccentric structure of the Roman calendar is based on historical happenstance ("custom or authority") rather than on a scientific rationale ("the logic of nature"). But for all its peculiarities, the Julian calendar was the official calendar of the Western Church, and the Christian faith had done much to redeem its vagaries. February, once the morbid month of fevers and propitiatory sacrifices to the subterranean gods, was now a feast of light and heavenly Purification. This apparent digression serves a number of purposes. It is not unlikely that Bede was a little embarrassed at trafficking in so much pagan mythology and religion through his long quotations from Macrobius; underscoring the contrast between the gruesome rites of the Roman month of the dead and the Purification allowed him to distance himself from this material. But Bede goes beyond the moralizing contrast between paganism and Christianity to stress that February's rituals now commemorate *a different kind of time*: not the repetitive cycles of worldly empires, but the eternal fulfilment of the heavenly kingdom. In this way, he evokes the larger themes of time and eternity which underpin the structure of *The Reckoning of Time*.

Chapter 13:

Chapter 13, on the divisions of the Roman months, steers the reader into the safer waters of rationalizing etymology. Bede's English pupils, of course, would have counted the days of the month as we do, sequentially. The Roman divisions had to be learned, not only in order to read the Julian calendar, but also to be able to add and subtract days within the Julian calendar. A practical application will be demonstrated in chapter 22.

Chapter 14:

Here Bede presents a complex and important argument for which the preceding chapters have set the stage. His interest in the "Greek" (really Macedonian) month names is focussed on one of the central documents of the Celtic Paschal reckoning, the *Liber Anatolii*. The crucial passage is the one which Bede discusses here, in which Anatolius identifies the starting point of the 19-year Paschal cycle according to the Egyptian, Macedonian and Roman calendars. The Celtic recension of the text rendered the Julian date as 8 kalends of April, or 25 March, while the Roman version gave it as 11 kalends of April, or 22 March. The distinction was, of course, a crucial one, because the Nicene Council, in Bede's view, had confirmed that the equinox fell on 21 March, and no longer on the date established in the Julian calendar at its inception, 25 March. Now it becomes plain why Bede was so concerned to establish a concordance of the Roman months with those of the Greeks and Egyptians: Anatolius' Egyptian date (26 Phamenoth) and his Greek one (22 Dystros) could only correspond to the 11th kalends of April, and not to the 8th. Therefore the Celtic computists were working with a corrupt text of Anatolius.[72]

Chapter 15:

Bede was not interested in ancient month-names for purely antiquarian reasons: indeed, he omitted completely Polemius Silvius' list of Athenian month-names, as they had no bearing on *computus*. It is an interesting irony, however, that later readers of *The Reckoning of Time*, to whom the controversy with the Irish was opaque, appreciated the discussion of the month-names precisely for their antiquarian value. They incorporated the names onto the pages of calendars,[73] expanded the repertoire

---

72 To make matters more complicated, the version of Anatolius found in Rufinus' translation of Eusebius says that Anatolius dated the equinox to 22 March. Bede, to avoid confusing his students, reserves this problem for his *Letter to Wicthed*, where he also revisits the issue of textual corruption of the *Liber Anatolii*: see translation in Appendix 3.3.

73 Concordances of Greek, Egyptian and Hebrew months are found in the calendars in PL 90.759–784 and of numerous medieval manuscripts, e.g. H.A. Wilson, *The Missal of Robert of Jumièges*, Henry Bradshaw Society Publications 11 (London: Harrison, 1896):9 *sqq.* ; London, British Library, Cotton Tiberius E.IV (Winchcombe, s. XII); Oxford, St John's College 17 (Thorney, *ca.*1110) fols. 16–22v; Cambridge, Trinity College R.15.32 (Winchester? s. X–XI). Partial concordances, and sometimes only the Greek or Hebrew month-names, are also found. Concordances could also be copied apart from the calendar,

with other month-names of regional interest,[74] and incorporated Bede's material into new tractates on the origins of the month-names.[75] In a sense, Bede is responsible for this development, for he indulges in a bit of patriotic antiquarianism himself in chapter 15, "The English Months". This chapter furnishes a unique record of the pagan Anglo-Saxon calendar. The calendar was evidently luni-solar, and operated on a fixed system of embolismic intercalation, though Bede does not say whether his ancestors used an 8- or 19-year cycle.[76] He essentially surveys the names of the months, and their meanings.

What delights the modern historian might have seemed merely trivial information to non-Englishmen contemporary with Bede, and even dangerous in the eyes of Englishmen anxious to set aside the pagan past. For Bede himself though, this account of the ancestral calendar was bound up with his larger concern to define the English as one of the *gentes* of Christendom. Chapter 15 also carries echoes of Bede's interest in Gregory the Great's policy of conversion: the pagan heritage of the English has been "translated" into Christian practice, not obliterated. Though Bede is generally taciturn about English paganism, he makes an exception for the months precisely because they illustrate the triumph of the conversion process. As Meaney observes:

> . . . we have to take into account that Bede here, for once, is setting up a deliberate contrast between the old times and the new: between the old pagan immolation of cattle and the new sacrifice of praise to Jesus; between the night of ceremonies for the Mother Goddesses and the celebration of the birth of the Lord. For once he does not wish to keep quiet about paganism, but to glory in its replacement.[77]

---

as in St Gall MS 251, p. 13, and Vatican City, Reg. Lat. 1260, fol. 107. This phenomenon is noted by Jones, *BOT* 136.

74 E.g. Vatican Library, Reg. lat. 123 (AD 1056, Ripoll) fol. 21, which gives the "Saracen" names for the months, and on fol. 22r, in a *rota*, lists Hebrew, Greek, Egyptian, English, and Saracen months.

75 E.g. the post-Bedan material of *De divisionibus temporum* (PL 90, 659–663), and the texts in Vatican Library, Pal. lat. 834 (Lorsch? *ca.*836), fols. 34r–35v, and Bamberg, Staatsbibliothek Msc. lit. 160 (Bamberg, s. XII) fols. 30v–33v.

76 For discussion, see Harrison, *Framework*, ch. 1.

77 A.L. Meaney, "Bede and Anglo-Saxon Paganism", *Parergon* n.s. 3 (1985):7. Cf. Bede, *HE* 1.30 (106–108).

The manner in which Bede records and explains these ancient Germanic month-names for an English audience indicates that they were already falling out of use in Bede's own day. As he himself points out, the word "Eostre" no longer denoted either a month or a goddess for the English, but rather the Christian feast of Easter.[78] Yet the subsequent use to which *The Reckoning of Time* was put lends an ironic twist to this story. While many Continental manuscripts dropped chapter 15 as irrelevant,[79] the old English month-names, like those of the Egyptians and Macedonians, did find a place on the pages of calendars, even outside the British Isles. Bede's antiquarian gesture (perhaps with Alcuin as the vector?) may lie behind Charlemagne's attempt to legislate native Frankish names for the months.[80] In short, the local and controversial material in *The Reckoning of Time* could be readily recycled as *doctrina*, shielded from accusations of being vain *curiositas* by the authority of the master. This is a parable of the history of *computus* as a whole. What begins as a highly focussed and practical exercise to devise a workable and theologically defensible Christian calendar comes eventually to include a broader spectrum of scientific information. When the polemics are over, the scientific information remains, and can be safely and innocently studied for its own sake because *computus* has been recognized as a branch of ecclesiastical learning.

Chapter 16:

The block of chapters on the zodiac signs (16–19) has no cognate in *On Times*, though elements of it can be found in *The Nature of Things*, chs. 17 and 21. Jones argues that it was inserted into *The Reckoning of Time*

---

78 On the problems surrounding the identification of the goddess Eostre as well as the "Hretha" of "Hrethmonath", and Bede's etymology of "Solmonath", see Meaney 6–7. Meaney thinks that Bede's knowledge of the pagan feasts' cycle was solid, but already a little remote; on the other hand, Harrison (*Framework*, ch. 1) argues that the circumstantial character of Bede's account is evidence for the persistence of the pagan calendar and religious practices in Bede's day.

79 This happened at a fairly early stage, e.g. Paris, Bibliothèque nationale lat. 7296 (s. IX[1]). Some manuscripts, however, reveal that the loss was recognized, and an attempt made to repair it: London, British Library Royal 15.B.19 (Reims s. IX[2]) inserts ch. 15 on a fly leaf (Jones, *BOT* 147); in Paris, Bibliothèque nationale 13403 (Corbie s. IX) ch. 15 is missing in the *capitula*, but included in the text, though from a different exemplar than the rest of the treatise: *ibid.* 148.

80 Einhard, *Vita Caroli* 29.

not for its computistical relevance, but because "the zodiac and planets fascinated early medieval students".[81] While early medieval students often indulged their curiosity about the zodiac and planets with impunity in the margins of the *computus* (and the evidence of *computus*-manuscripts confirms this), it seems very doubtful that this is what Bede is doing here, at least consciously. In the first place, he limits his discussion to the Sun and Moon, and barely alludes to the existence of the other planets.[82] Secondly, his discussion of the zodiac is clearly a means to an end. Bede's ultimate goal is to explain various ways of determining the age of the Moon, a topic with considerable direct relevance to *computus*. What Bede wants his students to grasp is that what we call the "age" of the Moon is its visible phase, which in turn is a function of its angular distance from the Sun along the ecliptic. This distance is compounded of the Moon's daily advance through the zodiac, minus the Sun's daily reduction of that advance. Bede devotes chs. 16–17 to an astronomical and mathematical explanation of the relationship between the Sun's position and progress along the ecliptic, the Moon's position and progress, and the lunar phase. The relationship between the Moon's phase and its position in the zodiac can be reduced to a crude mathematical formula (ch. 18), or for those who do not know how to calculate, read off a chart (ch. 19).

The formulae presented by Bede, albeit original, are highly approximate, even by medieval standards.[83] However, there was no need for Bede's students to know with extreme precision exactly which sign, and which degree of which sign, the Moon was in. What they did need to grasp, in order to comprehend how lunar computation works with respect to the Julian calendar, was roughly what is happening in the heavens: how the Moon's faster journey through the zodiac relates to the Sun's slower course. Were they merely curious about the speed of the Moon's passage, they could read *The Nature of Things* 21; but this was not enough for computistical purposes. To understand the relationship of lunar phase (synodic month) to celestial position (sidereal month), the student had to be able to conceive of (if not calculate with exceptional

---

81 *BOT* 351.

82 In fact, he only alludes to them in ch. 16 in the form of a brief cross-reference to ch. 8, where he mentions the relative speed of the other planets through the zodiac in order to explain the discrepancy between their natural order, and the order of the weekdays named after them.

83 Jones, *BOT* 353–354.

precision) the amount of zodiacal distance the Moon puts between itself and the Sun on each day of its lunation. Bede needs nothing more than an approximate formula here – indeed, if ch. 19 gives any indication, most of his readers could probably not cope with anything more complex – because his purpose is to demonstrate a principle.[84]

The originality of Bede's strategy in including this material in *The Reckoning of Time* is revealed by the difficulty of identifying his sources.[85] Much of the information can be traced back to Macrobius, Martianus Capella, Hyginus and the *Aratea*, but Bede seems to have known none of these works directly. He must, therefore, have used anonymous extracts and short texts in *computus* anthologies, but these can only be detected for the information on the zodiac in ch. 16.[86] The remaining material, including the formulae and possibly the table in ch. 19,[87] seem to be original to Bede.

Two main themes run through this group of chapters: Bede's concern to stress that the zodiac is a convention of measurement, and his paedagogical technique.

Introducing the zodiac in ch. 16, Bede is careful to describe it as an artifice of astronomical measurement. He is unconcerned with the mythology of the zodiac signs, and quotes Ausonius' poem only because it fixes the relationship between the signs and the months of the solar calendar.[88] The signs of the zodiac are each divided into 30 segments, representing a solar day; but the representation is only approximate, because the Sun actually passes through a sign in 30 days, $10\frac{1}{2}$ hours. The equinoxes and solstices have no meaningful relationship to the divisions of the zodiac, and indeed, different traditions locate these celestial events at different points (though Bede clearly favours the "Alexandrian" system).[89] Later in ch. 18, he defines a *pars* of a zodiac sign as

84 As usual, though, later generations interpreted Bede's inclusion of this material as an endorsement to study it for its own sake: see for example Alcuin's correspondence with Charlemagne (Introduction, n. 245) and the "Bobbio *computus*" (PL 129.1325).

85 Jones, *BOT* 351.

86 See *The Reckoning of Time*, ch. 16, n. 163.

87 See below, n. 91.

88 Ausonius' verses and others of the same type were frequently inscribed on the pages of the Julian calendar in *computus* manuscripts: see Wallis, "MS Oxford St John's College 17", 310–312, and John Henning, "Versus de mensibus", *Traditio* 11 (1955):65–90.

89 It is noteworthy that in ch. 16, Bede gives the conventional Roman dates for the solstices and equinoxes, mentioning that the Alexandrian date falls "a few days earlier". Only in the case of the spring equinox does he mention the precise Alexandrian date,

"nothing but the daily advance of the Sun in the heavens", as if to stress that these signs and their divisions have no natural existence or significance.

This emphasis is, I think, significant, for Bede plainly felt that in discussing the zodiac at all, he was handling something potentially dangerous. His reduction of the zodiac to a mere artifice of measurement, and his studious avoidance of any allusion to celestial influences or mythological connotations suggests a certain sensitivity to the threat of astrology. Indeed, one might ask whether his formulae for finding the Moon in the zodiac are deliberately crude and approximate to avoid their being used for any divinatory purpose.[90]

Chapters 17 and 18:

This section on the zodiac is also remarkable for Bede's display of his talents as a teacher. Indeed, one senses that he has inserted this entire section because students complained that they could not understand why the formulae for finding the age of the Moon were supposed to work. Hence Bede seems more than usually frustrated at the loss of face-to-face contact and feedback. In ch. 16, for example, he apostrophizes the *lector* frequently, and observes that "much can be said about [the Sun's passage through the zodiac], but it can be done to better effect by someone speaking than through the written word". We see him pause to explain terms like *bisse* in context (ch. 16), a good strategy for ensuring that the explanation will be retained; and in ch. 18, he will keep his students on their toes by reversing the equation, shifting at day five from correlating the Moon's position to the Sun's, to correlating the Sun's to the Moon's.

But most importantly, Bede undertakes throughout this section to provide numerous worked examples of the formulae he presents. The

---

because it is the only one of computistical significance: see Commentary on ch. 14. Bede may prefer the "scientific" dating of equinoxes and solstices to the traditional one, but he also recognizes that the calendar is a convention, and that it can tolerate inaccuracies as long as no essential computistical calculation is affected. This is the same practical logic which allows Bede to adopt an approximate but simple formula for the ratio of lunar and solar months in the interests of paedagogical clarity.

90 See remarks on Bede's knowledge of astrology in Commentary on chs. 3 and 7 above.

design of these worked examples is revealing: Bede begins with simple applications, then introduces more complex examples, and finally adds a reverse corroboration which works backwards from an incontestable axiom to demonstrate the soundness of the method. In ch. 17, his first example uses simple division without remainder: to find the position of the Moon on day 5, multiply by 4, which makes 20, and divide by 10. The quotient 2 means that the Moon and Sun are separated by two signs. He then proceeds to a slightly more complex example. This one involves a remainder, but a remainder expressed as a single unit of time-measurement: the quotient for the 8th day is 3, with 2 remaining: "Therefore the eighth Moon is separated from the Sun by three signs and two *puncti*. You know that two *puncti* equal six *partes*, that is, the distance covered by the Sun in its journey through the zodiac in six days." Level three of difficulty involves a remainder grouped into two registers of units: on day 19, the quotient will be 7, with 6 remaining: "Therefore, on the journey which it began away from the Sun, the Moon has parted company with it by seven signs, plus one hour, which is half a sign and a *punctus*, that is, three *partes*." Finally, there is the reverse corroboration. Everyone knows that the full Moon on day 15 is exactly opposite the Sun. Since there are 12 zodiac signs, this means it must be 6 signs away from the Sun, and $15 \times 4 = 60 \div 10 = 6$.

Chapter 19:

Bede is also willing, albeit somewhat grudgingly, to provide for the student who does not have enough mathematics to be convinced by any formula. In ch. 19, he introduces a system of key-letters attached to the dates in the Julian calendar. These key-letters refer to a table, the *pagina regularis*, which locates the Moon's position in the zodiac, and correlates this to the month in which the Sun will occupy the same position.[91] Bede's

91 Bede himself probably devised this table. It is modelled on an earlier version ascribed to Aldhelm, which was based on the synodic and not the siderial month: see Jones, *Bedae pseudepigrapha* 70. For discussion of the table and its history, see von Sickel, "Die Lunarbuchstaben", 181. Bede's interest in such tables was probably derived from Irish *computus* anthologies, which in turn inherited, through Spain, a strong north African tradition of computistical tables. Krusch (*Studien I*, 178–183) has reconstructed some tables based on the Carthaginian computus of 455, including a cycle of Dominical letters, a lunar-regular and lunar-letter table, and a table of Easters by ferial numbers – all types of calculation unknown to either Victorius or Dionysius; cf. Jones, *BOT* 76.

patronizing disdain for those who "do not know how to calculate" may be a bit of strategic sarcasm, designed to shame his students into making a greater effort. If so, it backfired miserably, for the *pagina regularis* enjoyed an enormous success in medieval *computus* collections and anthologies. Ironically, what Bede reluctantly included as a crutch for the mathematically challenged became the mainstay of *computus* pedagogy.[92]

Both Bede's reluctance and the subsequent popularity of the table are reflected in some peculiarities of the table itself, and the questions these peculiarities raised. The table translates the A–O letter series affixed to the solar calendar (see Appendix 1) into a position of the Moon within the zodiac. In the calendar, the letters A–O are separated by a single space, but there is no space after O. The letters and their spaces therefore account for 27 days, approximately the length of the sidereal lunar month (the time it takes the Moon to return to the same position in the zodiac as it started from at last conjunction). As Bede explains in ch. 16, it takes the Moon 2 days, 6 hours and $\frac{2}{3}$ of one hour to traverse each sign: multiplying 2 days and 6 hours by 12 signs yields 27 days. But what about the extra $\frac{2}{3}$ of an hour? Multiplied by 12, this excess produces an additional 8 hours, or over three months, a whole extra day. For this reason, every third A–O sequence in the calendar adds a space after O. If we turn to the *pagina regularis*, however, we find no space after any O. On the other hand, we discover that every fourth zodiac sign is given an additional space – three instead of two.

That Bede does not bother to explain these oddities indicates his disinclination to spend much time and attention on tables. The readers of *The Reckoning of Time*, however, would not rest content. The Carolingian computist who compiled the so-called "Byrhtferth Glosses"[93] answered their perplexity by pointing out that the space following the third O represents the extra day cumulated by the extra $\frac{2}{3}$ hour, and that the extended space accorded to every fourth zodiac sign is due to the fact that each line of the table is supposed to represent a day. It takes $2\frac{1}{4}$ days for the Moon to actually move through a zodiac sign, but the quarter-day cannot be represented in the table, and so it is cumulated and added to every fourth zodiac sign. But why, the students ask, is the

---

92 For examples of the *pagina*, both illustrating *The Reckoning of Time* and as a separate item, see Jones, *Bedae pseudepigrapha* 62–63, 68–70 and *BOT* 354.

93 See Introduction, section 6.

extra blank after the third O found in the calendar, but not in the table? One might answer that because each annual column must serve for all months in that year, the anomaly must be omitted. "Byrhtferth", however, takes a slightly different tack: "That constellation [i.e the zodiacal position of the Moon] between O and A can be known after the manner of the *bissextus* [i.e. leap-year day], for which no space is left blank in the calendar, but it is inserted between two solar days, that is, between the 7th and 6th kalends of March, in the place [assigned to] the 6th kalends of March."[94]

In short, the relationship between calendar and table is a complex one, involving adjustments in both directions, and an understanding of the *notional* character of the spatial strategies employed. We may conjecture that Bede felt that a student who was so weak as to require the use of a table would hardly be competent to handle such abstractions, and therefore omitted them.

Chapter 20:

Bede is now prepared to present his system for finding the age of the Moon on any given day of the Julian calendar. Since the system for finding the weekday is analogous, this will also be covered in this block of chapters on lunar and solar coordinates for every day of the month (20–23). Both are discussed in the context of the month because the computistical apparatus used to solve these problems is geared to the kalends or first day of each month, from which other dates are derived by counting.

---

94 *Ad hoc breviter respondere possumus, quod inter o et a illud sidus cognoscitur similitudine bissexti, cui nullus locus est vacuus in annali serie, sed inseritur inter duos dies solares, id est inter septimum et sextum Kalend. Mart., in loco VI Kalend. Mart.* (PL 90.391D). Similar questions about the table are addressed in the Old Welsh "*Computus* fragment", composed in the 10th century, and preserved in Cambridge University Library Add. 4543. The fragment was edited by E.C. Quiggan in *Zeitschrift für celtische Philologie* 8 (1911):407–410, and re-edited with annotations (including the "Byrhtferth" gloss) by Ifor Williams, "The Computus Fragment", *Bulletin of the Board of Celtic Studies* 3 (1927):245–272. The general computistical background to the fragment is presented by John Armstrong III, "The Old Welsh Computus Fragment and Bede's *Pagina regularis* [Part 1]", *Proceedings of the Harvard Celtic Colloquium* 2 (1982):187–273. The second part, which would have addressed the fragment itself, appears never to have been published. Armstrong's essay does not discuss the adjustment of table and calendar, and contains some misunderstandings and errors.

The problem of locating the age of the Moon, however, is certainly the more difficult of the two, and the more fraught with consequences for the Paschal reckoning. What made the business of calculating the Moon so difficult was the attempt to compromise between the customs of Rome and Alexandria – politically wise, perhaps, but computistically very confusing. Bede's system presupposes a 19-year cycle of lunations, beginning in a year when the Moon is new on a key date. The age of the Moon on that date in subsequent years is known as the annual *epact*. For the Alexandrians, the 19-year cycle began in the year when the epact was 0 on 22 March (c.f ch. 14). Bede adopted this system of epacts, but he was working with a Julian calendar which began on 1 January. This introduces an additional step: translating the epact into the age of the Moon on 1 January, so that the Moons of the other eleven months can be calculated in relation to it. Confusingly, another system of epacts beginning in years when the Moon was new on 1 January, also survived in the Middle Ages, and is enshrined in the *cyclus lunaris* column of the Paschal table (cf. discussion of ch. 56).

Since the Moon advances eleven days with respect to the Sun each solar year, the epacts increase by eleven, year by year, until a full lunar month of 30 days has accumulated, at which point this additional or embolismic lunar month is inserted into the reckoning. In the 19-year cycle, these embolisms are inserted in the third, sixth, eighth, eleventh, fourteenth, seventeenth, and nineteenth years. As Bede observed, it would make sense to insert the embolismic months at the end of the year, so as to cause minimal disturbance to the normal reckoning of the Moon, but in three years (8, 11 and 19), they are inserted in the spring, which necessitates additional adjustments for the other months. The *saltus lunae* in year 19 poses a similar problem. Bede will discuss both embolismic years and the *saltus* in greater detail in the context of the Paschal reckoning in chs. 42 and 45. At this point, he simply wants to alert his students to these exceptions to his rule.

The formula Bede presents for finding the Moon in ch. 20 seems to have been of his own devising, and proved so popular that it was adopted even by proponents of the Victorian *computus* which *The Reckoning of Time* would soon render obsolete.[95] It involves a relatively simple mathematical operation, plus some undemanding memorization. First, one should know the epact, or age of the Moon on 22 March, in

95 Jones, *BOT* 354.

the year in question: this can be calculated according to the formula presented in *The Reckoning of Time*, ch. 50, or simply looked up in the Paschal table. Secondly, one should memorize the monthly lunar "regulars". These regulars are numbers which represent the age of the Moon on the first day of each month, on a year when the epact is 0, i.e. year 1 of the cycle. Bede recommends simply memorizing them, but it is worth explaining how they are derived.

It will be recalled that medieval computists, to avoid working with fractions, alternated lunations of 29 and 30 days. The lunation of January (i.e. the one terminating in January) is calculated as "full" (30 days), the lunation of February as "hollow" (29 days) and so on through the year. In year 1, the Moon on 22 March is 0; counting backward, you will discover that the Moon on 1 January will be 9; this is January's "regular". January has 30 lunar days but 31 calendar days, so the Moon on 1 February will be one day older than it was on 1 January, or 10 days. The "regular" for February is therefore 10. February has 29 lunar days but 28 calendar days, so the Moon for 1 March will be one day younger than for 1 February, i.e. 9. The regular for March is, therefore, 9. The regulars for the other months are calculated in the same way. Add the annual epact to the monthly lunar regular, and you will have the age of the Moon on the first of the month.

Bede is conscious of the limitations of his system. It does not work, for example, in years 8 and 19, when the embolism is inserted in March, the same month when the epact changes, thereby pushing the lunar count in all subsequent months forward by one day (the embolismic month is always 30 days long, and thus interrupts the 29/30 day alternation of the lunar months). It also does not work in year 11, when the embolism is inserted exceptionally late in the month, on 5 December. This means that the embolismic month cannot end within the calendar month of December, as it ideally should: in fact, it ends on 3 January. The 30-day lunation proper to January does not end until 1 February, and February's 29-day lunation does not end until 2 March. That is why, as Bede observes, the Moon on 1 March in year 11 is 28 days old, when according to the formula, it ought to be 29 (epact 20 + lunar regular 9).

Bede also admits that the formula must be abandoned for the final months of the 19th year of the cycle, because of the *saltus lunae*. His system of monthly lunar regulars is adapted to the Julian calendar, and therefore begins in January. The *saltus lunae* (the shortening of one lunar month by one day, to bring the lunar and solar cycles together at the end of 19 years) is placed as close to the end of the cycle as possible, so as not to disturb calculations such as those explained in this chapter. That means that Bede's Roman system puts it in the penultimate month of the cycle, November of year 19. The Egyptian calendar began in September, and the Egyptians also inserted the *saltus* in the penultimate Egyptian month, i.e. July. Bede points out that if users of the Julian calendar follow this Alexandrian system, the epact-and-regular formula will not work for the last five months of the cycle.

The Alexandrian system of lunar regulars was also adapted to a calendar beginning in September; nonetheless, it had great authority and was in widespread use in the Middle Ages. Bede's system, though more "efficient" (as he himself remarks) for a Julian calendar, only slowly supplanted it, though it was certainly an advance over previous formulae, such as the one ascribed to Dionysius.[96] Though later computists preferred to work with the "Egyptian" regulars enumerated by Bede at the close of chapter 20, they retained the principles of his formula.

Chapter 21:

Bede now explains a cognate system for locating the weekday of any given calendar date. The *solar regulars* are analogous to the lunar regulars introduced in ch. 20, and the *concurrents* correspond to the epacts. The solar year of 365 days totals 52 weeks and one day. Hence a calendar date which this year falls upon a Sunday, will next year fall on a Monday, in the following year on a Tuesday, and so forth. However, the addition of an extra day at the end of February in leap year means that this sequence of weekdays "leaps" forward by one place after the leap-year day; e.g. if 1 March in the year preceding a leap year falls upon a Monday, it will fall on a Wednesday in leap year. The annual concurrent is a number representing the weekday of 24 March, a date chosen because it fell on the same weekday as 24 February, when the leap-year day was inserted, but within the period of the Easter reckoning; it was therefore a computistically handy reference point. The monthly regular represents the constant distance between the weekday of 24 March and the weekday of the first of each month. For example, the eight days

96 *Ibid.* 355.

between 24 March and 1 April are reduced to a week and one day, so the interval or regular for April is one; the thirty-eight days to 1 May are reduced to five weeks and five days, so the regular for May is five. Hence by adding the annual concurrent to the monthly regular, one can find the weekday of the first day of any month. Any weekday within the month may then be derived by simple counting.

Though the opening phrase of chapter 21 suggests that this system was not of Bede's invention, *The Reckoning of Time* certainly contributed to its well-deserved popularity. The solar regulars, like the lunar ones, were fairly simple to memorize, especially with the assistance of one of a number of mnemonic poems devised for this purpose.[97] Of course, both the monthly lunar ages and the weekdays could also be – and quickly were – reduced to a handy table.[98]

Chapter 22:

Bede somewhat diffidently presents a second method for finding the Moon or the weekday. Add up the number of days that have elapsed since 1 January, add in the age of the Moon on 1 January, and divide by 59 (one full plus one hollow lunar month). If the remainder is greater than 30, subtract 30. The result is the age of the Moon. To find the weekday, calculate the number of days elapsed since 1 January, add the number of the weekday of 1 January, and divide by 7. The remainder is the weekday.

97 The major collections of computistical verse are those edited by Paul de Winterfield in MGH Poetae 4.1 (1899):667–702, by Karl Strecker in MGH Poetae 6.1 (1951):186–208, and in P. Baehrens, *Poetae latini minores* (Leipzig: Teubner, 1883):5.349–356 and A. Riese, *Anthologia latina* (Leipzig: Teubner, 1870):1.258–259, 2.38, 91–92, 122–123, 140. The genre is discussed by John Hennig, "Versus de mensibus"; Strecker, "Zu den computistischen Rhythmen", *Neues Archiv* 36 (1911):317–342; and Gunter Berny, *Das lateinischen Epigramm in Übergang von der Spätantike zum frühen Mittelalter*, Münchner Beiträge zur Mediävistik und Renaissance-Forschung 2 (Munich: Arbeo-Gesellschaft, 1968):286–294. The most popular verse for solar regulars begins "Ianus et octimber binis regulantur . . .": see Dieter Schaller and Ehwald Köngsen, *Initia carminum latinorum saeculo undecimo antiquorum* (Göttingen: Vandenhoek & Ruprecht, 1977):341; Lynn Thorndike and Pearl Kibre, *Incipits of Medieval Scientific Writings in Latin*, 2nd ed. (Cambridge, Mass.: Mediaeval Academy of America, 1963):654.

98 E.g. Jones, *BOT* 356; lunar table in PL 90.753–754 (top); cf. Leofric Missal fol. 46r, ed. Robert Warren, *The Leofric Missal as Used in the Cathedral of Exeter . . .* (Oxford: Clarendon Press, 1883):37, and Wilson, *Missal of Robert of Jumièges* fol. 13v (p. 24); weekday table in *ibid.* fol. 14r (25).

To our modern eyes, this method might, at least at first glance, seem very much simpler than formulae involving epacts, concurrents and regulars; but in fact, this second method would have been much more challenging to Bede's students. Two factors contributed to its difficulty. The first is the difficulty of doing division with large numbers (up to 365), which we discussed in connection with *The Reckoning of Time*, ch. 1. In order to reduce the total number of days to a lunation, one must divide this figure by the length of a lunation, namely twenty-nine and one half days. However, division where the divisor is an integer and a fraction is very difficult, especially when the operation is done in one's head, or on one's fingers. Hence the reduction is effected by dividing first by a double lunation of fifty-nine days, and then reducing this double to a single lunation, if necessary, by subtracting thirty from the remainder. Bede suggests that the easiest expedient would be to memorize the products of 59 to 365. He does not explicitly advise memorizing the 7 times table (presumably because multiplication tables up to ten were memorized as a part of elementary education?),[99] but it seems that neither came easily to medieval students, for 59- and 7-times tables are commonly included in *computus* anthologies.

The second problem is the Roman calendar itself. Our method of designating dates by sequential count from the first of the month makes it fairly easy, assuming you know the length of each month, to add up the number of days since the beginning of the year. Roman dates, however, are numbered backwards from the kalends, nones and ides. Therefore one has to memorize the numerical position *vis-à-vis* 1 January of each of these fixed days for each month, and then subtract the number of days by which the date in question precedes the fixed day. All in all, epacts and concurrents were easier.

Why, then, did Bede include this second method? He claims to be doing so as a concession to tradition, but I think there is something more here. This old formula may be hard to apply in practice, but it offers a very clear illustration of the orderly progression of lunations and weekdays through the months. Since the Julian calendar in and of itself recorded neither lunations nor weekdays, these patterns were not as self-evident to the medieval student as they are to us. We are accustomed to calendars which are graphically organized as overlapping week-long strips; almost without thinking about it, we know that 1

99 This is Jones' explanation: *BOT* 358.

June, 8 June, and 15 June will all fall on the same day. Simply as a mental exercise, Bede's second formula would have impressed on the medieval student the fact that the Julian calendar can be conceived as successive blocks of seven days – an insight which his ordinary experience of the calendar as a document, and of the peculiar Roman method of naming dates, would not immediately suggest.

Chapter 23:

After a suitable display of schoolmasterly indignation, Bede once again furnishes a table for those "so lazy or slow-witted" that they cannot calculate the age of the Moon. Like the one described in ch. 19, it is based on a system of key-letters prefixed to each day of the solar calendar. The letters cover the 59 days of a full plus a hollow lunar month by using three sequential alphabets: the first alphabet has no distinguishing mark, the second has a dot following each letter, and the third has a dot preceding each letter. Hence these were commonly known as *litterae punctatae*. The third alphabet also ends with "t" rather than "u", so that the total number of letters in 59. The table translates these letters into the age of the Moon for every day of every year of the 19-year cycle. All one needs to know is what year of the cycle it is.[100]

Interestingly, Bede does not bother to explain that this table will also have to be tinkered with in years 8, 11 and 19 of the cycle. It is a crutch for dullards, who probably could not grasp the complexities of embolisms anyway. Bede may well have had other reasons for disapproving of tables, for this particular table was, surprisingly, often copied inaccurately in medieval *computus* manuscripts. Dots were omitted, letters skipped, or the scribe forgot that the third alphabet was shorter than the others. Rather than give up on tables, though, later computists tried other expedients to reduce the chances of scribal error, such as using a single alphabet with intervening spaces (as in the table in chapter 19) or re-ordering the tables to that the same letters were lined up in a row – proof against miscopying, but difficult to use.[101]

100 See the formula given in *The Reckoning of Time*, ch. 47. On the *litterae punctatae*, see von Sickel 159 *sqq.*

101 These tables were the specialty of the 10th-century computist Abbo of Fleury. For examples, see PL 90.802C–D and 805–806. I am presently preparing an article on this subject.

Chapter 24:

Bede closes his section on the month by examining a number of issues concerning the relationship of the Moon with the Sun, apart from the calendar. Chapters 24 to 29 are of exceptional interest, for they reveal Bede's qualities as an original scientific thinker and teacher. This block of chapters has no cognates in prior *computus* literature, and constitutes a genuine digression from the *computus*. Its purpose is very different, namely to set the record straight on a number of confusions and misconceptions about the behaviour of the Moon. It is, in fact, pure astronomy, with no connection to the calendar, save that a discussion of the month, which is etymologically and historically connected to the Moon, seems a logical place to "file" this material. This is an excellent illustration of how the *computus* could function as an occasion for scientific reflection as well as problem-solving, and as a haven for purely scientific knowledge with no fixed address in the clerical educational curriculum.

But Bede is usually restrained in the amount of extraneous encyclopaedic material he will admit to *The Reckoning of Time*. Why did he abandon this policy and permit himself this extensive digression? There are a number of reasons, which become clear as we proceed from chapter to chapter. In chapter 25, he wants to refute a commonplace error, found on the pages of virtually every ancient authority, and particularly in Isidore of Seville. In chapters 26 and 27, he seems to be responding to his students' demands for clarification. In chapters 28 and 29, he sets out the results of his own original researches into the relationship of the tides to the lunar month. Nowhere else in *The Reckoning of Time* do we have such a vivid picture of Bede's critical scientific mind, its pedagogical inventiveness, and spirit of rational inquiry. Terms like *natura, naturalis ratio, naturae ordo* are thickly strewn through these chapters, and the accent is on the conflict of reason with ignorance or superstition. These chapters, more than any other part of *The Reckoning of Time*, gave a scientific tone to the study of *computus*. Their importance for the history of science in the West is seriously underestimated. Bede's vision of what *computus* could encompass allowed monks like Gerbert and Abbo of Fleury, well before the arrival of Greek and Arabic astronomical texts, at least to pose questions and ponder problems which were purely scientific in character, independent of their applicability to the calendar.

Chapter 24 discusses the number of hours of moonlight, using a formula which, as Bede admits, will only really work at the equinoxes. This is, of course, not without computistical interest, given the symbolism of the Sun and Moon flooding the world with light at Easter (cf. *Reckoning of Time*, ch. 6), and the debate over the lunar limits of Easter. However, Bede does not bring up this computistical angle. The formula is, by his own admission, fairly useless for calculatory purposes.[102] But it is certainly not without its scientific and heuristic virtues, for a knowledge of the mathematics of the lunation, however crude, permits an understanding of the phenomenon of tides. In fact, Bede will invoke this formula in chapter 29, to support his explanation of the timing of tides.

Chapter 25:

Bede now turns his critical acumen upon a time-honoured superstition. The nightly variations in the Moon's shape and position are highly conspicuous, and in almost every culture and age have given rise to forms of prognostication, especially about weather. One common belief, enshrined in the works of a number of classical authorities, is that when the horns of the crescent Moon point upwards from the horizon, the weather will be stormy; when the Moon seems to be standing erect, it will be fair. Isidore of Seville recounts much in this vein in *De natura rerum* 38. This is typical of the kind of "untruths" that Bede did not want his spiritual sons to learn from the pages of Isidore, so he took this opportunity to explain why the tilt of the crescent Moon varies with the seasons.

Bede's explanation might be clearer if a short astronomical exegesis were provided, expressed in the geocentric terms he would have understood.

Once every day, the entire heavenly sphere appears to rotate around

---

102 Jones (*BOT* 359) argues that this formula should have been computistically embarrassing to Bede, as it would have supported the Irish claim, enunciated in ps.-Anatolius 4 (319–320), that Easter cannot be celebrated on the 21st day of the Moon, because at that point "darkness has overcome light". Jones interprets this phrase as meaning that on the 21st day of the lunation, the Moon will rise after midnight. But Bede's formula assigns 7 hours and 1 *punctus* to the 21st Moon (30−21=9x4=36 ÷ 5=7, with a remainder of 1) which means that it rises before midnight; in fact, even the 22nd Moon, at 6 hours and 2 *puncti*, is still within the limits by this criterion. Moreover, "darkness" has in a sense already overcome "light" by day 16, when there is already a gap of 4 *puncti* between sunset and moonrise.

Earth on an axis running through the north and south poles. That means that all the heavenly bodies travel across the sky in lines parallel to the celestial equator, which is a projection of the Earth's equator onto the celestial sphere. Wherever we stand on Earth, the celestial equator intersects our horizon at due east and due west. However, the *angle* at which it intersects varies with latitude. That angle is constant, and in northern latitudes, slants to the south.

The ecliptic, however, is set at a $23\frac{1}{2}$ degree angle to the celestial equator. Along with the rest of the heavens, it is carried around the Earth once each day, but its tilt to the equator means that the angle it forms with our horizon will vary throughout the day, as various parts of the ecliptic to the south and north of the equator come into view. If we could darken the sky for 24 hours so that we could actually see the constellations along the ecliptic, and set up a camera facing due west so that it would take a photograph at six-hour intervals, we would see the ecliptic in the following positions in succession:

    a. parallel to the equator to the north
    b. crossing the horizon at due west, but slanting to the south
    c. parallel to the equator to the south
    d. crossing the horizon at due west, but slanting to the north

This is illustrated in figure a.

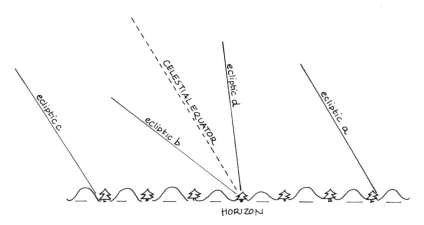

a. Positions of ecliptic with respect to celestial equator and horizon at six-hour intervals, looking west.

The ecliptic, then, seems to weave up and down the sky, to the north and south of the equator, and the angle it forms with the horizon changes hour by hour.

However, the daily timing of these angular changes varies throughout the year. In spring, the ecliptic at sunset will be crossing the equator and inclined at its maximum angle to the north. In northern latitudes, this means that the angle it forms with the horizon is very steep. The Moon, like the Sun, travels along the ecliptic.[103] Therefore the new Moon at sunset will appear high up in the sky towards the north, plunging almost straight down into the west. The setting Sun will illuminate the Moon from "below", producing two horns pointing away from the horizon, and up into the sky. Likewise at sunset at the autumn equinox, the ecliptic is at its maximum southward slant with respect to the equator. In northern latitudes, this means that the angle formed with the horizon is very shallow. Therefore the waxing crescent Moon, setting just after the Sun, and hugging the horizon, will appear to be "beside" rather than "above" the Sun (to borrow Bede's terms). The setting Sun will illuminate its northern flank, so that the Moon will seem to be standing upright, with its horns pointing south. At midwinter and midsummer on the other hand, the ecliptic is parallel to the celestial equator at sunrise and sunset, and as Bede observes, when new or very old the Moon is high in the north in the summer and low in the south in winter, because it is close to the Sun at these points in its monthly cycle. Therefore the waxing crescents at these seasons will be tilted midway between their spring and autumn extremes. These positions are illustrated in figure b.

Bede drives home his point with some force at the close of this chapter: the angle of the Moon is "natural and fixed"; it cannot portend anything. People have jumped to this conclusion simply because the season when the Moon is supine at sunset, i.e. spring, tends to have more unsettled weather than the season when the Moon is erect at sunset, autumn. There is a legitimate kind of weather prognostication based on the Moon, but it is based on atmospheric, not astronomical conditions.

---

103 It can, of course, be as much as 5 degrees to the north or south of the ecliptic, but this will not materially affect the phenomenon described here.

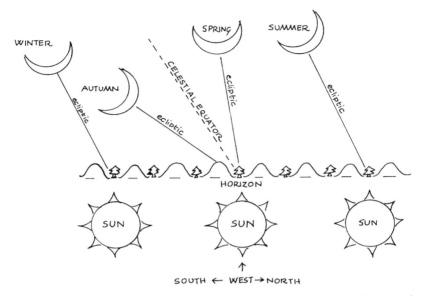

b. Angle and visible segment of crescent Moon at sunset in west, at solstices and equinoxes.[104]

Chapter 26:

Bede is at pains to correct another kind of false impression in ch. 26. A full Moon at midwinter will appear very far to the north – that is, it will be high up in the sky, towards the zenith – while the midwinter Sun remains in the south – low in the sky, close to the horizon. Apparently some of Bede's students confused the *elevation* of the Moon with its *distance* from Earth, as if "higher than the Sun in the sky" meant "above the Sun in the heavens". Bede seeks to eliminate this confusion by an ingenious experiment, using the hanging sanctuary lamps in a darkened church.

Jones claimed that this experiment could not have been actually tried by Bede, as it would require a church of far greater dimensions than any in Anglo-Saxon England.[105] But it hardly makes sense for Bede to offer

---

104 Figures a and b are adapted from Norman Davidson, *Astronomy and the Imagination* (London: Routledge and Kegan Paul, 1985):38, 73.

105 *BOT* 127; Jones does not actually cite the dimensions of Anglo-Saxon churches, nor explain how big a building he thinks would be required for the experiment, and why.

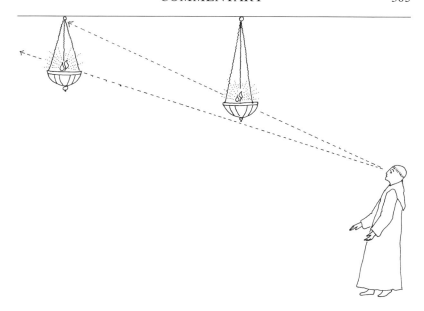

The above illustration may assist in visualizing Bede's experiment. Remember that the lamps themselves cannot be seen, but only the flames, for the church is dark. The nearer flame (the "Moon") will appear to be higher than the "Sun" flame.

such a concrete and circumstantial experiment if it could not be verified by his readers. In fact, the experiment does not require a large room at all. It is undoubtedly more evident in a large room, particularly as one can (as Bede indicates) more readily ignore the vessels and chains, and what they tell us about the true position of the lamps. Such a large space could very well have been available to Bede's readers.[106] But the principle of the relative apparent position of two lights, placed on the same plane as the viewer's advance, but at unequal heights, can be tested in a modern living room with two candles, or can even be successfully

106 Peter Hunter Blair, *The World of Bede*, 271, n. 18, discounts Jones' view that Northumbrian churches were too small for this experiment, though he does not explain why the experiment should work, or indeed, the fact that it can work even in a small room. As Bertram Colgrave observes, Anglo-Saxon churches may have been narrow, but they were also characteristically tall, which of course would suit this experiment perfectly: "Bede and His Times" (Jarrow Lecture 1958):16.

carried out with two lit matches, one held out at arm's length and high up, and the other closer to one's eyes, but lower down. The flame of the lower match can easily be made to appear above the farther, higher flame, as this figure shows.

Chapter 27:

Bede's concern to establish a natural explanation for eclipses is re-inforced by a rather ostentatious array of authorities, both pagan and Christian. Undoubtedly many of Bede's contemporaries in England regarded eclipses as uncanny omens of doom.[107] Nor was it only the uneducated who held these opinions; some within the Church probably thought the "darkness" at the time of Christ's Crucifixion was an eclipse. Hence Bede quotes Jerome to the effect that this darkness was a miracle, and not an eclipse. Christ was crucified at Passover, when the Moon is full and in opposition, and solar eclipses can only occur at conjunction. But Bede the scientist is also interested in what eclipses can reveal about the relative size of Sun, Moon and Earth. His source, Pliny, is unfortunately quite out of line with prevailing ancient opinion, which estimated the size of the Moon at one-sixth to one-half that of Earth.[108] But the fact that Bede saw fit to discuss this implication of eclipses bespeaks the range of his curiosity.

Chapter 28:

No ancient philosopher doubted for a moment that the heavenly bodies exert a natural influence upon life on earth, though they explained the physics of that influence in various ways. Medieval thinkers did not doubt it either, although to a greater degree than their ancient counter-parts they were careful to distinguish this natural or "environmental" influence from astrological determinism.[109] However in this chapter,

107 Cf. *The Reckoning of Time*, ch. 66, *s.a.* 664, and *HE* 3.27.

108 Jones, *BOT* 361.

109 For two balanced and sensible overviews of this subject, see Edward Grant, "Medieval and Renaissance Scholastic Conceptions of the Influence of the Celestial Region on the Terrestrial", *The Journal of Medieval and Renaissance Studies* 17 (1987):1–23 and J.D. North, "Medieval Concepts of Celestial Influence: A Survey", in *Astrology, Science and Society: Historical Essays*, ed. Patrick Curry (Woodbridge, Suffolk: Boydell, 1987):5–17.

Bede is not interested in the question of celestial influence in general. His focus is strictly on the Moon, for even in the midst of this scientific digression, Bede has not lost sight of his essentially computistical project.

As if to apologize for divesting the Moon of its control over the weather, Bede sets out to demonstrate that the Moon exerts a natural influence on growth and decay comparable to the Sun's. This influence is particularly marked in moist environments; as Ambrose observes, while the Sun dries out terrestrial moisture by day, the Moon replenishes it in the form of dew by night. Bede follows up Ambrose's statements with a corroborative passage from Basil, which leads him directly to the subject of the Moon's influence over channels and currents of open water. In short, this chapter is designed to set the stage for the following one, in which Bede will conclusively demonstrate that tides follow the course of the Moon.

Chapter 29:

This chapter on "The Harmony of the Moon and the Sea" has excited considerable admiration for its originality and critical clarity.[110] Here Bede digests, evaluates and offers significant corrections to previous information about tides, specifically the statements by Pliny in the *Historia naturalis*, and the more sophisticated discussions found in Philippus Presbyter and the Irish cosmographies of the seventh century.

Bede's sources were neither clear nor consistent on the subject of tides. The ancient texts came from the Mediterranean region, where the tides are not conspicuous. Only a few, for example, mention that there are two tides every day, or that the tides exhibit semi-monthly variations, which are especially marked at the equinoxes.[111] Most, however, mentioned a correlation between the tides and the phases of the Moon, and an attempt to coordinate tides with an 8-year calendar is reported

---

110 E.g. Wesley Stevens, "Bede's Scientific Achievement", 655–662, and T.R. Eckenrode, "Original Aspects in Venerable Bede's Tidal Theories with Relation to Prior Tidal Observations" (PhD dissertation, St. Louis University, 1970) and "Venerable Bede's Theory of Ocean Tides", *American Benedictine Review* 25 (1974):56–74.

111 In the sixth century Priscian of Lydus, in his *Solutionum ad Chosroem* described tidal action on the coasts of both the western Mediterranean and the Atlantic ocean, including spring and neap tides at conjunction and opposition; however, the only Latin translation of his treatise is rather later (though possibly derived from a Carolingian exemplar), and was not known to Bede: Stevens, "Bede's Scientific Achievement", 656.

by Pliny in *Historia naturalis* 2.99.215. Patristic exegesis, especially the commentary on Job by Philippus Presbyter, a disciple of Jerome, preserved the essence of classical learning on this subject.

Dwellers on the Atlantic coast, on the other hand, were much better situated to observe and calculate the times of tides, and it is in the realm of tide-lore that we find, for the first time, original contributions to Latin natural science coming from peoples and regions outside the confines of the old Roman empire.[112] These contributions were incorporated into two anonymous seventh-century tracts written in Ireland, both of which were known to Bede: the *Liber de mirabilibus sacrae scripturae* of ps.-Augustine, and the *Liber de ordine creaturarum* of ps.-Isidore.[113] To assess Bede's achievement in this chapter, we must examine first what his sources said – and did not say – about the tides.

Job 38.8 says that God "shut up the sea with doors, when it brake forth, as if it had issued out of the womb". This slants the exegesis of Philippus Presbyter towards the problem of the origin and cause of tides, and only incidentally to their periodicity or intensity. Neither he nor any ancient writer understood that tides were produced by the gravitational pull of the Moon, even though they recognized a correlation between lunar phases and the intensity of tides. Philippus knew that tides arrived twice each 24 hours, and was even aware that each tide was about $\frac{3}{4}$ of an hour later than the corresponding tide on the previous day (Bede quotes this passage in ch. 4), and yet he did not connect this to the fact that the Moon rises each day about $\frac{3}{4}$ of an hour later than the previous day. Philippus was more concerned with debates over whether the increase of water is due to an outpouring of waters from some sub-oceanic geyser, "the springs of the great abyss".[114] The alternative theory, explained though not endorsed by ps.-Augustine and ps.-Isidore, was that the actual quantity of ocean water increased under the influence of the Moon – a special case of the kind of power Bede described in ch. 28.[115] Both these theories rest on, or rather, encou-

112 See the illuminating discussion of Irish tide-lore in Smyth 241 *sq.*

113 See Introduction, section 2.

114 Philippus Presbyter, *Commentarium in Job* 38 (PL 26.752D–753A); see also critical edition of this passage by Dom Irénée Fransen, quoted by A.K. Brown, "Bede, a Hisperic Etymology, and Early Sea Poetry", *Mediaeval Studies* 37 (1975):423–424; cf. Smyth 247–248.

115 Cf. *De mirabilibus sacrae scripturae* 2.23 (2184), *De ordine creaturarum* 9.4–6 (148–150); Smyth 244–245.

rage the assumption that tides occur at the same time all over the world.[116]

Neither *De mirabilibus sacrae scripturae* nor *De ordine creaturarum* link the timing of the tides to the passage of the Moon. Indeed, though they both used Philippus, they did not repeat his information about the daily retardation of the tides. Both texts, however, connect the monthly variations in the intensity of tides to the phases of the Moon. They call these phases *ledo* and *malina*.[117] *Ledo* is customarily translated "neap tide" and *malina* "spring tide", but this is not exactly the meaning of the Irish sources. Ps.-Augustine says that each *malina* begins $3\frac{1}{2}$ days before new Moon and full Moon, and each *ledo* $3\frac{1}{2}$ days before the quarters. In short, they cover intervals of a week straddling the spring and neap tides; they denote, then, the peak above, and the trough below, the "average" tide, respectively.[118] Both authors state that *ledones* and *malinae* are characterized by specific durations: *ledones* flood and ebb for six hours, while *malinae* flood for five hours and ebb for seven.[119] *De mirabilibus sacrae scripturae* adds that there are six *malinae* in each season, and that a common lunar year of twelve lunar months has 24 *malinae*, while an embolismic year of thirteen lunar months has 26.[120] Both texts state that the tides at the equinoxes and solstices are especially high.[121]

Bede used all these texts, first in *The Nature of Things* and then in *The Reckoning of Time*, but in the later work he made numerous corrections. The first of these corrections involved clarifying the link between the daily passage of the Moon and the tide. For the first time, he connected lunar retardation to the daily period of tidal retardation, and he corrected the figures for the latter. Philippus Presbyter said that the retar-

116 There were other theories as well, such as the one recorded by Isidore of Seville, *DNR* 40.1, that the Ocean is an animal which breathes in and out.

117 The term *malina* appears in the work of the 4th-century medical writer Marcellus of Bordeaux to designate a phase of the Moon suitable for collecting herbs: *Marcelli de medicamentis liber* 36.49, ed. Max Niedermann and Eduard Leichtenhan, Corpus medicorum latinorum 5 (Berlin: Akademie Verlag, 1968): 612.3. He does not use the term *ledo*. The use of these terms to designate tidal phases was evidently quite current in 7th-century Ireland: see Michael Herren, ed. and trans., *Hisperica famina I–II* [the A-text]:399–411 (Toronto: Pontifical Institute of Mediaeval Studies, 1974–1987):1.94–97; Smyth 255.

118 Smyth 252–253.

119 *De mirabilibus sacrae scripturae* 1.7 (2159); *De ordine creaturarum* 9.4–6 (148–150).

120 1.7 (2159).

121 *Ibid.* 1.7 (2159); *De ordine creaturarum* 9.6 (150).

dation was a *dodrans* or $\frac{3}{4}$ of an hour. In Pliny, Bede found a more precise figure for the daily retardation of the Moon: a *dodrans* plus a *semiuncia*, or $\frac{3}{4}$ plus $\frac{1}{24}$ of an hour, or as we would say, $47\frac{1}{2}$ minutes,[122] and he reported that figure in *The Nature of Things*, ch. 39. In *The Reckoning of Time*, he revised that figure to 4 *puncti* or 48 minutes – still short of the correct figure of 50 minutes, but showing evidence of continuous reflection and research on this subject.

Secondly, Bede offered a new causal model for the tides in place of the "swelling" or "geyser" theories. The link between the passage of the Moon and the timing of the tides suggested to him that the Moon "drags" the waters of the ocean around the world with it. Bede borrows the word *protrahitur* from Ambrose, but gives it a new and literal meaning. He points out that in 59 days (2 synodic months), the Moon actually goes around the Earth 57 times, and there are twice this number of tides. He then scales this down to 29-day and 15-day periods, to show how a new Moon could appear at sunset in the west, and 15 days later appear at sunset in the east. Likewise, the tides, pulled around the Earth by the influence of the Moon, will appear at different times of day, but in a consistent pattern throughout the lunar month. In pointing out the exact correlation between lunar passage and tide, Bede also modifies and corrects ps.-Augustine's account of the number of *malina* tides each year. Since the Moon produces the tides, the number of times it circles the earth will equal the number of tides. In a common year of 354 days, the Moon actually circles the earth 342 times, so there are actually fewer tides than days. Needless to say, ps.-Augustine's statements about the number of tides in each season are based on a confusion of solar cycles (the seasons), with lunar ones; Bede does not allude to this at all.

Thirdly, Bede reproduces the Irish theory of *ledo* and *malina*, but corrects the notion that *ledo* and *malina* last exactly seven days each. Their periods are not equal, nor are they consistent, as strong winds or other forces can modify their course. Fourthly, Bede corrects a misunderstanding about the annual patterns of tides. This is something of a *retractatio* for Bede, because in his own *The Nature of Things*, he had copied the Irish cosmographers' statement that unusually high tides occur at the equinoxes and solstices. Here, he cites Pliny's statement that these exceptional tides occur at the equinoxes only.

122 *HN* 2.10.58 (206).

Finally, by linking the rise and fall of tides to the passage of the Moon around the earth, Bede refuted the assumption supported by the older theories that tides occurred all over the world at the same time. In showing that this was not the case, he enunciated for the first time the idea of a "rule of port". Dwellers along the North Sea coastline know, he says, that tides arrive earlier in some places than others.

How did Bede arrive at these ideas? Wesley Stevens suggests that he enlisted the aid of monastic and clerical friends up and down the coast of Britain as research assistants. Bede mentions the tides at Lindesfarne at *HE* 3.3. It stands to reason that the monks there would have had a good empirical notion of the tide-schedule, since the timing of visits and deliveries to their tidal island would depend on this. Bede certainly knew priests at Lindesfarne, and probably visited there in the course of preparing his versions of the life of St Cuthbert. Stevens also argues, albeit somewhat more speculatively, that Bede gathered information from Whithorn and the Isle of Wight (Bede does mention currents in the Solent at *HE* 9.16), and elsewhere. His intriguing reconstruction of a method for methodically measuring and timing tides over a significant period of time is attractive, but perhaps not totally convincing, given the demands of Bede's monastic schedule.[123]

So far, what Bede has presented is remarkable for its critical acumen and originality, but withal in harmony with his overall vision of nature as a sublimely regular system of laws, laws which are universal and reducible to physical and mathematical explanations. One has but to glance again at ch. 25 to understand afresh what he means by *naturalis ordo*. It is a vision of nature entirely consonant with Bede's aim as a computist, which was to submit time itself to a universal and mathematical *regula*. It is therefore all the more remarkable – and does credit to Bede's intellectual honesty – that his chapter on tides refuses to stop at this regularity, but rather, admits that no general theory can totally account for tidal phenomena.[124] The tides arrive sooner in the north than in the south, but Bede also recognized that every region had a *regula societatis* of the Moon and the tides, what we would now call a "rule of port". All of these rules might be upset by chance conditions such as weather. In short, though the tides are driven by the Moon, whose sway is exercised over all the Earth, their ultimate expression is eccentrically local. That a

123 Stevens, "Bede's Scientific Achievement", 657–660.
124 Stevens 662.

man of Bede's systematic mind, so passionately concerned with *regular-itas* and uniformity, should conduct such researches and express the results in this way is testimony to his critical spirit and freedom from dogmatism. The Irish cosmographers were likewise enchanted with the order of the universe, but that spell made them complacent about the information they found in their sources, particularly about the supposed simultaneity of tides and their precisely mathematical schedule.[125] Perhaps, however, Bede's vision was not so much at odds with his world-view as might at first seem to be the case. The unity of the Church never meant for him a mechanical uniformity of custom or observance, though variations which endangered community (such as divergent Easter reckoning) could not be tolerated. But every monastery had its particular customs and local ways of expressing the universal vision of the catholic Church, just as every place on the coast of England had its own tidal "rule", established at Creation, and which it follows faithfully every since. Unity of obedience could coexist with diversity of practice.

Chapter 30:

The final section of the first part of *The Reckoning of Time* (chs. 30–42) steps back to consider the solar year as a whole. The Moon names, shapes, and pervades the month; but the year is the Sun's domain. Its annual tour up and down the heavens defines the length of the year, and marks off its major phases, the seasons. Unlike lunar phases, these solar vicissitudes are slow and inconspicuous; it is easier to mark them by monitoring shadows cast by a pole stuck into the earth than by looking up at the Sun itself. Moreover, if the Moon's journey shifts the waters all over the globe, the Sun's advance and retreat is best read in the land – the seasons, and the latitudes which feel the Sun's power at different times and to different degrees. However, the natural course of both the Moon and the Sun is not readily reduced to the full days with which computists prefer to work. The leap year and the *saltus lunae* are devices to bring the conventional calendars of human society back into some kind of harmony with the transcendent calendar of the two great luminaries.

Much of the power and beauty of *The Reckoning of Time* as a theological vision of time – and indeed, much of the impulse medieval people

125 Smyth 262.

felt to study *computus*, even when there were no more problems left for it to solve – rests on Bede's artistry. His picture of the heavens is not only orderly, comprehensible and convincing, but full of life and meaning. The harmonious contrast of the swift, changeable, conspicuous course of the Moon to the stately, unvarying but more inscrutable motion of the Sun, suggests a dynamic balance between stability and change, swifter and slower rhythms of time.

The first subsection of this discussion of the solar year concerns the seasons (chs. 30–35). It is dominated by the theme of the distinction between natural time and calendar time. The seasons do not coincide with the divisions of the Julian solar calendar, which as Bede has been at pains to show, are to a large degree the result of Roman convention. On the other hand, they are the terrestrial manifestations of the Sun's annual journey north and south along the ecliptic. They are, therefore, linked to the solstices and equinoxes.

The section on the year therefore opens with this chapter on the solstices and equinoxes. It immediately lands Bede in the midst of a potentially troublesome problem, which he resolves in a manner both typical and instructive. At the time of the inception of the Julian calendar, the spring equinox fell on 25 March, the autumn equinox on 24 September, and the two solstices on 24 June and 25 December respectively. Much classical science and Christian symbolism came to be attached to those dates on the understanding that they represented the true dates of the equinoxes and solstices. Bede himself, at the time he wrote *On Times* ch. 7, still thought of them as such, and he opens ch. 30 of *The Reckoning of Time* by citing some ancient scientific and medical authorities who match the seasons to these dates. The ps.-Hippocratic citation is particularly interesting for what it reveals about Bede's medical reading,[126] but he may also have selected this passage because it highlights the relationship between the solstices and equinoxes on the one hand, and the seasons on the other; it also points out how the seasons vary in length. Bede then goes on to observe that the symbolism of many Christian feasts – including Christmas, coinciding with the old winter solstice – is based on an allegorization of the solstices and equinoxes.

126 Cf. M.L. Cameron, *Anglo-Saxon Medicine*, Cambridge Studies in Anglo-Saxon England 7 (Cambridge: Cambridge University Press, 1993): *passim*. Another side of Bede's medical interests will be explored in the context of ch. 35 below.

However, while avoiding any disparagement of either ancient learning or ecclesiastical traditions, Bede must assert that these conventional dates are not the true dates of the solstices and equinoxes. He invokes both scientific authority (the astronomical expertise of the Egyptians, since this is an Alexandrian doctrine) and ecclesiastical sanction, namely the decrees of the Council of Nicaea. According to both, the spring equinox must be on 21 March.[127] Though Bede claims that this date is supported by the evidence of the sundial (*horologica consideratione*) as well as conciliar authority, it is doubtful that he understood *why* the Alexandrians had moved the equinox back to 21 March.[128]

What is noteworthy about this chapter, however, is what it reveals about Bede's attitude toward fact and custom in matters relating to the calendar. Why does it *not* matter that the Lord's Nativity is not celebrated at the true winter solstice,[129] while the true spring equinox *does* matter in relation to his Resurrection? Though Bede does not pose this question, he does furnish its answer: the Mosaic Law (in his view) forbids the celebration of Passover before the equinox. Therefore the date of the equinox must be established with as much precision as possible. But the Bible furnishes only a general indication of when Jesus was born, and no injunctions are laid down as to when this event should be celebrated. The Church has assigned the Nativity to the conventional calendar date of the solstice for symbolic reasons, but no Biblical principle or precept is jeopardized if the date is off by a few days. Moreover, as far as Bede is concerned, there is only one date where the distinction between scientific and conventional dating of the equinoxes and solstices is significant, and that is the spring equinox, because of Easter. He is fairly casual about the others, merely pointing out that the true date falls before the conventional one by the same interval as applies in the case of the spring equinox.[130]

127 Anatolius, however, states that the equinox is on 22 March: Bede's solution to this problem is contained in the *Letter to Wicthed* (see Appendix 3.3).

128 See Introduction, section 3, and remarks on Bede's use of sundials in the Commentary to ch. 31.

129 Cf. Bede, *In Lucam* 1.26, ed. D. Hurst, CCSL 120 (1960):439–454, where Bede adds that it does not matter if Christ was born a little *after* the true solstice, because John's preaching of the Kingdom before Christ's official ministry began, constitutes the beginning of that ministry.

130 For the exception which proves this rule, see *Letter to Wicthed* 6, Appendix 3.3.

Whether dealing with tidal anomalies, dates for the solstices and equinoxes, or the discrepancy between the natural and computed age of the Moon, Bede is sensitive to the distinction between the idealized regularity of the calendar, the truth of nature, and the compromises of human society. Within certain limits, these divergences can be tolerated without scandal. But as Bede will later demonstrate, the religious significance of *computus* lies in its quasi-liturgical character as a model of heavenly regularity; if it contrasts with, instead of conforms to, the realities of nature, this may have its theological advantages. Through most of the Middle Ages, this sense that the calendar *had to* depart from nature in order to proclaim the triumph of eternity was understood; only in the waning centuries of that period did this anomaly become intolerable, and the call go out to reform the calendar to conform to nature, even if it meant the loss of its cyclical quality, and the symbolic values attached thereto.[131]

Chapter 31:

There is no disputing that the days are longer in summer than in winter, but one can imagine Bede's students asking him how one could be sure that today's daylight is longer or shorter than yesterday's. In other words, how would you actually know that the solstice had come? To answer questions like these, Bede composed four chapters (31–34) on the measurement of sunlight by sundial: they cover how the sundial registers time, and why the shadow of the gnomon on the same day is longer in some places than others.

Bede's familiarity with the sundial has been the subject of some controversy. He clearly invokes the evidence of the sundial in favour of revising the traditional date of the equinox in ch. 30, and in the *Letter to Wicthed*; in the letter of Abbot Ceolfrith to King Nechtan of the Picts in *HE* 5.21 – a letter commonly assumed to have been written by Bede himself – it is claimed that "we can confirm" that 21 March is the equinox "by consulting a sundial".[132] Bede does not describe in detail what this observation entailed, but it can be reconstructed fairly easily. Ancient and medieval sundials were circular or semicircular devices, usually mounted vertically on a wall, with a gnomon perpendicular to

---

131 Cf. Zélinsky, "Le calendrier".
132 *ut etiam ipsi horologica inspectione probamus*: *HE* 5.21 (542).

the support. The dial was marked like a fan with eight or twelve divisions, representing the hours.[133] As the solstices are relatively easy to observe (the Sun seems to "stand still" for several days at solstice) the outer edge of the semicircle could be marked with a line corresponding to the sweep of the gnomon's shadow on the longest day of the year; a smaller semicircle incised much closer to the gnomon would mark its path on the shortest day of the year. A line halfway between these two would mark the shadow's path at equinox.

But Jones argues that even had Bede made such an observation, it could not have produced the stated result, for the spring equinox in his day was falling around 17/18 March; if it did, it was because the sundials used in Anglo-Saxon England were very crude.[134] Harrison and Wesley Stevens, on the other hand, argue that Bede meant what he said. But even had he used a well-calibrated sundial, or improvised a more sophisticated experiment of his own – as he very well might have done, given that Pliny (*HN* 2.69) describes a way of confirming the equinox by means of "sights" (*dioptrae*), i.e. stakes or poles aligned with the rising and setting sun – the problems of local terrain and visibility would still have permitted a margin of error of more than one day. Hence Bede could at least prove that the old Roman equinox was not correct, even had his equipment and expertise not been sufficient to prove that the Alexandrian one was.[135] To put it another way, the sundial proved that the old Roman sources followed by the Celtic computists were wrong; therefore the Alexandrians authorities promoted by Bede must be right.

In short, Bede was familiar with sundials[136] and expected his readers to be as well, but there were limits to both the effectiveness of these instruments, and Bede's own motivation to press his researches much further. Nonetheless it is interesting to see how here, in the *Letter to Wicthed*, and in the letter of Ceolfrith to Nechtan, Bede insists that the sundial provides evident proof that so many ancient authorities were wrong about the equinox. Of course, he had other authorities – the Alexandrian

133 On Anglo-Saxon sundials, see Arthur Robert Green, *Sundials, Incised Dials or Mass-Clocks* (London: SPCK, 1926): ch. 2, and "Anglo-Saxon Sundials", *Antiquaries Journal* 8 (1928):489–516.

134 *BOT* 87.

135 Harrison, "Easter Cycles and the Equinox in the British Isles"; Stevens, "Bede's Scientific Achievement", 16 and nn. 48–49.

136 He alludes to them, for instance, in his *In Regum librum XXX quaestiones*, ed. D. Hurst, CCSL 119 (1962):316; cf. Harrison, "Easter Cycles", 5, n. 3.

computists – whom he was anxious to prove more reliable than Pliny or Isidore. But the implication that the contest of authorities can be settled by observations carried out with an instrument is unquestionably a bold one, and worthy of the name of science.

In chapter 31, Bede follows his policy of parallel quotations from classical and Christian writers to demonstrate that even though the length of the equinoctial day is the same everywhere on the globe, the shadow cast at noon will be much shorter in the tropics, where the sun is overhead, than in the north, where it is always to the south of the zenith. As Pliny points out, this means that sundials have to be calibrated to latitude. Moreover, on any day other than equinox, the length of the day itself will vary with latitude. The further north one goes, the longer the days will be at summer solstice, and the shorter they will be at midwinter.

Chapter 32:

In the previous chapter, Bede explained that the further north one is, the longer are the summer days. Here he explains why. The Earth is a sphere set in the middle of the universe. Once each day, the Sun revolves around the Earth from east to west. But from day to day, this orbit will shift southwards or northwards, depending on the time of year. Latitude determines length of day because the winter Sun in the southern sky is "blocked by the mass of the Earth" just as certain southern constellations cannot be seen by the inhabitants of a village on the northern slopes of a large mountain. But all locations at the same latitude (or as Bede puts it, "all who are placed facing one another in the same line under a northern or southern clime") will experience the same amount of sunlight on any given day.

Chapter 33:

All this is very well (we may imagine Bede's students replying), but how do you know what places "face one another" on the same parallel of latitude? Bede's response in ch. 33 is a long quotation from Pliny, showing how sundial readings at solstice, gathered from various locations, could be used to determine which places lie along the same latitude.[137]

---

137 Pliny named cities and prominent locations within the Roman Empire, and tried to assemble data about the hours of sunlight at solstice in each, to establish their relative latitude. Bede used some of this material in *The Nature of Things* 47–48. It is not, by our

Chapter 34:

Hitherto Bede's students have been looking downwards, so to speak, at the shadow cast by the Sun in a particular location. In ch. 34, he asks them to do something different: to stand outside the universe in their imagination, and visualize *why* the diurnal and annual motions of the universe produce these patterns of light and darkness on Earth, from east to west during the day, and from south to north to south again through the year. As Jones observes, this exposition of the world-zones and their climates is a very original and sophisticated piece of scientific writing for the early Middle Ages.[138] But what is particularly remarkable is the concrete, homely example with which Bede brings his students literally back down to earth again at the end. Imagine a rectangular open-air firepit on a cold night, with people sitting on either side. Nobody can sit right in the middle of the fire, because it is too hot; on the other hand, if you are too far away, you will freeze. If you are close, but not too close to the fire, you will be comfortable, no matter which side of the hearth you are on. Now imagine the rectangular firepit wrapped around the earth like a belt.[139] The two "comfort zones" represent the two habitable areas to the north and south of the equator, while the distant reaches stand for the Arctic and Antarctic polar zones. Both "comfort zones" are *habitable*, but Bede denies that both are actually *inhabited*. Pliny, after all, says that no one can cross the torrid zone. The missing minor premiss here is that mankind is descended from a single person, Adam, who began life in the northern hemisphere; hence none of his progeny could have ever reached the south. Note that the issue for Bede is not the possibility of life in the southern hemisphere, but the impossibility of travel to the southern hemisphere through the blazing equitorial zone. This means that if people lived there, the message of

---

standards, completely accurate: Meroe and Syene are too far south, the size of Sicily has been considerably inflated, and Narbonne is too far north: cf. Wesley Stevens, "Bede's Scientific Achievement", 652–653.

138 *BOT* 367–8.

139 This image suggests that Bede may have seen a zone-map, like the one illustrating Macrobius' *Commentary on the Dream of Scipio*. This showed the five climatic zones laid out across a disk-like earth. Such a map was sometimes imported into *The Reckoning of Time* as an illustration: see J.B. Harley and David Woodward, eds. *The History of Cartography. Volume 1: Cartography in Prehistoric, Ancient and Medieval Europe and the Mediterranean* (Chicago: University of Chicago Press, 1987):300–303.

salvation could never reach them, and there would be human beings for whom Christ had died in vain.[140]

Chapter 35:

Chapter 35, on the analogy of the four seasons, elements and humours, links the preceding chapters on the variable length of days to the following section on the year. But it is a linking chapter in other ways as well. At the exact mid-point of *The Reckoning of Time*, Bede pauses to reflect on the way in which time joins man to the universe. This connection is mediated by number, for the elements which compose the physical universe, the humours which constitute the human body, and the seasons of the natural year, are all four in number. The connection is also grounded in an ancient theory explaining the coincidence of stability and change in the cosmos. Four qualities – hot, cold, wet and dry – combine in pairs to make the elements and the humours; at each of the four seasons, one of these pairs dominates the world. Time, then, makes change rhythmic, not chaotic. This *syzygia* of *mundus*, *annus* and *homo* is derived from Isidore of Seville, *De natura rerum* 7, where it is illustrated by a much-copied and much-imitated diagram. Bede's language in describing the connection of the elements one to another suggests that he had seen the Isidorean diagram: he speaks, for instance, of each element "embracing what is on either side of it", and of the ensemble of elements looking like a sphere.[141]

The general scheme was a commonplace, but the exact dates when the seasons began was matter for debate. Bede's argument about these dates is a complex one, and since it has engendered some misunderstanding, it seems worthwhile to dissect it in detail.

Bede arranges the various opinions on this issue in an ascending hierarchy of authority. First, he reports the dates found in Isidore of Seville, but he plainly thinks little of them. Isidore's dates are certainly arbitrary

---

140 The patristic *locus classicus* is Augustine, *DCD* 16.9. The question was of considerable interest to Bede's Irish forerunners, who were sometimes a little uncertain as to whether the antipodeans dwelt on the far side of a spherical earth, or simply underground. The outrage expressed by Boniface at the Irish Bishop Virgil of Salzburg's suggestion that people might live "below the earth" shows how shocking the religious implications of the doctrine were felt to be, at least by some: see Smyth 285–290; J. Carey, "Ireland and the Antipodes: the Heterodoxy of Virgil of Salzburg", *Speculum* 64 (1989):1–10.

141 See ch. 35, n. 316.

and without scientific foundation, but Bede's criterion for dismissing them is that they are *provincial*, a local Spanish reckoning. This is not exactly the case, but the fact that Bede makes this point is, as we shall shortly demonstrate, of some significance.

The Greek and Roman dates rate higher in his estimation, and Bede does take note of their astronomical foundation. However, he passes over these dates as well, though without comment. The reason for doing so, however, can be deduced from what follows. Bede, as we shall see, has a special interest, for computistical reasons, in determining the beginning of *spring*. The Greco-Roman system furnishes clear criteria for the beginning of summer and winter (the heliacal rising and setting of the Pleiades), but spring and summer are only vaguely marked by their rising and setting "about midnight".

Ranking very much higher in Bede's estimation is the method of ps.-Anatolius, which takes the solstices and equinoxes as the mid-points of the seasons. There is a serious problem here, however. The passage from ps.-Anatolius does not actually date the spring equinox, but the dates provided for the autumn equinox and the solstices indicate that he was using the old Roman dates, and that his spring equinox would be on 25 March. Bede is once again using the Celtic recension of this text. If spring has 90 days (cf. ch. 33) and the mid-point is 25 March, it will begin on 8 February. Most medieval *computus* texts which date the seasons according to this method use either 8 February or 6 February as the beginning of spring.

Jones states that Bede approved, adopted, and popularized this system, but "moved the seasons back from v id. [9 February] to vii id. [7 February]" to adjust for the 21 March equinox.[142] But careful reading of this chapter will reveal that Bede never explicitly states when the seasons actually begin. Moreover, moving the beginning of spring from 9 February to 7 February would by no means adjust for the four-day interval between 25 March and 21 March; spring, using Anatolius' system but Bede's equinox, would come on 5 February.

Bede adds yet another rung to his hierarchy of authorities on the dates of the seasons: the Bible. The quotation from Deuteronomy indicates that spring begins in the Passover month; this is a lunar month, but determined by the equinox. The most orthodox computistical autho-

142 *BOT* 369.

rities, the Egyptians (represented here by Proterius), also use this criterion.

Bede in fact at no point fixes calendar dates for the beginnings of the seasons. Rather, he sets out in ascending order of authority the various schemes of which he knew, ending with the Hebrew/Egyptian method. Given Bede's reverence for both biblical authority and Egyptian computistical expertise, this is clearly the formula of choice. It is also the only one of any computistical consequence, and as we have seen earlier, Bede's interest in taking sides on these contentious questions is only roused by computistical consequences. Moreover, even if he did think Anatolius' system plausible, to have committed himself to it would have confused his students, since Anatolius used Roman dates for the equinoxes, and his spring began before the earliest acceptable calendar date for the Passover month, Nisan.

Finally, Bede makes the interesting observation that the etymologies of the Latin names for the seasons only make sense in terms of the climate Latin-speaking peoples know. Unlike solstices and equinoxes, which are astronomical phenomena, seasons (like hours of daylight) are local conditions, different in India, Egypt and England. They are sublunary and terrestrial, and hence widely variable and inconsistent. This may be another reason why Bede refrained from any dogmatic statement about their starting dates.

Chapter 36:

Chapters 36–43 of *The Reckoning of Time*, devoted to the year, are also dominated by the theme of the distinction between natural and computed time. The "year" which the student sees laid out before him in the *annalis libellus* is not precisely equivalent to the natural "year", for it is six hours shorter than the time the Sun actually takes to make a complete circuit of the zodiac. Since calendars have to the constructed out of whole days, the six extra hours can only be "counted" when they have cumulated into a full day of 24 hours; this happens every four years. For somewhat analogous reasons, the calculated Moon will occasionally not quite match the visible Moon in the sky.

This distinction was an important one for Bede, and one which he was at pains to impress upon his students. Its implications are complex, for though time-reckoning was an activity steeped in theological significance, the Julian calendar itself was only a human convention, one of the many calendars employed by mankind through the centuries, and in fact, a pagan invention. God's time was the time marked by the Sun and Moon themselves, the exemplar-time which good calendars strove to replicate as closely as possible. Though all calendars ultimately had to sacrifice scientific perfection to the human need for simplicity and convenience, there is a sort of *felix culpa* here: the mathematical cycles by which the calendar compensated for its loss of natural accuracy transformed it into a symbol of eternity.

Significantly, Bede begins ch. 36 by destabilizing his readers' notions of what the word "year" means.[143] *Annus* denotes merely a "circuit". Therefore the length of the "year" depends on what is making the circuit: the Sun, the Moon, the planets, or even all of the above. When applied to the Moon, this "circuit" can have no fewer than four meanings: a siderial lunar month, a synodic lunar month, a "common" year of 12 lunar months, or an embolismic year of 13. Where the latter two begin and end also varies, depending on whether one is following the "Hebrew" or Roman system. The Sun's year is the length of time required for it to return to the same point in the zodiac, and planetary years are defined in the same way. A Great Year is the length of time it takes for all the planets to resume the same position in relation to one another. It is interesting to note that this definition is rather vague. Bede seems not to have known much about the ancient debates concerning the length of the Great Year, the nature of the grand planetary conjunction, the notion that all the events of previous Great Years would be repeated, or the theories of alternating floods and conflagrations. This information was conveyed to the Middle Ages largely by Macrobius' *Commentary on the Dream of Scipio*,[144] a book which Bede apparently did not know. Hence the only clue that his library yielded as to the length of the Great Year was the quotation from Josephus. Josephus' "Great Year" is actually the Babylonian *naros*, a cycle designed to

---

143 Jones speculates (*BOT* 371) that Bede's source for this chapter was a short tract in his "Irish computus", whose incipit reads *Quot modis soleat annus nominari. Primus modus est de luna . . . ("The number of ways in which one can designate a year. The first way is according to the Moon . . .")*: see Introduction, section 5. It survives in a number of Carolingian computus manuscripts, e.g. Berlin Staatsbibliothek 128, fol. 117r; Besançon 186, fol. 40r; Geneva 50, fols. 165v–166r; Rouen 26, fol. 150; Paris, B.N. lat. 2341, fol. 6r; Paris, B.N. nouv. acq. lat. 1613, fol. 3v; and Munich, CLM 210, fols. 72v–73r. Analogous passages appear in the "Bobbio *computus*", PL 129.1306.

144 2.11.8–17, ed. J. Willis (Stuttgart and Leipzig: Teubner, 1994):128.31–129.20.

predict eclipses. It is far too short a period to bring all the planets together in the fourth degree of Aries, which is the definition of the grand conjunction Bede proposes in ch. 6.[145]

Chapter 37:

Not only do natural years vary in length, but so the conventional years of human time-reckoning. Bede illustrates this fact in ch. 37 with an apt quotation from St Augustine. Augustine's point is that ancient pagan peoples used the term "year" to denote a number of different measures of time. Anti-Christian polemicists exploit this fact to cast doubt on the Old Testament's statements about the lifespans of the patriarchs. What Bede omits, and presumably was confident his readers would understand, is that Augustine goes on to argue that the "years" of Genesis are true years, and not synonyms for shorter periods of time.

Chapter 38:

Leap-year day was a subject of perennial and seemingly inexhaustable interest for computists, both in Bede's day and later. One 11th-century anthology, Vatican Reg. lat. 123, contains no fewer than 14 tracts on leap year, and it is far from unique. Chapters 38 and 39 of *The Reckoning of Time* in fact were originally composed as a separate treatise on leap year in the form of a letter to a certain Helmwald.[146] It was in turn excerpted in whole or in part into later *computus* anthologies.

Bede's goal in these two chapters is to explain leap-year day without overcomplicating the issue. For the sake of students with little scientific background or mathematical aptitude, he omits elaborate calculations of the leap-year increment,[147] and focusses on more basic issues: why an extra day accumulates over four years, and why it is important to intercalate it accurately.

---

145 Godefroid de Callatäy, *Annus platonicus: A Study of World Cycles in Greek, Latin and Arabic Sources* (Louvain-la-Neuve: Université Catholique de Louvain, Institut Orientaliste, 1996):98–99. Bede's remarks on the Great Year elaborate on his statements in *On Times* 9. The Macrobian tradition about the Great Year was certainly known to Carolingian computists, e.g. Paris, B.N. nouv. acq. lat. 1615, fols. 166v–167v: cf. Jones, *BOT* 371.

146 The part of the letter which is not reproduced in *The Reckoning of Time*, is translated in Appendix 3.2.

147 Bede reserves some of these more subtle details for his *Letter to Wicthed* 7 (see Appendix 3.3).

Chapter 39:

Here is a concrete illustration of why Bede wanted to keep his exposition of the leap-year day as simple and clear as possible. In Bede's day, there was a misconception abroad that the bissexile increment was 3 hours per year. Apparently, many thought that the "quarter-day" was one-fourth of a conventional day of 12 hours, rather than of a full day of 24 hours.[148] Typical of the numerology used to justify this "fact" is an argument found in ch. 34 of the "Bobbio *computus*". The world was created in 7 days, but the Bible does not say that the seventh day had an evening. This is why the year is a little longer than expected. 365 days × 24 hours = 8,760 hours in the year. 8,760 divided by the 7 days of the week = 1,251, with a remainder of three; these three hours were the annual leap-year increment.[149]

That this error was so deeply entrenched, and furnished with arguments such as these, explains both Bede's monotonous insistence that the true length of the annual increment is six hours, and his production of a lengthy quotation from Augustine's *De Trinitate*, which discusses the leap-year increment in the context of the symbolic perfection of the number six.

Chapter 40:

Since much of Augustine's disquisition on the number six in relation to leap year alludes to the day on which the extra day is intercalated – the

---

148 Jones, *BOT* 372–374, argued that this was an "Irish" error which arose through a confusion between the three *unciae* accumulated every year by the Sun ($\frac{3}{12}$ or $\frac{1}{4}$ of a day), and three *hours*. The perpetrator, in his view, was an addendum, which he identifies as Celtic in origin, to the *Prologus Cyrilli* (see Introduction, p. lxxiv). The Celtic character of this addendum is proven by elimination. Since the text invokes *unciae* or duodecimal fractions, which were of no computistical use to the proponents of the 19-year cycle, it must have come from an 84-year cycle environment. The Irish are certainly not immune to this mistake, for it appears in MS Vatican Reg. lat. 39, as an appendix to the spurious *Acts of the Synod of Caesarea* (ed. Wilmart, 27), a "Celtic forgery" (see Introduction, section 5); it also surfaces in the "Bobbio *computus*" (see n. 149 below). However, Krusch found this identical error in some texts rather remote from Celtic sphere, e.g. Merovingian Gaul ("Das älteste fränkische Lehrbuch der dionysianischen Zeitrechnung", *Mélanges offerts à Emile Châtelain* (Paris: Champion, 1910): 235), and a Victorian *computus* dating from 727 found in MS Bern 611 (ed. Krusch, *Studien II*, 55–56). Indeed, the same mistake is found in ps.-Cyprian, *De pascha computo* 19, so it must be much older.

149 "Bobbio *computus*", ch. 43 (PL 129.1296C). Cf. Jones, *BOT* 374.

6th kalends of March (24 February) – Bede was probably asked by his students to explain why this date was chosen. Macrobius' *Saturnalia* furnishes the answer, but one senses that Bede is uncomfortable with this account, since the reason for the choice was pure pagan superstition. Hence he follows up his quotation with a somewhat moralizing tirade. This effort to distance himself abruptly from the pagan content of Macrobius might be compared with Bede's remarks on February in ch. 12.

Chapter 41:

This is a very original explanation of a subject which had received no recorded notice before Bede's time: the effect of the leap-year day on the calculation of the age of the Moon. The leap-year day may not appear as a marked place in the Julian calendar (the way 29 February is a marked place in our calendars) but it is a real day in the real world, and its effect is to make the Moon one day older that it was before the intercalation. This will affect the date of the Easter full Moon.

But it is the final, somewhat cryptic paragraph in this chapter which is the most intriguing. Although he does not enter into the details, Bede is acknowledging here that thanks to the leap years, the 19-year cycle is not exactly cyclic. The 19-year lunar cycle is incommensurable with a fourth-year leap-year intercalation. Each 19-year cycle will accumulate 4.75 leap-year days for both the Sun and Moon (.25 × 19), which means that cycle will not contain a whole number of days. Therefore the lunar phases may be slightly off their computed calendar dates from one 19-year cycle to the next.

The total number of lunar and solar days in a 19-year cycle is actually 6,939.75, computed as follows:

Solar years:	
19 solar years	6,939.75 days (365.25 × 19)
Lunar years:	
12 common years	4,248 days (354 × 12)
7 embolismic years	2,688 days (384 × 7)
Bissextile lunar days	4.75 days (.25 × 19)
Total lunar days	6,940.75
Subtract *saltus*	1
	6,939.75

So from a computistical perspective some 19-year cycles will be "hollow" and contain 6,939 lunar days, while others will be "full" and contain 6,940. The variation will be regularized over the course of the great Paschal cycle of 532 years, which is the product of the 19 years of the lunar cycle, and the 28 years it takes to move the leap-year days through each of the days of the week.[150] This means that the computist must be careful to track the course of the Moon each leap year, so that he can ascertain its correct age on the critical date of the spring equinox.

Computists before Bede hardly noticed the problem, because they calculated the age of the Moon with annual epacts and lunar regulars. It only surfaces when the issue becomes the establishing and demonstration of a *cycle*; indeed, the unique pre-Bedan allusion comes from a critique of Victorius of Aquitaine's cycle by a Spanish writer named Leo.[151] It had little impact on the *computus* as a whole, but it illustrates how profoundly Bede had grasped the problem of constructing a luni-solar cycle.

Chapter 42:

This chapter and the next concern anomalies of lunar reckoning. They constitute a kind of appendix to the section of *The Reckoning of Time* devoted to the solar calendar, and serve as a bridge to the Paschal table.

In plotting the phases of the Moon against the grid of the solar calendar, anomalies are bound to occur due to the fundamental incommensurability of lunar and solar cycles. Over 19 solar years, the calculated Moon will gradually build up an advance of one day, which will have to be deducted at some stage of the cycle to bring the cycle back into phase with reality. Thus the Moon will have to "jump over" one day. But the issue of this *saltus lunae* also raises broader questions about natural and calculated time – between what things *are* and how they are *named*. For in fact, Bede was aware that calendrical discrepancies were occuring fairly often, and not only in connection with the *saltus*.

As Bede points out in the opening sentences of ch. 42, the lunar *saltus* is the leap-year problem turned on its head. In the latter case, the calendar has to "slow down" every four years by adding a day to "let the Sun catch up"; in the case of the Moon, the planet starts to run ahead of the calendar, and eventually "jumps over" a day in the count

---

150 *BOT* 374.
151 Ed. Krusch, *Studien I*, 299. Cf. Jones, *BOT* 374–375.

of lunar days. Bede admits that it is not easy to see this happening, since the advance amounts to only one day over 19 years; but it is not hard to measure it.

Bede's calculation of the annual increment which eventually produces the *saltus* is a silent correction of his own earlier error in *On Times* 12, which was in turn a critique of a persistent misconception, namely that the *saltus* accumulated at the rate of $\frac{1}{12}$ of a day per year. In *On Times*, Bede rendered the *mensura* as 1 hour, 10 *momenta*, $\frac{1}{2}$ of a *momentum*, and $\frac{1}{19}$ of $\frac{1}{2}$ of a *momentum*. This figure is corrected in *The Reckoning of Time*, and the statement simplified by grouping the 10 *momenta* into one *punctus*.[152]

But where should the "leap" take place? Bede argues that it should properly take place at its "natural" time, which is the spring equinox. What Bede means by "natural" is the will of God as expressed in the act of Creation.[153] Since God created the Moon and Sun at the spring equinox (cf. ch. 6), then their cycles begin and end at that point, and it would be logical to correct the lunar anomaly there before starting a fresh cycle. But Bede acknowledges that the conventions of the calendar tend to outweigh the claims of "nature", and that computists prefer to insert the *saltus* at the end of the calendar year – November in the case of the Roman calendar, July for the Egyptians – so that the anomaly can be absorbed by the final lunation of the year.

Bede himself declares no rule, and does not tell us which system he uses,[154] though he seems to favour the practice of inserting the *saltus* at

152 Cf. Jones, *BOT* 377. The sources for chapter 42 as a whole are fairly obscure, and Jones lists a number of anonymous *saltus* tracts which Bede might have used. The most important is the one whose incipit is *De saltu lunae pauca dicimus* (i.e. ps.-Alcuin, *De saltu lunae* 1, PL 101.984–989), and which appears in the "Sirmond manuscript" and its cognates; Jones argues that it is a displaced chapter of the Irish *computus* textbook found in this MS: *BOT* 375–6. Bede's approach to the calculation of the *saltus* increment resembles somewhat that found in ps.-Columbanus, *De saltu lunae*, ed. Walker, *Sancti Columbani opera* 212–214.

153 Cf. Augustine, *DCD* 21.8 (773.104 *sq.*) where he argues that portents are not "contrary to nature" because they happen by the will of God, which *is* "the nature of every created thing".

154 Dionysius himself laid down no rule for positioning the *saltus*. Ps.-Alcuin, *De saltu lunae* (see n. 152 above) lists several possibilities: Victorius and the "Latins" place it on 25 November, the Greeks on 22 March, the Egyptians on 26 September, and Dionysius on 17 April: cf. van Wijk 19–22. It is little wonder that Bede, normally quite clear and decisive on matters computistical, refused to declare his position on the *saltus*.

the very end of the 19-year cycle, or rather, at the beginning of the new cycle, on 21 March, so that the epacts on 22 March change from 18 (year 19) to 0 (year 1), and not to 29. This is why Bede refers to the *saltus* as happening in the *first* year of the 19-year cycle. Normally the "year" in this sense means the period between the Easter new Moon of year 19, and the Easter new Moon of year 1 (cf. ch. 45), but in this case it means "on 22 March" since the new epact of year 1 comes into effect on that date. This is also why he claims that the *saltus* produces a lunar year of 353 days; year 19 of the cycle is embolismic, and has 384 days, but year 1 is common.[155]

Chapter 43:

Chapter 43, "Why the Moon sometimes appears older than its computed age", is one of the most complex and revealing passages in *The Reckoning of Time*. It shows Bede attempting to account for a discrepancy in the computistical system made evident by the direct observation of nature. The explanation, although purely computistical, cannot win assent unless it is shored up by authority. But the authorities are not as convincing as they might be.

Bede admits from the outset that the observed Moon in the sky is sometimes older than its computed age. There are four reasons for this. First, the gradually accruing *saltus* increment will mean that by the end of the 19-year cycle, a discrepancy of about a day will be fairly normal. Secondly, we may be fooled into thinking the Moon is older than its calculated day if we mistakenly count a crescent Moon appearing *just*

---

155 The question of the *saltus* seems to have been a matter of some confusion in Bede's day: cf. the rather obscure riddle by "Eusebius" (i.e. Bede's abbot Hwætbert) on this subject. *De aetate et saltu. Rite uicenis cum quadragies octies una/ Quaeque sororum formatur de more mearum/ Nempe momentis; tunc ego sola peracta uidebor,/ Cicli nondecimus cum deficit extimus annus.* ("*Of the age [of the Moon] and the leap [of the Moon].* According to rite, of exactly twenty times forty-eight / *momenta* each one of my sisters is habitually / formed: but I, the unique, apparently [have already] run my course / the moment the cycle's nineteenth and last year ends.") Ed. and trans. Maria Di Marco, *Variae Collectiones Aenigmatum Merovingicae Aetatis*, CCSL 133 (1968):239. The "sisters" are the lunar days, each of which is 24 hours long (20×48 *momenta* = 960 *momenta*, divided by 40 *momenta* per hour = 24. But the final lunar day of the 19-year cycle is in fact already over by the time the final calendar day of the cycle rolls around, since it represents the advance which the Moon has accumulated throughout 19 years. That is why it is dropped out of the reckoning, or "leaped over".

after sunset as the first Moon: lunar count begins *at* sunset (c.f. ch. 5), so only a Moon visible *by* sunset counts as the first. Thirdly, a Moon visible at sunset may actually be a new Moon, i.e. in day 30 of its previous lunation, not in day 1 of its new lunation. It can happen that we see the thin crescent of the old Moon in the eastern sky at dawn, and the thin crescent of the new Moon in the western sky at sunset on the same day – i.e. day 30 of the old lunar month. Such sightings are unusual, but they can happen, and especially in the spring, for reasons explained by Bede in ch. 25. Due to the shifting angle formed by the ecliptic and the horizon, the old Moon in spring will rise at dawn "beside" the Sun, and will be visible as a faint arc. Conjunction will take place during the daytime. By sunset, however, the Moon will be plunging almost straight downwards towards the horizon along the ecliptic; hence the setting Sun will light up a thin arc of the Moon from beneath.[156] But the Moon is still technically in day 30.

The fourth reason for anomaly is the most complex and interesting case of all. Bede evokes an imaginary observer – not a lone individual, but one who can summon witnesses to affirm what he has seen. This observer claims that on 4 April in year 19, when the table says that the Moon will be new, he has seen a Moon two days old in the sky. One day might be accounted for by any of the three reasons given above. But how to account for two?

We now know, though Bede did not, what has caused this gap. Nineteen solar years total 6,939 days and 18 hours, but the 235 lunations which computists consider the equivalent of 19 solar years actually total 6,939 days, 16 hours and 31 minutes (235×29.530588 days). Therefore, every 19 years the Moon gets about 1 hour and 29 minutes ahead of its computed age. After 308 years, the discrepancy amounts to one day, so by Bede's day, assuming the calibration was correct at the time of the Council of Nicaea, the slippage was about $1\frac{1}{2}$ days.[157]

Since this discrepancy would be visible any night of the year, Jones assumes that Bede's case involving the new Moon in April of year 19 is not an instance of first-hand observation, but an echo of a controversy which had occurred in 550. In that year, the 19th of the Alexandrian cycle, the Victorian and Alexandrian tables disagreed, and it looked as though the Victorian tables were more accurate. Victorius fixed Easter

156 See figure b in Commentary to ch. 25.
157 Jones, *BOT* 378.

on 17 April, the 15th day of the Moon, while the Alexandrian table declared it to be 24 April, the 21st day of the Moon. Therefore Victorius' lunation began on 3 April and the Alexandrian on 4 April. In his defence of the Alexandrian reckoning, Victor of Capua indicates that the Victorian camp pointed out that the new Moon was indeed visible on 3 April; he responds by invoking the first reason rehearsed above, namely the accumulation of the *saltus* increment. But "whether in 550 or 725, the discussion is highly academic. Bede shows no knowledge of the actual observed Moon".[158]

This argument can be contested on the following grounds. First, Victor is actually discussing the generic issue of the calendrical *terminus ad quem* of Easter. The date in question is not 24 April, but 25 April, and Victor nowhere alludes to the *saltus*.[159] Secondly, Bede's explanation for this anomaly is completely different from the one which Jones ascribes to Victor. He points out that in year 19, the last lunation before the Easter Moon is not a hollow one of 29 days, as is usual, but a full one of 30 days, because an embolismic month is inserted in March in year 19. It begins on 5 March, and ends 3 April. Therefore to the day accumulated by the *saltus* increment, one must add an additional day to the lunar reckoning because of the embolism.[160] For this reason, all the arguments invoked by Bede focus on the authority of the 19-year cycle, and particularly on its pattern of common and embolismic years (see Commentary on ch. 46, below).

He begins with a common-sense assumption. Given that the Council of Nicaea (in Bede's view) approved the 19-year cycle, they must have approved its sequence of common and embolismic years. Surely the Fathers at Nicaea observed this discrepancy; but they chose to subordinate nature to the higher goal of maintaining the proper sequence of

158 Jones, *BOT* 378.

159 Victor of Capua, *De pascha* 11 (300).

160 See commentary on ch. 45 below. Because the embolismic lunation is 30 days long, it temporarily – but only temporarily – pushes the calculated Moon ahead of astronomical reality, as the following comparison of calculated and true new Moons for AD 664 (see n. 163 below) shows:

	CALCULATED	TRUE
March ("full")	3 Feb.	3 Feb.
Embolism ("full")	5 March	3 March
April ("hollow")	4 April	2 April
May ("full")	3 May	1 May
June ("hollow")	31 May	31 May

common and embolismic years, so that the cycles of Moon and Sun would coincide over a period of 19 years (cf. ch. 46). But as there is no documentation for such an assumption, Bede feels obliged to back it up with further evidence. The evidence is certainly colourful, but not exactly Bede-like, being a string of wonder-tales and religious legends, none of them with impeccable credentials.

The first is the legend that St Pachomius, the founder of cenobitic monasticism, had received the rule for the 19-year cycle from an angel. This story was to have a long medieval career,[161] but it is noteworthy that Bede emphasizes that the saint established the sequence of the *common and embolismic years.*

Reinforcements are still required, so Bede invokes a rather elliptical reference from the *Epistola Cyrilli.* It is in fact not much of a proof, since it is a negative hypothesis: had the Council of Nicaea not established the proper Easter cycle, the true new Moons could have been tested by observing the selenite stone, which changes colour with the waxing and waning Moon (c.f. ch. 28). This is a flamboyant rhetorical defence of the absolute accuracy of the Nicene system, not a laboratory test, but Bede at this stage is ready to take all the help he can get. He then cites a passage from Paschasinus of Lilybaeum which tells of how a miraculous font settled a dispute over the date of Easter in the 19th year of a cycle – a dispute occasioned by failing to account for the embolismic Moon of March.[162] Finally, Bede hints in a somewhat vague way that the Pope had sanctioned the Alexandrian system.

Bede is intent on proving, almost at any cost, that the 19-year cycle, and especially its sequence of common and embolismic years, is sanctioned by natural, ecclesiastical and divine authority. Given the shakiness of the case, Bede must have had very pressing reasons for wanting to discuss the issue of this lunar anomaly in the first place. A number of people in Bede's circle must have actually observed the discrepancy; if not, why bring it to the attention of his readers and risk confusing them?

The answer lies buried elsewhere in *The Reckoning of Time,* namely in the chronicle in ch. 66, under the year 4622, or in AD reckoning, 664.

---

161 Jones, "A Legend of St Pachomius".

162 Miraculous fonts are frequently resorted to in such cases; see Gregory of Tours, *History of the Franks* 10.23, ed. B. Krusch and W. Levison, MGH SS Rer. Mer. 1 (1951):514.22–515.6, where he describes how in 590, a year when Victorius' tables offered both a "Latin" and a "Greek" Easter, a miraculous spring confirmed that Gregory's choice of the "Latin" date was correct; for another example from 577, see 5.16 (215.4–8).

664 happens to be year 19 of the Alexandrian cycle. It was also the year in which the synod of Whitby took place, though Bede does not mention this fact in the chronicle. In that year, a solar eclipse took place on 1 May, which is recorded in various contemporary chronicles, notably Irish ones. However, according to Dionysius' table, the new Moon would not have appeared until 3 May, for the full Moon of Easter fell on 17 April. Bede dates the eclipse to 3 May. In short, the situation in 664 was exactly the one described in this chapter: the Moon on 3 May was two days older in the tables than it was in the heavens. If Bede was aware of the true date of the eclipse, then his position would have been extremely delicate. If he referred explicitly to this observed anomaly – one which cast doubt on the accuracy of the very table endorsed by the synod of Whitby – he jeopardized the still fragile allegiance of the Insular Churches to the Dionysian system. He also risked putting himself in opposition to men like Victor of Capua and Paschasinus of Lilybaeum, who had fought for the acceptance of the late Alexandrian Easters of year 19 against the traditional Roman calendar limit of 21 April. But he had to explain to his own satisfaction, and to the satisfaction of his more astute and well-informed readers, how a two-day gap could have transpired. So he recorded the Dionysian date for the eclipse in the chronicle, where it would do no computistical harm, and devoted this singular and difficult chapter to elucidating such discrepancies in lunar reckoning. It is always possible that Bede was in fact not aware of the true date of the 664 eclipse – it might not have been dated in his sources, and he himself might have extrapolated the 3 May date from a Dionysian table – but in any event, his admission that the true new Moon in April of year 19 appears on 2 April wittingly or unwittingly hinted at the true date of the 664 eclipse.[163]

---

163 In her article "Doubts About the Calendar. Bede and the Eclipse of 664", *Isis* 89 (1998):50–65, Jennifer Moreton also argues that ch. 43 is implicitly about the 664 eclipse. However, she does not attend to Bede's argument about how the embolism, in conjunction with the *saltus* contributes to a two-day anomaly. Rather, she claims that Bede changed the date of the eclipse to cloak his dissatisfaction with Dionysius' AD reckoning, particularly in regard to the date of the historic Resurrection. While I agree that Bede had doubts about Dionysius' chronology – doubts which he scarcely disguises in chs. 47 and 61 – it seems plain that Bede does offer a purely computistical solution to the problem of the 664 eclipse in this chapter.

Chapter 44:

Having completed his exposition of the solar calendar, its adnexa and problems, Bede now turns to the second major graphic document of *computus*, the Paschal table. He will expound this table column by column, from left to right.[164] As the Laon-Metz glossator remarks, the backbone of the Paschal table is the 19-year cycle; this is a list of the full Moons of Easter which will repeat in a cycle of 19 years, and the rest of the table – its indictions, lunar cycle, concurrents, etc. – was simply added on by Dionysius.[165] Hence, though the 19-year cycle proper does not appear until column six of the Dionysian Paschal table, Bede begins his exposition with an introductory section on the cycle itself (chs. 44–46).

Chapter 44 is therefore of signal importance, for Bede's task was as much to establish the credibility of the cycle as to demonstrate its workings. Its architect, Dionysius Exiguus, was unfortunately not an "authority", but a mere monk, an Eastern refugee in Rome about whom little was known. For his computistical system to gain a hearing, it needed a patristic pedigree, the patina of antiquity, and above all, the sanction of the Apostolic See. Furnishing such credentials was not an easy task.

Bede begins by diverting attention from the obscure Dionysius to a recognized patristic figure, Eusebius of Caesarea.[166] He adds that the system underlying the table is even older, and was well known and widely accepted throughout the Eastern Church. But using this system to calculate Easter year by year was a cumbersome business, causing delays in communication of the date of Easter to the Churches, and occa-

---

164 Bede's Paschal table is reproduced in Appendix 2. This form of exposition seems to have become standard in the medieval schools, as witness a mnemonic jingle on the order and content of the several columns, beginning *Linea Christi . . .*, ed. P. de Winterfeld, MGH Poetae 4 (1899):153.

165 *Per decennouenalem ciclum inuenitur XIIIIa luna, et dies paschae, et praeter hos inueniuntur anni incarnationis domini, et indictiones, reliqua. Hoc tamen addidit Dionisius ad ipsum ciclum.* ("The 14th Moon, and the date of Easter, are found by means of the 19-year cycle. In addition, one can find the years of Our Lord's Incarnation, the indictions, and so on. However, Dionysius added this to the cycle"; 418, *ad lineam* 3).

166 Jones argues that though medieval writers only knew of this Eusebian cycle through Jerome's *De uiris illustribus*, it is likely that Eusebius did indeed compose such a table, copies – or reconstructions – of which can be found in a few Carolingian *computus* manuscripts (Jones, *BOT* 24, n. 2). However, since the differences between these tables and the one of Dionysius are minimal, it is more probable that the Carolingian compilers were ascribing the Dionysian table to Eusebius on the strength of Bede's implication.

sioning disputes. A cycle was both easier to learn, and less contentious. Bede claims that the responsibility in this matter was *delegated* by the Pope to the bishop of Alexandria; the bishop reported back to Rome, who issued an announcement to the Churches. Most of the rest of this chapter is devoted to affirming Rome's control over the calendar. Even when Pope Leo the Great referred the Paschal dispute of 454 to Alexandria,[167] he pronounced his judgement in advance; and when the response was received from Bishop Proterius, Leo congratulated the Egyptians on their fidelity to the doctrine which Peter, founder of the Roman Church, taught to his disciple Mark, the first bishop of Alexandria. No more impressive *imprimatur* for the 19-year cycle could be found.

This apostolic sanction is, unfortunately, a fabrication, though probably not a deliberate one. The particular letter of Leo to Proterius which Bede quotes here concerns a dispute about the theology of the Incarnation, not the Paschal question. Krusch thought the mistake unintentional,[168] MacCarthy, deliberate.[169] Jones judged it both honest, and in a sense, intelligent. The letter from Proterius to Leo in Bede's "Sirmond manuscript" is undated, but Leo's letter is dated 10 March, 454, so close to Easter that Bede could easily have jumped to the conclusion that its opening sentence (the passage cited here) referred to the receipt of a response to the Paschal question.[170]

Chapter 45:

Before launching into the separate columns of the Paschal table, Bede wants to clarify two structural elements of the 19-year cycle: the sequence of common and embolismic years, and the division of the cycle into eight-year and eleven-year segments (*ogdoas* and *hendecas*). The common and embolismic years is the theme of chapter 45, and has been discussed at some length above.[171] Here again, Bede's aim is to justify as much as to explain. He insists on the point that the insertion of embolismic months was a Hebrew practice, enshrined in their "natural" lunar calendar. They inserted embolisms as needed, but the Romans, with their fixed solar months, preferred to insert them in locations where the extra

---

167 See Introduction, section 3.
168 *Studien I*, 136.
169 *Annals of Ulster* 4.cxlv.
170 *BOT* 379–80.
171 See Introduction, section 3, and Commentary on ch. 43.

month could be absorbed in its entirety within a calendar month. This makes the embolismic lunation "invisible", a lunation which belongs to no month.

However, in years 8 and 19 of the cycle (third and seventh embolisms), it is impossible to find a suitable month for an "invisible" embolism. The embolism in both instances is inserted in early March (cf. commentary on ch. 43, above), but ends after the first of April. Hence in years 8 and 19, the Paschal reckoning must be closely monitored, because the lunation beginning 5 March and ending 3 April will, contrary to normal practice, be a "full" one – the embolismic one.[172]

Chapter 46:

The *ogdoas* (8-year period) and *hendecas* (11-year period) have little significance for computation, but they were frequently marked on Paschal tables, and are directly linked to the embolisms. Notably, they serve to remind the computist that in years 8 and 19 (the last years of the *ogdoas* and *hendecas* respectively) the embolism follows upon a single common year, not two, as is otherwise the case. But Bede finds the subject interesting for more than mnemonic reasons. *Ogdoas* and *hendecas* demonstrate the truly cyclical character of the 19-year period. Though it was commonly assumed by many ancient peoples that eight solar years equal eight lunar years, this was eventually proved not to be the case; an additional 11 years were necessary for a proper luni-solar cycle. But it is the way in which Bede demonstrates this that catches our attention, with its concrete, even violent language, and interactive classroom manner. Students "hit upon" the answer and "gut the bowels" of the problem; the Sun is a debtor imploring the help of his friends; and finally, Bede makes clever tactical use of a red herring, the leap-year day.

If leap year is left out of the reckoning, the Moon will be two days ahead of the Sun at the end of eight years. How can the Sun catch up? Can he "call in" the extra leap-year days? If we turn to the *hendecas*, though, we find exactly the inverse situation. The Sun has gained two

---

172 Jones (*BOT* 380) seriously misinterprets this passage, and claims that the embolisms are shifted from their prescribed places in these two years. This is not the case, and Bede's source, the tract *De ratione embolismorum* (printed in *BOD* 685–689) has been misread at this point: it is *only* by inserting an embolismic month of 30 days in March that one can bring about a situation where the Moon of 1 April and the Moon of 1 May are the same age (688.10–689.19).

days over the Moon in this period. What about the leap-year days then? In fact, the Sun does not need them, for the two days won in the *hendecas* compensate for the two days lost in the *ogdoas*. Over 19 years, the number of lunar and solar days, setting aside leap-year days, is equal. And it is right to set aside leap-year days: they are irrelevant, a red herring. As Bede proved in ch. 41, they affect both Moon and Sun, so they cannot be used to even up either planet's score – a point which Bede will raise again to discomfit Victorius in ch. 51. Secondly, the number of leap years in any 19-year cycle will vary between four and five, but the overall synchronicity of the lunar and solar cycles will not be affected (cf. ch. 41).

Chapter 47:

From this point onwards, Bede follows the order of the eight columns of the Dionysian Paschal table. He explicitly situates each column (*uersus, linea, ordo* . . .) in numerical sequence, from left to right. One is always conscious of the physical presence of the table in this section, and this is reinforced by Bede's literary finger, so to speak, pointing to the graphic page. In this chapter, for example, he indicates the Paschal data for the year of Christ's Incarnation "here", and later invites the reader, "with the cycles of Dionysius open before you", to look for the traditional date for Easter in Dionysius' year of the Passion. The irony is that he won't find it there – but to appreciate the irony, the reader must actually be looking at the table.

Bede's relationship to *annus Domini* reckoning, and the reasons for his use of the more conventional *annus mundi* era in his chronicle of world history, have been discussed in the Introduction to this volume (section 4, "*Computus* and history"), and will be considered again in the Commentary to ch. 66 below. But it is essential to consider his historiographical choices in the light of this very curious chapter of *The Reckoning of Time*. If ch. 43 was designed to confront an observed anomaly in the Dionysian system and show how it was an exception which proved the rule, ch. 47, in an extraordinary turnabout, seems to do just the opposite. Here Bede, brandishing his weapons against Victorius and the 84-year Celtic cycle, grounds the accuracy of the Dionysian *computus* in its chronology, and then apparently without even noticing it, shows that chronology to be completely untenable.

Bede begins by clarifying the relationship between the great Paschal cycle of 532 years (the product of the 19 years of the lunar cycle and the 28 years of the cycle of weekdays) and the year of the Incarnation. Since Dionysius began his table in AD 532, it follows that a similar cycle began in BC 1,[173] and that Christ was born in the second year of this Great Cycle. This is implied by Dionysius' computistical formulae or *argumenta*; for example, to find the year of the 19-year cycle, take the AD year and add 1 before dividing by 19. That being established, Bede now turns to the year of Christ's Passion, which is his real interest. The Gospels seem to agree that Christ had passed this 33rd birthday when he died, or to put it another way, the year of the Passion was AD 34. Bede anticipates some resistance here, as most people were unfamiliar with the Dionysian era. He recounts the voyage of the Wearmouth monks to Rome in 701. There the pilgrims saw candles inscribed with the *annus Passionis* 668, i.e 701. The year of the Passion was therefore AD 34, which holds the same position within this first great Cycle as does the year 556 within the second cycle, beginning in 532. Now, says Bede, just look in your Dionysian tables for Easter in 556, and if you can find the traditional date of the Resurrection, 27 March – well, good for you!

As the Laon-Metz glossator remarks, Bede is being heavily ironic here, because the date for Easter in Dionysius' table for AD 34/556 is 28 March.[174] Bede's target here is the *idée fixe* that the historic date of the Resurrection must have been 27 March.[175] Nor will the statement of ps.-Theophilus that the Resurrection occurred on 25 March pass muster. The only year of the Dionysian cycle which produces an Easter on that date is AD 31, which was, as Bede observes, the beginning and not the end of Jesus' ministry. But he has unwittingly let an embarrassing

173 See ch. 47, n. 22, and cf. Harrison, *Framework*, 37. It should be noted that Bede nowhere states when the year begins. R.L. Poole claims that *The Reckoning of Time* favours Christmas, but there is no evidence for this: "The Beginning of the Year in the Middle Ages", in *Studies in Chronology and History* (Oxford: Clarendon Press, 1934):8. Indeed Bede at several junctures indicates that 1 January is the beginning of the year (e.g. ch. 22). However, the Church throughout the early Middle Ages strove to fix the beginning of the year at Christmas, to counter the traditional Roman revelries of 1 January. Cf. Harrison, *Framework*, 38, and "The Beginning of the Year in England", *Anglo-Saxon England* 2 (1963):51–71.

174 Laon-Metz glossator: *AGE. Sc. hironice dixit hoc, dum nullo modo inuenitur.* ("GIVE THANKS. He is speaking ironically here, because it cannot be found at all." *BOD* 432, *ad lineam* 89).

175 See Commentary on ch. 61 below.

cat out of the bag. The Paschal data in Dionysius' table for AD 34 not only does not conform to tradition, but it does not even match the facts recounted in the Gospel. If Christ was crucified on the 15th day of the Moon, a Friday, he would have risen on Sunday, the 17th day of the Moon. But the Easter Moon of AD 34/556 is unmistakably 21 days old.

Dionysius, of course, never noticed this problem, because unlike Hippolytus or ps.-Cyprian, it never occured to him to test the validity of his computational system by chronology. Chronology was of incidental interest to him, and he borrowed his Passion chronology from traditional Roman sources.[176] But the fact that Bede did not notice this problem is startling, especially since he thrusts it in his reader's face, so to speak, by telling him to check out the data in Dionysius' table. Certainly his successors remarked it, and from the Carolingian period onwards, began to propose various alternative solutions; by the 11th century, a veritable chronological reform movement was under way.[177]

How can we interpret this paradox? There seem to be only two possibilities. Either Bede was focussing so myopically on the date of Easter Sunday in AD 34 that he ignored completely the issue of the age of the Moon, which flatly contradicted the biblical data; or he did indeed notice the problem, and used sarcasm to divert attention away from it. But perhaps there is a third possibility, and that is that Bede himself doubted the historical foundation of Dionysius' AD chronology. His admonition that the reader should be sceptical about chronography, but place absolute faith in the Gospel, reveals his own dilemma. Dionysius' chronology was wrong, but in the end the error was negligible, for AD 36 produces an Easter on the 17th day of the Moon. Had Bede pointed out the error and proposed a correction, he would have put at risk his main project of securing acceptance for the computistical system to which this chronology was attached. Dionysius' system *was* this table, and if there was a small mistake in a computistically irrelevant part of the table, that would have to be tolerated for the sake of promoting orthodox Easter reckoning. Chronology had no bearing on salvation and the order of the Church; the Paschal *computus* most definitely did.

---

176 Dionysius' dating of the Passion corresponds, for example, to that of the *Chronograph of 354*. Victorius dates it to AD 28, but it appears that his chronology never found favour: Jones, *BOT* 63.

177 See Introduction, section 6, "Use by Later Computists".

Chapter 48:

The second column of the Paschal table records the indictional year. The indictions were a Roman bureaucratic cycle of 15 years, instituted in the reign of Diocletian and Constantine for taxation purposes.[178] From the time of Constantine, the cycle began on 1 September, which became the beginning of the Byzantine fiscal year; under Justinian, the indiction became a compulsory feature of official dating. Indictions were included in the Alexandrian Paschal tables, from whence they were taken over by Dionysius Exiguus.[179] Though familiar to men like Ambrose who were connected with imperial administration (see below), they were not in common use in the West; Isidore of Seville, for instance, does not mention them. The earliest Latin computist to mention them is Dionysius Exiguus, and though the Papal chancery was using them in dating clauses by 584[180] this was probably because it was using Dionysius' tables.

For this reason, Bede was at somewhat of a loss to explain what indictions were, or why they were important. He guessed (wrongly) that their purpose was to help historians fix dates in the past, as it had helped him to confirm the date on the Paschal candle his brothers had seen in Rome.[181] But his lack of direct experience of indictions also led him to date the beginning of the cycle to 24 September, rather than the correct date of 1 September. Jones argued that this error stemmed from a statement Bede encountered in Ambrose's *De Noe et Arca* 17.60. Ambrose's point is that the year really has its beginning at the spring equinox, the time of the Hebrew "first month", even though "as our use of indictions shows", we tend to think of September as the first month. Ambrose's parallel encouraged Bede to conclude that the indictional cycle began at the analogous point of the year to the "first month", i.e. at the autumn equinox.[182] Harrison doubts that Bede made this discovery on the pages of Ambrose, but he agrees that the autumn equinox may have attracted Bede to this date. Harrison also points out that the age of the Moon on

---

178 A.H.M. Jones, *The Later Roman Empire A.D. 284–602* (Baltimore: Johns Hopkins University Press, 1986):1.448 *sqq.*

179 Jones, *BOT* 384.

180 Franz Rühl, *Chronologie des Mittelalters und der Neuzeit* (Berlin: Reuter & Reichard, 1897):171.

181 But see ch. 47, n. 31.

182 Jones, *BOT* 383–384.

24 September will match that at 1 January in ordinary years,[183] though it
is not clear of what possible computistical use this could be. But if the
autumn equinox was indeed the attraction, the fact that Bede fixed the
start of the indictional cycle at 24 September is further evidence of his
relative indifference to the dates of the solstices and equinoxes, apart
from the spring equinox.

Indictions continued to be used in dating formulae throughout the
Middle Ages, but by the beginning of the ninth century, the indictional
year was assumed to overlap with the calendar year, suggesting that the
indictions were simply being copied out of a Paschal table. Bede's
eccentric indiction of 24 September was never used by chanceries.[184]

Chapter 50:

Bede spends little time on the mechanics of lunar epacts, since they have
already been explained in ch. 20. Instead, he takes the opportunity they
afford to launch a full-scale attack against the *computus* of Victorius of
Aquitaine. He loads his ammunition in ch. 50, and discharges it with
stunning éclat in ch. 51.

The real theme of ch. 50 is not epacts at all, but the calendar limits for
the Paschal lunation. The lunar epact is the age of the Moon on 22
March; as Bede insistently points out, this is the earliest possible date
for Easter. The epact is therefore a handy index for locating the Paschal
lunation: if it is 16 or less, 22 March is within the Paschal lunation, but if
it exceeds 16, the Paschal lunation will be the following lunar month.
This is because the Paschal full Moon must not precede the equinox, for
the Moon was created on the evening of the equinox (cf. ch. 6). But this
is exactly the error of Victorius, who held the equinox to be the *terminus
a quo* of Easter itself, not of the Easter full Moon. Bede sets his sights on
his target when he criticizes those computists who begin the Paschal
lunation on 5 March, for this is precisely the doctrine of Victorius.

Chapter 51:

In this chapter, Bede removes the gloves and starts naming names. The
casus belli would seem to have been the Easter coming up in 729. In

---

183 *Framework*, 43–44.
184 *Ibid.* 115–119.

Victorius' table, AD 729 and 976 share the same data, save that 729 is an ordinary year, and 976 a leap year. In both years, 1 January is a Saturday, and the Moon is 27 days old. However, Easter in 729 is scheduled for 24 April, the 21st Moon, while Easter in 976 is on 23 March, because of the leap year. In his preface to his table, Victorius notes that years like 729 present a problem for Roman traditionalists, because 24 April was outside the ancient Roman calendar limits for Easter. The only alternative would be to move Easter back to the previous lunation, but there was simply no available Sunday in this lunation. The closest one was Sunday, 27 March, but 27 March fell on the 23rd day of the Moon, which was outside everyone's lunar limits. Moving the "Latin" date back a week to 20 March will not help, *even though the Moon agrees* (it will be 16 days old on that date), because 20 March precedes the equinox. According to Victorius, the equinox is the *terminus a quo* for Easter, so he advises celebrating later with the Alexandrians.[185]

What touched off Bede's wrath was Victorius' claim that a Moon 16 days old on 20 March was a "true" Paschal Moon, even though it produced a manifestly uncanonical date for Easter. Obviously, Victorius' "Latin" Paschal lunation begins too early: a full Moon before the equinox is not the full Moon of Easter. Victorius has not grasped that it is *Passover*, i.e. the Paschal full Moon, which is linked to the equinox, for astronomical reasons; Easter follows Passover for *historical* reasons. Finally, Victorius is guilty of bad faith. In leap years like 976, when the 16th Moon falls on Monday, 20 March, Victorius' "Latin" system should allow him to celebrate on Sunday, 26 March; yet even in this case, his table postpones Easter to 23 April, even though 26 March is within Victorius' own lunar limits of 16–22. So Victorius tacitly admits that the Alexandrian system is the correct one; why then does he not follow it?

By citing Victor of Capua's critique of Victorius, Bede hopes to demolish the major defence of the Victorian system. As was pointed out earlier,[186] Victorius' tables, for all their evident flaws, seemed to carry papal sanction because they were commissioned by a man who later became pope, at the request of the then Pope, Leo. In Bede's view, support from Victor, who was a bishop writing at papal behest, trumps Victorius, who is demoted to *quidam* – "a certain person".

185 This will only work, of course, if the computist has remembered to add an extra day to the February lunation, to include the leap-year day: see ch. 41 above.
186 Introduction, section 3.

Chapter 53:

Solar epacts or concurrents, like lunar epacts, have already been introduced in connection with customizing the Julian calendar to the shifting
pattern of weekdays (ch. 21). In this chapter, Bede elects to focus on the
28-year weekday cycle itself, because in conjunction with the 19-year
cycle, it creates the Great Paschal Cycle of 532 years.

The solar epact or concurrent is the weekday of 24 March, a date
chosen, as Bede remarks rather casually, because it is "quite close to the
Paschal festivity". No further reason is given for this choice, which has
puzzled many commentators, and provoked a number of ingenious explanations. Jones, however, offers the most sensible reading. 24 March falls
on the same weekday as 24 February, the leap-year day, but it falls within
the Paschal lunation. Hence it furnishes a useful reference point for
locating the weekday of the Paschal full Moon, in ordinary or in leap
years.[187] Notice that Bede advises his pupils to memorize the other four
dates within the Paschal period which share this same weekday. This
operation is hardly necessary for us, or for the Alexandrian computists,
since we both number the days of the month sequentially, and can therefore just keep adding seven to find the next corresponding date. But the
non-sequential Roman dates, based on reverse counting from kalends,
nones and ideas, made it obligatory for Bede and his readers.

Chapter 55:

The remainder of this section is devoted to expounding the cyclical character of the solar epact cycle, and how it can be used to determine these
data for any year in the past or future. There are basically only three
numbers one needs to know: 5, 6 and 11. Any given year will share the
same solar epact with years in the past or future according to this pattern:

	PAST	FUTURE
leap year	-5	+6
year 1	-11	+6
year 2	-6	+11
year 3	-6	+5

But what if you want to find the concurrent for a year much further in the
past or future? Bede's example is 300 years. Divide this by 30, and multiply

187 Jones, *BOT* 387–388.

the quotient (10) by 2: the result is 20. Therefore the solar epacts 300 years from now will be the same as they will be 20 years from now. Since 20 is a small number, it can easily be reckoned up using the pattern given above. Lunar epacts can also be projected into the past or future, but much more simply, since every 20th year will have the same lunar epact.

Bede then decribes how to use the human hand as a form of modular table for the 19-year lunar cycle and the 28-year weekday cycle. If we include the fingertips, the hand contains nineteen joints, on which the 19 epacts of the 19-year cycle can be inscribed in sequence, from the base to the tip of each finger. One can then count off the number of years in the past or future which one wishes, and find its position within the cycle. The joints of the fingers of both hands, excluding fingertips this time, number 28, and are therefore suitable as a grid for the solar cycle. But here the numbering is not sequential, because what you will be looking for is the position of the year within the four-year leap-year sequence. Therefore the first six leap-year cycles are inscribed in rows horizontally across the joints, so that the little finger contains all the leap years, the ring finger all the years immediately following leap years, and so forth. The seventh leap-year sequence of the 28-year cycle is consigned to the four joints of the thumbs.

To understand how this portable modular table would have worked, it is helpful to visualize it. An illustration inserted into the margin of this chapter in the 11th-century manuscript Oxford St John's College 17, fol. 98r–v has been adapted here to make this clearer (see figs. 1 and 2).

The lunar epact hand is fairly unproblematic, as the epacts rise steadily by 11 each year. It is the solar epact hands that shows the system's real worth. No memorization is required. If you know the position of the present year within the four-year leap-year cycle, and its solar epact, you can find where it belongs on your hand by continuously adding 5 to the base number on the lowest rank of joints on the left hand, numbered 1–4. For example, suppose the present year is a leap year with solar epact 2. Begin with the leap-year finger of the left hand. The lowest joint is solar epact 1. To find the epact of the joint above, add 5 (4 days for each year of the cycle + extra day for leap year). The solar epact corresponding to this joint is 6 (1+5); that for the topmost joint of the little finger is 4 (6+5=11–7=4). Then shift to the little finger of the right hand. The bottom joint yields 4+5=9–7=2: this is the year you are looking for. You can then count forward or backward any number of years you wish, and read off the solar epact in the same way. Forward counting

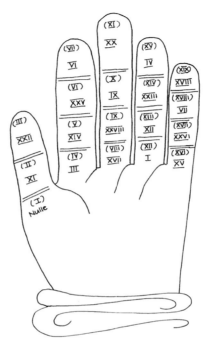

*Fig. 1. Lunar cycle hand, showing lunar epacts for each year of the 19-year cycle.
The year of the cycle is noted in parentheses above.*

moves horizontally in ascending rows, backward counting proceeds in
descending rows. To continue our example: suppose you want to know
what the solar epact will be 25 years from now. Count forward twenty-
five places from your "present year", that is, starting from the bottom
joint of the right hand. Remember to include the four joints of the left
and right thumbs in your count, when you have finished with the fingers
of the right hand. You will land on the topmost joint of the middle
finger of the left hand, i.e. the second year after the leap year. The
bottom joint of this finger has the concurrent of 3. The middle joint of
this finger therefore represents 8 (3+5), and the top joint – the year you
are looking for – will have a solar epact of 6 (8+5=13–7=6). In short,
once you have counted forward or backward to the year you want, you
can "read off" its solar epact by adding up from the base of the "table",
i.e. the bottom joints of the left hand.

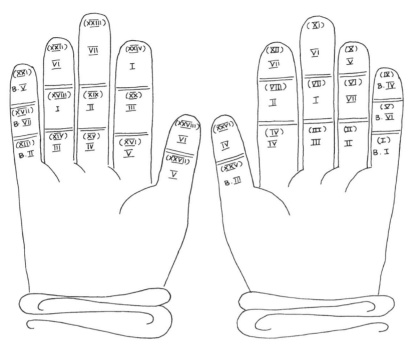

Fig. 2. Solar cycle hands, showing concurrents (solar epacts) for each year of the 28-year weekday cycle. The year of the cycle is noted in parentheses.

Bede lays no claim to inventing this device, but he is almost the only computist to describe it.[188] It certainly needs an illustration to explain it, and St John's 17 is, as far as I am aware, the only manuscript of *The Reckoning of Time* to contain such an illustration. Nonetheless, it is the first in a long line of ingenious mnemonic and calculating systems for the calendar based on the hand, known generically as *computus manualis* or *computus chirometralis*. The history of this intriguing genre of *computus*-literature awaits a fuller exposition, but the fact that a

188 The only contemporary analogue is in the "Bobbio *computus*", PL 129.1322. Ch. 76 of Rabanus Maurus' *De computo* is copied directly from Bede: see Florence Yeldham, "An Early Method of Determining Calendar Dates by Finger-Reckoning", *Archeion* 9 (1928):325–326.

13th-century version in Bodleian Digby 56 fol. 165v bears the legend *manus bede* indicates that Bede's paternity was acknowledged.[189]

Chapter 56:

The lunar cycle in column 5 must have been something of a puzzle to Bede, as it plays no role at all in the computation of Easter. It replicates the structure of the 19-year cycle, but begins in year 3 of the 19-year cycle, when the Moon is new on 1 January. Because of this 1 January epact, Bede took a guess that it was a Roman lunar cycle. This is not the case, but its true origin remains obscure. It arrived in the West with the ps.-Cyrillan Paschal table, and it was obviously meant to serve as a concordance to some other form of reckoning.[190] Based on his guess, Bede therefore created a table patterned after the one Dionysius made for the 19-year cycle. The lunar cycle does have one use, however; it makes it easy to locate the age of the Moon on 1 January (ch. 57).

Chapter 59:

Column 6 of the Paschal table contains the marrow of the Dionysian-Alexandrian reckoning: the dates, recurring over 19 years, of the 14th Moon of "the first month". The *terminus a quo* is the spring equinox, and the last possible date is 29 days later (because April's lunar month is "hollow"). For Bede, this is the only scheme which fulfils the criteria both of the Old Testament Passover, and of the history of Christ's Passion and Resurrection. He makes a careful distinction (to be developed more thoroughly in ch. 63) between the Passover proper, and the Feast of Unleavened Bread. The former begins at sundown on the day of the 14th Moon, and symbolizes Christ's sacrificial death. That is why the rite of baptism, a symbolic inclusion in that death, is celebrated at the great vigil of Easter on this night. Easter proper corresponds to the Feast of Unleavened Bread, from the 15th to the 21st Moon. Because Christ historically rose on a Sunday, Christians celebrate on whichever Sunday falls within this period.

These dates for the 14th Moon are also the crux of the controversy

---

189 Illustrated in John E. Murdoch, *Album of Science: Antiquity and the Middle Ages* (New York: Charles Scribner's Sons, 1984):80, no. 75.

190 Jones (*BOT* 388) suggests that the system was Byzantine, but furnishes no evidence.

between the Dionysian reckoning and its rival systems (to be explained in ch. 60). The Celtic Churches which celebrate Easter Day on the 14th Moon must sometimes break their fast on the 13th, which is too early; at the same time, they reject the 21st Moon, which Moses includes within the time of Unleavened Bread. Victorius on the other hand rejects the 15th Moon, which is also within this Feast of Unleavened Bread, while including the 22nd, which is manifestly outside it. The passage from Theophilus which Bede quotes particularly addresses the Celtic reckoning. If the 14th Moon falls on a Sunday, we cannot celebrate Easter then, for that would mean either breaking our fast on the 13th, or (to avoid this) fasting until sundown on Sunday, both of which are prohibited.

Chapter 60:

The formula for finding these 14th Moons deserves our close attention for the insight it affords into Bede's pedagogy. If the pupil has the Dionysian table before him – and the whole latter half of *Reckoning of Time* is structured on the assumption that this is the case – why learn a formula for what can be seen plainly on the page? No table, however, explains how or why the pattern of the full Moons unfolds as it does, nor can it bring home, as does this formula, the crucial link between the full Moon and the equinox. In Bede's hierarchy of certitude, memory is the most reliable; formulae are less so, because people make mistakes in calculation especially when doing mental arithmetic; tables are the least reliable of all, since it is easy to introduce mistakes when copying them. But only the formula has a heuristic value.

Chapter 61:

The culminating point of the Dionysian Paschal table is the seventh column, containing the dates for Easter Sunday. Bede sees this as an opportunity to recapitulate the criteria for determining this date: it must be after the equinox, in the "first [lunar] month", in the third week, i.e. between the 15th and 21st day of the Moon, and on a Sunday. The last criterion was an innovation of the New Law, and Bede felt it was necessary, through the words of Theophilus, to emphasize that varying the age of the Moon to accommodate Sunday did not violate the Old Testament rules. After all, even the first Easter fell on the 17th Moon.

This last point obliges Bede to venture out once more onto the thin ice of the debate over the historic calendar date of the Resurrection, first raised in ch. 47. The traditional possibilities are 25, 27 or 28 March, but if one follows the Dionysian system, none of these dates will match years within the historic period of Christ's ministry. Resurrection on 25 March needs a 14th Moon on 22 March, which will happen in AD 42, but no earlier: 25 March is the date of Easter in AD 31, but the 14th Moon falls on March 24 that year, or Holy Saturday. 27 March requires a 14th Moon on 24 March, and the closest one falls in AD 12,[191] while an Easter on 28 March with 14th Moon on 25 March will not happen until AD 191! It should be noted, however, that Bede implies, but does not explicitly state, the signal criterion of any historic date for Easter, namely that it would have to fall on the 17th day of the Moon. He avoids drawing attention to this issue by expressing his data in terms of the date of the 14th Moon, and the concurrent. This subtle ellipsis veils the fact that Dionysius' table itself cannot produce an Easter on the 17th Moon for any year close to the traditional year of the Resurrection. The closest years with Easters on the 17th Moon are AD 26 and AD 36. However, the Easters in these years are not on the traditional calendar dates. So Bede must reject the traditional dates, without drawing too much attention to the fact that Dionysius himself offers nothing better. This is evidently embarrassing, for ever since Hippolytus and ps.-Cyprian, congruence of a Paschal table with the data of known Easters had been invoked as a foolproof test of validity. Bede's ploy is to distract the reader's attention by changing the subject to a critique of certain computists who argued that the historic Resurrection fell on the 16th Moon.

Chapter 62:

The final column of the Paschal table records the age of the Moon on Easter Day, ranging from 15 to 21. Here Bede takes the opportunity to revisit the arguments against Victorius' *computus* enunciated in ch. 51. Jones found it curious that Bede should do so, given that the examples

---

191 Considering Bede's dismissal of 25 March, it is strange that he seems to accept this date in ch. 66, where in the preface to the chronicle of the First Age, he dates the creation of Adam and the Passion to 23 March. But he only does so very tentatively, "unless some stronger argument should prevail". For a similar endorsement of 27 March, see *Letter to Wicthed* 12. On Bede's ambivalence regarding the date of the historic Resurrection, see Moreton, "Doubts", 62–64.

of erroneous calculation discussed in ch. 62 would not be applicable to any year in the near future from 725, and therefore could hardly have been raising any immediate concern. Hence he speculates that what Bede is actually doing in this chapter is recounting the conflict between Pope Leo and the Alexandrians over the Easter of 444, which hinged on a similar issue.[192]

But *The Reckoning of Time* is neither an occasional polemic nor an antiquarian essay: it is a textbook, and the atmosphere of ch. 62 is that of Bede's class-room. First, our teacher asks the class a riddle, adopting the *persona* of a Victorian straw man. What after all is wrong with celebrating Easter on 26 March when the Moon is 20 days old? Is not 26 March a legitimate date? Is not a Moon 20 days old within the legal limits? The answer, of course, is that a Moon 20 days old on 26 March would have been full before the equinox. The crucial error of the Victorians is to base their calculation on the relationship of *Easter* to the equinox, when it is the relationship of the *full Moon* to the equinox which counts. By the same token, we cannot have an Easter on 22 March with a Moon 16 days old, or one on 23 March with a Moon 17 days old, etc. To be sure, some of the inspiration and ammunition for ch. 62 may indeed have come from the disputes of AD 444, but Bede's aim here is to help his students grasp the relationship of the Moon on Easter Day to the lunation of the first month, not to resurrect forgotten battles. In any event, Bede thought that Leo endorsed the Alexandrian system;[193] his real target here is Victorius.

Chapter 63:

How does one "prove" the correctness of a system of Paschal computation? Nature does not furnish a complete answer; on the other hand, authority is malleable and Scripture open to interpretation. In the case of Scripture, it is not enough to cite the Old Testament regulations for Passover, or the New Testament records of the Resurrection. It is also necessary to determine how these two are related to one another, historically and theologically. This is Bede's task in chapters 63 and 64 of *The Reckoning of Time*. Bede adopts a very different approach and tone here, one strongly redolent of his biblical commentaries and homilies. These are

192 *BOT* 389–390.
193 Cf. ch. 44.

the most personal chapters of the treatise, and also the ones which reveal most clearly the character of *computus* as *doctrina christiana*.

Ch. 63 draws on ancient Christian traditions[194] to weave a uniquely subtle argument about the relationship between the Old Testament type and its New Testament antitype. The ancient Hebrew feast comprises two elements: the Passover proper, celebrated on a single night, and the Feast of Unleavened Bread, spread over seven days. As we have seen, this fact was a crucial weapon in Bede's contest with the Celtic *computus*, which conflated the two feasts by celebrating on the 14th Moon. But it is noteworthy that Bede does not want to endorse a strong dichotomy between Passover/Passion and Unleavened Bread/Resurrection, for that would play into the hands of the Victorians, who rejected the 15th Moon because it belonged to the Passion, not the Resurrection. Bede's argument is, in fact, more subtle and profound. The Christian Easter incorporates both Passover and Unleavened Bread, without effacing the distinction between the two. The single night of Passover typifies both Christ's sacrifice as the Lamb of God, and his Resurrection or "passing over" from death to life. As the Gospels bear out, Passover and the Unleavened Bread are in fact a single entity: the disciples in Mark 14.12 speak of the Passover on the 14th Moon as if it were the Unleavened Bread, while John 18.28 refers to Good Friday on the 15th Moon as "Passover". Each contains the other – an argument already invoked by the quotations from Theophilus in the preceding chapters, justifying the celebration of Easter on any Sunday from the 15th to the 21st Moon. Sliding into homiletic mode, Bede draws a moral lesson: just as Easter is both a moment in time, and an event unfolding over time, so also our individual experience of Easter is both a discrete moment – our baptism – and a grace which unfolds "for the whole time of our pilgrimage" through the seven "days" of our life.

---

194 Though Bede does not cite these directly, his predecessors include Gaudentius, *Tractatus paschales* 1.9–10, ed. Ambrosius Glueck, CSEL 58.20 (1886):20.50–62, and Origen, whose Homily 9.5 on Leviticus on the subject of Passover and Unleavened Bread is quoted by ps.-Anatolius (317) and Cummian (62–64): *Origines Werke 6: Homilien zum Hexateuch in Rufins Übersetzung*, ed. Wilhelm Baehrens, Griechischen christlichen Schriftsteller 29 (Leipzig, 1920):426. The older opinion that "Origen" was a lost or spurious authority, represented by Jones (*BOT* 390), has been refuted by Ó Cróinín, *Cummian's Letter*, 62–63, n. 52. Jerome's Greek nickname for Origen, *chalcenterus* ("bronze entrails" – apparently a compliment on his indefatigable scholarship!), came to mean "computist" in 7th-c. Ireland: see D. Ó Cróinín, "Hiberno-Latin *calcenterus*", *Peritia* 1 (1982):296–297.

The idea, then, is that the Christian computist acknowledges both the distinction between Passover and Unleavened Bread, and their reciprocity. As Bede says, we must "eat unleavened bread in the Passover season, and in the days of Unleavened Bread ... make a spiritual passover". The Celts fail to honour the distinction; the Victorians, to acknowledge the reciprocity.

Chapter 64:

Ch. 64 is another highly original "sermon" whose "text" is the four computistical criteria for Easter enunciated in ch. 61. Though it draws heavily on Augustine's *Letter* 55 to Januarius, its structure and aim are very different. Augustine is concerned to defend the Church against the charge that by observing the astronomical criteria for Easter, it in fact worships the Sun and Moon; Bede, on the other hand, takes the Sun and Moon as his "scripture". The parallels between this chapter and the letter of Ceolfrid to King Nechtan included in *HE* 5.21 are very striking.

We celebrate Easter at the vernal equinox because Christ, the Sun of Righteousness, has won His victory over darkness; because He has "made all things new", we celebrate in the first month; because His Resurrection inaugurated a third age of grace (succeeding the ages of natural and written Law), Easter falls on the third week of this first month. We are like the Moon, shining in His borrowed light. After the 15th day of the lunation, the Moon turns its illuminated side away from Earth and towards the heavens; so also Easter, falling on the 15th day or later, teaches us to spurn earthly things for the sake of heavenly ones. But the Moon is also Christ himself, increasing in visibility to earthly eyes up until the time of His Passion at the full Moon, and then returning to heaven after the Resurrection. Finally, we celebrate on Sunday, the first day of the week, in honour of Creation, re-made in Christ's triumph. Easter Sunday can fall on any one of seven days of the lunar month; this symbolizes the universality of the Church, just as John in his Apocalypse addressed the same admonition to each of the seven Churches of Asia. Bede then returns to the beginning of his theme, and rewrites the entire allegory in a moral mode, admonishing his readers to "pass over" from sin to virtue, bring forth the "first fruits" of redemption, shine like the Moon both for others and for God alone, and finally, trust in the sevenfold gifts of the single Spirit.

In treating *computus* as *figura*, Bede has followed much the same method as he used in his highly original expositions of the Tabernacle and Temple, where he turned the nuts and bolts of architectural detail into theological signs. But within the economy of *The Reckoning of Time*, this homily also serves a strategic purpose. By colouring the Alexandrian rules with mystical meaning, Bede has raised them above the sordid plane of computistical controversy, and transformed them into a kind of sacred text. In his exposition of the Paschal table, he had to defend and justify the Alexandrian reckoning line by line; now he can celebrate its apotheosis. But Bede has done more than simply close the books on the Paschal question. He has engraved the Alexandrian *computus* on his pupils' memory, not just as a body of rules to be followed – or contested – but as theological truth and moral imperative.

Chapter 65:

The note of sublimity struck in ch. 64 sets the stage for the final topics of *The Reckoning of Time*, namely the relationship of *computus* to time as a whole, and in its largest sense. Three motifs articulate this theme: the Great Cycle by which the Easters will repeat over 532 years; how *computus* structures all of past time, right back to Creation; and how it illuminates future time, carrying it over the threshold into eternity.

Bede was the first computistical writer to expound the cyclical character of the Alexandrian reckoning correctly. Ps.-Cyril believed, incorrectly, that his Paschal data would repeat after 95 years. Victorius noticed that his data began to repeat after 532 years, but did not know why. Dionysius knew that Cyril was mistaken, but apparently either did not know, or did not appreciate, the significance of Victorius' discovery. Bede does not claim to have discovered the 532-year cycle, and he does not seem to expect his readers to be struck with its novelty.[195] But there

195 Ceolfrith's letter to Nechtan (*HE* 5.21 [546]) claims that many in England were capable of producing Easter cycles, but whether he meant Great Cycles or not is unclear. Since the extension to Dionysius' table expired in 721, someone would have had to project them further into the future, but this does not necessarily imply the creation of a Great Cycle. However, it should be noted that the author of *De mirabilibus sacrae scripturae* realized that the Victorian 532-year cycle was the product of the 19-year lunar cycle and the 28-year solar cycle (2.4, 2176); cf. Smyth 151–152; Ó Cróinín, "Early Irish Annals from Easter Tables: a Case Re-Stated", *Peritia* 2 (1983):74–86. Cummian drew the same conclusion (88.225–226), and *De ratione conputandi* 14 applies the same logic within the context of the Dionysian system (124.1–6). It is therefore more than likely that a Dionysian Great Cycle was in circulation before *The Reckoning of Time* was composed.

is no question that he was amongst the first to explain why the data recurred after 532 years, and definitely the first to publish a Great Cycle. Bede indicates that his exemplar of *The Reckoning of Time* contained such a cycle; we also have included it, as an appendix to the present translation.

At a stroke, the Great Cycle solved the problem of *computus* forever. All the elaborate explanations, formulae and arguments of *The Reckoning of Time* are in a sense totally unnecessary, for by means of this table, true Easters can be found for as long as the world lasts. Bede's table would run out in 1063, but the data could then simply be recycled for the next 532 years. All that would need to be adjusted were the indictions (an arbitrary count of years, without astronomical basis or computational significance) and the years of the Incarnation.

The years of the Incarnation are, of course, not cyclic, but linear, and their steady, objective march into the future serves to identify and label all time to come. But it is not this possibility which primarily intrigues Bede here. It is rather the potential of the table, and its count of years, to plot past time. It can be used to "clarify an ancient text" and "clearly identify all the years"; in short, it can be a framework for history. Bede's *computus* has hitherto been exclusively prospective in its intentions. Its object was to determine dates for Easter, and the only such dates of interest were future ones. The link between *annus Domini* and the Paschal table, however, gives it an alternative, retrospective orientation. Bede is, of course, justly famous for the way in which he used *annus Domini* in his own historical writings. When he looks at past time in his *computus* treatise, however, he adopts a rather different approach.

Chapter 66:

As we have seen, one of Bede's most remarkable innovations was to solemnize the link between chronology and *computus*. In his hands, chronology became applied *computus*.[196] The initial inspiration for this came from Isidore and the Irish, who suggested a framework for a treatise on *computus* based on an ascending hierarchy of units of time. In *De natura rerum* and the *Etymologiae*, Isidore stops at the year; his next category is *mundus* ("world" or "universe"), which bridges time into

---

196 Jones, "Bede as Early Medieval Historian", 32; von den Brinken 102–108; Tristram 26. See Introduction, section 4.

cosmos. The Irish *De computo dialogus* goes from the year to the cycle, from the cycle to the *aetas* ("age"), thence to the *saeculum* (literally "generation", but here used in its extended sense of "era"), and finally to *mundus*.[197] Bede followed the Irish lead by adding the Great Cycle and the World-Age to the divisions of time. *On Times* chs. 1–9 follows the conventional Isidorean divisions up to the year; chs. 10–15 deal with the Paschal cycle; and finally, chs. 16–22 describe the Six World-Ages through the medium of a chronicle. This format was retained and expanded in *The Reckoning of Time*. But in his later work, Bede chose to treat the World-Ages twice: once as a type of week (ch. 10) and a second time as the extension of the Great Paschal Cycle (ch. 66).

Bede's two world chronicles are unusual in a number of respects. Not only are they embedded in *computus* treatises, but they propose a radical revision of the received chronology of world history. To understand the form, content and role of ch. 66 within *The Reckoning of Time*, we must therefore first situate it with respect to its major sources.

The Christian concept of time is remarkable for its linear focus, its boundedness, and its exclusivity. Time began at Creation, and will end at the Last Judgement. It goes in only one direction, forwards: before Christ, its goal was Christ himself; after Christ, its goal is the Parousia. Finally, there is no time outside this sacred time of God's Creation: there is only one God, only one world, and only one history for all peoples. These three characteristics underpin the unique Christian genre of universal history. God's plan of salvation embraces all nations, whether they are aware of it or not. He foreordained that the Romans would rule the world at the time of the Incarnation, just as He chose the Jews as His special people. The single thread of divine purpose draws all particular histories into one continuum of time.

The linear direction and unity of this world-history is secured by chronology, the precise establishing and synchronizing of all historical dates within one time-frame. The father of Christian chronology was Eusebius of Caesarea (263–339).[198] Basing himself on the pioneering Christian chronology of Julius Africanus (*fl.* 221), and the Hellenistic

---

197 *De divisonibus temporum* 633B.

198 Croke; Tristram 19–22; Alden A. Mosshammer, *The Chronicle of Eusebius and Greek Chronographic Tradition* (Lewisburg: Bucknell University Press and London: Associated University Presses, 1979); Molly Miller, "The Chronological Structure of the Sixth Age in the Rawlinson Fragment of the 'Irish World Chronicle'", *Celtica* 22 (1991):79–83.

tradition of synchronized histories of the Mediterranean world, Eusebius contructed a pair of seminal chronological works, the *Chronographia* and the *Chronikoi Kanones*. The former survives only in a doubtful Armenian translation and a Syriac fragment, and seems to have had little influence. The impact of the latter, however, was enormous. The *Kanones* is a time-line chronicle, beginning with Abraham. Its chronological backbone is the regnal lists of the world-empires of Assyria, Persia, Macedonia and Rome, supplemented after 776 BC by the Olympiads. As peoples and empires appear on the stage of world history, they are granted their own parallel columns within the table; as they are absorbed into the world-empires, these columns gradually disappear, until the Roman empire occupies the entire frame. This unusual format stresses the coincidence of the histories of various peoples, and the key role of chronology in tying these stories together. Indeed, Eusebius' signal achievement was to synchronize the dates and events of the Bible with those of the pagan kingdoms within a format that furnished continuity, parallelism, and chronological exactitude.[199] The *Kanones* had an extraordinary career, particularly in the West, where they were translated and edited by Jerome. The original *Kanones* ended at AD 303; Jerome's continuation went until 378, that of Prosper (*ca.*390–463) to 445, and that of Marcellinus Comes to 534. Bede had access to this full edition of the *Kanones*.

Eusebius' *Kanones* synchronizes eras, but does not synthesize them, save in the form of a running total of years since the birth of Abraham in the outer margin. In the preface of the work, however, Eusebius discusses the periodization of world-history before Christ. He divides the time between Abraham and the beginning of Jesus' ministry into four periods – (1) Abraham–Moses (2) Moses–First Temple (3) First Temple–Second Temple (4) Second Temple–Ministry – and adds that the time prior to Abraham can be divided into two periods, namely (1) Adam–Flood, and (2) Flood–Abraham.[200] The dates for the period after Abraham are established by the regnal lists, for Abraham was born in the 43rd year of King Nilus. The first two Ages are much more problematic, because the years and generations in the Septuagint translation of the Old Testament are quite different from those in the Hebrew text. Eusebius preferred the Septuagint version, on the grounds that

199 Tristram 20; Miller 80.
200 Helm's introduction to his ed. of Jerome's *Chronicon*, 14–17.

modern Hebrew codices are more likely to be corrupt than the ancient exemplars used by the Seventy Translators, but he probably began his chronicle with Abraham precisely to avoid this controversy.[201] Eusebius dated the beginning of Christ's ministry to AM 5228. If Christ was 30 years of age at this time, he would have been born in 5197 or 5198; Jerome preferred 5199, which was the conventional date in the West.

Augustine was enormously dependent on Eusebius, and it was through Augustine's eyes that Bede read the Greek chronographer. Augustine was less interested in exact chronology than in the mystical significance of the World-Ages, whose allegory and number symbolism he explored in many of his works.[202] He reduced Eusebius' six pre-Christian Ages to five, which with the addition of the Christian era, made six. He also reordered the last three Ages before Christ to match the three periods of fourteen generations enunciated in Matthew 1.17: Abraham to David, David to the Babylonian Captivity, Babylonian Captivity to Christ.

The number six was highly significant for Augustine, because it linked the Ages of the world to the days of Creation. This analogy was commonplace in patristic literature,[203] but Augustine elaborated it considerably: each Age, like a day, had a morning, a noontide, and an evening – in other words, a bright beginning, a zenith, and a troubled period of decline. Bede, as we have seen, borrowed this model for his exposition of the World-Ages in ch. 10 of *The Reckoning of Time*. Augustine also proposed a second analogy between the Six Ages of the world and the six ages of human life: indeed, he deliberately selected *aetas* as his term for the six eras or epochs of world-history, by analogy with the ages (*aetates*) of human life. The key text here is *De Genesi contra Manichaeos* 1.23.35 *sqq.*, a work which Bede certainly knew,[204] but the analogy is also found in other Augustinian works, notably *De civitate Dei* 16.43. The First Age is like infancy, for the world emerged into the

201 Miller 80.

202 For a survey of these, see Tristram 23; cf. Luneau pt. 4.

203 Siniscalco 316–317. As Siniscalco later observes, this analogy has the advantage of contrasting the linearity of historic time with the cycles of nature and the calendar; history, like Creation, is unrepeatable: *ibid.* 327.

204 He mentions it as a source for his own commentary on Genesis, composed in 720: *In Gen., praef.* (1). Bede's much more skeletal treatment of the analogy between the World-Ages and the ages of man in *On Times* seems to be based on Isidore, *Etym.* 5.38; cf. Siniscalco 323.

light of Creation just as man emerges into the light of life. The Second Age corresponds to childhood, the Third to *adolescentia*, the Fourth to *juventus*, and the Fifth to "a decline from youth into old age".[205] Bede borrowed this scheme for his prologue to ch. 66, but with some significant changes, as we shall see.[206]

Augustine was interested in the Six Ages as a theology of history, not as a chronological framework. Indeed, when he touched on the length of the World-Ages at all, he tended to employ generations rather than years, vacillated between the Incarnation and the Ministry as the starting point for the Sixth Age, and confessed agnosticism or lack of interest with regard to the discrepancies between Septuagint and Hebrew chronologies.[207] Significantly, his protégé Orosius did not use the Six Ages scheme in his own *Historia*.

A different approach was taken by Isidore of Seville, one which had a formative effect on Bede. Isidore took the momentous step of fusing Eusebius' chronology to Augustine's division of the Six Ages, to produce the first world chronicle dated according to *annus mundi*.[208] In other words, he abandoned Eusebius' parallel eras for a single chronological backbone based on the date of Creation. His figures, however, were derived entirely from the Septuagint chronology outlined in Eusebius' prologue. Isidore organized the history of the first three Ages according to the generations of the Hebrew patriarchs, and then according to the reigns of the Judges and kings. After the fall of the Jewish kingdom, the chronological frame was furnished by the heathen empires which ruled Judaea: Macedon, Ptolemaic Egypt, and finally Rome. Isidore dated reigns according to their closing year, not their inception.[209] Finally, he also innovated by fixing the beginning of the Sixth Age definitively at Christ's birth.[210] All these features were taken over by Bede. Indeed Bede's debt to Isidore for

205 *declinatio a juventate ad senectutem*: Augustine, *De Genesi Contra Mani-chaeos* 1.23.40 (PL 35.191).

206 On this theme in general, see Elizabeth Sears, *The Ages of Man*, ch. 3, "The Ages of Man and the Ages of World History"; Paul Archambault, "The Ages of Man and the Ages of the World: a Study of Two Traditions", *Revue des études augustiniennes* 12 (1966):193–228.

207 *DCD* 18.43 (639.22–28). Cf. Tristram 24.

208 Eusubius and Jerome, it will be recalled, used the "era of Abraham"; Prosper used the *annus Passionis*.

209 This policy is also followed by the *Chronicon Paschale*.

210 Tristram 24–25.

the substance as well as the form of the chronicle in *On Times* is very significant, and if Isidore is somewhat eclipsed by Eusebius in *The Reckoning of Time*, the formal influence remains.

To summarize the argument thus far, Bede was the heir to a complex tradition of world-chronicles, based on Eusebius' synchronization of biblical and secular time, and overlaid with a particularly Augustinian reading of the World-Ages. His immediate source was Isidore's chronicle in *Etymologiae* 5.29, but he reached back to Augustine for the analogy with the six days of Creation, and with the six ages of man. However, he introduced a fundamental modification of this tradition. In *On Times*, Bede replaced the Eusebian-Septuagint chronology of the first two World-Ages with a new chronology based on Jerome's translation of the Hebrew text of the Old Testament. As had been recognized long since, even by Eusebius himself, the two vary considerably in their chronology of the first two Ages. Where the Septuagint assigns 2,242 years to the First Age and 942 to the Second, Bede, following the Vulgate, gives 1,656 years to the First Age and 292 to the Second Age. The upshot is that the birth of Christ, which Isidore, following Eusebius, dated to AM 5197, is dated by Bede to AM 3952. This revised chronology drew down accusations of nothing less than heresy, for there was a common conception that each of the Six Ages was 1,000 years in length; Bede's new chronology, therefore, did not place the birth of Christ in the Sixth Age. Bede refuted these accusations in his *Letter to Plegwin*, and devoted ch. 66 of *The Reckoning of Time* to restating and elaborating his case.

Central to his argument is the idea that not all the World-Ages are of equal length, and by implication, the duration of the Sixth Age cannot be known, a point he will expand upon in chs. 67 and 68. At the outset of ch. 66, Bede sets the stage for this larger argument by recasting slightly Augustine's analogy between the Six World-Ages and the six ages of man's life. Though Augustine explores this theme in *De Genesi contra Manichaeos*, Bede's text corresponds more closely to *De vera religione* 26.48: for example, the First Age is likened to childhood because childhood is a time which is lost in oblivion, as the First Age was wiped out by the Flood.[211] But what is particularly significant is Bede's comment on the Sixth Age, where he steps outside the Augustinian framework

---

211 *De vera religione*, ed. J. Martin, CCSL 32 (1961):217. Contrast *De Genesi contra Manichaeos* (PL 35.190) which simply connects the Creation in the First Age to human birth.

completely. The sixth age of human life is characterized not only by decline, but by the fact that we do not know the time of death. Unlike the other five ages of life, it has no predetermined end-point. Similarly the Sixth Age, unlike the other five, has no fixed number of years, for its end – and hence the end of the world – is known to God alone. Isidore, to be sure, closed his chronicle by averring that "the rest of the Sixth Age is known to God alone",[212] and Augustine noted in *De civitate Dei* 23 that there is no number of generations to be assigned to the Sixth Age, but Bede's integration of this warning into Augustine's anthropological analogy is quite original. It reinforces the whole message of the world-chronicle, which is a polemic against chiliasm.

The choice of a Six Ages scheme and an AM reckoning for the chronicle in ch. 66 has puzzled some scholars, who wonder why Bede discarded both in his *Ecclesiastical History*.[213] It might also be remarked that the AM reckoning is a little unexpected in the context of the *Reckoning of Time* itself, since the chronicle is a kind of gloss on the Paschal table, in which years are numbered according to Dionysius' AD scheme. Why did Bede take this route? A second question is related to this first one: why did Bede revise the standard chronology of the *annus mundi* so radically in *On Times*, shaving nearly 1200 years off the conventional age of the world, and why, when this new chronology provoked criticism, did he proclaim it afresh, and in even more emphatic terms, in *The Reckoning of Time*?

These questions are linked to the issue of the relationship between *computus*, history, chronology, and the end of time. In a recent and very penetrating essay, Richard Landes has pointed out that the fusion of world-chronicle and the scheme of the Six Ages is not only unique to the West, but fraught with theological and pastoral implications.

Though subject to variation, most AM chronologies dated the Incarnation to the fifth millenium: Julius Africanus to 5500, and Eusebius to 5197/8 (or in Jerome's version, 5199). When the schema of the Six Ages is laid over this chronology, it immediately suggests that each of the six World-Ages is about 1,000 years long, for Christ's Incarnation inaugurated the Sixth Age. This chronological coincidence tapped into an

---

212 *Residuum sextae aetatis tempus Deo soli est cognitum*, *Etym.* 5.29.42; cf. *Chron.* 418 (ed. Th. Mommsen, p. 481): *residuum saeculi tempus humanae inuestigationis incertum est.*

213 Introduction, section 4.

ancient piece of Christian eschatological folklore, namely that there are six Ages, each Age one thousand years in length. The cue here is II Peter 3.8 "... one day is with the Lord as a thousand years, and a thousand years as one day" (cf. Psalm 90.4/89.4). This notion of the divine thousand-year day, when joined to the analogy between the World-Ages and the days of Creation, produced the idea of the thousand-year World-Age. The logical conclusion is that the world will end in *annus mundi* 6000, but if that is the case, it means that one can predict the time of the end of the world if one knows the date of its beginning. And if each of the Six Ages is 1,000 years long, then the thousand-year reign of the saints after the general resurrection (Apoc. 20.4) must be a seventh World-Age, a cosmic sabbath corresponding to the seventh day of the Creation week, but still within historic time. Chronology, then, becomes unwittingly the accomplice of a both a heretical chiliasm and an explosive millenarianism.[214] Millenarianism was the target of Augustine's redefinition of the Seventh Age as the duration of the Church Expectant, from the time of Abel until the Last Judgement, a position which Bede will expound in ch. 71. Chiliasm, however, was less easy to uproot.

The reaction of the Church to chiliasm was paradoxical, for though it strenuously condemned it, it hesitated to take the obvious measure to uproot it, namely jettisoning the notion of the Six World-Ages. Moreover, the nature of Christian historiography prevented the Church from downplaying chronology in the hopes of avoiding speculation about the date of the Last Judgement. The result was that Western writers revised the *annus mundi* chronology at regular intervals to "rejuvenate" the world and postpone the millenium, without abandoning the basic framework of the Six World-Ages. As Landes has observed, these revisions always took place in the final century leading up to the year 6000, and usually rejuvenated the world by only a few hundred years: enough to allay apocalyptic terrors, but not enough to push the end over the horizon of visibility. First, Julius Africanus' chronology was replaced by Eusebius' in the late fourth century, and as Eusebius' year 6000 approached in the middle of the eighth century, fresh revisions of the AM reckoning were aired. Bede's radical revision of AM was one of these. Jerome's Vulgate inadvertently offered Bede a way of eluding the impending Eusebian deadline.[215]

---

214 Landes 142–149; von den Brincken 50–60.
215 Cf. Landes 176; von den Brinken 110.

Bede's new chronology drew fire because unlike previous revisions, it postponed the Last Judgement for a very long time – almost 1,200 years. The captious critics at Bishop Wilfred's table were not "rustics" (to borrow Bede's scornful epithet in the *letter to Plegwin*) but ordinary educated clerics, and their reaction reveals the problems engendered by the Western Church's ambivalence with regard to chiliasm. Deeply as it deplored these speculations, the Church could not abandon the Six Ages chronology or the equation of age and millenium sanctioned by so many authoritative texts. Instead, it elected, in Landes' phrase, to "mortgage the future" for the sake of peace and good order in the present, by continually pushing the Parousia a little bit ahead, but not so far ahead as to rupture the implicit connection between millenia and World-Ages.

We may hypothesize, then, that in *On Times* Bede revised Eusebius' chronology in the interests of what he saw as historic accuracy. But after the Plegwin affair, he defended his innovation for a new reason: because it offered a radical challenge to the chronic heresy of chiliasm.[216] As Landes points out, chiliasm was far from moribund in Bede's day, nor was it merely a feature of "popular religion". The Plegwin episode shows that a purely chiliastic reading of the Six Ages of the world could be aired in the presence of an English bishop without any hue and cry of heresy being raised. On the contrary, it was Bede, whose revised chronology was radically anti-chiliastic, who was called a heretic. Bede may also have been responding to the aberrant chronology of the Byzantine writer John Malalas, introduced into England by Archbishop Theodore of Canterbury, and promulgated through his *Laterculus Malalianus*, a historical exegesis of the life of Christ. In the *Laterculus*, Theodore complains about the stupidly and quarrelsomeness of the Irish, who insist on rejecting what Theodore regards as the orthodox and accepted chronology of Christ's life. That chronology actually dated the Crucifixion to AM 6000, which meant the world was already into its seventh millenium. The peevish Irish were probably defending the standard Eusebian chronology, but they were also demonstrating the force of chiliasm, which continued to regard the year 6000 as the scheduled end of the Age.[217]

---

216 Cf. Siniscalco 309–311, who cites evidence of Bede's increasing use of this motif in the biblical commentaries written between *On Times* and *The Reckoning of Time*.

217 See the Irish text of AD 645 on this subject edited by Ó Cróinín in "Early Irish Annals from Easter Tables: a Case Re-Stated", *Peritia* 2 (1983):79–81; *ibid.*, "Irish Provenance", 235; and the anonymous 7th-century Irish commentary on the First Epistle of John, ed. Robert E. McNally, *Scriptores Hiberniae minores*, CCCM (1973):1.40; cf. Smyth 107.

Certainly the *Laterculus* is frank about its belief that each Age of the world lasts one thousand years – a position Bede deplores.[218] For all these reasons, Bede saw fit to reassert his position in *The Reckoning of Time*.[219]

Ironically, Bede unwittingly solved the problem of chiliasm by popularizing *annus Domini* reckoning, both through his *Ecclesiastical History*, and through his Dionysian Paschal tables.[220] Taken up by the Carolingians, the *annus Domini* era eventually ended the reign of the Creation era in the West, and drew the millennial sting out of the schema of the Six World-Ages.[221]

Landes argues that Bede's new *annus mundi* chronology was almost universally rejected, but von den Brinken shows that many Carolingian

218 On the *Laterculus Malalianus* see Jane Stevenson, *The Laterculus Malalianus and the School of Archbishop Theodore*, Cambridge Studies in Anglo-Saxon England 14 (Cambridge: Cambridge University Press, 1995): esp. ch. 3 of text (p. 124), and *idem*, "Theodore and the *Laterculus Malalianus*", in Michael Lapidge, ed., *Archbishop Theodore: Commemorative Studies on his Life and Influence*, Cambridge Studies in Anglo-Saxon England 11 (Cambridge: Cambridge University Press, 1995):204–221. There is no evidence that Bede knew the *Laterculus Malalianus* directly, but he certainly had access to other literature from the school of Theodore and Hadrian. Benedict Biscop, after all, had guided the newly consecrated archbishop and his deacon to England, and had lived in close proximity to them as abbot of St Augustine's for several years after. Bede revised the *Passio S. Anastasii* which Theodore probably brought to England: see Carmela Vircillo Franklin, "Theodore and the *Passio S. Anastasii*", in *Archbishop Theodore*, 175–203. Bede also had access to another rare Greek text associated with the Canterbury mission, the lapidary of Epiphanius: see Peter Kitson, "Lapidary Traditions in Anglo-Saxon England: Part II, Bede's *Explanatio Apocalypsis* and Related Works". Had Bede known of the *Laterculus*, he might not necessarily have associated it with Theodore, as he seems to have known little about Theodore's career and learning. In *The Reckoning of Time*, ch. 66, *s.a.* 4591, he mentions the Roman monastery of "Ad Aquas Salvias", but does not remark that Theodore was the abbot of this house: cf. Franklin 203.

219 Jones also argues that the doctrine of the Six Ages was of minimal importance to Bede, a mere "teaching device" without connection to historical reality. Indeed, he claims that the Plegwin episode made Bede almost embarrassed by the concept: "Bede as Early Medieval Historian", 31–32. This position is untenable, and Jones seems to have modified his views in later life: see the section on the Six Ages in "Some Introductory Remarks", 191–198. Not only does Bede invoke the concept of the Six Ages with great enthusiasm in many of his works (for a list of these, see Plummer xli–xlii, n. 6, and Siniscalco 311, n. 30), but the Plegwin affair made him bolder than ever in asserting its relevence, for he wanted, above all, to detach it from the pernicious notion that each Age is one thousand years long. This is why he devotes the final seven chapters of *The Reckoning of Time* to the issue of the World-Ages.

220 Landes 178.

221 *Ibid.* 140–141.

and post-Carolingian historians used it without demur,[222] and Tristram argues that the later success of the Six Ages doctrine was largely due to Bede.[223] Indeed, Bede's eccentric *annus mundi* reckoning was taken up by some for exactly the reason that Bede devised it. MS Leiden Scaliger 28 is a 9th-century *computus* manuscript from Autun, with Bedan Easter cycles extending from AD 1 to AD 1006. The closing date is significant: the tables should have gone to 1063, the end of the second 532-year cycle. Did the scribe imagine that the world might end in AD 1000, and terminate his table at the completion of the 19-year cycle closest to the fatal year? Certainly one of the readers of Scaliger 28 entertained analogous apocalyptic anxieties. He collated the AD years to their AM equivalent. He used both the Eusebian and the Bedan chronology, but discontinued the Eusebian one after AD 789, the year 5990 in the Eusebian scheme. Thereafter, he continued up to his own time (AD 806) only with Bede's chronology, for the other chronology had already put him over the frontier of *annus mundi* 6000.[224]

The Great Chronicle of ch. 66, and indeed the whole block of chapters 66–71, forms a thematically distinct section within *The Reckoning of Time*, but its position within the work has often seemed somewhat tenuous, even to the first generations of its readers. While Bede's extension of the divisions of time to the cycle and the age might seem logical, the somewhat controversial tone and the predominantly historical and theological content of these chapters sets them apart from the rest of *The Reckoning of Time*. As a result, from very early on in the manuscript tradition, there was a tendency to detach this section from the rest of the book. One very early codex, the Reichenau manuscript Karlsruhe, Aug. 167 (s. IX[1]),[225] drops the chronicle completely. So do other Carolingian manuscripts like Munich CLM 14725 (Regensberg? s. IX[1]) and Paris, Bibliothèque nationale n.a.l. 1632 (prov. Orléans, s. IX[2]). Dropping the chronicle sometimes led to dropping everything after the chronicle, e.g. in Berlin, Staatsbibliothek lat. 130 (Phill. 1832) (Metz, 873?).[226]

---

222 Cf. von den Brincken 113–120 and Appendix, Table IV. For its use by Irish chroniclers, see Miller, and Ó Cróinín, *The Irish 'Sex Aetates Mundi'* (Dublin: Institute for Advanced Sciences, 1983). For use by Freculf of Lisieux and by St Gall historians, see Allen, "Bede and Freculf", 78–80.

223 Tristram 30.

224 Landes 188–189.

225 See Introduction, section 6.

226 Other examples include Brussels 9932–9934 (Liège, s. XI), Cambrai 925 (prov. Cambrai Cathedral, s. IX); Ivrea 42 (813?); St Gall 248 (St Gall, s. IX); Vat. lat. 644 (St Gall,

Occasionally one encounters a curiously exclusive interest in the chronicle and closing chapters. Rome, Biblioteca Vallicelliana E 26 (Lyon, s. IX[1]) fols. 43v–71v is "possibly a half of *DTR* rearranged in an accurate and careful text",[227] followed (fols. 91r–136v) by the full text of the chronicle and closing chapters. Troyes 1071 (s. X) fols. 46r–62 is an example of a manuscript where the chronicle and last chapters (incomplete) are included as a separate work.[228] Some codices bear witness to efforts to repair a truncated text from another exemplar: Vatican Pal. lat. 1449 (Mainz s. IX), for example, contains chs. 1–65 on fols. 27v–104r, and the rest on fols. 121r–145v; Vatican Reg. lat. 1038 (s. X) also contains chs. 1–65, but with chs. 69–72 added later.

It is hard to pinpoint any single reason for these amputations. The chronicle is certainly a long chapter which would consume much parchment and not be very useful, since its dating scheme was eccentric; the closing chapters on the last days probably seemed of little practical use to Carolingian schoolmasters bent on using *The Reckoning of Time* as a classroom text. But it is not beyond possibility that a more overt hostility may lie behind this phenomenon. Bede's radical postponing of AM 6000 and his impeccably Augustinian policy on the unpredictability of the last days would be unwelcome both to millenarians and to anti-millenarians. Millenarians would object to the explicit message, anti-millenarians to the implicit one, namely that by harping so insistently on the dangers of forecasting the Last Day, Bede was admitting that the Church's war against such speculation was still far from won.[229] *On Times* had already been compromised in some eyes by this new chronology, and *The Reckoning of Time* was far too useful to place at risk.

The content of Bede's world-chronicle has until recently not inspired much admiration. It has generally been assumed that Bede assembled information from prior chronicles without any conscious plan or

---

s. X); Paris lat. 5239 (Limoges, s. X); lat. 5543 (Fleury s. IX[2]); lat. 7297 (Blois, s. X); St Gall 459 (St Gall, s. X[1]); also later MSS like Oxford Rawlinson C 308 (prov. Hereford, s. XIII).

227 Jones, *BOT* 156.

228 *Ibid.* 157.

229 Landes comments on the final chapters of *The Reckoning of Time* that "[o]ne could not ask for less ambiguous testimony to the survival of the most dangerous kind of sabbatical millenarianism in the early 8th century" (177).

overarching purpose.[230] Jones remarked that Bede "selected items for his chronicle which illustrated the generalizations about time recorded in the theoretical part of the volume",[231] by which he seems to mean that he selected events more or less at random to fill out the designated periods of each Age. Bede wants to demonstrate the theory of the Six Ages, not write a history.

More recent studies of the tradition of the Christian world-chronicle, however, provide evidence that Bede was following a coherent tradition in selecting material for ch. 66. In selecting the Eusebian model, and dating by *annus mundi*, Bede was consciously choosing a genre which conveyed a specific message to his readers.[232] Jones argues that Bede was constrained to use AM by his source, Eusebius; when he came to write the *Historia ecclesiastica*, his sources were annals in Paschal tables, and he could therefore use AD.[233] This argument founders on the fact that ch. 66 contains much of the same material found in the *Historia ecclesiastica*, but dated according to AM. The choice, it would seem, was deliberate, not compulsory. Bede's purpose in ch. 66 is to demonstrate the sequence and articulation of the World-Ages, and also to refute the claims of chiliasm. To have begun the count of years afresh at the birth of Christ – in other words, to have introduced a separate count for the Sixth Age – would have interrupted this sequence and worked against Bede's purpose.

As we noted above, Bede's main chronological "policies" in ch. 66 are largely derived from Isidore: the continuous *annus mundi* chronology, the articulation of Hebrew history according to patriarchs, judges and kings, the dating of reigns at their close rather than their inception, and the choice of the Incarnation as the beginning of the Sixth Age.

In the first five Ages and the beginning of the Sixth, the guiding thread is essentially the biblical narrative. It is only in the Sixth Age where we see Bede selecting certain themes for inclusion. His choices were limited by the basic formula for the world-chronicle established in Hellenistic

---

230 The only major study of the chronicle, Georg Wetzel, *Die Chronicen des Baeda Venerabilis* (Halle: Plötz'sche Buckdruckerei, 1878), opens by describing the chronicles in *On Times* and *The Reckoning of Time* as "an sich unbedeutend und Compilationen" (3). Wilhelm Levison dismissed the chronicle as "a compressed Jerome": "Bede as Historian", in *Bede: His Life, Times and Writings*, 121.

231 "Bede's Place", 32.

232 See Introduction, section 4.

233 "Bede's Place", 50.

Alexandria, notably by Eratosthenes: a world-chronicle should include battles, victories, political events, games and festivals, literary figures, and natural phenomena.[234] For "literary figures", Bede could substitute Church Fathers and ecclesiastical writers, culled from the collective biographies of Jerome and Gennadius. Political events were expanded to include heresies, a subject of intense importance to Bede. The capacity of the world-chronicle to absorb the histories of new peoples as they appear on the global scene[235] permits Bede to include much material about Britain, derived largely from Gildas. But there are two kinds of information which seem to be peculiar to Bede, and not accounted for by his sources and models: architecture and *computus*. Bede combed the *Liber pontificalis* for items of general ecclesiastical interest, but building projects exercised a disproportionate fascination for him. Constantine's churches are listed in loving detail, while the Council of Nicaea is dismissed in a mere sentence (*s.a.* 4290); the building projects of Damasus (*s.a.* 4328), Innocent (*s.a.* 4362) Sixtus (*s.a.* 4403) and others are carefully chronicled. So are the milestones in the history of *computus*: the writings of Theophilus (*s.a.* 4338), Paschasinus of Lilybaeum (*s.a.* 4403), Victorius (*s.a.* 4427), and Dionysius and Victor of Capua (*s.a.* 4518). Above all, Bede marks out the bounds of the Paschal controversy in the British Isles: the detection of the Celtic error (*s.a.* 4591), and the conversion of Iona by Egbert (*s.a.* 4670) – the latter event, significantly, is also dated according to *annus Domini*.[236] In short, Bede stretched the conventions of world history to accommodate some very personal interests, such as the history of his own nation, and some subjects especially pertinent to *The Reckoning of Time*: *computus* itself, and architecture, whose permanence and order made it an allegory of *computus*.[237]

234 Croke 119.

235 Miller 81.

236 Cf. *HE* 5.22 (552–554). What Bede omitted from ch. 66 may also have been conditioned by his focus on *computus*. Thacker argues that Aidan and Oswald are not mentioned in the chronicle because Bede was reluctant "to give prominence to two saints who, however much esteemed, nevertheless celebrated Easter incorrectly, in a work designed to illustrate the proper reckoning of time and the triumph of its orthodox exponents" (57).

237 Cf. Faith Wallis, "Images of Order in the Medieval Computus", in *ACTA XIV: Ideas of Order in the Middle Ages*, edited by Warren Ginsberg (Binghamton: State University of New York Press, 1990):45–67.

Chapter 67:

In *On Times*, Bede did not discuss future time, or the end of time, or the notion of the Seventh and Eighth Ages, nor did his model, Isidore, in *Etymologiae* 5.[238] Why did he deem it essential to do so in *The Reckoning of Time*? In one respect, this coda, and particularly chapter 67 on "the remainder of the Sixth Age", is a natural extension of the chronicle of the Six Ages. These Six Ages articulate the time of this fleeting world: they are *sex aetates ... mundi labentis*. *Labens* in a temporal context means "passing away", but in fact its primary meaning is spatial: gliding, sliding, slipping in a gentle and almost insensible manner. The flavour of the word is slightly sinister, as of the silent beginnings of an avalanche, or the first careless footstep into quicksand: hence the extended meanings "to sink into ruin", "to fall into error". To Bede, the *saeculum* began with the Fall, and the "passage" of time describes a decline, a decline which will be played out to the end of the Age.

But the closing chapters of *The Reckoning of Time* are also a continuation of the argument against chiliasm begun in ch. 66. These pages may contain little of the burning outrage of the *Letter to Plegwin*, but the episode is still present *en filigrane*. At the outset of ch. 67, Bede strikes a pose of cool detachment: there may be those who object to his chronology, but if so, they are under no obligation to read it, or they can use the same sources to construct one that pleases them better. Details of chronology are not matters of principle, and it is to matters of principle that Bede wishes to direct his reader's attention. These are three. First, no matter what chronology one proposes, it is not licit to use it to speculate about the timing of the end of the Age, for that is known to God alone. Bede may well have regretted not following the example of Isidore's *Chronicle* by ending his *On Times* chronicle with an admonition against calculating the remainder of the Age. If so, he made up for it in *The Reckoning of Time* by a lengthy demonstration of the impossibility and inadvisibility of such computations. The problem was all the more acute in that the widespread availability of Paschal tables equipped with an era (Victorius' *annus Passionis* or Dionysius' *annus Domini*) made it very easy to track the progress of the sixth millenium: all one had to know was the *annus mundi* of the initial year of the era, and one could find out

---

238 Isidore does, however, bring up the Seventh Age in *Quaestiones in Vetus Testamentum* (PL 83.213); cf. Tristram 25.

readily where one stood with respect to the end. *Computus*, ironically, helped undermine the official anti-chiliasm of the Church.[239]

Secondly, Bede wishes to remind his reader that the belief that each of the Six World-Ages is one thousand years long has no biblical or patristic authority. This is the burden of his dialogue with the millenarians, where he takes his revenge for the Plegwin episode by concluding that their chiliasm is heretical. Thirdly, Bede attacks millenarianism, the belief that there will be a post-Judgement seventh historical World-Age, in which the saints will reign with Christ on Earth. Like the chiliasts, proponents of this view based their contention on an analogy between the World-Ages and the week of Creation. Bede, on the other hand, is the spokesman for Augustine's concept of the Seventh Age, enunciated principally in *De civitate Dei*, as the "sabbath rest" of the souls of the righteous in heaven. This sabbath runs in parallel to the World-Ages. It began with the division of humanity into two "cities" when Cain murdered his brother Abel.

Bede's conception of the end of time and the Seventh and Eighth Ages is deeply rooted in this Augustinian model of time, and is reiterated in many of his other homiletic and exegetical writings. It is announced in various passages of *The Reckoning of Time* as well, notably in his comment in the Chronicle about the construction of Solomon's Temple: "[The Temple] was finished in seven years, and dedicated in the seventh month of the eighth year, as a symbol of the totality of time in which the Church of Christ, which is made perfect in the future [Age], is built up in this world." Seven is the number of this Age, eight of the Age to come.

Millenarianism, the heretical twin of chiliasm, is also a target of patristic arguments. But at the same time as they were trying to stifle millenarianism, the Fathers developed the idea of the "sabbatical millenium", where the millenium predicted in John's Apocalypse becomes a seventh World-Age. All this did, of course, was encourage the persistence of the notion that the other six Ages would likewise be 1,000 years in length.[240] To save the idea of the "sabbatical millenium" while avoiding this invidious corollary, Augustine devised an ingenious interpretation: the Seventh Age was not an Age like the others, but a way of designating the duration of the Church Expectant, the totality of the elect from the time of Abel to the end of the world. It is, then, simulta-

239 Landes 168–169.
240 Landes 144.

neous with the Six World-Ages, and lasts far longer than one thousand years. Bede's resistance to millenarianism is measured by his enthusiasm for this Augustinian Seventh Age, which he not only unfolds in *The Reckoning of Time*,[241] but discusses in other theological and homiletic writings.[242]

Bede, then, is the heir to a complex and paradoxical situation: like Jerome and Augustine, he is intent on pressing home the gospel message that no one can calculate the Second Coming, but at the same time, he is shackled to the concept of the Six Ages. This scheme was not only approved by generations of Fathers, and notably Augustine, but had a special appeal for Bede as a computist. The concept of the World-Age as a "week" evidently held considerable charm for Bede, who plays with it not only in *The Reckoning of Time*, but in other works: though potentially dangerous, it also demonstrated the applicability of *computus* to biblical exegesis.

Chapter 68:

Bede also draws upon Augustine in this chapter to disarm any notion that we can speculate about the time of the end. Augustine's *Letter* 199, itself a miniature treatise on the Day of Judgement, is the moral high ground from which Bede will reproach potential critics. Our first consideration in speaking about the end of time should be the ethical implications of such discourse. If we say that the Lord will come soon, we risk, however sincere our motives, bringing discredit upon the faith if we are wrong. But if we remain agnostic, or acknowledge that His coming may be delayed, we can only be pleasantly surprised if we are wrong, and we put no one's faith in peril.

Chapter 69:

There are, indeed, only two certain signs that the end is coming, though even these do not tell us precisely *when* it will come. These are

---

241 Cf. *The Reckoning of Time*, ch. 10 and ch. 71.

242 E.g. *Hom.* 2.19 (Vigil of St John the Baptist) describes the Seventh Age as a time of tranquillity for the soul *in the next life* (320.86.92); *Hom.* 1.24 tells how the Transfiguration occurs "after six days", which Bede interprets as a figure of the beatific vision after the Six Ages of the world (172.82–88); *Hom.* 2.6 (Easter) describes Christ's repose in the tomb on Holy Saturday as representing the sabbath rest of the saints in heaven (225.17–226.26).

the appearance of Antichrist, and the conversion of the Jews. The subject of the Antichrist, and indeed the entire prospective or prophetic dimension of the Apocalypse, is not one with which Bede was sympathetic. His own commentary on the Apocalypse, albeit written many decades before *The Reckoning of Time*, takes a strongly Augustinian approach to this work, interpreting it not as a blueprint of time to come, but as an allegory of the continuous troubles of the Church in this world.[243]

The "biography" of Antichrist in ch. 69 finds no analogue in Bede's commentary on the Apocalypse, and draws on a source which Bede did not use in this work, namely Jerome's commentary on Daniel. On the whole, Bede is not interested in Antichrist's character or career. He only brings him up because the prophet Daniel, the evangelist John and their patristic interpreters seem to have provided a precise chronological framework for his reign. But as Bede points out, these figures are highly ambivalent. Daniel says that Antichrist's reign will be 1290 days, and yet he also says that the faithful will have to wait 45 days more. So even the appearance of Antichrist does not give us licence to speculate on the exact timing of the Last Judgement.

Chapter 70:

This chapter seems like a curious, even eccentric digression. Its sole point is to refute the idea that the "new heavens" which will appear at the Last Judgement will entail the destruction of the present heavens, with their planets and stars. Drawing on Augustine, Bede argues that the "heavens" to be destroyed at the Judgement are only those which were destroyed at the Flood, namely the lower heavens or atmosphere. The darkening of the heavenly bodies foretold by Christ is nothing more than the occultation of their light by the overwhelming splendour of the Lord's advent; it does not mean that they will be annihilated. Even if the stars fell, their "place", the upper heaven itself, would have to remain, or there would be no place for them to fall *from*. The "new heavens" are "renewed by fire and glorified by a certain power of resurrection", not annihilated. Bede evidently thought much of this argument, for it is substantially repeated in his commentary on II Peter

---

243 See Gerald Bonner, "Saint Bede and the Tradition of Western Apocalyptic Commentary" (Jarrow Lecture, 1966).

3.5–6 and 10.[244] Though not formally developed in his early *Explanatio Apocalypsis*, it is already implicitly present, for Bede asserts that the

244 *'Per quae ille tunc mundus aqua inundatus periit'. 'Per quae', per caelos et terram dicit quae prius nominauerat. Per haec enim perdita mundus qui in his constiterat periit. Nam superiores mundi partes diluuium minime tangit . . . Perierunt et caeli 'secundum quantitatem et spatia aeris huius. Excreuit enim aqua', ut sanctus Augustinus ait, 'et totam istam capacitatem ubi aues uolitant occupauit; ac si utique caeli perierunt propinqui terris secundum quod dicuntur aues caeli. Sunt autem', inquit, 'et caeli caelorum superiores in firmamento' et utrum et ipsi perituri sunt igni, an hi soli qui etiam diluuio perierunt, disceptatio est aliquanto scrupulosior inter doctos*: 277.35–48. (*"Throughout these then that world was deluged with water and perished. Throughout these,* he means the heavens and the land which he [had] just named. For it was throughout these [things which perished] that the world which was made up of them perished. For the Flood did not touch the upper parts of the world at all . . . The heavens also perished, as Saint Augustine said, *according to the extent and breadth of this air. For the water increased and occupied the entire space where the birds fly. And so the heavens that are near the land, which people refer to when they say "The birds of heaven",* certainly perished. There are, however (he says) *also the higher heavens of heavens in the firmament.* But as to whether these also will perish by fire or only those that also perished in the Flood is a question that is somewhat more carefully disputed among the learned": trans. David Hurst, *The Commentary on the Seven Catholic Epistles of Bede the Venerable*, Cistercian Studies 82 (Kalamazoo: Cistercian Publications, 1985):147–148). Also: *'Adveniet autem dies domini ut fur in quo caelo magno impetu transiet.' Illos proculdubio caelos dicit qui in diluuio transierunt, hoc est aerem hunc terrae proximum qui igni perdendus est, tantum, ut recte creditur, spatii tenentem quantum aqua tenebat diluuii. Alioquin si quis caelos superiores ubi sol et luna et astra sunt posita transituros asseruerat, quomodo uult illam domini sententiam intelligere qua dicitur tunc 'sol obscuratur et luna non dabit lumen suum et stellae cadent de caelo'; si enim locus siderum transierit, id est caelum, qua ratione potest dici in eodem die domini et sidera uel obscurari uel cadere et ipsorum siderum loca quo fixa tenentur igne absumente transisse? 'Elementa uero calore soluentur.' Quattuor sunt elementa quibus mundus iste consistit, ignis, aer, aqua, et terra, quae cuncta ignis ille maximus absumet. Nec tamen cuncta in tantum consumet, duo uero in meliorem restituet faciem*: 279.110–125. ("*For the day of the Lord will come as a thief; on it the heavens will pass away with great violence.* He is undoubtedly speaking of those heavens that passed away with the Flood, that is, this air nearest the earth which is to be destroyed by fire, and which occupies only so much space, as is properly believed, as the water of the Flood occupied. Otherwise, if anyone maintains that the upper heavens where the Sun and the Moon and the stars are located will pass away, how does he wish to understand that thought of the Lord wherein it is said that then *the Sun will be darkened and the Moon will not give its light and the stars will fall from heaven*? For if the place of the constellations, that is, the heaven, has passed away, by what reason can it be said that on the same day of the Lord constellations are either darkened or fall and that the place of these constellations where they are held in place has passed away because of the consuming fire? *The elements in fact will be dissolved by heat.* There are four elements from which this world is formed, fire, air, water, and earth. That very great fire will lay waste all these. But yet it will not consume them that they will completely cease to exist, but it will consume two entirely, two in fact it will restore to a better appearance": trans. Hurst 150–151.)

new heavens and earth will be a transformation of the old world, not its destruction;[245] moreover, when he comes to comment on Apoc. 21.23 ("And the city had no need of the Sun, neither of the Moon ..."), he allegorizes the verse rather than comment on the fate of the physical Sun and Moon.[246]

Bede's concern with this detail – it is a mere polemical aside in Augustine's *De civitate Dei* – is indeed curious. In his commentary on Genesis, Bede explains that the Sun and Moon belong to "our heaven", as distinct from the "upper heaven" of God and his angels, created on the first day.[247] "Our heaven" is the firmament, which "God called ... heaven", (Gen.1.8), and in which the heavenly bodies are stationed.[248] The world of the elements, the Earth and lower atmosphere will be transmuted by fire; the Sun, Moon and stars will remain unscathed, but not because they are outside the world of time.

Since the Sun and Moon are within the world of time, and essentially the reckoners of time, Bede's insistence on their survival must say something about the relationship of time to eternity. If the Sun and Moon were destroyed, the reckoning of time would be destroyed as well, for these are the Creator's signs of times and seasons. Indeed, the Irish *De mirabilibus sacrae scripturae* deduced that because time was not destroyed in the Deluge, the waters of the Flood could not have reached

245 *Tunc figura hujus mundi supernorum ignium conflagratione praeteribit, ut, caelo et terra in melius commutatis, incorruptioni et immortalitati sanctorum corporum congrua utriusque commutationis qualitas conveniat*: PL 93.194C on Apoc. 21.1. ("Then the fashion of this world will pass away in the conflagration of the heavenly fires so that, when heaven and Earth have been changed into something better, the fitting quality of this double transformation may accord with the incorruption and immortality of the bodies of the saints.")

246 *Quia non lumine aut elementis mundi regitur Ecclesia, sed Christo aeterno sole deducitur per mundi tenebras*: PL 93.203C. ("Because the Church is not ruled by the light or the elements of the world, but she is led through the shadows of the world by Christ the eternal Sun.")

247 *Ipsum est enim caelum superius quod, ab omni huius mundi uolubilis statu secretum, diuinae gloria praesentiae manet semper quietum (nam de nostro caelo, in quo sunt posita luminaria huic saeculo necessaria, in sequentibus scriptura uel quomodo uel quando sit factum declarat)*: *In Gen.* 1.1 (4.37–41). ("For the heaven [created on the first day] is that upper heaven which, hidden from the condition of this fleeting world, remains ever at peace in the glory of the divine presence (for Scripture relates in what follows how and when our heaven, in which are placed the luminaries needful for this world, was made.")

248 *In Gen.* 1.1.6–8 (10.247–248).

the heavenly bodies.[249] Bede's concern, I believe, derives its meaning from the chapter which follows, the closing one of *The Reckoning of Time*, and its religious capstone. The meaning of Resurrection – Christ's at Easter, our own at the end of the Age – is not annihilation but transfiguration. Christ who is the same yesterday and today, is also the same forever. So also the Creation, and even time, are caught up into eternity, not destroyed by it.[250]

Chapter 71:

Easter is the *figura* of both the Seventh and the Eighth Ages, the sabbath rest of the saints, and the new Age of the world to come.[251] Bede reinforces this by self-consciously using a bit of *computus* jargon to describe this new age as "the endless *ogdoad* to come".[252] Here, once again, Bede reveals his deep indebtedness to the Irish cosmographers of the seventh century. The author of the *Liber de ordine creaturarum* envisages the new heaven and new Earth as an improved version of the old. They will share in the glorification of man at the resurrection, transformed and perfected, but not destroyed. The new Earth will have a spiritual nature, impervious to change; the Sun and Moon will shine at all times.[253] In short, what is now

249 1.8 (2160); cf. Smyth 164–165.

250 Cf. Bede's commentary on II Peter 3.11: '*Praeterit enim figura huius mundi', [I Cor. 8.31] non substantia, sicut et carnis nostrae non substantia perit sed figura immutabitur quando quod 'seminatur corpus animale surgit corpus spiritale'[I Cor. 15.44]*: 280.134–136 ("*For the shape of this world passes away* not its substance, just as with our bodies too, the shape will be changed, [but] the substance [will] not perish when *what is sowed as a physical body rises as a spiritual body*": trans. Hurst 151.)

251 On the patristic tradition of the analogy between Holy Week and the Six Ages, see Siniscalco 319–320; A. Quacquarelli, *L'ogdoade patristica e suoi reflessi nella liturgia et nei monumenti* (Bari: Adriatica, 1973): W. Rordorf, *Sabbat et dimanche dans l'Eglise ancienne* (Neuchâtel: Delachaux & Niestle, 1972).

252 Jerome also refers to the Eighth Age as the *ogdoad* in *Ep.* 140.8 (PL 22.1172). The computistical context of *The Reckoning of Time*, however, lends a different dimension to the allusion.

253 *De ordine creaturarum* 5.4 (114) and 11.4–7 (170–172); Smyth 292–3 and 307. Cf. *De ordine creaturarum* 5.5–7 (114–116), where the author invokes the prophecy of Isaiah 30.25–26 and 65.17 that in the new Age the Moon will outshine the Sun, and the Sun be seven times brighter than it now is, to prove that the luminaries will not be destroyed, but restored to the brilliance they lost at Adam's fall: cf. Smyth 174. The cosmographers' emphasis on transformation and glorification is in strong contrast to the images of annihilation conveyed by other Irish depictions of the end-time, such as the *Altus Prosator*, or ps.-Hilary,

experienced as time, will then be experienced as eternity. The Eighth Age is still an age, but it is an age without end, an age which, unlike those of this *mundus labens,* will never decline and fall. Time will not cease to *exist,* but it will cease to *run out,* and the "fleeting and wave-tossed" course of the ages will be redeemed in "eternal stability and stable eternity".

Bede's library contained many works by Augustine, the Father closest to Bede himself in his interest in time. We do not know whether it also contained the *Soliloquies* – Bede never cites this work – but Bede would surely have embraced the image of the reckoning of time expressed therein:

> O God, by whose laws the poles revolve, the stars follow their courses, the Sun rules the day, and the Moon presides over the night; and all the world maintains, as far as this world of sense allows, the wondrous stability of things by means of the orders and recurrences of the seasons: through the days by the changing of light and darkness, through the months by the Moon's progressions and declines, through the years by the succession of spring, summer, autumn and winter, through the cycles by the completion of the Sun's course, through the great eras of time by the return of the stars to their starting-points.
>
> O God, ... by whose ever enduring laws the varying movement of movable things is not suffered to be disturbed, and is always restored to a relative stability by the controls of the encompassing ages ...[254]

or even by Bede's own *De die judicii*: see the critical edition of this poem in *Opera rhythmica,* ed. J. Fraipont, CCSL 122 (1955):439–444, and discussion by L. Whitbread, "A Study of Bede's *Versus de die iudicii*", *Philological Quarterly* 23 (1944):193–221; "The Old English Poem *Judgement Day II* and its Latin Source", *ibid.* 45 (1966):635–656; and "Bede's Verses on Doomsday: A Supplementary Note", *ibid.* 51 (1972):485–486.

254 *Deus cui serviunt omnia, quae serviunt; cui obtemperat omnis bona anima. Cujus legibus rotantur poli, cursus suos sidera peragunt, sol exercet diem, luna temperat noctem: omnisque mundus per dies, vicissitudine lucis et noctis: per menses, incrementis decrementisque lunaribus; per annos, veris, aestatis, autumni et hiemis successionibus; per lustra, perfectione cursus solaris; per magnos orbes, recursu in ortos suos siderum, magnam rerum constantiam, quantum sensibilis materia patitur, temporum ordinibus replicationibusque custodit. Deus cujus legibus in aevo stantibus, motus instabilis rerum mutabilium perturbatus esse non sinitur, frenisque circumeuntium saeculorum semper ad similitudinem stabilitatis revocatur...: Soliloquies* 1.1.4 (PL 32.871); trans. Thomas F. Gilligan, Fathers of the Church 5 (New York: Cima, 1948):347–348.

Computus anticipates the Eighth Age of eternal stability and stable eternity through a hierarchy of units of temporal measurement, and an order based on rhythm and reiteration. Its ultimate product is the flawless mechanism of the Paschal table, which weaves the erratic courses of Sun and Moon into a perfectly paced ballet, whose figures never jar, and which can be repeated without variation or error until the end of the Age. The Paschal table embraces all time to come, and all time past, on which it bestows unambiguous dates. Within that system, the histories of all the peoples of the Earth are synchronized, and subordinated to the truth of Scripture and the Church. In the tables and formulae of the *computus*, time "moves" but "is not suffered to be disturbed"; the stability it achieves is "relative" and yet it is grounded in "enduring laws". *Computus*-time is, in short, a sacrament of eternity: not merely the shadow or symbol, but the pledge and foretaste of the everlasting Age to come.

APPENDICES

APPENDIX 1: BEDE'S SOLAR CALENDAR

This appendix contains a reconstruction in English of the solar calendar described by Bede in *The Reckoning of Time*. Its format is that of the standard ancient and medieval Julian calendar. The days are arranged in a vertical column, with the key letters to the left, and the martyrology and other notes to the right.

This reconstruction is based on Jones' in *BOT* 567 *sq.*, except that I have omitted the martyrology, and corrected the dates for the entry of the Sun into the various zodiac signs, to agree with Bede's main source, *De causis quibus nomina acceperunt duodecim signa* (see ch. 16). Bede himself only states that the Sun enters each zodiac sign "in the middle" of the month (ch. 16). The month-names are taken from chs. 11–15, the beginnings of the seasons from ch. 35, the lunar-zodiac letters from ch. 19, and the *litterae punctatae* from ch. 23.

JANUARY

The Hebrew month Tevet; the Egyptian month Tybi; the Greek month Eudymios; the English month Giuli.

a	a	[1]	kalends
	b	[2]	4 nones
b	c	[3]	3 nones
	d	[4]	2 nones
c	e	[5]	nones
	f	[6]	8 ides
d	g	[7]	7 ides
	h	[8]	6 ides
e	i	[9]	5 ides
	k	[10]	4 ides
f	l	[11]	3 ides
	m	[12]	2 ides
g	n	[13]	ides
	o	[14]	19 kalends of February
h	p	[15]	18 kalends
	q	[16]	17 kalends
i	r	[17]	16 kalends SUN IN AQUARIUS
	s	[18]	15 kalends
k	t	[19]	14 kalends
	u	[20]	13 kalends
l	a.	[21]	12 kalends
	b.	[22]	11 kalends
m	c.	[23]	10 kalends
	d.	[24]	9 kalends
n	e.	[25]	8 kalends
	f.	[26]	7 kalends
o	g.	[27]	6 kalends
a	h.	[28]	5 kalends
	i.	[29]	4 kalends
b	k.	[30]	3 kalends
	l.	[31]	2 kalends

FEBRUARY

The Hebrew month Shevat; the Egyptian month Mecheir; the Greek
month Peritios; the English month Solmonath.

c	m.	[1]	kalends
	n.	[2]	3 nones
d	o.	[3]	2 nones
	p.	[4]	nones
e	q.	[5]	9 ides
	r.	[6]	8 ides
f	s.	[7]	7 ides SPRING BEGINS ACCORDING TO GREEKS AND ROMANS
	t.	[8]	6 ides
g	u.	[9]	5 ides
	.a	[10]	4 ides
h	.b	[11]	3 ides
	.c	[12]	2 ides
i	.d	[13]	ides
	.e	[14]	16 kalends of March
k	.f	[15]	15 kalends
	.g	[16]	14 kalends SUN IN PISCES
l	.h	[17]	13 kalends
	.i	[18]	12 kalends
m	.k	[19]	11 kalends
	.l	[20]	10 kalends
n	.m	[21]	9 kalends
	.n	[22]	8 kalends SPRING BEGINS ACCORDING TO ISIDORE
o	.o	[23]	7 kalends
a	.p	[24]	6 kalends LOCATION OF LEAP-YEAR DAY
	.q	[25]	5 kalends
b	.r	[26]	4 kalends
	.s	[27]	3 kalends
c	.t	[28]	2 kalends

Remember in leap year to compute the Moon of February as one of 30
days, so that the Moon of March will still be one of 30 days, as usual,
lest the Easter reckoning waver. [cf. ch. 41]

MARCH

The Hebrew month Adar; the Egyptian month Phamenoth; the Greek month Dystros; the English month Hrethmonath.

	a	[1]	kalends
d	b	[2]	6 nones
	c	[3]	5 nones
e	d	[4]	4 nones
	e	[5]	3 nones
f	f	[6]	2 nones
	g	[7]	nones
g	h	[8]	8 ides
	i	[9]	7 ides
h	k	[10]	6 ides
	l	[11]	5 ides
i	m	[12]	4 ides
	n	[13]	3 ides
k	o	[14]	2 ides
	p	[15]	ides
l	q	[16]	17 kalends of April
	r	[17]	16 kalends
m	s	[18]	15 kalends SUN IN ARIES. FIRST DAY OF CREATION
	t	[19]	14 kalends (cf. ch. 6)
n	u	[20]	13 kalends
	a.	[21]	12 kalends EQUINOX ACCORDING TO THE EGYPTIANS (chs. 6, 30, 38)
o	b.	[22]	11 kalends LOCATION OF EPACTS (cf. chs. 20, 41)
	c.	[23]	10 kalends
a	d.	[24]	9 kalends LOCATION OF CONCURRENTS (cf. ch. 53)
	e.	[25]	8 kalends EQUINOX ACCORDING TO THE ROMANS (cf. ch. 30)
b	f.	[26]	7 kalends
	g.	[27]	6 kalends
c	h.	[28]	5 kalends
	i.	[29]	4 kalends
d	k.	[30]	3 kalends
	l.	[31]	2 kalends

APRIL

The Hebrew month Nisan; the Egyptian month Pharmouthi; the Greek month Xanthicos; the English month Eosturmonath.

e	m.	[1]	kalends
	n.	[2]	4 nones
f	o.	[3]	3 nones
	p.	[4]	2 nones
g	q.	[5]	nones
	r.	[6]	8 ides
h	s.	[7]	7 ides
	t.	[8]	6 ides
i	u.	[9]	5 ides
	.a	[10]	4 ides
k	.b	[11]	3 ides
	.c	[12]	2 ides
l	.d	[13]	ides
	.e	[14]	18 kalends of May
m	.f	[15]	17 kalends
	.g	[16]	16 kalends
n	.h	[17]	15 kalends SUN IN TAURUS
	.i	[18]	14 kalends
o	.k	[19]	13 kalends
a	.l	[20]	12 kalends
	.m	[21]	11 kalends
b	.n	[22]	10 kalends
	.o	[23]	9 kalends
c	.p	[24]	8 kalends
	.q	[25]	7 kalends
d	.r	[26]	6 kalends
	.s	[27]	5 kalends
e	.t	[28]	4 kalends
	a	[29]	3 kalends
f	b	[30]	2 kalends

MAY

The Hebrew month Iyyar; the Egyptian month Pachons; the Greek month Artemisios; the English month Thrimilchi.

	c	[1] kalends	
g	d	[2] 6 nones	
	e	[3] 5 nones	
h	f	[4] 4 nones	
	g	[5] 3 nones	
i	h	[6] 2 nones	
	i	[7] nones	
k	k	[8] 8 ides	
	l	[9] 7 ides	SUMMER BEGINS ACCORDING TO GREEKS AND ROMANS
l	m	[10] 6 ides	
	n	[11] 5 ides	
m	o	[12] 4 ides	
	p	[13] 3 ides	
n	q	[14] 2 ides	
	r	[15] ides	
o	s	[16] 17 kalends of June	
a	t	[17] 16 kalends	
	u	[18] 15 kalends	SUN IN GEMINI
b	a.	[19] 14 kalends	
	b.	[20] 13 kalends	
c	c.	[21] 12 kalends	
	d.	[22] 11 kalends	
d	e.	[23] 10 kalends	
	f.	[24] 9 kalends	SUMMER BEGINS ACCORDING TO ISIDORE
e	g.	[25] 8 kalends	
	h.	[26] 7 kalends	
f	i.	[27] 6 kalends	
	k.	[28] 5 kalends	
g	l.	[29] 4 kalends	
	m.	[30] 3 kalends	
h	n.	[31] 2 kalends	

JUNE

The Hebrew month Sivan; the Egyptian month Payni; the Greek month Daisios; the English month Litha.

	o.	[1]	kalends
i	p.	[2]	4 nones
	q.	[3]	3 nones
k	r.	[4]	2 nones
	s.	[5]	nones
l	t.	[6]	8 ides
	u.	[7]	7 ides
m	.a	[8]	6 ides
	.b	[9]	5 ides
n	.c	[10]	4 ides
	.d	[11]	3 ides
o	.e	[12]	2 ides
	.f	[13]	ides
a	.g	[14]	18 kalends of July
	.h	[15]	17 kalends
b	.i	[16]	16 kalends
	.k	[17]	15 kalends SUN IN CANCER
c	.l	[18]	14 kalends
	.m	[19]	13 kalends
d	.n	[20]	12 kalends
	.o	[21]	11 kalends
e	.p	[22]	10 kalends
	.q	[23]	9 kalends
f	.r	[24]	8 kalends SOLSTICE ACCORDING TO ANATOLIUS (ch. 35) AND ROMANS (ch. 30)
	.s	[25]	7 kalends
g	.t	[26]	6 kalends
	a	[27]	5 kalends
h	b	[28]	4 kalends
	c	[29]	3 kalends
i	d	[30]	2 kalends

JULY

The Hebrew month Tammuz; the Egyptian month Epieph; the Greek month Panemos; the English month Litha.

	e	[1] kalends
k	f	[2] 6 nones
	g	[3] 5 nones
l	h	[4] 4 nones
	i	[5] 3 nones
m	k	[6] 2 nones
	l	[7] nones
n	m	[8] 8 ides
	n	[9] 7 ides
o	o	[10] 6 ides
a	p	[11] 5 ides
	q	[12] 4 ides
b	r	[13] 3 ides
	s	[14] 2 ides
c	t	[15] ides
	u	[16] 17 kalends of August
d	a.	[17] 16 kalends
	b.	[18] 15 kalends SUN IN LEO
e	c.	[19] 14 kalends
	d.	[20] 13 kalends
f	e.	[21] 12 kalends
	f.	[22] 11 kalends
g	g.	[23] 10 kalends
	h.	[24] 9 kalends
h	i.	[25] 8 kalends
	k.	[26] 7 kalends
i	l.	[27] 6 kalends
	m.	[28] 5 kalends
k	n.	[29] 4 kalends
	o.	[30] 3 kalends
l	p.	[31] 2 kalends

AUGUST

The Hebrew month Av; the Egyptian month Mesore; the Greek month Loios; the English month Weodmonath.

	q.	[1]	kalends
m	r.	[2]	4 nones
	s.	[3]	3 nones
n	t.	[4]	2 nones
	u.	[5]	nones
o	.a	[6]	8 ides
a	.b	[7]	7 ides AUTUMN BEGINS ACCORDING TO GREEKS AND ROMANS
	.c	[8]	6 ides
b	.d	[9]	5 ides
	.e	[10]	4 ides
c	.f	[11]	3 ides
	.g	[12]	2 ides
d	.h	[13]	ides
	.i	[14]	19 kalends of September
e	.k	[15]	18 kalends
	.l	[16]	17 kalends
f	.m	[17]	16 kalends
	.n	[18]	15 kalends SUN IN VIRGO
g	.o	[19]	14 kalends
	.p	[20]	13 kalends
h	.q	[21]	12 kalends
	.r	[22]	11 kalends
i	.s	[23]	10 kalends AUTUMN BEGINS ACCORDING TO ISIDORE
	.t	[24]	9 kalends
k	a	[25]	8 kalends
	b	[26]	7 kalends
l	c	[27]	6 kalends
	d	[28]	5 kalends
m	e	[29]	4 kalends
	f	[30]	3 kalends
n	g	[31]	2 kalends

SEPTEMBER

The Hebrew month Elul; the Egyptian month Thoth; the Greek month Gorpiaios; the English month Halegmonath.

	h	[1]	kalends	
o	i	[2]	4 nones	
	k	[3]	3 nones	
a	l	[4]	2 nones	
	m	[5]	nones	
b	n	[6]	8 ides	
	o	[7]	7 ides	
c	p	[8]	6 ides	
	q	[9]	5 ides	
d	r	[10]	4 ides	
	s	[11]	3 ides	
e	t	[12]	2 ides	
	u	[13]	ides	
f	a.	[14]	18 kalends of October	
	b.	[15]	17 kalends	
g	c.	[16]	16 kalends	
	d.	[17]	15 kalends	SUN IN LIBRA
h	e.	[18]	14 kalends	
	f.	[19]	13 kalends	
i	g.	[20]	12 kalends	
	h.	[21]	11 kalends	
k	i.	[22]	10 kalends	
	k.	[23]	9 kalends	
l	l.	[24]	8 kalends	EQUINOX ACCORDING TO ROMANS (ch. 30) AND ANATOLIUS (ch. 35). INDICTIONS BEGIN (ch. 48)
	m.	[25]	7 kalends	
m	n.	[26]	6 kalends	
	o.	[27]	5 kalends	
n	p.	[28]	4 kalends	
	q.	[29]	3 kalends	
o	r.	[30]	2 kalends	

OCTOBER

The Hebrew month Tishri; the Egyptian month Phaophi; the Greek month Hyperberetaios; the English month Winterfilleth.

a	s.	[1]	kalends
	t.	[2]	6 nones
b	u.	[3]	5 nones
	.a	[4]	4 nones
c	.b	[5]	3 nones
	.c	[6]	2 nones
d	.d	[7]	nones
	.e	[8]	8 ides
e	.f	[9]	7 ides
	.g	[10]	6 ides
f	.h	[11]	5 ides
	.i	[12]	4 ides
g	.k	[13]	3 ides
	.l	[14]	2 ides
h	.m	[15]	ides
	.n	[16]	17 kalends of November
i	.o	[17]	16 kalends
	.p	[18]	15 kalends SUN IN SCORPIO
k	.q	[19]	14 kalends
	.r	[20]	13 kalends
l	.s	[21]	12 kalends
	.t	[22]	11 kalends
m	a	[23]	10 kalends
	b	[24]	9 kalends
n	c	[25]	8 kalends
	d	[26]	7 kalends
o	e	[27]	6 kalends
a	f	[28]	5 kalends
	g	[29]	4 kalends
b	h	[30]	3 kalends
	i	[31]	2 kalends

NOVEMBER

The Hebrew month Marheshvan; the Egyptian month Hathyr; the Greek month Dios; the English month Blodmonath.

c	k	[1]	kalends
	l	[2]	4 nones
d	m	[3]	3 nones
	n	[4]	2 nones
e	o	[5]	nones
	p	[6]	8 ides
f	q	[7]	7 ides

WINTER BEGINS ACCORDING TO GREEKS AND ROMANS

	r	[8]	6 ides
g	s	[9]	5 ides
	t	[10]	4 ides
h	u	[11]	3 ides
	a.	[12]	2 ides
i	b.	[13]	ides
	c.	[14]	18 kalends of December
k	d.	[15]	17 kalends
	e.	[16]	16 kalends
l	f.	[17]	15 kalends

SUN IN SAGITTARIUS

	g.	[18]	14 kalends
m	h.	[19]	13 kalends
	i.	[20]	12 kalends
n	k.	[21]	11 kalends
	l.	[22]	10 kalends
o	m.	[23]	9 kalends

WINTER BEGINS ACCORDING TO ISIDORE

	n.	[24]	8 kalends
a	o.	[25]	7 kalends
	p.	[26]	6 kalends
b	q.	[27]	5 kalends
	r.	[28]	4 kalends
c	s.	[29]	3 kalends
	t.	[30]	2 kalends

DECEMBER

The Hebrew month Kislev; the Egyptian month Choiac; the Greek month Apellaios; the English month Giuli.

d	u.	[1]	kalends
	.a	[2]	4 nones
e	.b	[3]	3 nones
	.c	[4]	2 nones
f	.d	[5]	nones
	.e	[6]	8 ides
g	.f	[7]	7 ides
	.g	[8]	6 ides
h	.h	[9]	5 ides
	.i	[10]	4 ides
i	.k	[11]	3 ides
	.l	[12]	2 ides
k	.m	[13]	ides
	.n	[14]	19 kalends of January
l	.o	[15]	18 kalends
	.p	[16]	17 kalends
m	.q	[17]	16 kalends
	.r	[18]	15 kalends SUN IN CAPRICORN
n	.s	[19]	14 kalends
	.t	[20]	13 kalends
o	a	[21]	12 kalends
a	b	[22]	11 kalends
	c	[23]	10 kalends
b	d	[24]	9 kalends
	e	[25]	8 kalends SOLSTICE ACCORDING TO THE ROMANS (ch. 30) AND ANATOLIUS (ch. 35)
c	f	[26]	7 kalends
	g	[27]	6 kalends
d	h	[28]	5 kalends
	i	[29]	4 kalends
e	k	[30]	3 kalends
	l	[31]	2 kalends

APPENDIX 2: BEDE'S 532-YEAR PASCHAL TABLE

[A.D.]	[Indiction]	[lunar epact]	[concurrent]	[lunar cycle]	[14th Moon]	[Easter]	[Moon on Easter Day]
[-1] 532	10	0	4	17	5 April	11 April	20
[1] 533	11	11	5	18	25 March	27 March	16
[2] 534	12	22	6	19	13 April	16 April	17
[3] 535	13	3	7	1	2 April	8 April	20
[4] 536	14	14	2	2	22 March	23 March	15
[5] 537	15	25	3	3	10 April	12 April	16
[6] 538	1	6	4	4	30 March	4 April	19
[7] 539	2	17	5	5	18 April	24 April	20
[8] 540	3	28	7	6	7 April	8 April	15
[9] 541	4	9	1	7	27 March	31 March	18
[10] 542	5	20	2	8	15 April	20 April	19
[11] 543	6	1	3	9	4 April	5 April	15
[12] 544	7	12	5	10	24 March	27 March	17
[13] 545	8	23	6	11	12 April	16 April	18
[14] 546	9	4	7	12	1 April	8 April	21
[15] 547	10	15	1	13	21 March	24 March	17
[16] 548	11	26	3	14	9 April	12 April	17
[17] 549	12	7	4	15	29 March	4 April	20
[18] 550	13	18	5	16	17 April	24 April	21
[19] 551	14	0	6	17	5 April	9 April	18
[20] 552	15	11	1	18	25 March	31 March	20
[21] 553	1	22	2	19	13 April	20 April	21
[22] 554	2	3	3	1	2 April	5 April	17
[23] 555	3	14	4	2	22 March	28 March	20
[24] 556	4	25	6	3	10 April	16 April	20
[25] 557	5	6	7	4	30 March	1 April	16
[26] 558	6	17	1	5	18 April	21 April	17
[27] 559	7	28	2	6	7 April	13 April	20
[28] 560	8	9	4	7	27 March	28 March	15
[29] 561	9	20	5	8	15 April	17 April	16
[30] 562	10	1	6	9	4 April	9 April	19
[31] 563	11	12	7	10	24 March	25 March	15
[32] 564	12	23	2	11	12 April	13 April	15
[33] 565	13	4	3	12	1 April	5 April	18
[34] 566	14	15	4	13	21 March	28 March	21
[35] 567	15	26	5	14	9 April	10 April	15
[36] 568	1	7	7	15	29 March	1 April	17
[37] 569	2	18	1	16	17 April	21 April	18

[A.D.]	[Indiction]	[lunar epact]	[concurrent]	[lunar cycle]	[14th Moon]	[Easter]	[Moon on Easter Day]
[38] 570	3	0	2	17	5 April	6 April	15
[39] 571	4	11	3	18	25 March	29 March	18
[40] 572	5	22	5	19	13 April	17 April	18
[41] 573	6	3	6	1	2 April	9 April	21
[42] 574	7	14	7	2	22 March	25 March	17
[43] 575	8	25	1	3	10 April	14 April	18
[44] 576	9	6	3	4	30 March	5 April	20
[45] 577	10	17	4	5	18 April	25 April	21
[46] 578	11	28	5	6	7 April	10 April	17
[47] 579	12	9	6	7	27 March	2 April	20
[48] 580	13	20	1	8	15 April	21 April	20
[49] 581	14	1	2	9	4 April	6 April	16
[50] 582	15	12	3	10	24 March	29 March	19
[51] 583	1	23	4	11	12 April	18 April	20
[52] 584	2	4	6	12	1 April	2 April	15
[53] 585	3	15	7	13	21 March	25 March	18
[54] 586	4	26	1	14	9 April	14 April	19
[55] 587	5	7	2	15	29 March	30 March	15
[56] 588	6	18	4	16	17 April	18 April	15
[57] 589	7	0	5	17	5 April	10 April	19
[58] 590	8	11	6	18	25 March	26 March	15
[59] 591	9	22	7	19	13 April	15 April	16
[60] 592	10	3	2	1	2 April	6 April	18
[61] 593	11	14	3	2	22 March	29 March	21
[62] 594	12	25	4	3	10 April	11 April	15
[63] 595	13	6	5	4	30 March	3 April	18
[64] 596	14	17	7	5	18 April	22 April	18
[65] 597	15	28	1	6	7 April	14 April	21
[66] 598	1	9	2	7	27 March	30 March	17
[67] 599	2	20	3	8	15 April	19 April	18
[68] 600	3	1	5	9	4 April	10 April	20
[69] 601	4	12	6	10	24 March	26 March	16
[70] 602	5	23	7	11	12 April	15 April	17
[71] 603	6	4	1	12	1 April	7 April	20
[72] 604	7	15	3	13	21 March	22 March	15
[73] 605	8	26	4	14	9 April	11 April	16
[74] 606	9	7	5	15	29 March	3 April	19
[75] 607	10	18	6	16	17 April	23 April	20
[76] 608	11	0	1	17	5 April	7 April	16
[77] 609	12	11	2	18	25 March	30 March	19
[78] 610	13	22	3	19	13 April	19 April	20
[79] 611	14	3	4	1	2 April	4 April	16
[80] 612	15	14	6	2	22 March	26 March	18

[A.D.]	[Indiction]	[lunar epact]	[concurrent]	[lunar cycle]	[14th Moon]	[Easter]	[Moon on Easter Day]
[81] 613	1	25	7	3	10 April	15 April	19
[82] 614	2	6	1	4	30 March	31 March	15
[83] 615	3	17	2	5	18 April	20 April	16
[84] 616	4	28	4	6	7 April	11 April	18
[85] 617	5	9	5	7	27 March	3 April	21
[86] 618	6	20	6	8	15 April	16 April	15
[87] 619	7	1	7	9	4 April	8 April	18
[88] 620	8	12	2	10	24 March	30 March	20
[89] 621	9	23	3	11	12 April	19 April	21
[90] 622	10	4	4	12	1 April	4 April	17
[91] 623	11	15	5	13	21 March	27 March	20
[92] 624	12	26	7	14	9 April	15 April	20
[93] 625	13	7	1	15	29 March	31 March	16
[94] 626	14	18	2	16	17 April	20 April	17
[95] 627	15	0	3	17	5 April	12 April	21
[96] 628	1	11	5	18	25 March	27 March	16
[97] 629	2	22	6	19	13 April	16 April	17
[98] 630	3	3	7	1	2 April	8 April	20
[99] 631	4	14	1	2	22 March	24 March	16
[100] 632	5	25	3	3	10 April	12 April	16
[101] 633	6	6	4	4	30 March	4 April	19
[102] 634	7	17	5	5	18 April	24 April	20
[103] 635	8	28	6	6	7 April	9 April	16
[104] 636	9	9	1	7	27 March	31 March	18
[105] 637	10	20	2	8	15 April	20 April	19
[106] 638	11	1	3	9	4 April	5 April	15
[107] 639	12	12	4	10	24 March	28 March	18
[108] 640	13	23	6	11	12 April	16 April	18
[109] 641	14	4	7	12	1 April	8 April	21
[110] 642	15	15	1	13	21 March	24 March	17
[111] 643	1	26	2	14	9 April	13 April	19
[112] 644	2	7	4	15	29 March	4 April	20
[113] 645	3	18	5	16	17 April	24 April	21
[114] 646	4	0	6	17	5 April	9 April	18
[115] 647	5	11	7	18	25 March	1 April	21
[116] 648	6	22	2	19	13 April	20 April	21
[117] 649	7	3	3	1	2 April	5 April	17
[118] 650	8	14	4	2	22 March	28 March	20
[119] 651	9	25	5	3	10 April	17 April	21
[120] 652	10	6	7	4	30 March	1 April	16
[121] 653	11	17	1	5	18 April	21 April	17
[122] 654	12	28	2	6	7 April	13 April	20
[123] 655	13	9	3	7	27 March	29 March	16

[A.D.]	[Indiction]	[lunar epact]	[concurrent]	[lunar cycle]	[14th Moon]	[Easter]	[Moon on Easter Day]
[124] 656	14	20	5	8	15 April	17 April	16
[125] 657	15	1	6	9	4 April	9 April	19
[126] 658	1	12	7	10	24 March	25 March	15
[127] 659	2	23	1	11	12 April	14 April	16
[128] 660	3	4	3	12	1 April	5 April	18
[129] 661	4	15	4	13	21 March	28 March	21
[130] 662	5	26	5	14	9 April	10 April	15
[131] 663	6	7	6	15	29 March	2 April	18
[132] 664	7	18	1	16	17 April	21 April	18
[133] 665	8	0	2	17	5 April	6 April	15
[134] 666	9	11	3	18	25 March	29 March	18
[135] 667	10	22	4	19	13 April	18 April	19
[136] 668	11	3	6	1	2 April	9 April	21
[137] 669	12	14	7	2	22 March	25 March	17
[138] 670	13	25	1	3	10 April	14 April	18
[139] 671	14	6	2	4	30 March	6 April	21
[140] 672	15	17	4	5	18 April	25 April	21
[141] 673	1	28	5	6	7 April	10 April	17
[142] 674	2	9	6	7	27 March	2 April	20
[143] 675	3	20	7	8	15 April	22 April	21
[144] 676	4	1	2	9	4 April	6 April	16
[145] 677	5	12	3	10	24 March	29 March	19
[146] 678	6	23	4	11	12 April	18 April	20
[147] 679	7	4	5	12	1 April	3 April	16
[148] 680	8	15	7	13	21 March	25 March	18
[149] 681	9	26	1	14	9 April	14 April	19
[150] 682	10	7	2	15	29 March	30 March	15
[151] 683	11	18	3	16	17 April	19 April	16
[152] 684	12	0	5	17	5 April	10 April	19
[153] 685	13	11	6	18	25 March	26 March	15
[154] 686	14	22	7	19	13 April	15 April	16
[155] 687	15	3	1	1	2 April	7 April	19
[156] 688	1	14	3	2	22 March	29 March	21
[157] 689	2	25	4	3	10 April	11 April	15
[158] 690	3	6	5	4	30 March	3 April	18
[159] 691	4	17	6	5	18 April	23 April	19
[160] 692	5	28	1	6	7 April	14 April	21
[161] 693	6	9	2	7	27 March	30 March	17
[162] 694	7	20	3	8	15 April	19 April	18
[163] 695	8	1	4	9	4 April	11 April	21
[164] 696	9	12	6	10	24 March	26 March	16
[165] 697	10	23	7	11	12 April	15 April	17
[166] 698	11	4	1	12	1 April	7 April	20

[A.D.]	[Indiction]	[lunar epact]	[concurrent]	[lunar cycle]	[14th Moon]	[Easter]	[Moon on Easter Day]
[167] 699	12	15	2	13	21 March	23 March	16
[168] 700	13	26	4	14	9 April	11 April	16
[169] 701	14	7	5	15	29 March	3 April	19
[170] 702	15	18	6	16	17 April	23 April	20
[171] 703	1	0	7	17	5 April	8 April	17
[172] 704	2	11	2	18	25 March	30 March	19
[173] 705	3	22	3	19	13 April	19 April	20
[174] 706	4	3	4	1	2 April	4 April	16
[175] 707	5	14	5	2	22 March	27 March	19
[176] 708	6	25	7	3	10 April	15 April	19
[177] 709	7	6	1	4	30 March	31 March	15
[178] 710	8	17	2	5	18 April	20 April	16
[179] 711	9	28	3	6	7 April	12 April	19
[180] 712	10	9	5	7	27 March	3 April	21
[181] 713	11	20	6	8	15 April	16 April	15
[182] 714	12	1	7	9	4 April	8 April	18
[183] 715	13	12	1	10	24 March	31 March	21
[184] 716	14	23	3	11	12 April	19 April	21
[185] 717	15	4	4	12	1 April	4 April	17
[186] 718	1	15	5	13	21 March	27 March	20
[187] 719	2	26	6	14	9 April	16 April	21
[188] 720	3	7	1	15	29 March	31 March	16
[189] 721	4	18	2	16	17 April	20 April	17
[190] 722	5	0	3	17	5 April	12 April	21
[191] 723	6	11	4	18	25 March	28 March	17
[192] 724	7	22	6	19	13 April	16 April	17
[193] 725	8	3	7	1	2 April	8 April	20
[194] 726	9	14	1	2	22 March	24 March	16
[195] 727	10	25	2	3	10 April	13 April	17
[196] 728	11	6	4	4	30 March	4 April	19
[197] 729	12	17	5	5	18 April	24 April	20
[198] 730	13	28	6	6	7 April	9 April	16
[199] 731	14	9	7	7	27 March	1 April	19
[200] 732	15	20	2	8	15 April	20 April	19
[201] 733	1	1	3	9	4 April	5 April	15
[202] 734	2	12	4	10	24 March	28 March	18
[203] 735	3	23	5	11	12 April	17 April	19
[204] 736	4	4	7	12	1 April	8 April	21
[205] 737	5	15	1	13	21 March	24 March	17
[206] 738	6	26	2	14	9 April	13 April	18
[207] 739	7	7	3	15	29 March	5 April	21
[208] 740	8	18	5	16	17 April	24 April	21

[A.D.]	[Indiction]	[lunar epact]	[concurrent]	[lunar cycle]	[14th Moon]	[Easter]	[Moon on Easter Day]
[209] 741	9	0	6	17	5 April	9 April	18
[210] 742	10	11	7	18	25 March	1 April	21
[211] 743	11	22	1	19	13 April	14 April	15
[212] 744	12	3	3	1	2 April	5 April	17
[213] 745	13	14	4	2	22 March	28 March	20
[214] 746	14	25	5	3	10 April	17 April	21
[215] 747	15	6	6	4	30 March	2 April	17
[216] 748	1	17	1	5	18 April	21 April	17
[217] 749	2	28	2	6	7 April	13 April	20
[218] 750	3	9	3	7	27 March	29 March	16
[219] 751	4	20	4	8	15 April	18 April	17
[220] 752	5	1	6	9	4 April	9 April	19
[221] 753	6	12	7	10	24 March	25 March	15
[222] 754	7	23	1	11	12 April	14 April	16
[223] 755	8	4	2	12	1 April	6 April	19
[224] 756	9	15	4	13	21 March	28 March	21
[225] 757	10	26	5	14	9 April	10 April	15
[226] 758	11	7	6	15	29 March	2 April	18
[227] 759	12	18	7	16	17 April	22 April	19
[228] 760	13	0	2	17	5 April	6 April	15
[229] 761	14	11	3	18	25 March	29 March	18
[230] 762	15	22	4	19	13 April	18 April	19
[231] 763	1	3	5	1	2 April	3 April	15
[232] 764	2	14	7	2	22 March	25 March	17
[233] 765	3	25	1	3	10 April	14 April	18
[234] 766	4	6	2	4	30 March	6 April	21
[235] 767	5	17	3	5	18 April	19 April	15
[236] 768	6	28	5	6	7 April	10 April	17
[237] 769	7	9	6	7	27 March	2 April	20
[238] 770	8	20	7	8	15 April	22 April	21
[239] 771	9	1	1	9	4 April	7 April	17
[240] 772	10	12	3	10	24 March	29 March	19
[241] 773	11	23	4	11	12 April	18 April	20
[242] 774	12	4	5	12	1 April	3 April	16
[243] 775	13	15	6	13	21 March	26 March	19
[244] 776	14	26	1	14	9 April	14 April	19
[245] 777	15	7	2	15	29 March	30 March	15
[246] 778	1	18	3	16	17 April	19 April	16
[247] 779	2	0	4	17	5 April	11 April	20
[248] 780	3	11	6	18	25 March	26 March	15
[249] 781	4	22	7	19	13 April	15 April	16
[250] 782	5	3	1	1	2 April	7 April	19
[251] 783	6	14	2	2	22 March	23 March	15

[A.D.]	[Indiction]	[lunar epact]	[concurrent]	[lunar cycle]	[14th Moon]	[Easter]	[Moon on Easter Day]
[252] 784	7	25	4	3	10 April	11 April	15
[253] 785	8	6	5	4	30 March	3 April	18
[254] 786	9	17	6	5	18 April	23 April	19
[255] 787	10	28	7	6	7 April	8 April	15
[256] 788	11	9	2	7	27 March	30 March	17
[257] 789	12	20	3	8	15 April	19 April	19
[258] 790	13	1	4	9	4 April	11 April	21
[259] 791	14	12	5	10	24 March	27 March	17
[260] 792	15	23	7	11	12 April	15 April	17
[261] 793	1	4	1	12	1 April	7 April	20
[262] 794	2	15	2	13	21 March	23 March	16
[263] 795	3	26	3	14	9 April	12 April	17
[264] 796	4	7	5	15	29 March	3 April	19
[265] 797	5	18	6	16	17 April	23 April	20
[266] 798	6	0	7	17	5 April	8 April	17
[267] 799	7	11	1	18	25 March	31 March	20
[268] 800	8	22	3	19	13 April	19 April	20
[269] 801	9	3	4	1	2 April	4 April	16
[270] 802	10	14	5	2	22 March	27 March	19
[271] 803	11	25	6	3	10 April	16 April	20
[272] 804	12	6	1	4	30 March	31 March	15
[273] 805	13	17	2	5	18 April	20 April	16
[274] 806	14	28	3	6	7 April	12 April	19
[275] 807	15	9	4	7	27 March	28 March	15
[276] 808	1	20	6	8	15 April	16 April	15
[277] 809	2	1	7	9	4 April	8 April	18
[278] 810	3	12	1	10	24 March	31 March	21
[279] 811	4	23	2	11	12 April	13 April	15
[280] 812	5	4	4	12	1 April	4 April	17
[281] 813	6	15	5	13	21 March	27 March	20
[282] 814	7	26	6	14	9 April	16 April	21
[283] 815	8	7	7	15	29 March	1 April	17
[284] 816	9	18	2	16	17 April	20 April	17
[285] 817	10	0	3	17	5 April	12 April	21
[286] 818	11	11	4	18	25 March	28 March	17
[287] 819	12	22	5	19	13 April	17 April	18
[288] 820	13	3	7	1	2 April	8 April	20
[289] 821	14	14	1	2	22 March	24 March	16
[290] 822	15	25	2	3	10 April	13 April	17
[291] 823	1	6	3	4	30 March	5 April	20
[292] 824	2	17	5	5	18 April	24 April	20
[293] 825	3	28	6	6	7 April	9 April	16
[294] 826	4	9	7	7	27 March	1 April	19

[A.D.]	[Indiction]	[lunar epact]	[concurrent]	[lunar cycle]	[14th Moon]	[Easter]	[Moon on Easter Day]
[295] 827	5	20	1	8	15 April	21 April	20
[296] 828	6	1	3	9	4 April	5 April	15
[297] 829	7	12	4	10	24 March	28 March	18
[298] 830	8	23	5	11	12 April	17 April	19
[299] 831	9	4	6	12	1 April	2 April	15
[300] 832	10	15	1	13	21 March	24 March	17
[301] 833	11	26	2	14	9 April	13 April	18
[302] 834	12	7	3	15	29 March	5 April	21
[303] 835	13	18	4	16	17 April	18 April	15
[304] 836	14	0	6	17	5 April	9 April	18
[305] 837	15	11	7	18	25 March	1 April	21
[306] 838	1	22	1	19	13 April	14 April	15
[307] 839	2	3	2	1	2 April	6 April	18
[308] 840	3	14	4	2	22 March	28 March	20
[309] 841	4	25	5	3	10 April	17 April	21
[310] 842	5	6	6	4	30 March	2 April	17
[311] 843	6	17	7	5	18 April	22 April	18
[312] 844	7	28	2	6	7 April	13 April	20
[313] 845	8	9	3	7	27 March	29 March	16
[314] 846	9	20	4	8	15 April	18 April	17
[315] 847	10	1	5	9	4 April	10 April	20
[316] 848	11	12	7	10	24 March	25 March	15
[317] 849	12	23	1	11	12 April	14 April	16
[318] 850	13	4	2	12	1 April	6 April	19
[319] 851	14	15	3	13	21 March	22 March	15
[320] 852	15	26	5	14	9 April	10 April	15
[321] 853	1	7	6	15	29 March	2 April	18
[322] 854	2	18	7	16	17 April	22 April	19
[323] 855	3	0	1	17	5 April	7 April	16
[324] 856	4	11	3	18	25 March	29 March	18
[325] 857	5	22	4	19	13 April	18 April	19
[326] 858	6	3	5	1	2 April	3 April	15
[327] 859	7	14	6	2	22 March	26 March	18
[328] 860	8	25	1	3	10 April	14 April	18
[329] 861	9	6	2	4	30 March	6 April	21
[330] 862	10	17	3	5	18 April	19 April	15
[331] 863	11	28	4	6	7 April	11 April	18
[332] 864	12	9	6	7	27 March	2 April	20
[333] 865	13	20	7	8	15 April	22 April	21
[334] 866	14	1	1	9	4 April	7 April	17
[335] 867	15	12	2	10	24 March	30 March	20
[336] 868	1	23	4	11	12 April	18 April	20
[337] 869	2	4	5	12	1 April	3 April	16

[A.D.]	[Indiction]	[lunar epact]	[concurrent]	[lunar cycle]	[14th Moon]	[Easter]	[Moon on Easter Day]
[338] 870	3	15	6	13	21 March	26 March	19
[339] 871	4	26	7	14	9 April	15 April	20
[340] 872	5	7	2	15	29 March	30 March	15
[341] 873	6	18	3	16	17 April	19 April	16
[342] 874	7	0	4	17	5 April	11 April	20
[343] 875	8	11	5	18	25 March	27 March	16
[344] 876	9	22	7	19	13 April	15 April	16
[345] 877	10	3	1	1	2 April	7 April	19
[346] 878	11	14	2	2	22 March	23 March	15
[347] 879	12	25	3	3	10 April	12 April	16
[348] 880	13	6	5	4	30 March	3 April	18
[349] 881	14	17	6	5	18 April	23 April	19
[350] 882	15	28	7	6	7 April	8 April	15
[351] 883	1	9	1	7	27 March	31 March	18
[352] 884	2	20	3	8	15 April	19 April	18
[353] 885	3	1	4	9	4 April	11 April	21
[354] 886	4	12	5	10	24 March	27 March	17
[355] 887	5	23	6	11	12 April	16 April	18
[356] 888	6	4	1	12	1 April	7 April	20
[357] 889	7	15	2	13	21 March	23 March	16
[358] 890	8	26	3	14	9 April	12 April	17
[359] 891	9	7	4	15	29 March	4 April	20
[360] 892	10	18	6	16	17 April	23 April	20
[361] 893	11	0	7	17	5 April	8 April	17
[362] 894	12	11	1	18	25 March	31 March	20
[363] 895	13	22	2	19	13 April	20 April	21
[364] 896	14	3	4	1	2 April	4 April	16
[365] 897	15	14	5	2	22 March	27 March	19
[366] 898	1	25	6	3	10 April	16 April	20
[367] 899	2	6	7	4	30 March	1 April	16
[368] 900	3	17	2	5	18 April	20 April	16
[369] 901	4	28	3	6	7 April	12 April	19
[370] 902	5	9	4	7	27 March	28 March	15
[371] 903	6	20	5	8	15 April	17 April	16
[372] 904	7	1	7	9	4 April	8 April	18
[373] 905	8	12	1	10	24 March	31 March	21
[374] 906	9	23	2	11	12 April	13 April	15
[375] 907	10	4	3	12	1 April	5 April	18
[376] 908	11	15	5	13	21 March	27 March	20
[377] 909	12	26	6	14	9 April	16 April	21
[378] 910	13	7	7	15	29 March	1 April	17
[379] 911	14	18	1	16	17 April	21 April	18

[A.D.]	[Indiction]	[lunar epact]	[concurrent]	[lunar cycle]	[14th Moon]	[Easter]	[Moon on Easter Day]
[380] 912	15	0	3	17	5 April	12 April	21
[381] 913	1	11	4	18	25 March	28 March	17
[382] 914	2	22	5	19	13 April	17 April	18
[383] 915	3	3	6	1	2 April	9 April	21
[384] 916	4	14	1	2	22 March	24 March	16
[385] 917	5	25	2	3	10 April	13 April	17
[386] 918	6	6	3	4	30 March	5 April	20
[387] 919	7	17	4	5	18 April	25 April	21
[388] 920	8	28	6	6	7 April	9 April	16
[389] 921	9	9	7	7	27 March	1 April	19
[390] 922	10	20	1	8	15 April	21 April	20
[391] 923	11	1	2	9	4 April	6 April	16
[392] 924	12	12	4	10	24 March	28 March	18
[393] 925	13	23	5	11	12 April	17 April	19
[394] 926	14	4	6	12	1 April	2 April	15
[395] 927	15	15	7	13	21 March	25 March	18
[396] 928	1	26	2	14	9 April	13 April	18
[397] 929	2	7	3	15	29 March	5 April	21
[398] 930	3	18	4	16	17 April	18 April	15
[399] 931	4	0	5	17	5 April	10 April	19
[400] 932	5	11	7	18	25 March	1 April	21
[401] 933	6	22	1	19	13 April	14 April	15
[402] 934	7	3	2	1	2 April	6 April	18
[403] 935	8	14	3	2	22 March	29 March	21
[404] 936	9	25	5	3	10 April	17 April	21
[405] 937	10	6	6	4	30 March	2 April	17
[406] 938	11	17	7	5	18 April	22 April	18
[407] 939	12	28	1	6	7 April	14 April	21
[408] 940	13	9	3	7	27 March	29 March	16
[409] 941	14	20	4	8	15 April	18 April	17
[410] 942	15	1	5	9	4 April	10 April	20
[411] 943	1	12	6	10	24 March	26 March	16
[412] 944	2	23	1	11	12 April	14 April	16
[413] 945	3	4	2	12	1 April	6 April	19
[414] 946	4	15	3	13	21 March	22 March	15
[415] 947	5	26	4	14	9 April	11 April	16
[416] 948	6	7	6	15	29 March	2 April	18
[417] 949	7	18	7	16	17 April	22 April	19
[418] 950	8	0	1	17	5 April	7 April	16
[419] 951	9	11	2	18	25 March	30 March	19
[420] 952	10	22	4	19	13 April	18 April	19
[421] 953	11	3	5	1	2 April	3 April	15
[422] 954	12	14	6	2	22 March	26 March	18

[A.D.]	[Indiction]	[lunar epact]	[concurrent]	[lunar cycle]	[14th Moon]	[Easter]	[Moon on Easter Day]
[423] 955	13	25	7	3	10 April	15 April	19
[424] 956	14	6	2	4	30 March	6 April	21
[425] 957	15	17	3	5	18 April	19 April	15
[426] 958	1	28	4	6	7 April	11 April	18
[427] 959	2	9	5	7	27 March	3 April	21
[428] 960	3	20	7	8	15 April	22 April	21
[429] 961	4	1	1	9	4 April	7 April	17
[430] 962	5	12	2	10	24 March	30 March	20
[431] 963	6	23	3	11	12 April	19 April	21
[432] 964	7	4	5	12	1 April	3 April	16
[433] 965	8	15	6	13	21 March	26 March	19
[434] 966	9	26	7	14	9 April	15 April	20
[435] 967	10	7	1	15	29 March	31 March	16
[436] 968	11	18	3	16	17 April	19 April	16
[437] 969	12	0	4	17	5 April	11 April	20
[438] 970	13	11	5	18	25 March	27 March	16
[439] 971	14	22	6	19	13 April	16 April	17
[440] 972	15	3	1	1	2 April	7 April	19
[441] 973	1	14	2	2	22 March	23 March	15
[442] 974	2	25	3	3	10 April	12 April	16
[443] 975	3	6	4	4	30 March	4 April	19
[444] 976	4	17	6	5	18 April	23 April	19
[445] 977	5	28	7	6	7 April	8 April	15
[446] 978	6	9	1	7	27 March	31 March	18
[447] 979	7	20	2	8	15 April	20 April	19
[448] 980	8	1	4	9	4 April	11 April	21
[449] 981	9	12	5	10	24 March	27 March	17
[450] 982	10	23	6	11	12 April	16 April	18
[451] 983	11	4	7	12	1 April	8 April	21
[452] 984	12	15	2	13	21 March	23 March	16
[453] 985	13	26	3	14	9 April	12 April	17
[454] 986	14	7	4	15	29 March	4 April	20
[455] 987	15	18	5	16	17 April	24 April	21
[456] 988	1	0	7	17	5 April	8 April	17
[457] 989	2	11	1	18	25 March	31 March	20
[458] 990	3	22	2	19	13 April	20 March	21
[459] 991	4	3	3	1	2 April	5 April	17
[460] 992	5	14	5	2	22 March	27 March	19
[461] 993	6	25	6	3	10 April	16 April	20
[462] 994	7	6	7	4	30 March	1 April	16
[463] 995	8	17	1	5	18 April	21 April	17
[464] 996	9	28	3	6	7 April	12 April	19
[465] 997	10	9	4	7	27 March	28 March	15

[A.D.]	[Indiction]	[lunar epact]	[concurrent]	[lunar cycle]	[14th Moon]	[Easter]	[Moon on Easter Day]
[466] 998	11	20	5	8	15 April	17 April	16
[467] 999	12	1	6	9	4 April	9 April	19
[468] 1000	13	12	1	10	24 March	31 March	21
[469] 1001	14	23	2	11	12 April	13 April	15
[470] 1002	15	4	3	12	1 April	5 April	18
[471] 1003	1	15	4	13	21 March	28 March	21
[472] 1004	2	26	6	14	9 April	16 April	21
[473] 1005	3	7	7	15	29 March	1 April	17
[474] 1006	4	18	1	16	17 April	21 April	18
[475] 1007	5	0	2	17	5 April	6 April	15
[476] 1008	6	11	4	18	25 March	28 March	17
[477] 1009	7	22	5	19	13 April	17 April	18
[478] 1010	8	3	6	1	2 April	9 April	21
[479] 1011	9	14	7	2	22 March	25 March	17
[480] 1012	10	25	2	3	10 April	13 April	17
[481] 1013	11	6	3	4	30 March	5 April	20
[482] 1014	12	17	4	5	18 April	25 April	21
[483] 1015	13	28	5	6	7 April	10 April	17
[484] 1016	14	9	7	7	27 March	1 April	19
[485] 1017	15	20	1	8	15 April	21 April	20
[486] 1018	1	1	2	9	4 April	6 April	16
[487] 1019	2	12	3	10	24 March	29 March	19
[488] 1020	3	23	5	11	12 April	17 April	19
[489] 1021	4	4	6	12	1 April	2 April	15
[490] 1022	5	15	7	13	21 March	25 March	18
[491] 1023	6	26	1	14	9 April	14 April	19
[492] 1024	7	7	3	15	29 March	5 April	21
[493] 1025	8	18	4	16	17 April	18 April	15
[494] 1026	9	0	5	17	5 April	10 April	19
[495] 1027	10	11	6	18	25 March	26 March	15
[496] 1028	11	22	1	19	13 April	14 April	15
[497] 1029	12	3	2	1	2 April	6 April	18
[498] 1030	13	14	3	2	22 March	29 March	21
[499] 1031	14	25	4	3	10 April	11 April	15
[500] 1032	15	6	6	4	30 March	2 April	17
[501] 1033	1	17	7	5	18 April	22 April	18
[502] 1034	2	28	1	6	7 April	14 April	21
[503] 1035	3	9	2	7	27 March	30 March	17
[504] 1036	4	20	4	8	15 April	18 April	17
[505] 1037	5	1	5	9	4 April	10 April	20
[506] 1038	6	12	6	10	24 March	26 March	16
[507] 1039	7	23	7	11	12 April	15 April	17
[508] 1040	8	4	2	12	1 April	6 April	19

[A.D.]	[Indiction]	[lunar epact]	[concurrent]	[lunar cycle]	[14th Moon]	[Easter]	[Moon on Easter Day]
[509] 1041	9	15	3	13	21 March	22 March	15
[510] 1042	10	26	4	14	9 April	11 April	16
[511] 1043	11	7	5	15	29 March	3 April	19
[512] 1044	12	18	7	16	17 April	22 April	19
[513] 1045	13	0	1	17	5 April	7 April	16
[514] 1046	14	11	2	18	25 March	30 March	19
[515] 1047	15	22	3	19	13 April	19 April	20
[516] 1048	1	3	5	1	2 April	3 April	15
[516] 1049	2	14	6	2	22 March	26 March	18
[518] 1050	3	25	7	3	10 April	15 April	19
[519] 1051	4	6	1	4	30 March	31 March	15
[520] 1052	5	17	3	5	18 April	19 April	15
[521] 1053	6	28	4	6	7 April	11 April	18
[522] 1054	7	9	5	7	27 March	3 April	21
[523] 1055	8	20	6	8	15 April	16 April	15
[524] 1056	9	1	7	9	4 April	7 April	17
[525] 1057	10	12	1	10	24 March	30 March	20
[526] 1058	11	23	3	11	12 April	19 April	21
[527] 1059	12	4	4	12	1 April	4 April	17
[528] 1060	13	15	5	13	21 March	26 March	19
[529] 1061	14	26	6	14	9 April	15 April	20
[530] 1062	15	7	1	15	29 March	31 March	16
[531] 1063	1	18	2	16	17 April	20 April	17

APPENDIX 3: BEDE'S LETTERS ON *COMPUTUS*

APPENDIX 3.1: LETTER TO PLEGWIN

Translator's note: The circumstances surrounding Bede's letter to Plegwin have been sketched in the Introduction, section 2. The main themes – chronology and eschatology – are dealt with in the commentary to the Preface, and to ch. 66. Nothing is known of Plegwin except what can be inferred from Bede's *Letter*, namely that he was a cleric in the entourage of Wilfred.

LETTER TO PLEGWIN

/617/ To his dearest brother, worthy of honour *in the bowels of Christ*,[1] Plegwin, Bede sends greeting in the Lord.

1. Two days ago, beloved brother, a messenger from your Sanctity came to me, bearing gladsome words of peaceful salutation from you. But soon thereafter he threw these into disorder by adding something very unfortunate, namely that you had heard it babbled out by lewd rustics in their cups that I was a heretic. I confess I was aghast; blanching, I asked of what heresy I was accused. He replied that it was because I had denied that our Lord and Saviour had come in the flesh in the Sixth Age of the world. So I began to enquire how this had been expressed – whether because the Lord had not come in the flesh when it was already the Sixth Age, or because it was not yet the Sixth Age when the Lord came, or that the Seventh Age had indeed already come, although patently the Sixth Age could not have begun save with His Incarnation. I discovered that whether accused of one thing or the other, I shared on both points the faith and consensus of the Church.

2. For if I had denied that Christ had come, how could I be a priest in Christ's Church? Or how could it follow that I, believing the evangelical and apostolic words, could fail to believe that He was incarnated in the

---

1 Philippians 1.8.

Sixth Age, when the evangelist Matthew in his book of the genealogy of Jesus Christ discerns the four final Ages by very particular distinctions of time through fourteen generations?[2] The Apostle Peter, indeed, distinguishes the first two [Ages][3] by such an arrangement, asserting that the Earth together with the heavens were destroyed by the Flood and replaced by others,[4] so that we could call these not so much two Ages of the world as two worlds, were not the Valentinian heresy to be feared.[5]

3. So when I perceived that I was innocent of denying this very thing, I began very carefully to consider how this calumny could have been launched against me. At length I called to mind that I once showed to one of your number /**618**/ my short work *On Times* which I wrote five years ago.[6] In this work, the sequence of the years was given according to the Hebrew Truth;[7] this is far shorter than the Septuagint, so that up to the Advent of the Saviour in the flesh, five thousand years were not completed. And I believe that I advised, in consideration of fraternal charity and truth itself, that credence be given to Holy Scripture as it is translated by our Christian interpreter, rather than to Jewish translators, or the ignorance of chronographers, pointing out how Eusebius in his designation of times followed neither Hebrew Truth nor the version of the Seventy in every [instance]. I will undertake to do this very thing here in writing. To be sure, he, whoever he was to whom I intimated these things, twisted the office of brotherly love into the loathing of envy, and the high Judge transformed the light of patent truth into the darkness of blind error, as he deserved.

4. But lest you think, beloved, that now that my heresy has been discovered, I prefer to make my escape under the protection of denial rather

2 Matthew 1.1–18.

3 I.e. the first two Ages of the world, from Adam to Noah, and from Noah to Abraham. Matthew's genealogy of Christ began with Abraham.

4 II Peter 3.5–12.

5 Both Prosper 634 (425) and Rufinus' trans. of Eusebius, *HE* 4.14, mention the Valentinian heresy, but do not provide details about its content. Bede may have acquired this information from Irenaeus, *Adversus haereses*, Books 1–2, ed. Adelin Rousseau and Louis Doutreleau, Sources chrétiennes 263–266 (Paris: Cerf, 1979), or from Tertullian's *Adversus Valentinianos*, ed. Jean-Claude Fredaille, Sources chrétiennes 280–281 (Paris: Cerf, 1980).

6 In 703; therefore the *Letter to Plegwin* was composed about 708.

7 I.e. Jerome's Vulgate.

than receive the grace of truth, listen to what I wrote concerning the Ages in that little book I mentioned. For after I touched upon moments and hours, day and night, weeks and months, the year and what pertains to the year, the nineteen-year cycle and its distinctive points (as far as I believed it sufficed for me and my [readers]), it happened that something was also briefly set down concerning the Ages. I began the sixteenth chapter as follows: I said, *The times of the world are divided into Six Ages. The First Age from Adam to Noah contained ten generations and 1,656 years; this all perished in the Flood, just as infancy should be drowned in forgetfulness. The Second, from Noah to Abraham, likewise contains ten generations, and 292 years. In it language – that is, the Hebrew language – was discovered, for in childhood a person learns to speak after infancy, which derives its name from the fact that he cannot 'fari', that is, speak. The Third, from Abraham to David, contains 14 generations and 942 years, and because man starts to be capable of begetting at adolescence, Matthew takes up the beginning of the generations [of Christ] from Abraham, who was established as the father of nations. /619/ The Fourth extends from David to the Babylonian exile in 14 generations likewise, according to Matthew, and 473 years, from which the time of kings began, for the dignity of a young man is apt for ruling. Then the Fifth Age stretches up to the coming of our Saviour in the flesh likewise in 14 generations and 589 years, in which, as if weary from a burdensome old age, the Hebrew folk were crushed by many misfortunes. The Sixth Age, in which we are now, will end, not in a fixed sequence of generations and times, but, like extreme old age itself, with the death of the whole world.*[8]

5. Considering these matters in a prefatory note on the Ages in the afore-mentioned short work, I immediately added thereto another chapter in which the course and order of the whole World-Age was to be laid open, beginning as follows: *Therefore the First Age contains, according to the Hebrews, 1,656 years, and according to the Seventy Translators 2,242. Adam at the age of 130 begat Seth, who was born in place of Abel. Seth at the age of 105 begat Enos*, and so on, down to the Flood.[9] Then I added: *The Second Age contains 292 years according to the Hebrews, and*

---

8 *On Times* 16 (600–601). The main source for the division of the Ages here is Isidore, *Etym.* 5.38–39. On the sources for the parallel between the six ages of man and the six World-Ages, see Commentary on ch. 66.

9 *Ibid.* 17 (601–602).

according to the Seventy Translators 942, or with Cainan added in, 1072. In the Second year after the Flood Shem begat Arfaxat, from whom the Chaldeans [are descended]. Arfaxat at the age of 35 begat Sela.[10] Here the Seventy Translators, whom Luke follows, insert Cainan, who had begotten Sela when he was 30 years old. Again, concerning the Third Age, I say, *The Third Age contains 942 years.*[11] Again, when it came to the Fourth [Age], I say, *The Fourth Age contains 473 years according to the Hebrews, and the Seventy Translators add on 12.*[12] *The Fifth Age*, I say, *contains 589 years.*[13] When therefore in the course of the work I arrived at the Nativity of our Lord and Saviour, I wrote thus: *Octavianus reigned 56 years. In his 42nd year, our Lord was born, 3,952 years after Adam, and according to others 5,199.*[14]

6. These are the things which, according to the faith of Holy Scripture, I was solicitous to summarize for myself, and for those of my circle who asked me, in a concise and simple manner, as I believe and think, offering not the slightest contradiction to such an erudite and praiseworthy chronographer,[15] /620/ at least in those things in which he himself did not contradict Holy Scripture. Dreading the rod of Origen's *Hexapla*, he removed the generation of Cainan. Also he called into doubt [*fidem derogauit*] both the Gospel of Luke and the Seventy Translators, and erased 130 years from their translation.[16] He corrected the generations according to the Hebrew books, but he did not take care likewise to emend the number of years according to [their reckoning], and so he removed 1,032 years in the sum total of his computation of these years. Thus observing neither road, he proceeded by [the way] he wished.

7. In the case of the First and Second Age, there is a discrepancy between the codices transmitted to us (that is, from the Hebrew source) and those of the Greeks. As Jerome the translator of sacred history relates in his *Hebrew Questions, Before the birth of the aforementioned son, we find that the age of his father is 100 years less in the Hebrew version; but*

10 *Ibid.* 18 (602.1–3).
11 *Ibid.* 19 (603.1).
12 *Ibid.* 20 (604.1–2).
13 *Ibid.* 22 (607.1–3).
14 *Ibid.* 22 (607.2–5).
15 I.e. Eusebius.
16 Cf. *The Reckoning of Time*, ch. 66, *s.a.* 1693.

after [the son] is born, we find that he lives 100 years less in the Greek than in the Hebrew [version], and thus the grand total [of years] agrees in both.[17] *From this corruption of the Greek manuscripts arises that notorious question about Methuselah, who is reckoned to have lived fourteen years after the Flood, although it is certain, according to the Hebrew manuscripts, that he died in the same year as the Flood.*[18]

8. *Should any say that the Hebrew books were falsified later on by the Jews and that they, when they begrudge authority to us,*[19] *deprive themselves of the truth, let them listen to Origen, who in the seventh book of his commentary on Isaiah, solves this trivial problem, [saying] that our Lord and the Apostles, who accuse the scribes and Pharisees of other crimes would never have refused to speak about this crime, which is a very grave one.*[20] Let him listen to Josephus the historian of the Jews who in the first book ἀρχαιότητος [of the *Antiquities*] against the grammarian Apion[21] openly and lucidly demonstrates that the lesser number of years is correct, and that Holy Scripture was in no way /**621**/ falsified by the Jews. He says, *Nor are there among us innumerable books which disagree with one another, but there are 22 books which contain the entire sequence of time and which are justly believed to be divinely inspired, of which five are by Moses containing the laws of life and tracing the lineage of human succession down to the death of Moses himself, encompassing slightly less than 3,000 years.*[22] And a little later: *The great respect with which we treat our Scriptures is evident from this fact. For although so many ages have intervened, no one has ever dared to add or delete or change anything. But for all the people of our nation, this faith is implanted and, as it were, inborn: to believe these [Scriptures] to have been inspired by God, and to remain steadfastly attached to them, and should the occasion demand, willingly to die for them.*[23]

---

17 This section of the quotation is not from Jerome, but is slightly adapted from Augustine, *DCD* 15.10 (466.6–10). The father is Adam, the son is Seth.

18 *Ibid.* 15.11; cf. Jerome, *Hebraicae quaestiones in Geneseos* 5.25–27 (8.18–9.17).

19 "... when they begrudge authority to us": cf. Augustine, *DCD* 15.11 (468.33–35).

20 Jerome, *In Isaiam* 3.6.9.10, ed. M. Adriaen, CCSL 73 (1963):92.49–54.

21 *Against Apion* is a separate treatise, and not the first book of the *Antiquities*.

22 Josephus, *Contra Apionem* 1.38–39 (11.7–13).

23 *Ibid.* 1.42 (11.20–12.4).

9. Let him listen to the blessed Jerome who says that the Greek, and not the Hebrew manuscripts are false. For when he was compelled to translate the Book of Chronicles, he for this very reason wrote the following preface to the book: *If the version of the Seventy Translators had remained pure and as it had been translated by them into Greek, O Chromatus, most holy and learned of bishops, it would not have been necessary for you to impel me to translate the Hebrew volumes into Latin for you. For it was just that [the translation] which once had captured the ears of men and strengthened the faith of the new-born Church should also be confirmed by our silence.*[24] *But now that many proofs have appeared from various corners of the world that the genuine and ancient translation is corrupted and violated, you consider our choice [to be] either to judge which of the many [versions] is true, or to hammer out a new work in place of the old work . . .*[25] and so on, in the same vein.

10. The sainted Bishop Augustine agrees with this viewpoint. Since he did not wish to ascribe the cause of this discrepancy either to the malice of the Jews or to the error or jealously *of the Seventy* Translators, *who were Jews themselves*, lest he indiscriminately accuse of falsehood those men whom the spirit of God had filled in their seventy cells, he therefore says: *So he is more worthy of credence* /**622**/ *who said that when these [Scriptures] first began to be transcribed from the library of Ptolemy, then some [error] of this sort could have happened in one codex, and from that original transcription could have been disseminated at large, where a scribal error could have happened. This could plausibly be the case in the question of [the length of] Methuselah's life.* And later on: *For even nowadays, whenever numbers contribute nothing to ease of comprehension, or useful knowledge, they are carelessly written down and carelessly corrected. For who would consider it worth learning how many thousand men were in each of the tribes of Israel, when it is not thought to be of any benefit.* And a little later on: *But however this is interpreted, I should certainly in no way doubt that when some divergence is found in both codices [i.e. Septuagint and Hebrew], and when it is impossible for both to be true to the historical facts, greater faith should be placed in [the version] in the*

---

24 The Old Testament of the *Vetus latina*, the primitive Latin version of the Scriptures, was translated from the Septuagint.

25 Jerome, *In libros Paralipomenon praefatio*, PL 28.1325B–1326B.

language from which the translation was made into another language by interpreters.[26]

11. Read Book 15 of *The City of God*: *We ought not to listen to those who think that in [Old Testament] times, years were computed in a different way, that is, who believe that they were so short that one of our years equals ten of theirs*, or in other words, that they comprised 36 days. They use the argument *that it is found in many historians that the Acarnians* had a year *of six months, the Arcadians of three, and the Egyptians of four months, and that these people* sometimes *closed* their *year with the end of the lunation.*[27] Scripture itself directly refutes any such meaning. Where the Hebrew [version] reads: *And Adam lived 130 years and begot Seth, and the days of Adam after he had begotten Seth were 800 years and he begat sons and daughters; and all the days that Adam lived were 930 years, and he died,*[28] the Seventy say: *And Adam lived 230 years and begot Seth, and the days of Adam after he had begotten Seth were 700 years; and all his days were 930 years*, and so on in the same vein. What then? Is it possible that they posited shorter years in the first part of each verse, longer ones in the middle, and equal ones at the end? So to [conclude] is both to fall into error and to be suspected of stupidity.

12. Some people wonder why the great and famous translator of Holy Scripture Jerome – of whom it is justly said, *Jerome, translator most learned in diverse languages, Bethlehem /**623**/ extols you, the whole world resounds [with] your [praise]*[29] – why, I say, although he translated the book of the *Chronicles* [of Eusebius], he did not wish to pass on to the Latins the truth he had learned from the Hebrews? Whoever is disturbed by this should know that, just as he himself stated in the preface to the same book on times [i.e. the *Chronicle* of Eusebius], he wished either to be a translator or the creator of a new work.[30] And perhaps [it was] because he did not wish, where there was no pressing

26 Augustine, *DCD* 15.13 (471.17–22, 37–42).

27 *Ibid.* 15.12 (468.1–3, 469.37–39, 50). Cf. *The Reckoning of Time*, ch. 37.

28 Genesis 5.3–5.

29 Isidore, *Versus in bibliotheca* 7.1–2, ed. C.H. Beeson, *Isidorestudien* (Munich: Beck, 1913):160.

30 Bede was unaware of the fact that Jerome translated Eusebius' *Chronicle* before undertaking the Vulgate. The *Chronicle* was translated in 380, the Vulgate begun in 383: cf. J.N.D. Kelly, *Jerome* (London: Duckworth, 1975):32, 86.

necessity, to provoke the fury of the insane against himself on his own initiative, but to leave this matter to posterity, as he says to Eustochium the virgin in a certain place: *To wish to buy shuttles and heddles for oneself, but to leave to others the trouble of weaving.*[31] Indeed, on account of such a necessary translation of Holy Scripture, he was pelted with stones by both the Latins and the Hebrews, almost at the same time – by the Hebrews, because he had robbed them of the opportunity to mock and revile the Christians on account of their fallacious books, but by the Latins because he had introduced new and unfamiliar, albeit better, things in place of old and familiar ones.

13. St Augustine, in letters transmitted across so great a space of land and sea with friendly brotherly love, took care to insert the following: *But I would not wish you to labour at turning the canonical books into the Latin language, save in that manner in which you translated Job, so that what is divergent between your translation and the Septuagint (which is a very weighty authority) might be manifest through editorial marks. I would be extremely surprised if anything were to be found in the Hebrew exemplars which escaped such expert translators from that tongue. But I exempt the Seventy: when it comes to their decisions, and their greater agreement of mind, just as if they had been one man, I dare not pass categorical judgement on any score, save that I think that there can be no doubt that they are the pre-eminent authority in this domain. I am more disturbed by those later translators who, although they stuck more closely to the pattern of Hebrew words and locutions (as it is said), not only did not agree amongst themselves, but also left many things which remained long after to be unearthed and discovered. For if these matters are obscure, you likewise could be mistaken concerning them; if they are evident, it is not to be believed that those men would have been able to make a mistake in this regard.*[32]

14. On the subject of times and years, /**624**/ I warn your simplicity, dearest brother, lest seduced by vulgar opinion you should expect that this present world will endure 6,000 years, as it were. According to a book of I know not what heretic which I remember seeing as a lad, written in old-fashioned script, though the day and hour of Judgement

---

31 The source of this quotation has not been traced.
32 Augustine, *Ep.* 66.2 (to Jerome) CSEL 54.497–498.

cannot be known, the year can.[33] In fact you think that man can foresee this, because the Lord said, *But of that day and hour knoweth no man*,[34] and yet said nothing about the year. And again when he said, *It is not for you to know the times or the seasons*,[35] he did not add *the years* thereto. And again this heresiarch chronographer exerted himself to add that 5,500 had passed before the Incarnation, and from thence to the Day of Judgement only 500 remained, of which 300 and some had passed at the time when he uttered his ravings.[36] Accumulating the sum of this number partly from Genesis, partly from the Book of Judges, but attributing forty years each to Samuel and Saul,[37] he used the argument that in the Lord's vineyard, the workers hired last will work one hour,[38] and that John had said, *Little children, it is the final hour*,[39] as if 500 years out of 6,000 ought to be as one hour out of a day, when evening falls after twelve hours of daylight. So when 12 times 500 years had passed, that is 6,000 years, the eternal retribution of the just Judge would come to pass.

15. On this matter I confess I am quite grieved, and often irritated to the limit of what is permissible, or even beyond, when every day I am asked by rustics how many years are left in the final millenium of the world, or learn from them that they know that the final millenium is in progress, when our Lord in the Gospel did not testify that the time of His advent was near at hand or far off, but commanded us to keep watch *with our loins girded and our lamps lit*,[40] and to *wait for Him* until *He should come*. For I notice that when in conversation with the brothers the occasion arises for us to dispute concerning the Ages of the world, certain of

33 Jones (*BOT* 135, n. 2) identifies this text with the Cologne Prologue, ed. Krusch, *Studien I*, 227 *sq.* Cf. Augustine, *Ep.* 199.6 (256 *sq.*). Though this idea was not unknown amongst the Fathers, Bede's identification of the heretic as a chronographer suggests that the book was computistical, not exegetical.

34 Matthew 24.36, Mark 13.32.

35 Acts 1.7.

36 Cologne Prologue 5 (231–232). This is the chronology of Hippolytus and Julius Africanus: Landes 138.

37 The lengths of Samuel's tenure as judge and of Saul's reign are not given in Scripture. In *The Reckoning of Time*, ch. 66, *s.a.* 2870 and 2890 Bede supplies them from Josephus: 12 years for Samuel and 20 for Saul.

38 Cf. Matthew 20.1–16.

39 I John 2.18.

40 Cf. Luke 12.35–37.

the less learned ones allege that we are speaking of 6,000 years, and there are those who think that this world will end at 7,000 years because it unfolded in seven days. /625/ The said doctor Augustine in his explanation of Psalm 6 often openly accused these people of temerity, saying amongst other things: *For if this day is to come after 7,000 years, everybody can add these to the computed years of his Advent. How will it be that the Son does not know [the time]? It means, of course, that men cannot know this through the Son, not that He does not know it in himself, according to the saying, "The Lord your God proveth you, to know",*[41] *and "Arise, O Lord",*[42] *that is, make us arise. Therefore although it is said that the Son does not know the day, [it is] not that He does not know, but that He does not cause them to know, that is, He does not show this to them. What does this strange presumption seek, which hopes [to find] the most certain day of the Lord by counting 7,000 years?*[43]

16. I send these letters, signs of charitable friendship, dearest brother, as pledges of my innocence, lest you judge me deeply lacking in secular or sacred learning. Rather, [we shall] state openly that our Lord and Saviour took mortal flesh for our sake in the Sixth Age of the world, according to the Scriptures or acknowledging the faithful and catholic voice of the Fathers. But the course of the world is not defined for us in any fixed number of years, and is known only to the Judge himself. And if anyone were to say to me, "Lo, Christ is here", or "Lo, He is there",[44] that is, [that Christ is] to come to Judgement at such and such a time, I would not listen to or follow him at all. For I know that *as the lightning, that lightneth under heaven shineth on those things which are under heaven; so shall the Son of Man be in his day.*[45] For you know by what opinion of the multitude it will come in six or seven thousand years, and by what authority I build the assertion of my computation: namely by the Hebrew Truth, recorded by Origen, published by Jerome, praised by Augustine, confirmed by Josephus. I have found none more learned in such matters than these. Nor is it to be wondered at that that praiseworthy man Eusebius, although he was able, as they say, to bind iron

41 Deuteronomy 13.3.
42 Psalm 3.7.
43 Augustine, *Enarrationes in Psalmos* 6.1 (27.15–26).
44 Cf. Luke 17.23.
45 Luke 17.24.

and brick by his marvellous talent in speaking and thinking, nevertheless could not do what he had not learned to do, that is, to know the Hebrew language. By his justly reverend fear, he himself had no fear of corrupting what he knew, as we showed above. /**626**/

17. I beseech you to present this account of my exoneration to our religious and very learned brother David,[46] that he may read it before our venerable lord and father Bishop Wilfred, so that, because I was earlier assailed in his presence and hearing by the abusive talk of the foolish, it may now be made plain to him in his hearing, and with him as the final judge, how unworthily I suffer this same abuse. Also I ask this same David, above all others, to follow the example of the boy whose namesake he is, and to exert himself sedulously to expel the madness of spirit from the unreasonable brother by the exhortation of healthful words, as if by the sweet modulation of psalmody.[47] In any event, the man who, tipsy with drink, strove to incriminate me in that banquet – who ought rather to have made himself blameless by paying heed to his reading – was unable to bring it off, for he did not realize that he was extolling my own meaning and my very words. For truly it is said that *if a serpent will bite in silence, there is no wealth for the enchanter.*[48] May the almighty Lord deign to preserve your fraternity in safety.

---

46 Jones thought that David was the perpetrator of the accusation against Bede, but as the remainder of this section makes plain, David is being asked to clear Bede's name before Wilfred, and to reproach the nameless accusor: see Dieter Schaller, "Der verleumdete David: zum Schlusskapitel von Bedas *Epistola ad Pleguinam*", in *Literatur und Sprache im europäischen Mittelalter: Festschrift für Karl Langosch* (Darmstadt: Wissenschaftliche Buchgesellschaft, 1973):39–43.

47 The allusion is to the biblical David's capacity to soothe the rage of King Saul through his music: I Samuel 16.14–23.

48 Ecclesiastes 10.11.

APPENDIX 3.2: LETTER TO HELMWALD

Translator's note: This appendix contains the opening paragraphs of Bede's letter to Helmwald. The remainder of the text was incorporated by Bede without change into chs. 38–39. Who Helmwald was, or where he lived, is not known.

LETTER TO HELMWALD

/**629**/ To Helmwald, his most beloved brother in Christ, Bede the servant of Christ [sends] greeting.

I confess that I rejoiced greatly, dearest brother in Christ, to learn that in the place where, at the will of God, you joined the long-desired pilgrimage, you begin to pursue peace and virtue, and even to devote some effort to reading. And therefore I shall not delay to expound plainly to you, as I see it, those things which you inquired of me in [your] letters, desiring with grateful expectation that I may deserve to see you restored to the [heavenly] homeland in the company of the eminent persons of our age, at once endued with the light of spiritual knowledge, and ministering the radiant joys of celestial teaching and life. So you asked to have explained to you, briefly and plainly, the calculation of the fourth year which is called *bissextus*, in so far as this is possible, and, if I may so express it, to have opened up to the full view of free and serene understanding the depths of more secret nature by the keys of straightforward writing. Nor do you who ask such things deserve to be refused by my insignificance, whom you decided should be appealed to by [your] letters, and consulted concerning matters of importance to the Church, though such vast distances of land and sea lie between us . . .

APPENDIX 3.3: LETTER TO WICTHED

Translator's note: This letter was evidently a response to a query raised by Bede's discussion of the equinox and the leap-year day in *The Reckoning of Time*. Its close connection to *The Reckoning of Time* is revealed by the manuscripts; except when it is included in a collection of Bede's letters, the *Letter to Wicthed* always accompanies *The Reckoning of Time*. Occasionally it is incorporated into the text of *The Reckoning of Time*.[1]

At various points in *The Reckoning of Time*, Bede insists that the spring equinox falls on 21 March, and cites Anatolius as his authority. However, the text of Anatolius which he and his readers held to be authentic clearly stated that the equinox was on 22 March. In this letter, Bede explains that the equinox can fall on 22 March in the final year or two of the leap-year cycle, but that this should not inhibit Christians from celebrating Easter on that date.

The date of composition is unknown, but since it follows *The Reckoning of Time*, and is found in the autobio-bibliography in *HE*, it must have been written between 725 and 731.

LETTER TO WICTHED

/635/ To his most reverend and holy brother Wicthed, priest, Bede [sends] longed-for greeting in the Lord.

1. Beloved brother in Christ, I gladly received the letters of your benevolence, and I made haste to copy out promptly the chapters which you requested and to send them to you, recalling the friendliness and kindness with which you received me when I came thither.[2] But now I have undertaken to explain at greater length that justly famous question about the *Ecclesiastical History* [of Eusebius] which you asked me when I was with you, and to which I then replied briefly, as I was able. This concerns the spring equinox, which in the aforementioned book the most reverend

---

1 Jones, *BOT* 138–139.
2 The location of Wicthed's monastery is not known, but Bede evidently visited there.

Bishop Anatolius is held to have placed on the 11th kalends of April [22 March],[3] although the other masters of the Egyptians decreed that it should rather be placed on the 12th kalends of the same month [21 March]. For the holy Proterius, bishop of the church of Alexandria, writing to the most blessed Pope Leo on the subject of Easter, says: *The equinox is plainly identified according to the course of the Sun as the 25th day of [the Egyptian month] Phamenoth, which is the 12th kalends of April [21 March]. But we should not start the first month according to the Moon straightaway from this equinox, though in all other respects the course of the Moon ought to harmonize with the cycle of the Sun.*[4] St Cyril, prelate of the same church, also said: *For the Sun itself terminates in land and sea at the close of the day, and at the beginning of the day is unveiled in like manner, and the Sun finishes the course of the whole year on the 12th kalends of April.*[5]

2. For this reason Anatolius disturbs you, just as he disturbs other scholars, when he writes, in contradiction to the Egyptian teachers (although he is by birth and education an Egyptian) that the equinox is wont to occur on the 11th kalends of April [22 March], saying: *Therefore in the first year, which is the beginning of the nineteen-year cycle, the beginning of the first month, according to the Egyptians, /636/ is the 26th day of the month Phamenoth; according to the Macedonians the 22nd day of the month of Dystros; and according to the Romans the 11th kalends of April [22 March]. On this day the Sun is found not only to have climbed up to the first part, but to have in that day a quarter, that is, in the first of twelve parts. For this little part [particula] is the spring equinox, and it is the beginning of months and the starting-point of the cycle and the completion of the course of the stars called "planets" (that is, "wandering") and the end of the twelfth little part is the close of the whole cycle.*[6]

3. But I think that it is possible to demonstrate by a very simple reckoning that he is not opposed to the other teachers of Egypt and the East on this

3 Anatolius 2, as cited in Rufinus' trans. of Eusebius, *HE* 7.32.14 (723.10–14). Cf. *The Reckoning of Time*, ch. 14.

4 Proterius, *Ep. ad Leonem* 8, ed. Krusch, *Studien I*, 276–277.

5 *Ep. Cyrilli*, ed. Krusch, *Studien I*, 346.

6 Anatolius 2, as cited in Rufinus' trans. of Eusebius, *HE* 7.32.14–15 (723.10–19).

point. For the calculation of the quarter-day which is called *bissextus* is made so that the Sun reaches the equinoctial point of its cycle in the zodiac now at dawn, now at noon, now at sunset, and now in the middle of the night.[7] And whenever the equinox happens at dawn or at midday, it is on the 12th kalends of April [21 March], and when it happens at sunset or midnight, it pertains to the 11th kalends of April [22 March]. For it is beyond dispute that the entire night on which our Lord rose from the dead is moved forward to the time of the following day, and not placed with the preceding [day].[8]

4. Therefore Anatolius prudently does not forbid Easter to be celebrated *on* the 11th, but rather *before* the 11th kalends of April [22 March], for immediately after those statements of his which we set down [above], he adds: *And so we declare that those who think that Easter ought to be celebrated before the beginning of the new year commit no small offence.*[9] So he does not say that it is forbidden to celebrate Easter *on* that beginning which is the equinox, but rather *before* that beginning of the new year, since he himself in the same book assiduously affirms, from the writings of ancient and modern Fathers alike, that Easter cannot happen before the equinox has passed. For he is his own witness when he writes that he indicates not the earliest time of the seat of the equinox, but the latest, after which, he knew, the celebration of Easter could begin. For although Easter Sunday can fall on the 11th kalends of April [22 March] in the second or third year after the leap year, /**637**/ it is nevertheless agreed that the Paschal season begins with the equinox. For its ceremonies are partly celebrated in the beginning of the night, although no one doubts that the equinox can take place in the middle [of that night] or even at the beginning [of the day], because undoubtedly the hour of the Lord's Resurrection is rightly to be celebrated at dawn.

5. And so that you may plainly observe that Anatolius writes specifically about an evening equinox, after which Easter can begin, and that he regards the 11th kalends of April [22 March] according to the rule [*regulariter*], notice what he adds in the following, [drawn] from the opinion of the ancient Fathers: *For since there are two equinoxes, spring and*

7 Cf. Victor of Capua, *De pascha* 2 (297).
8 Cf. *The Reckoning of Time*, chs. 5, 6, 71.
9 Anatolius 2, as cited in Rufinus' trans. of Eusebius, *HE* 7.35.15 (723.19–21).

autumn, separated by equal lengths [of time], and since the solemnity is prescribed for the fourteenth day of the first month, when the Moon is detected in the region opposite the Sun after sunset, as can be verified with the eyes, the Sun is without doubt found reaching the position of the spring equinox, and the Moon, the autumn [equinox].[10]

6. Is it any wonder that he says that the equinox happens on the 11th kalends of April [22 March]? After all, when he speaks about this hour, he declares that in the evening when the Sun sets, the Moon on the other hand begins to rise, while at the same time it is understood that he said that the equinoxes of spring and autumn are separated by equal lengths [of time]. So it follows that we should compute half a year from equinox to equinox, and that the autumn [equinox] is fixed on the 13th kalends of October [19 September], separated from the spring equinox by 182 days. Inspection of the sundial demonstrates this, especially since [Anatolius] adduces this perfectly equal separation of the year between the equinoxes in the words of the ancient and most learned doctor Aristobolus, who was one of those famous Seventy Translators of Holy Scripture.[11] /**638**/ The same Anatolius says: *On this day the Sun is found not only to have climbed up to the first part, but to have in that day a quarter, that is, in the first of twelve parts.* The twelve parts denote the twelve signs of the zodiac, each of which holds the Sun for thirty days and a few hours longer. Here Maro says, on the same subject:

> *thus the golden Sun holds sway over the sphere,*
> *divided into fixed sections, by the twelve stars.*[12]

7. The first of these parts begins according to nature [*iuxta naturam*] from the place of the spring equinox. This first of twelve [parts] is the spring equinox, because without doubt the beginning of this part (that is, where the Sun was positioned at the beginning [of the world]) contains the equinox itself.[13] [Anatolius] correctly stated by way of preface that "the Sun on that day not only climbs to the first part but also already has a quarter on that same day", because whenever the equinoctial time according to the aforementioned calculation falls on

10 *Ibid.* 7.32.18 (725.10–14).
11 *Ibid.* 7.32.17 (725.8).
12 Vergil, *Georgics* 1.231–232. Bede quotes the same passage in *The Reckoning of Time*, ch. 16.
13 Cf. *The Reckoning of Time*, ch. 6.

the 11th kalends of April [22 March], then at that same moment of time the fourth part of the day which ought to accrue annually is recognized to be complete according to nature [*secundam naturam*]. For when the equinox falls on the 11th kalends [22 March], then the quarter-day is finished on the very same day, and in the same equinoctial hour. We add it thus according to nature, because according to human custom, it is [added] at various times of the year, as it pleases each people to insert it. But according to natural computation [*naturalis . . . rationis*] it is to be added at the completion of the solar cycle. Cyril also indicates this in that statement which we cited above, saying: *And the Sun finishes the course of the whole year on the 12th kalends of April* [21 March].[14]

8. And should someone perhaps object that we have not [respected] what Anatolius himself /639/ understood in this statement, I wish to add what the learned Victor, bishop of the city of Capua, wrote on the subject of the parts and of the equinox: *The celestial circle, through which the Sun and Moon and the stars called planets are borne with their proper motion, against the impulse of the whole heaven,[15] is divided into twelve parts according to the judgement of wisdom. These are completed in 365 days, in which sum of time the year is completed. When the Sun regains the beginning of the first part of the aforementioned circle, the year begins, and this happens from the 12th kalends of April [21 March] to the 11th [22 March]. [The equinox] is ascertained to take place at evening, sometimes at night, and occasionally on the day of 11th kalends of April [22 March] itself. On the 25th or 26th day of the month of March according to the Alexandrians[16] – that is, according to the Latins, the 12th or 11th kalends of April – the beginning of the first month according to the course of the Sun takes place. Lest anyone wish to accuse us of inconsistency because we do not define the day of the first month, using (so to speak) the ambiguous phrases of a waverer, let this fault-finder, or rather, inquisitor know that the most skilful and very subtle investigation of the Egyptians teaches that the first hour of the latter day*

14 Cf. n. 5 above.

15 For explanation, see *The Reckoning of Time*, ch. 34, n. 298.

16 Victor means, of course, the Egyptian month corresponding to March, i.e. Phamenoth, not the Roman month of March: 25/26 Phamenoth = 21/22 March (cf. *Reckoning of Time*, ch. 11). This is a rather unfortunate expression, because it might suggest that Victor supported the 25 March equinox.

begins in the seventh hour of the day.[17] *And although the 25th day of March*[18] *is the 12th kalends of April [21 March], if the Sun lights up the space of the first part on the evening of the same day, it is already ascribed to the 26th day, as we said, following the subtle and without doubt probable Egyptian tradition.* And a little later he says: *Therefore if the 14th Moon is found, with the Sun established in the first part of its cycle, it is rightly to be assigned to the first month. But if the 14th Moon should have shone forth with the full light of its orb before the Sun has reached the aforementioned first part of its cycle, it is considered to belong to the [preceding] twelfth month. And therefore every time this happens, the solemnity of Easter is postponed until the next full Moon, which must take place while the Sun is still established in the first part of the twelve parts of the circle.*[19] It has pleased us to include these words of the blessed Victor, **/640/** seeing that the statement of the holy father Anatolius, which in his own work on Easter and in the *Ecclesiastical History* is, for many people, obscure, may through his statement be rendered more clear to you.

9. But it should not be overlooked that there are those who contend that Anatolius did not write "the 11th" but "the 8th kalends of April" [25 March]. They say that Eusebius, when he quoted this statement in the *Ecclesiastical History*, substituted the former date for the latter. He saw that the rest was well and philosophically said, and so he wished to correct one word which he perceived had been expressed less perfectly, lest by introducing into his histories what [Anatolius] originally said, he openly expose to ridicule a man whom he was disposed to praise. But it would be astounding if Eusebius, a writer circumspect in word and meaning, had been so eager to praise others as to pretend that they had said things which they did not say when he excerpted their writings, and yet not be afraid that he would be criticized when his readers came upon the complete works of these men, and took umbrage that what he himself had taken from these men was found to be otherwise in his authorities. It would be astounding had such an imposture escaped the notice of Victor, whose statement we have quoted above, a man of equally exceptional learning

17 This is a somewhat obtuse way of saying that in the second year of the 4-year leap-year cycle, the Sun will reach the equinoctial point one quarter of a day, or 6 hours, later than it did in the previous year.

18 See n. 16 above.

19 Victor of Capua, *De pascha* 2–4 (297).

who, writing about Easter in another work, inserted this same statement of Anatolius from the *Ecclesiastical History*,[20] adopting it, so to speak, as justly praiseworthy and memorable, when it might readily have been refuted. And he himself wrote "11th kalends of April" and not 8th.[21]

10. But it is also to be wondered why Dionysius, surnamed Exiguus, a man outstanding in knowledge, would have pressed into service in his letters on Easter the support of Anatolius from the *Ecclesiastical History*, if he thought his statements had been falsified. The watchfulness of one who was very expert in the Greek language would make it impossible to conceal from him how these [statements] had first been published: *But because this month, whence it starts or where it ends, /***641***/ could not be inferred there*, that is, in the writings of Moses, *the aforementioned 318 venerable bishops,*[22] *investigating in a resourceful manner the observance of ancient custom, and what was handed down thence by holy Moses, as is recounted in the seventh book of the Ecclesiastical History, stated that a new Moon from the 8th ides of March [8 March] up to the nones of April [5 April] defines the beginning of the first month, and the fourteenth Moon should be skilfully sought for from the 12th kalends of April [21 March] up to the 14th kalends of May [18 April].*[23]

11. Therefore it seems likely that Eusebius faithfully inserted into his histories what he found in the Greek authority. But that same treatise by Anatolius afterwards became corrupted in certain Latin exemplars, by whose considerable fraud those who did not know the true time of Easter would eagerly desire to defend their erroneous observance with the authority of so great a Father. But they who prefer to think that Eusebius emended one sentence, rather than that certain others falsified all of Anatolius, enquire how it could be written in the same treatise: *But is it any wonder that they err about the 21st Moon, who add three days before the equinox in which, they proclaim, the Pasch can be sacrificed, which surely is utterly absurd?*[24] To them it ought to be replied that Anatolius

---

20 What "other work" by Victor Bede might be referring to is not clear. No quotation from Anatolius is found in the fragments of *De pascha* printed in *Spicilegium Solesmense*.

21 See quotation in section 8 above. Cf. *The Reckoning of Time*, ch. 14.

22 I.e. the Council of Nicaea.

23 Dionysius Exiguus, *Ep. ad Petronium*, ed. Krusch, *Studien II*, 65.

24 Ps.-Anatolius 6 (321).

could have known many who believed one thing or another about Easter, but of which we, however, have no knowledge. On the contrary, they should be asked how it could be written in the same treatise: *The beginning of the first month according to the Egyptians, is the 26th day of the month Phamenoth, according to the Macedonians the 22nd day of the month of Dystros, and according to the Romans the 11th kalends of April [22 March]*, when the 26th day of the Egyptian month Phamenoth and the 22nd day of the Macedonian month Dystros are not the 8th kalends of April [25 March] but the 11th kalends of April [21 March]. /**642**/ When the person who changed the 8th kalends of April to the 11th kalends forgot to change the dates of the Greek and Egyptian months, but left them as they were because they were less familiar, and changed only that which was evident to himself and his [readers], does it not seem highly probable that the statement could be falsified?

12. Let those who can read Anatolius in Greek see which of these is more true. Indeed, if Eusebius altered one statement, or if someone else altered the entire treatise from its [original] state, it nevertheless remains beyond doubt, although great numbers of ancient [authorities] affirm [it], that the equinox cannot be found on the 8th kalends of April [25 March], which is proved both by inspecting the sundial and by transparent reasoning.[25] The rule of ecclesiastical observance, which is confirmed both by the edicts of the Fathers of old and more clearly by the Nicene Council, is that Easter Sunday is to be sought from the 11th kalends of April [22 March] to the 8th kalends of May [25 April]. Indeed the rule of catholic instruction prescribed that Easter should not be celebrated before the spring equinox has passed. Whoever thinks that the 8th kalends of April [25 March] is the equinox is either obliged to state that it is licit for Easter to be celebrated before the equinox, or to deny that it is licit for Easter to be celebrated before the 8th kalends of April. Also, he will confirm that the Passover which our Lord kept with His disciples either was not on the 9th kalends of April [24 March],[26] or was before the equinox.

Beloved brother, may you always fare well in the Lord.

25 Cf. *The Reckoning of Time*, ch. 30.

26 Given Bede's rejection of 25 March as a possible date for the historic Good Friday (cf. *The Reckoning of Time*, chs. 47 and 61), it is curious that he should invoke it here as evidence for the 21 March equinox.

APPENDIX 4: A NOTE ON THE TERM *COMPUTUS*

The word *computus*, whatever meaning is assigned to it, is post-classical; classical Latin prefers *computatio*. The *Oxford Latin Dictionary* defines *computare* (383) as primarily a mathematical or counting term, though it also has an extended or metaphorical sense of "to work out (a problem etc.)" which links it to some of the possible connotations of *ratio* (see Introduction, section 2). The earliest use of *computus* in connection with time seems to have been astrological, e.g. Firmicius Maternus 1.12, which may be another reason for its slow adoption as a term for Christian time-reckoning. For the most part, Bede uses *computus* only in its broader mathematical sense of "calculation", e.g. in the title of ch. 1 of *Reckoning of Time*, *De computo uel loquela digitorum*, though he certainly uses *computare* in the narrower sense of time-reckoning: e.g. in the same chapter, *computando seriem temporum* (268.6–7). On only one occasion does he use the term to refer to a work on Easter-reckoning, and that is in ch.66, *s.a.* 4338, where he notes that Theophilus "wrote his Paschal *computus*" (*paschalem compotum scribit*). However, a volume containing calendar tables, texts and rules is referred to as a *computus* by Bede's younger contemporary Egbert, archbishop of York, in the prologue of his *Penitential* (ed. D. Wilkins, *Concilia Magnae Britanniae et Hiberniae*, London, 1737, vol. 2, p. 417). Aelfric[1] echoes Egbert in insisting that priests should have, amongst their *divini libri*, a *computus*. By the time of Alcuin, *computus* apparently was accepted as referring specifically to time-reckoning: *cur in quibusdam mensibus ad eosdem dies in anno secundo uel tertio uel quarto computus cursus solis non peruenisset* (*Ep.* 171, ed. Dümmler, *Epistulae Karolini Aevi*. MGH Ep. 4. Berlin, 1985). This usage expanded in the Carolingian period, e.g. *Computus hic alphabeto confectus. . .* the incipit of the metrical textbook on time-reckoning by Agius de Corvey, ed. by Karl Strecker, *Agii versus computistici*, MGH Poetae 4.3 (1914):937–943; and by the twelfth century was generalized: e.g. the anonymous treatises beginning *Computus est scientia rationis temporum . . .* the earliest

---

[1]*Epistola* 2.137, ed. Bernhard Fehr, *Die Hirtenbriefe Aelfrics in altenglischer und lateinischer Fassung*, Bibliothek der angelsächsichen Prosa, 9 (Hamburg: Grand, 1914).

manuscript of which (Berlin Staatsbibliothek F.307 fols.8v–10v) dates from the twelfth century, and *Computus est computatio temporum secundum cursum solis et lune* ... in Montpellier 322, fols. (3v)–(18v), also from the twelfth century. By the time of Sacrobosco and Grosse-teste, *computus* meant almost exclusively time-reckoning: cf. the incipit of Sacrobosco's textbook, *Computus est scientia considerans tempora ex solis et lune motibus* ... The noun *computista* or *compotista* is also twelfth-century coinage; Bede prefers *computator* (e.g. *Reckoning of Time*, ch. 38, where he warns that the *peruersus computator* who forgets to intercalate the leap-year day will eventually be horrified to discover that the seasons are out of phase with the calendar: 400.38). For further citations, see R.E. Latham, *Dictionary of Medieval Latin from British Sources*, fasc. IIC (London: Oxford University Press for the British Academy, 1981):415. On the evolution of computistical terms in general, see Faith Wallis, "Chronology and Systems of Dating", in *Medieval Latin Studies: An Introduction and Bibliographic Guide*, edited George Rigg and Frank A.C. Mantello (Washington: Catholic University of America Press, Washington D.C., 1996):383.

APPENDIX 5: BRIEF GLOSSARY OF
COMPUTISTICAL TERMS

bissextus: Leap year or leap-year day. In the Julian calendar, the leap-year day was inserted on the 6th kalends of March (24 February). The intercalated day was thus the "second 6th (kalends)" or *bissextus*. A leap year is an *annus bissextilis*.

common year: A year in which there are 12 lunar months, in contrast to an embolismic year (q.v.).

concurrent: In the context of the 19-year cycle used by Bede, the concurrent is the number of the weekday of 24 March, counted from Sunday. Since the tropical year of 365 days contains 52 weeks plus 1 day, the concurrent will advance by one each year. However, the insertion of leap-year day will advance the concurrent by two every fourth year. It therefore requires 28 years for the cycle of concurrents to be complete: 7 weekdays, times 4 for the leap years.

cycle: see "luni-solar cycle", "Paschal cycle".

embolism: Because the Julian solar calendar is 365 days long, and 12 lunar months of $29\frac{1}{2}$ days total 354 days, the age of the Moon on any given calendar date increases by 11 days each year. When the accumulation tops 30 days, the maximum length of a calculated lunar month, an additional 13th lunar month is deemed to be inserted within the calendar year. This additional intercalated lunation is called an "embolism" and a year with 13 lunations is "embolismic".

embolismic year: A year in which there are 13 lunar months, due to the insertion of the embolism (q.v.). Contrasts with common year (q.v.).

epact: Generally, the age of the Moon on any given day. Specifically, and in the context of Dionysius Exiguus' 19-year cycle, "epact" refers to the age of the Moon on 22 March. In 84-year cycles, and in the tables of Victorius, the epact is the age of the Moon on 1 January.

"full" lunar month: A synodic lunar month or lunation is slightly more than $29\frac{1}{2}$ days. However, in order to harmonize solar and lunar cycles (see "luni-solar cycle" and "luni-solar calendar" below) a normalized lunar month expressed as a number of whole solar calendar days must be adopted. Medieval computists consequently adopted the

fiction that lunations are alternately 29 and 30 days long. 29-day lunations are "hollow", and 30-day lunations "full". Full and hollow lunations alternate throughout the year, January's being full, February's
hollow, etc. All embolisms (q.v.) are full.

hendecas: The final 11 years of the 19-year cycle, where the pattern of
common and embolismic years (q.v.) is CCECCECCECE. See also
"ogdoas".

"hollow" lunar month: see "full" lunar month.

leap of the Moon: see "*saltus lunae*".

lunation: The period between one new Moon and the next, now calculated as approximately 29.5306 days. Also known as "synodic lunar
month". Computists consider a lunation to "belong" to the solar
month in which it ends, i.e. the lunation of January is the lunation
which ends within the calendar month of January.

luni-solar calendar: A lunar calendar adjusted to maintain the relationship of lunar months to solar seasons, by the insertion of embolisms
(q.v.).

luni-solar cycle: A whole number of solar years into which a whole
number of lunar months can be divided, so that the lunar phases fall
on the same solar calendar dates after the end of the cyclic period.

Nisan: A month in the Jewish lunar calendar, corresponding to the first
lunation of spring. Bede identifies Nisan with April.

ogdoas: The first 8 years of the 19-year cycle, where the pattern of
common and embolismic years (q.v.) is CCECCECE. See also
"*hendecas*".

Quartodecimans: Early Christian communities in Syria and Asia Minor
who celebrated Easter on the fourteenth day of the first lunation of
spring, i.e. at the same time as the Jewish Passover.

Paschal cycle: A luni-solar cycle (q.v.) modified to permit repeated
projection of the dates of Easter into the future, normally by the incorporation of a third cycle to accommodate the shifting date of Sunday
in the Julian calendar (see "concurrent").

Paschal table: A table projecting the dates of Easter for a discrete
number of years into the future. A Paschal table may or may not be
based on a Paschal cycle.

regular: Bede uses the term "regular" to denote a number of computistical figures, but the most important and commonly used are the solar
and lunar regulars. The solar regular is a number assigned to each
month which represents the interval between the weekday of 24

March (the concurrent: q.v.) and the weekday of the first day of the month in question. Solar regular plus annual concurrent yields the weekday of the first day of the month in the year in question. The lunar regular is the age of the Moon on the first day of the month in year 1 of the 19-year cycle. The lunar regular, when added to the annual epact (q.v.), will yield the age of the Moon on the first of the month in that year.

saltus lunae: All luni-solar cycles gradually accumulate an additional *calculated* lunar day, due to the fact that the average lunation is slightly longer than the notional month of $29\frac{1}{2}$ days. To bring the lunar count back into phase with reality, the calculated age of the Moon must "jump over" a day. This is called the *saltus lunae*. In the 19-year cycle used by Bede, there is one *saltus* in each cycle, and it occurs in the final year of the cycle.

year: The tropical or solar year is the period of time required for the Earth to orbit the Sun, or in medieval terms, the time required for the Sun to complete its journey through the zodiac. The modern length of the tropical year is 365.2422 days. A lunar year is 12 or 13 lunations, depending on whether the year is common (q.v.) or embolismic (q.v.).

BIBLIOGRAPHY

1. BIBLIOGRAPHIC NOTE: LITERATURE ON *COMPUTUS*

The most recent and accessible survey of *computus* is Arno Borst's essay "Computus: Zeit und Zahl im Mittelalter", *Deutsches Archiv* 44 (1988):1–82 [9], and its expanded version, *Computus: Zeit und Zahl in der Geschichte Europas* (Stuttgart: Verlag Klaus Wagenbach, 1990); Eng. trans. *The Ordering of Time* (Chicago: University of Chicago Press, 1994). This book furnishes useful orientation, but contains numerous errors and is somewhat superficially conceived: see the review by Bruce Eastwood in *Speculum* 71 (1996):692–693. Supplementing Borst are the following: Olaf Pedersen, "The Ecclesiastical Calendar and the Life of the Church", in *The Gregorian Reform of the Calendar. Proceedings of the Vatican Conference to Commemorate its 400th Anniversary, 1582–1982*, ed. G.V. Coyne, M.A. Hoskin and O. Pedersen (Vatican City: Pontificia Academia Scientiarum; Specola Vaticana, 1983):17 *sqq.*; Alfred Cordoliani, "Comput, chronologie, calendriers", in *L'Histoire et ses méthodes*, ed. Charles Samaran (Paris: Gallimard, 1961):37–51 (very brief); A.N. Zélinsky, "Le calendrier chrétien avant la réforme grégorienne", *Studi medievali* ser. 3, 23 (1982):529–597; and Faith Wallis, "Images of Order in the Medieval Computus", in *ACTA XIV: Ideas of Order in the Middle Ages*, edited by Warren Ginsberg (Binghamton: State University of New York Press, 1990):45–67, and "The Church, the World, and the Time", in *Normes et pouvoirs à la fin du moyen âge*, edited by Marie-Claude Deprez-Masson, Inedita et rara 7 (Montreal: Ceres, 1990):15–29. More specialized literature on the history of *computus* before and during the time of Bede is cited below in the Select Bibliography. I am presently preparing two publications designed to provide a broad synthesis of the literature and history of *computus*: the first is a volume entitled *Computus: Manuscripts, Texts and Tables* for the series *Typologie des sources du Moyen Age occidental*; the second is a cultural and intellectual history of *computus*, tentatively entitled *A Diagram of Time*. For an overview of the secondary literature on *computus* in English, see Wallis, "Chronology and Systems of Dating", in *Medieval Latin Studies: An Introduction and Bibliographic Guide*, edited by

George Rigg and Frank A.C. Mantello (Washington: Catholic University of America Press, Washington D.C., 1996):383–387.

There is a formidable body of technical literature on the history of calendar construction. Most pertinent to the medieval *computus* are: F.K. Ginzel, *Handbuch der mathematischen und technischen Chronologie*, 3 vols. (Leipzig: J.C. Hinrichs'sche Buchhandlung, 1906–1914): esp. vol. 3, ch. 14, "Die Zeitrechnung des Mittelalters"; Franz Rühl, *Chronologie des Mittelalters und der Neuzeit* (Berlin: Reuter & Reichard, 1897); Bartholomew MacCarthy's introduction to his edition of the *Annals of Ulster*, vol. 4 (Dublin: His Majesty's Stationery Office, 1901); W.E. van Wijk, *Le Nombre d'or. Étude de chronologie technique suivi du texte de la Massa compoti d'Alexandre de Villedieu* (La Haye: Martinus Nijhoff, 1936); and Vénance Grumel, *La Chronologie*, Traité des études byzantines 1 (Paris: Presses universitaires de France, 1968). Grumel's bibliography is a very well organized guide to the older technical literature. A useful modern summary of the mathematics involved in the medieval *computus* is furnished by Werner Bergmann, "Easter and the Calendar. The Mathematics of Determining a Formula for the Easter Festival to [*sic*] Medieval Computing," *Journal for General Philosophy of Science* 22 (1991):15–41.

2. SELECT BIBLIOGRAPHY

This bibliography excludes works already cited in the Abbreviations list in the front of this volume, and items cited only once in the notes.

PRIMARY SOURCES

[(Spurious) *Acts of the Synod of Caesarea*]. *Acta synodi Caesareae*. Ed. Krusch, *Studien II*, 303–310, and from a different recension by A. Wilmart, "Un nouveau texte du faux concil de Césarée sur le comput pascal", in his *Analecta reginensia*, Studi e testi 59. Vatican City: Bibliotheca Apostolica Vaticana, 1933. Pp. 19–27.

Aldhelm, *Opera*. Ed. R. Ehwald. MGH AA 13 (1913). Trans. Michael Lapidge and Michael Herren, *Aldhelm: The Prose Works*. Cambridge: D.S. Brewer and Totowa, N.J.: Rowman and Littlefield, 1979.

Ambrose, *De Noe et Arca*. Ed. C. Schenkl. CSEL 32.1 (1897).
   Hexaemeron 4.4. Ed. C. Shenkl. CSEL 32.1 (1897).

Ps.-Anatolius, *De ratione paschali.* Ed. Krusch, *Studien I*, 316–327.

Arnobius, *In Psalmos.* Ed. Klaus.-D. Daur. CCSL 25 (1990).

Augustine, *Adnotationes in Iob.* Ed. I. Zycha. CSEL 28.2 (1895).

    Contra Faustum. Ed. J. Zycha. CSEL 25.1 (1891).

    De consensu Evangelistarum. Ed. F Weihrich. CSEL 43 (1904).

    De doctrina christiana. Ed. Joseph Martin. CCSL 32 (1961).

    De Genesi ad litteram. Ed. I. Zycha. CSEL 28.1 (1894).

    De sermone Domini in monte. Ed. A. Munzenbecker. CCSL 35 (1967).

    De Trinitate. Ed. W.J. Mountain and F. Glorie. CCSL 50 (1978). 2 vols.

    Enarrationes in Psalmos. Ed. E. Dekkers and J. Fraipont. CCSL 39 (1961). 3 vols.

    Epistolae. Ed. W. Goldbacher. CSEL 34 (1898), 44 (1904), 57 (1911), 67 (1911).

    Quaestiones evangeliorum. PL 35.1321–1364.

    Quaestiones in Heptateuchum. Ed. J. Fraipont. CCSL 33 (1958).

Ps.-Augustine, *De mirabilibus sacrae scripturae.* PL 35.2149–2200.

Ausonius, *Eclogae.* Ed. Sextus Prete. Leipzig: Teubner, 1978.

Basil, *Hexaemeron.* Trans. Eustathius. PL 53.965–966.

Bede, *Explanatio Apocalypsis.* PL 93.129–206.

    Historia abbatum. Ed. C. Plummer, *Baedae opera historica.* Oxford: Clarendon Press, 1896. Vol. 2, pp. 364–387.

    In Epistolas VII Catholicas. Ed. David Hurst. CCSL 121 (1983). Trans. David Hurst, *Bede the Venerable. Commentary on the Seven Catholic Epistles.* Cistercian Studies 82. Kalamazoo: Cistercian Publications, 1985.

    In Lucam. Ed. D. Hurst. CCSL 120 (1960).

    In Regum librum XXX quaestiones. Ed. D. Hurst. CCSL 119 (1962).

"Bobbio *computus". Liber de computo.* PL 129.1273–1372.

[Bonifatius Primicerius]. Ed. Bruno Krusch, "Ein Bericht der päpstlichen Kanzlei an Papst Johannes I. von 526 und die Oxforder HS Digby 63 von 814", in *Papstum und Kaisertum. Forschungen zur politischen Geschichte und Geisteskultur des Mittelalters Paul Kehr zum 65. Geburtstag dargebracht.* Ed. Albert Brackmann. Munich: Verlag der Münchner Drucke, 1926. Pp. 48–58.

Capitularia regum francorum. Ed. A. Boretius. MGH Leges (Quarto), 2 (1883).

"Carthaginian computus of 455". Ed. Krusch, *Studien I*, 279–297.

Cassian, *De incarnatione Domini contra Nestorium.* Ed. M. Petschenig. CCSL 17 (1886).

Cassiodorus, *Computus paschalis*. Ed. Paul Lehmann, "Cassiodorusstudien II: Die Datierung der Institutiones und der Computus paschalis", *Philologus* 71 (1912):278–299 (text is on 297–299).

Chronicon Paschale 284–628 AD. Trans. Michael Whitby and Mary Whitby. Translated Texts for Historians 7. Liverpool: Liverpool University Press, 1989.

Chronograph of AD 354. Ed. Theodore Mommsen. MGH AA 9:13–148.

Ps.-Clement, *Recognitiones*. Trans. Rufinus. PG 1.1157–1454.

"Cologne Prologue". Ed. Krusch, *Studien I*, 227–235.

Columbanus, *Epistolae*. Ed. W. Gundlach. MGH Ep. 3 (1892). Ed. and trans. G.S.M. Walker, *Sancti Columbani Opera* (Dublin, 1970).

Constantius, *Vita Germani*. Ed. W. Levison. MGH Script. rerum merov. 7 (1920).

Cummian's Letter 'De controversia paschali' and the 'De ratione conputandi'. Ed. and trans. M. Walsh & D. Ó Cróinín. Studies and Texts 86. Toronto: Pontifical Institute of Mediaeval Studies, 1988.

Ps.-Cyprian, *De pascha computus*. Ed. Wilhelm Hartel. CSEL 3.3 (1871):248–271. Trans. George Ogg. *The Pseudo-Cyprianic De pascha computus*. London: S.P.C.K., 1955.

Ps.-Cyril, *Prologus Cyrilli*. Ed. Krusch, *Studien I*, 337–343.

De causis quibus nomina acceperunt duodecim signa. Ed. Jones, *BOD* 665–667.

De computo dialogus. PL 90.647–652.

De divisionibus temporum. PL 90.653–664.

Dionysius Exiguus, *Argumenta titulorum paschalium*. Ed. Krusch, *Studien II*, 75–81; J.W. Jan, PL 67.453 *sq.*; Joan Gómes Pallarès, "Hacia una nueva edición de los 'Argumenta titulorum paschalium' de Dionisio el Exiguo", *Hispania sacra* 46 (1994):13–31.

Ep. ad Bonifacium et Bonum. Ed. Krusch, *Studien II*, 82–86.

Ep. ad Petronium. Ed. Krusch, *Studien II*, 63–68.

Einhard, *Vita Caroli*. Ed. Louis Halphen. 3rd ed. Paris: Les belles lettres, 1947.

Eusebius, *De vita Constantini*. Ed. I.A. Heikel, *Eusebius Werke* 1. Leipzig: Hinrich'sche Buchhandlung, 1902. Pp. 3–148.

Historia ecclesiastica. See Rufinus of Aquileia, below.

Eutropius, *Breviarium ab Urbe condita*. Ed. C. Santini. Leipzig: Teubner, 1979. Trans. H.C.S. Bird. Translated Texts for Historians 14. Liverpool: Liverpool University Press, 1993.

Gennadius, *De uiris illustribus*. Ed. W. Herdin. Leipzig: Teubner, 1879.

Gildas, *De excidio et conquestu Britanniae.* Ed. T. Mommsen. MGH AA 13 (1898):25–85.

Gregory, *Dialogi.* Ed. A. de Vogüé. Sources chrétiennes 251. Paris: Cerf, 1978.

Moralia in Job. Ed. M. Adriaen. CCSL 143A (1979). 3 vols.

Registrum. Ed. P. Ewald and L.M. Hartman. MGH Ep. 1 (1887).

Hegesippus, *Hegesippi qui dicitur libri quinque.* Ed. V. Ussani. CSEL 66 (1932).

Ps.-Hippocrates, *Ad Antigonum regem.* Ed. E. Liechtenhan, J. Kollesch and D. Nickel, *Marcelli de medicamentis liber.* Corpus medicorum latinorum 5. Berlin: Akademie-Verlag, 1968.

Historia abbatum auctore anonymo. Ed. C. Plummer, *Baedae opera historica* (see above). Vol. 2, pp. 388–404.

Isidore of Seville, *Chronica maiora.* Ed. T. Mommsen. MGH AA 11 (1894).

Ps.-Isidore of Seville, *Liber de ordine creaturarum: Un anónimo irlandés del siglo VII.* Ed. Manuel C. Diaz y Diaz. Santiago de Compostela: Universidad de Santiago, 1972.

Jerome, *Adversus Jovinianum.* PL 23.211–338.

Apologia aduersus libros Rufini. PL 23.397–492.

Chronicon. Ed. Rudolf Helm, *Eusebius Werke* 7.1. Griechischen christlichen Schriftsteller 24. Leipzig: J.C. Hinrichs, 1913.

De viris illustribus. Ed. W. Herding. Leipzig: Teubner, 1879.

Epistolae. Ed. I. Hilberg. CSEL 54–56 (1996). 4 vols.

Hebraicae quaestiones in Geneseos. Ed. P. de Lagarde. CCSL 72 (1959).

In Danielem. Ed. F. Glorie. CCSL (1964).

In Evangelium Matthaei. Ed. D. Hurst and M. Adriaen. CCSL 77 (1959).

In Ezechielem. Ed. F. Glorie. CCSL 75 (1964).

In Isaiam. Ed. M. Adriaen. CCSL 73 (1963).

In Zachariam. Ed. M. Adraien. CCSL 76A (1970).

Liber interpretationis hebraicorum nominum. Ed. P. de Lagarde. CCSL 72 (1959).

Josephus, *Antiquitates* [Books 1–6]. Ed. Franz Blatt, *The Latin Josephus.* Copenhagen: Munksgaard, 1958.

Antiquitates, [*versio latina*]. Cologne: ex aedibus Eucharii Cervicorni, 1534.

Contra Apionem. Ed. C. Boysen. CSEL 37 (1898).

Liber pontificalis. Ed. L. Duschesne. Paris: E. de Boccard, 1955.

Macrobius, *Saturnalia.* Ed. J. Willis. Leipzig: Teubner, 1970.

Mansi, J.D., *Sacrorum conciliorum amplissima collectio.* Paris: Hubert Welter, 1901.

Marcellinus Comes, *Chronicon.* Ed. T. Mommsen. MGH AA 11 (1894):37–108. Trans. Brian Croke, Byzantina Australiensia 7. Sydney: Australian Association for Byzantine Studies. 1995.

Marius Aventicensis, *Chronica.* Ed. Th. Mommsen. MGH AA 11 (1892):255–239.

Martin of Braga, *Opera omnia.* Ed. Claude W. Barlow. New Haven: Yale University Press, 1950.

"Merovingian *Computus* of 727". Ed. Krusch, *Studien II*, 53–57.

Orosius, *Historiarum adversos paganos libri quinque.* Ed. C. Zangemeister. CSEL 5 (1882).

Paschasinus of Lilybaeum, *Epistola ad Leonem papam.* Ed. Krusch, *Studien I*, 247–250.

Paulinus, *Vita Ambrosii.* Ed. M. Pellegrino. Verba seniorum n.s. 1. Rome, 1961.

Philippus Presbyter, *Commentarii in librum Iob.* PL 26.619–802.

[Pliny]. Ed. Karl Rück. *Auszüge aus der Naturgeschichte des C. Plinius Secundus in einem astronomisch-komputistischen Sammelwerke des achten Jahrhunderts.* Programm des Königlichen Ludwigs-Gymnasiums für das Studienjahr 1887–88. Munich: F. Straub, 1888.

Poetae latini minores. Ed. P. Baehrens. Leipzig: Teubner, 1883.

Polemius Silvius, *Laterculus.* Ed. Theodore Mommsen, *Corpus Inscriptionum Latinarum* 1, 2nd ed. Berlin: Georg Remer, 1893. Pp. 335–357.

Possidius, *Vita Augustini.* Ed. M. Pellegrino. Verba seniorum 4. Alba: Edizioni Paolini, 1955.

Priscian, *De figuris numerorum.* Ed. Heinrich Keil. *Grammatici latini.* Leipzig: Teubner, 1855–1880. Vol. 3, pp. 406–417.

Prosper of Aquitaine, *Epitoma chronicorum.* Ed. Th. Mommsen. MGH AA 9 (1892):385–485.

Proterius, *Epistola ad Leonem.* Ed. Krusch, *Studien I*, 269–278.

Riese, A. (ed.). *Anthologia latina.* Leipzig: Teubner, 1870.

Romana computatio. Ed. Jones, *BOD* 671–672.

Rufinus of Aquileia, trans. of Eusebius, *Historia ecclesiastica.* Ed. Th. Mommsen, *Eusebius Werke* 2.1–2. Griechischen christlichen Schriftsteller 9.1–2. Leipzig: J.C. Hinrichs, 1903–1908.

Solinus, *Collectanea rerum memorabilium*. Ed. Th. Mommsen. Berlin: Weidemann, 1895.

Theophilus of Alexandria, *Prologus Theophili*. Ed. Krusch, *Studien I*, 220–226.

Vegetius, *Epitoma rei militaris*. Ed. K. Lang. Leipzig: Teubner, 1869. Trans N.P. Milner. Translated Texts for Historians 16. 2nd ed. Liverpool: Liverpool University Press, 1996.

Vergil, *Aeneid*. Ed. R.A.B. Mynors. Oxford: Clarendon Press, 1969.

Eclogues. Ed. Wendell Clausen. Oxford: Clarendon Press, 1994.

Georgics. Ed. R.A.B. Mynors. Oxford: Clarendon Press, 1990.

Victor of Capua, *De pascha*. Ed. J.B. Pitra. *Spicilegium Solesmense*. Paris: Firmin Didot, 1852. Vol. 1, pp. 296–301.

Victorius of Aquitaine, *Calculus*. Ed. G. Friedlein, *Zeitschrift für Mathematik und Physik* 7 (1871):42–97.

Prologus ad Hilarium archidiaconum. Ed. Krusch, *Studien II*, 17–26.

Vindicianus, *Epistola ad Pentadium*. Ed. V. Rose. *Theodori Prisciani Euporiston libri III*. Leipzig: Teubner, 1894. Pp. 485–492.

de Winterfeld, P., "Rhythmi computistici". MGH Poetae 4.1 (1899). Pp. 667–702.

SECONDARY SOURCES

Allen, Michael Idomir, "Bede and Freculf at Medieval St. Gallen", in *Beda Venerabilis* (see below), 61–80.

Amos, T.L., "Monks and Pastoral Care in the Early Middle Ages", in *Religion, Culture and Society in the Early Middle Ages: Studies in Honor of Richard E. Sullivan*. Ed. T.F.X. Noble and J.J. Contreni. Studies in Medieval Culture 23. Kalamazoo: Medieval Institute Publications, 1987. Pp. 165–180.

Anscombe, A., "The Paschal Canon Attributed to Anatolius of Laodicea", *English Historical Review* 10 (1895):515–535.

Armstrong, John, III, "The Old Welsh Computus Fragment and Bede's *Pagina regularis* [Part 1]", *Proceedings of the Harvard Celtic Colloquium* 2 (1982):187–273.

Beda Venerabilis: Historian, Monk and Northumbrian. Ed. L.A.J.R. Houwen and A.A. MacDonald. Groningen: Egbert Forsten, 1996.

Bede: His Life, Times and Writings. Ed. A.H. Thompson. Oxford: Clarendon Press, 1935.

Bischoff, Bernhard, "Zur Kritik der Heerwagenschen Ausgabe von

Bedas Werken (Basel 1563)", in his *Mittelalterliche Studien*. Stuttgart: Hiersemann, 1966–1981. Vol. 1, pp. 112–117.

Borst, Arno, "Alkuin und die Enzyklopädie von 809", in *Science in Western and Eastern Civilization in Carolingian Times*. Ed. P.L. Butzer and D. Lorhmann. Basel: Birkhäuser Verlag, 1993. Pp. 53–78.

Bouchier, Gilles [Bucherius, Aegidius], *De doctrina temporum*. Antwerp: ex officina Plantiniana Balthasari Moreti, 1633.

Brincken, Anna-Dorothee von den, *Studien zur lateinischen Weltchronistik bis in die Zeitalter Ottos von Freising*. Düsseldorf: Michael Triltsch, 1957.

Brown, G.H., *Bede the Venerable*. Twayne's English Authors Series 443. Boston: Twayne, 1987.

Cantalamessa, Rainero, *Easter in the Early Church: An Anthology of Jewish and Early Christian Texts*. Ed. and trans. James M. Quigley and Joseph T. Leinhard. Collegeville, Minn.: Liturgical Press, 1993.

Casel, Odon, *La Fête de Pâques dans l'église des Pères*. Lex orandi 37. Paris: Cerf, 1963.

Contreni, John J., *The Cathedral School of Laon from 850 to 930: Its Manuscripts and Masters*. Münchner Beiträge zur Mediävistik und Renaissance-Forschung 29. Munich: Arbeo-Gesellschaft, 1978.

Cordoliani, Alfred, "Les computistes insulaires et les écrits pseudo-alexandrins", *Bibliothèque de l'École des Chartes* 106 (1945–1946):5–34.

Croke, Brian, "The Origins of the Christian World Chronicle", in *History and Historians in Late Antiquity*. Ed. Brian Croke and Alanna M. Emmett. Sydney and Oxford: Pergamon Press, 1983. Pp. 116–131.

Cross, J.E., "Bede's Influence at Home and Abroad", in *Beda Venerabilis* (see above), 17–30.

Davidse, Jan, "On Bede as Christian Historian", in *Beda Venerabilis* (see above), 1–15.

"The Sense of History in the Works of the Venerable Bede", *Studi medievali*, ser. 3, 23 (1982):647–695.

Diesner, H.-J., "Das christliche Bildungsprogramm des Beda Venerabilis (672/3–735)", *Theologische Literatur-Zeitung* 106 (1981):865–872.

Dillon, Miles, "The Vienna Glosses on Bede", *Celtica* 3 (1956):340–344.

Eckenrode, Thomas, "The Growth of a Scientific Mind: Bede's Early and Late Scientific Writings", *Downside Review* 94 (1976):197–212.

"Original Aspects in Venerable Bede's Tidal Theories with Relation to Prior Tidal Observations". PhD dissertation, St. Louis University, 1970.

"The Venerable Bede and the Pastoral Affirmation of the Christian Message in Anglo-Saxon England", *Downside Review* 99 (1981):258–278.

"Venerable Bede as a Scientist", *American Benedictine Review* 22 (1971):486–507.

"Venerable Bede's Theory of Ocean Tides", *American Benedictine Review* 25 (1974):56–74.

Englisch, Brigitte, *Die Artes liberales im frühen Mittelalter (5.–9. Jh.): Das Quadrivium und der Komputus als Indikatoren für Kontinuität und Erneuerung der exacten Wissenschaften zwischen Antike und Mittelalter.* Sudhoffs Archiv, Beiheft 33. Stuttgart: Franz Steiner, 1994.

"Realitätsorientierte Wissenschaft oder praxisferne Traditionswissen? Inhalte und Probleme mittelalterlicher Wissenschaftsvorstellungen am Beispiel von *De temporum ratione* des Beda Venerabilis", in *Dilettanen und Wissenschaft. Zur Geschichte und Aktualität eines wechselvollen Verhältnisses.* Ed. Elisabeth Strauss. Philosophie und Repräsentation / Philosophy and Representation 4. Amsterdam: Rodopi, 1996. Pp. 11–34.

Famulus Christi. Essays in Commemoration of the Thirteenth Centenary of the Birth of the Venerable Bede. Ed. Gerald Bonner. London: S.P.C.K., 1976.

Fordyce, C.J., "A Rhythmical Version of Bede's *De ratione temporum*", *Archivum latinitatis medii aevi* 3 (1927):59–73, 129–141.

Ginzel, F.K., *Handbuch der mathematischen und technischen Chronologie.* Leipzig: J.C. Hinrichs'sche Buchhandlung, 1906–1914. 3 vols.

Gómez Pallarés, Juan, "Los *excerpta* de Beda (*De temporum ratione*, 23–25) en el MS. ACA, Ripoll 225", *Emerita* 59 (1991):101–122.

The Gregorian Reform of the Calendar. Proceedings of the Vatican Conference to Commemorate its 400th Anniversary, 1582–1982. Ed. G.V. Coyne, M.A. Hoskin and O. Pedersen. Vatican City: Pontificia Academia Scientiarum; Specola Vaticana, 1983.

Grosjean, Paul, "La date de Pâques et le Concile de Nicée", *Académie royale de Belgique. Bulletin de la classe des sciences*, 5th ser., 48 (1962):55–66.

"Recherches sur les débuts de la controverse paschale chez les Celtes",

Analecta Bollandiana 64 (1946):200–244.

Grumel, Vénance, *La Chronologie.* Traité des études byzantines 1. Paris: Presses universitaires de France, 1968.

"Le problème de la date paschale aux IIIe et IVe siècle", *Revue des études byzantines* 18 (1960):161–178.

Halton, T., "The Coming of Spring: A Patristic Motif", *Classical Folia* 30 (1976):150–164.

Harley, J.B. and David Woodward, (eds.), *The History of Cartography. Volume 1: Cartography in Prehistoric, Ancient and Medieval Europe and the Mediterranean.* Chicago: University of Chicago Press, 1987.

Harrison, Kenneth, "Easter Cycles and the Equinox in the British Isles", *Anglo-Saxon England* 7 (1978):1–8.

"Epacts in Irish Chronicles", *Studia Celtica* 12–13 (1977–8):17–32.

"Episodes in the History of Easter Cycles in Ireland", in *Ireland in Early Medieval Europe. Studies in Memory of Kathleen Hughes.* Ed. Dorothy Whitelock, Rosamond McKitterick and David Dumville. Cambridge: Cambridge University Press, 1982. Pp. 307–319.

The Framework of Anglo-Saxon History to A.D. 900. Cambridge: Cambridge University Press, 1975.

"A Letter from Rome to the Irish Clergy", *Peritia* 3 (1984):222–229.

Henning, John, "Versus de mensibus", *Traditio* 11 (1955):65–90.

History and Historians in Late Antiquity. Ed. Brian Croke and Alanna M. Emmett. Sydney and Oxford: Pergamon Press, 1983.

Hunter Blair, Peter, *The World of Bede.* London: Secker and Warburg, 1970.

Illmer, Detlef, *Formen der Erziehung und Wissenvermittlung im frühen Mittelalter.* Münchner Beitäge zur Mediävistik und Renaissance-Forschung 7. Munich: Arbeo-Gesellschaft, 1971.

Jones, C.W., *Bedae pseudepigrapha: Scientific Works Falsely Attributed to Bede.* Ithaca and London: Cornell University Press, 1939.

"Bede and Vegetius", *The Classical Review* 46 (1932):248–249.

"Bede as Early Medieval Historian", *Mediaevalia et Humanistica* 4 (1946):26–36.

"The Byrhtferth Glosses", *Medium Aevum* 7 (1938):88–97.

"An Early Medieval Licensing Examination", *History of Education Quarterly* 3 (1963):19–29.

"A Legend of St Pachomius", *Speculum* 18 (1943):198–210.

"The 'Lost' Sirmond Manuscript of Bede's Computus", *English Historical Review* 51 (1937):204–219.

"Materials for an edition of Bede's *De temporum ratione*". PhD dissertation, Cornell University, 1932.

"Polemius Silvius, Bede, and the Names of the Months", *Speculum* 9 (1934):50–56.

Saints' Lives and Chronicles. Ithaca: Cornell University Press, 1947.

"Some Introductory Remarks on Bede's *Commentary on Genesis*", *Sacris Erudiri* 19 (1969–1970):115–198.

"The Victorian and Dionysiac Tables in the West", *Speculum* 9 (1934):408–421.

"Two Easter Tables", *Speculum* 13 (1938):204–205.

Kenney, J.F., *Sources for the Early History of Ireland: Ecclesiastical. An Introduction and Guide*, 2nd ed. (1968) rpt. Dublin: Four Courts Press, 1993.

Killion, Stephen B., "Bede's Irish Legacy: Knowledge and Use of Bede's Works in Ireland from the Eighth through the Sixteenth Century". PhD dissertation, University of North Carolina at Chapel Hill, 1992.

King, Vernon, "An Unreported Early Use of Bede's *De natura rerum*", *Anglo-Saxon England* 22 (1993):85–91.

Kitson, Peter, "Lapidary Traditions in Anglo-Saxon England: Part I, the Background; the Old English Lapidary", *Anglo-Saxon England* 7 (1978):9–60; "Part II, Bede's *Explanatio Apocalypsis* and Related Works", *Anglo-Saxon England* 12 (1983):73–123.

Krusch, Bruno, "Chronologisches aus Handschriften", *Neues Archiv* 10 (1885):83–94.

"Die Einführung der griechischen Paschalritus im Abendland", *Neues Archiv* 9 (1884):101–169.

Landes, Richard, "Lest the Millenium Be Fulfilled: Apocalyptic Expectations and the Pattern of Western Chronography 100–800 CE", in *The Use and Abuse of Eschatology in the Middle Ages*. Ed. Werner Verbeke, Caliel Verhelst and Andries Welkenhuysen. Mediaevalia Lovaniensia Series I/Studia XV. Leuven: Leuven University Press, 1988. Pp. 137–209.

Lapidge, Michael (ed.), *Bede and His World*. Aldershot: Variorum, 1994. 2 vols.

Lejbowicz, Max, "Postérité médiévale de la distinction isidorienne *astrologia/astronomia*: Bède et le vocabulaire de la chronometrie", in *Documents pour l'histoire du vocabulaire scientifique*, no. 7. Paris: Editions du CNRS, 1985. Pp. 1–41.

Lipp, Frances Randall, "The Carolingian Commentaries on Bede's *De natura rerum*". PhD dissertation, Yale University, 1961.

Lowe, E.A., *Codices latini antiquiores*. 11 vols. + suppl. Oxford: Clarendon Press, 1934–72.

Luneau, Auguste, *L'Histoire du salut chez les Pères de l'église: La doctrine des âges du monde*. Théologie historique 2. Paris: Beauchesne, 1964.

MacCarthy, Bartholomew (ed.), *Annals of Ulster*, vol. 4. Dublin: His Majesty's Stationery Office, 1901.

McCarthy, Daniel, "The Chronological Apparatus of the Annals of Ulster AD 431–1131", *Peritia* 8 (1994):46–97.

"Easter Principles and a Fifth-Century Lunar Cycle Used in the British Isles", *Journal of the History of Astronomy* 24 (1993):204–224.

"The Origin of the *Latercus* Paschal Cycle of the Insular Celtic Churches", *Cambrian Medieval Celtic Studies* 28 (1994):25–49.

McClure, Judith, "Bede and the Life of St Ceolfrid", *Peritia* 3 (1984):71–84.

McCluskey, Stephen C., *Astronomies and Cultures in Early Medieval Europe*. Cambridge: Cambridge University Press, 1998.

Meaney, A.L., "Bede and Anglo-Saxon Paganism", *Parergon* n.s. 3 (1985):1–29.

Menninger, Karl, *Number Words and Number Symbols: A Cultural History of Numbers*. Trans. Paul Broneer. Cambridge, Mass.: M.I.T. Press, 1969.

Miller, Molly, "The Chronological Structure of the Sixth Age in the Rawlinson Fragment of the 'Irish World Chronicle'", *Celtica* 22 (1991):79–83.

Moreton, Jennifer, "Doubts about the Calendar. Bede and the Eclipse of 664", *Isis* 89 (1998):50–65.

Neugebauer, O., "On the *Computus paschalis* of 'Cassiodorus'", *Centaurus* 25 (1982):292–302.

North, J.D., "Monasticism and the First Mechanical Clocks", in *The Study of Time*. Proceedings of the Second Conference of the International Society for the Study of Time, 1973. New York: Springer, 1973. Pp. 381–398.

O'Connell, D.J., "Easter Cycles in the Early Irish Church", *Journal of the Royal Society of Antiquaries of Ireland* 66 (1936):67–106.

Ó Cróinín, D., "The Irish Provenance of Bede's *Computus*", *Peritia* 2 (1983):238–242.

Ó Cróinín, D. and McCarthy, D., "The 'Lost' Irish 84-year Easter Table Rediscovered", *Peritia* 6–7 (1987–88):227–242.

Petau, Denis [Dionysius Petavus], *De doctrina temporum*. Paris: Sebastien Cramoisy, 1727.

Petersohn, Jürgen, "Neue Bedafragmente in Northumbrischer Unziale Saec. VIII", *Scriptorium* 20 (1966):215–247 and Pl. 17–18.

"Die Bückeburger Fragments von Bedas *De temporum ratione*", *Deutsches Archiv für Erforschung des Mittelalters* 22 (1966):587–597.

di Pilla, Alessandra, "Cosmologia e uso delle fonte nel *De natura rerum* di Beda", *Romanobarbarica* 11 (1991):128–147.

Poole, R.L., *Studies in Chronology and History*. Oxford: Clarendon Press, 1934.

Ray, Roger, "Bede, the Exegete, as Historian", in *Famulus Christi* (see above), 125–140.

"Bede's *vera lex historiae*", *Speculum* 55 (1980):1–21.

Richard, Marcel, "Comput et chronologie chez saint Hippolyte [part 1]", *Mélanges de sciences religieuses* 7 (1950):237–268; [part 2], *Mélanges de sciences religieuses* 8 (1951):19–50.

"Le comput pascal par octaétéris", *Muséon* 87 (1974):307–339.

Rühl, Franz, *Chronologie des Mittelalters und der Neuzeit*. Berlin: Reuter & Reichard, 1897.

Saints, Scholars and Heroes. Ed. Margot King and Wesley Stevens. Collegeville, Minn.: Hill Monastic Manuscript Library, 1979. 2 vols.

Samuel, Alan E., *Greek and Roman Chronology. Calendars and Years in Classical Antiquity*. Handbuch der Altertumswissenschaft I,17. Munich: Beck, 1972.

Santosuosso, Alma Colk, "Music in Bede's *De temporum ratione*: An 11th-Century Addition to MS London, British Library, Cotton Vespasian B. VI", *Scriptorium* 43 (1989):255–259.

Schaller, Dieter, "Der verleumdete David: zum Schlusskapitel von Bedas *Epistola ad Pleguinam*", in *Literatur und Sprache im europäischen Mittelalter: Festschrift für Karl Langosch*. Darmstadt: Wissenschaftliche Buchgesellschaft, 1973. Pp. 39–43.

Schneiders, Marc, "Zur Datierung und Herkunft des Karlsruher Beda (*Aug.* CLXVII)", *Scriptorium* 43 (1989):247–252.

Schwartz, Eduard, *Christliche und jüdische Ostertafeln*. Abhandlungen der königlichen Gesellschaft der Wissenschaften zu Göttingen, phil.-hist. Kl. n.f. 8,6 (1905).

Science in Western and Eastern Civilization in Carolingian Times. Ed. P.L. Butzer and D. Lorhmann. Basel: Birkhäuser Verlag, 1993.

Sears, Elizabeth, *The Ages of Man. Medieval Interpretations of the Life Cycle.* Princeton: Princeton University Press, 1986.

Siniscalco, P., "Le età del mondo in Beda", *Romanobarbarica* 3 (1978):297–331.

Smyth, Marina, *Understanding the Universe in Seventh-Century Ireland.* Studies in Celtic History 15. Woodbridge: Boydell, 1996.

Staub, Kurt Hans, "Ein Beda-Fragment des 8. Jahrhunderts in der Hessischen Landes- und Hochschulbibliothek Darmstadt", *Bibliothek und Wissenschaft* 17 (1983):1–7.

Stevens, Wesley, "Bede's Scientific Achievement", Jarrow Lecture, 1985.

Strachan, John, "The Vienna Fragments of Bede", *Revue celtique* 23 (1902):40–49.

Strecker, Karl, "Zu den computistischen Rhythmen", *Neues Archiv* 36 (1911):317–342.

Strobel, August, *Texte zur Geschichte des frühchristlichen Osterkalendars.* Münster in Westfalen: Achendorffsche Verlagsbuchhandlung, 1983.

   Ursprung und Geschichte des frühchristlichen Osterkalendars. Texte und Untersuchungen 121. Berlin: Akademie Verlag, 1977.

Teres, Gustav, "Time Computations and Dionysius Exiguus", *Journal of the History of Astronomy* 15 (1984):177–188.

Thacker, A.T., "Bede and the Irish", in *Beda Venerabilis* (see above), pp. 31–57.

Tristram, Hildegard L.C., *Sex aetates mundi. Die Weltzeitalter bei den Angelsachsen und den Iren. Untersuchungen und Texte.* Heidelberg: Carl Winter, 1985.

Turner, C.H., "The Paschal Canon of 'Anatolius of Laodicea'", *English Historical Review* 10 (1895):699–710.

van de Vyver, André, "L'évolution du comput alexandrin et romain du IIIe au Ve siècle", *Revue d'histoire ecclésiastique* 52 (1952):5–25.

van Wijk, W.E., *Le nombre d'or. Étude de chronologie technique suivi du texte de la Massa compoti d'Alexandre de Villedieu.* La Haye: Martinus Nijhoff, 1936.

Viereck, Wolfgang, "Beda in Bamberg", in *Einheit in der Vielfalt: Festschrift für Peter Lang zum 60. Geburtstag.* Ed. Gisela Quast. Bern: Peter Lang, 1988. Pp. 556–569.

von Sickel, Theodore, "Die Lunarbuchstaben in der Kalendarien des

Mittelalters", *Sitzungsberichte der Akademie der Wissenschaften zu Wien*, Phil.-hist. Kl. 38 (1861).

Wallis, Faith, "Chronology and Systems of Dating", in *Medieval Latin Studies: an Introduction and Bibliographic Guide*. Ed. George Rigg and Frank A.C. Mantello. Washington: Catholic University of America Press, Washington D.C., 1996. Pp. 383–387.

"The Church, the World, and the Time", in *Normes et pouvoirs à la fin du moyen âge*. Ed. Marie-Claude Deprez-Masson. Inedita et rara 7. Montreal: Ceres, 1990. Pp. 15–29.

"Images of Order in the Medieval Computus", in *ACTA XIV: Ideas of Order in the Middle Ages*. Ed. Warren Ginsberg. Binghamton: State University of New York Press, 1990. Pp. 45–67.

"Medicine in Medieval Computus Manuscripts", in *Manuscript Sources of Medieval Medicine*. Ed. Margaret Schleissner. New York: Garland, 1995. Pp. 105–143.

"MS Oxford, St John's College 17: A Medieval Manuscript in its Contexts". PhD dissertation, University of Toronto, 1985.

Ward, Benedicta, *The Venerable Bede*. London: Geoffrey Chapman, 1990.

Warren, Robert, *The Leofric Missal as Used in the Cathedral of Exeter*... Oxford: Clarendon Press, 1883.

Wetzel, Georg, *Die Chronicen des Baeda Venerabilis*. Halle: Plötz'sche Buckdruckerei, 1878.

Williams, Burma P. and Williams, Richard S., "Finger Numbers in the Greco-Roman World and the Early Middle Ages", *Isis* 86 (1995):587–608.

Wilson, H.A., *The Missal of Robert of Jumièges*. Henry Bradshaw Society Publications 11. London: Harrison, 1896.

Wormald, Francis, *English Kalendars Before A.D. 1100*. Henry Bradshaw Society Publications 72. London: [The Society], 1934.

Yeldham, Florence, "Notation of Fractions in the Earlier Middle Ages", *Archeion* 8 (1927):313–329.

Zélinsky, A.N., "Le calendrier chrétien avant la réforme grégorienne", *Studi medievali* ser. 3, 23 (1982): 529–597.

INDEX OF SOURCES

All references are to the chapter of *The Reckoning of Time*, unless otherwise indicated. References to chapter 66 are further subdivided by *annus mundi* (in brackets).

25.3	8
25.8–31	8
25.10	2

Numbers

9.10	61
9.10–11	51
14.34	66 (2493)
28.11	13

Deuteronomy

13.3	Pleg. 15
16.1	35

Joshua

4.19	66 (2519)

Judges

3.8	66 (2559)
3.15–30	66 (2639)
4.2	66 (2697)
4.23–24	66 (2697)
6.3, 14	66 (2719)
10.1–2	66 (2745)
10.3–5	66 (2767)
11	66 (2773)
11.26	66 (2773)
12.8–9	66 (2780)
12.11	66 (2790)
12.14	66 (2798)
13.1	66 (2798)

I Kings

6.31	66 (2979)
10.1	66 (2979)
12.20	66 (2979)
15.2	66 (2990)
16.24	66 (3031)
16.34	66 (3031)
17.1	66 (3056)
19.16	66 (3056)
22	66 (3056)

II Kings

2	66 (3064)
8.17	66 (3064)
8.18	66 (3034)
8.20	66 (3064)
8.24–25	66 (3065)
10.15	66 (3065)
12.1	66 (3111)
13.3	66 (3031)
13.20	66 (3140)
13.22	66 (3140)
14.2	66 (3140)
15.1–2	66 (3192)
15.33	66 (3208)
15.35	66 (3208)
16.2	66 (3224)
16.5–9	66 (3224)
18.9	66 (3253)
21.19	66 (3310)
22.1	66 (3341)
23.36	66 (3352)
24.8	66 (3352)
24.10–12	66 (3352)
25.1	66 (3352)
25.2–12	66 (3363)
25.26	66 (3363)
25.27–28	66 (3389)

I Chronicles

15.21	71

II Chronicles

12.2	66 (2987)
13.17–18	66 (2990)
16.4	66 (3031)
16.13	66 (3031)
22.10–11	66 (3071)
22.12	66 (3071)
24.21	66 (3111)
29.1	66 (3253)
33.11–13	66 (3308)
33.24	66 (3310)

Zechariah

1.12	66 (3468)
7.5	66 (3468)

Malachi

4.6	69

I Esdras

2.13	66 (3423)

I Maccabees

1.7	66 (3629)

Matthew

1.1–18	Pleg. 2
3.16	9
5.8	71
5.16	64
5.17	9
5.26	4
7.6	9
12.38–40	5
20.1–16	Pleg. 14
24.2	9
24.29	70
24.36	67, Pleg. 14
24.48	68
25.10	12
28.1	5, 71

Mark

1.10	9
13.2	9
13.32	Pleg. 14
13.35	70
14.12	63
16.1	5

Luke

2.52	64
3.22	9

3.23	47
3.36	66 (1693)
10.1, 17	5
12.35–36	67
12.35–37	Pleg. 15
17.23	Pleg. 16
17.24	Pleg. 16
21.6	9
23.43	71
24.13	66 (4175)
24.39	66 (4536)

John

1.9	6
1.29	9
1.32	9
2.20	66 (3468)
3.30	30
8.25	40
11.9	3
12.1	71
13.1	47
14.6	71
16.28	64
18.28	63
19.14	4
19.34	10
20.27	5
21.12–13	5

Acts of the Apostles

1.7	Pleg. 14
1.8	64
3.2	66 (3208)
8.1	66 (3984)
11.28	66 (4007)
12.23	66 (3966, 3993)

Romans

8.11	64

2. Classical, patristic and medieval authors

4.3.11	7
4.4	3
4.5.23	31
4.5.24	16
4.7	29
4.7.29	28
4.8.32	6
6.2.8	7

Anatolius, *De ratione paschali*

("Insular forgery" version, ed.
Krusch, *Studien I*, 316–327).

6	Wic. 11
10	22
14	35

(as quoted by Rufinus in his trans. of
Eusebius, *Historia ecclesiastica* 7.32)

2	14, 30, 42, Wic. 1, 2, 4, 5, 6

Arnobius, *In psalmos*

104	66 (1757)

Augustine

Adnotationes in Iob

6

Contra Faustum

12.14.15	66 (1656)
18.5	12

De civitate Dei

3.15	27
11.30	8
15.10	Pleg. 7
15.11	66 (874)
15.11	Pleg. 7
15.12	37, Pleg. 11
15.13	66 (1656), Pleg. 10
15.17	66 (687)
15.19	66 (687)
15.23	66 (622)

16.9	34
16.10	66 (1948)
16.43	66 (Prologue)
18.36	66 (3548)
18.45	66 (3548)
20.13	69
20.24	70
20.26	70
21.5.1	28
22.30	71

De consensu evangelistarum

3.13.41	4

De doctrina christiana

3.35.50–51	5

De Genesi ad litteram

1.10	5
3.2	70
5.23	5
6.10	5

De Genesi contra Manichaeos

1.23.35–41	66 (Prologue)

De sermone Domini in monte

1.11.30	4

De Trinitate

1.3.25–28	Pref.
4.4	39

Enarrationes in Psalmos

6.1.15–26	Pleg. 15
6.1–2	10
10.3	6, 25
101 (II).13	70

Epistolae

55	25, 64
55.5	64

Ps.-Isidore, *De ordine creaturarum*

Etymologiae

Jerome

Adversus Jovinianum

Apologia adversus libros Rufini

Chronicon

189.24–26	66 (4049)	214.8–9	66 (4175)
191.8	66 (4069)	214.20–24	66 (4175)
192.1–4	66 (4049)	215.1–2	66 (4188)
192.16–18	66 (4049)	215.22	66 (4188)
192.25–26	66 (4050)	215.25–26	66 (4188)
193.3–8	66 (4050)	216.8–9	66 (4191)
193.18–19	66 (4069)	216.18–19	66 (4197)
194.19–22	66 (4069)	217.9–13	66 (4204)
194.24–26	66 (4069)	218.7–8	66 (4205)
195.21–23	66 (4069)	218.11–13	66 (4205)
195.26	66 (4069)	218.16–18	66 (4205)
196.12–26	66 (4069)	218.21–23	66 (4207)
197.9–12	66 (4090)	220.1–2	66 (4222)
197.21–22	66 (4090)	220.9–11	66 (4222)
200.18–19	66 (4069)	220.12–14	66 (4222)
201.10–12	66 (4090)	221.22–23	66 (4224)
201.15–17	66 (4069)	221.26–222.3	66 (4224)
202.1–3	66 (4113)	222.8–9	66 (4229)
203.13–18	66 (4113)	223.7, 9–10	66 (4229)
203.25	66 (4090)	223.12	66 (4230)
204.7–9	66 (4132)	223.13–14	66 (4230)
204.10–12	66 (4132)	223.16	66 (4236)
205.5–6	66 (4132)	223.20–22	66 (4236)
205.9–10	66 (4132)	223.21–22	66 (4230)
205.13	66 (4132)	223.23–24	66 (4236)
206.1–3	66 (4132)	223.25–26	66 (4230)
206.4–6	66 (4132)	224.19–20	66 (4238)
207.18–19	66 (4132)	225.7–8	66 (4258)
208.20–21	66 (4145)	225.13–14	66 (4258)
209.16–18	66 (4145)	225.18–23	66 (4258)
210.4–5	66 (4146)	225.26–226.3	66 (4258)
210.9–10	66 (4146)	227.2–3	66 (4258)
210.16–19	66 (4146)	228.5–6	66 (4258)
210.20–21	66 (4163)	228.12–14	66 (4258)
211.1–4	66 (4163)	228.18–19	66 (4259)
211.5–8	66 (4163)	228.20–21	66 (4258, 4259)
212.7–9	66 (4163)	228.26–229.1	66 (4290)
212.23–25	66 (4170)	229.1–3	66 (4290)
213.4–5	66 (4170)	229.6	66 (4290)
213.8–10	66 (4170)	231.22–25	66 (4290)
213.23–24	66 (4171)	234.13–15	66 (4314)
214.5–6	66 (4171)	234.24–25	66 (4314)

Paschasinus of Lilybaeum, *Epistola ad Leonem papam*

Paulinus, *Vita Ambrosii*

Philippus Presbyter, *Commentarii in librum Iob*

Pliny, *Historia naturalis*

Polemius Silvius, *Laterculus*

GENERAL INDEX

Note: All references are to page number.

"Aachen encyclopaedia of 809": xci–xcii
Aaron (martyr): 211
Abbo of Fleury: xciv
Abdon: 170
Abel: 159
Abgar: 205
Abijah: 173
Abimelech: 169
Abraham: 25, 165, 275
Acephalite heresy: 229–232
Achilleus (usurper): 211
Adam: 159
Aefric: xcvii
Aegialius: 165
Aegyptus Silvius: 173
Aelle (king of Northumbria): 226
Aeneas: 170
Aeneas Silvius: 171
Aethelberht (king of Kent): 226
Aethelfrith (king of Northumbria): 226
Aethelthryth, St: 232
Aetius: 220, 222
Africa: 220, 223, 224, 232
Agamemnon: 170
Agatho (pope): 231
ages of man: 157–159, 407
(St) Agnes, basilica of (Rome): 213
Agrippa (son of Herod Agrippa): 197
Agrippa Silvius: 174
Ahab: 173–174
Ahasuerus: 182, 186
Ahaz: 176
Alans: 216, 218, 220
Alaric: 218

Alba: 171
Alban (martyr): 211
Albinus: 198
Albinus, Clodius (emperor): 204
Alboin (king of the Lombards): 226
Alba Silvius: 172
Alchymus: 191
Aldhelm: lviii, lxi, lxii
Alethis: 171
Alexander (bp. of Jerusalem): 204, 207
Alexander (high priest): *see* Janaeus Alexander
Alexander (pope): 200
Alexander (son of Herod the Great): 194
Alexander Severus (emperor): 205
Alexander the Great: 187
Alexandra (wife of Janaeus Alexander): 192
Alexandria: 187, 193; lighthouse of: 188
alphabet, Athenian: 186
Amaziah: 175
Ambrose, St: lxxxii, lxxxiii–lxxxiv, 216, 217
Ambrosius Aurelianus: 223
Amon: 177
Amulius Silvius: 176
Ananias, Azarias and Misael: 179
Anastasius, St: 227–228
Anastasius I (emperor): 223–224
Anastasius II (emperor): 235, 236
Anatolius of Laodicea: xlvii, 209–210
(ps.)-Anatolius of Laodicea, *Liber Anatolii*: lvi–lix, lxxii, lxxv